Blue-Green Infrastructure for Sustainable Urban Settlements

P. K. Joshi · K. S. Rao · Rahul Bhadouria ·
Sachchidanand Tripathi · Rishikesh Singh
Editors

Blue-Green Infrastructure for Sustainable Urban Settlements

Implications for Developing Countries Under Climate Change

Springer

Editors
P. K. Joshi
School of Environmental Sciences
Jawaharlal Nehru University
New Delhi, Delhi, India

K. S. Rao
Department of Botany
University of Delhi
New Delhi, Delhi, India

Rahul Bhadouria
Delhi College of Arts and Commerce
University of Delhi
New Delhi, Delhi, India

Sachchidanand Tripathi
Department of Botany
Deen Dayal Upadhyaya College
University of Delhi
New Delhi, Delhi, India

Rishikesh Singh
Amity School of Earth & Environmental Sciences
Amity University Punjab
Mohali, Punjab, India

ISBN 978-3-031-62292-2 ISBN 978-3-031-62293-9 (eBook)
https://doi.org/10.1007/978-3-031-62293-9

This Springer imprint is published by the registered company Springer Nature Switzerland AG
The registered company address is: Gewerbestrasse 11, 6330 Cham, Switzerland

Foreword

In the face of unprecedented urbanization and the escalating impacts of climate change, the quest for sustainable urban development has become an urgent global imperative. The anthology you hold in your hands, *Blue-Green Infrastructure for Sustainable Urban Settlements—Implications for Developing Countries Under Climate Change*, is a beacon of critical inquiry and innovative solutions. It illuminates the path toward resilient and sustainable urban futures, particularly in the context of developing nations.

As we navigate the complexities of urban resilience and sustainability, the chapters by distinguished cadre of scholars and practitioners assembled within this edited book offer a multifaceted exploration of blue-green infrastructure—a transformative paradigm that integrates natural elements into the urban fabric to enhance resilience, mitigate environmental risks, and foster sustainable development. This volume stands as a testament to the urgency and importance of embracing innovative solutions that harmonize human activity with the natural world.

In Part One, "Urban Resilience and Sustainability," the foundational principles and overarching implications of blue-green infrastructure are meticulously examined. The issues addressed by the part range from the role of the infrastructure in mitigating the urban heat island effect to its contribution toward achieving Sustainable Development Goals. The chapters in this part elucidate the manifold benefits and challenges of implementing blue-green infrastructure in the context of developing countries.

Part Two, "Management and Development of Blue-Green Infrastructure," delves into the practicalities of adopting and nurturing such innovative systems. Through invaluable insights garnered from real-world experiences, case studies and best practices across diverse geographical and cultural contexts, the contributors illuminate the challenges and opportunities inherent in integrating blue-green infrastructure into the urban landscape. In particular, they emphasize the crucial role of community engagement and vertical greening for sustainable planning.

The imperative for scaling up and assessing the impact of blue-green infrastructure is the focus of Part Three, "Scaling and Assessment for Urban Planning." From modeling climate change-resilient urban agriculture to addressing spatial accessibility disparities including redeveloping green spaces, the chapters in this part underscore the importance of evidence-based planning and inclusive governance in realizing the full potential of blue-green infrastructure.

Finally, in Part Four, "Policy Issues and Sustainable Development," the nexus between blue-green infrastructure and policy frameworks is explored. The contributors use case studies from diverse geographical contexts in order to highlight the governance challenges and multifunctional patterns inherent in implementing blue-green infrastructure. In doing so, they underscore the need for strategic planning and policy innovation to drive sustainable urban development.

It is with great appreciation and admiration that I acknowledge the alignment of this forthcoming edited volume with the broader objectives of the URGENT (Urban Resilience and Adaptation for India and Mongolia) Project. This initiative exemplifies the collaborative spirit and transdisciplinary approach necessary to address the complex challenges facing urban settlements in the twenty-first century. By supporting blue and green infrastructure initiatives, the edited book contributes to the advancement of sustainable urban development agendas across the globe.

As we confront the challenges of an uncertain future, may the insights and perspectives presented within these pages inspire collective action and transformative change toward more resilient, equitable, and sustainable urban settlements for generations to come.

February 2024

Anton Shkaruba, Ph.D.
Senior Researcher—Estonian
University of Life Sciences
Environmental Protection
and Landscape Management
Tartu, Estonia
Director—Erda Research
Technology, Education
The Hague, The Netherlands
anton@erda-rte.eu
http://www.erda-rte.eu/

Preface

It is with great pleasure and anticipation that we present this comprehensive volume. As the urbanization trend continues to surge globally, the importance of sustainable development, particularly in the context of developing countries, becomes increasingly paramount. This book aims to shed light on the critical role of Blue-Green Infrastructure (BGI) in fostering urban resilience, sustainability, and climate adaptation, with a particular focus on the unique challenges and opportunities faced by cities in the developing world.

Part One of this book delves into the foundational aspects of urban resilience and sustainability, setting the stage for understanding the implications of climate change on developing countries. Chapters "Strength and Limitation of Nature-Based Solutions Towards Adaptation and Mitigation of Climate Change in Developing Countries" through "Role of Blue-Green Infrastructure in Achieving Sustainable Development Goals in Urban Centres of Developing countries" offer a comprehensive overview of the concept of BGI and its potential in mitigating the adverse effects of climate change in urban settlements. From exploring the role of BGI in addressing the Urban Heat Island (UHI) effect to its contribution toward achieving Sustainable Development Goals (SDGs), this part provides a holistic understanding of the subject matter. Chapter "Strength and Limitation of Nature-Based Solutions Towards Adaptation and Mitigation of Climate Change in Developing Countries" highlights the potential of natural methods in cities to combat climate change by promoting blue and green urban areas. It suggests using Nature-based Solutions (NbS) to mitigate climate impacts and enhance resilience. Key points include developing metrics to measure efficacy, addressing knowledge gaps, overcoming obstacles, and prioritizing socio-environmental justice in NbS implementation. Ultimately, NbS can serve as tools for climate adaptation while improving societal well-being and sustainable urban development.

Chapter "Blue-Green Infrastructure for Urban Resilience and Sustainability in Developing Countries" examines the significance of BGI in enhancing urban resilience and sustainability, particularly in developing countries. It outlines the challenges being faced, such as rapid urbanization and limited resources, and proposes BGI as a solution to promote resilience, sustainability, and social equity. It also addresses implementation challenges, governance issues, and emphasizes the importance of community involvement and education for the long-term success of BGI initiatives.

Chapter "Integrating Blue-Green as Next-Generation Urban Infrastructure in Developing Countries" highlights the increasing importance of NbS like BGI for sustainable urban design, especially in the face of climate change and rapid urbanization. The chapter aims to review the benefits, applications, and scientific knowledge surrounding BGI to support its integration into urban planning for sustainable development, thus addressing priorities and differences in BGI development globally.

Chapter "Understanding the Role of Blue-Green Infrastructure in Abatement of Urban Heat Island Effect" discusses the effectiveness of BGI in mitigating UHI effects, showing a temperature reduction of 2 °C to 4 °C compared to built-up areas. For example, a study in Singapore reveals that an 8% increase in BGI can lower temperatures by 1.21 °C. The findings highlight the need for BGI planning in developing regions to create sustainable built environments with distributed BGI networks.

Chapter "Role of Blue-Green Infrastructure in Achieving Sustainable Development Goals in Urban Centres of Developing countries" emphasizes the role of BGI in promoting sustainable living through urban planning, ecological restoration, and fostering a connection with nature. BGI contributes to various sectors—economic, social, and environmental—by controlling ecosystems, reducing pollution, and lowering urban temperatures. It reviews BGI's applicability in achieving SDGs and discusses its implications, benefits, and challenges, highlighting its effectiveness in meeting targets for sustainable development.

Moving on to Part Two, the focus shifts toward the management and development of BGI. Chapters "Challenges and Opportunities in Adopting Blue-Green Infrastructure in Cities of the Developing World" through "Blue-Green Infrastructure for Flood Resilience: Case Study of Indonesia" delve into the practical challenges and opportunities in adopting BGI in cities of the developing world. From examining the eco-cultural legacy of Feng Shui to exploring the role of community engagement in building sustainable BGI, this part offers valuable insights and case studies to inform policymakers, urban planners, and practitioners alike. Chapter "Challenges and Opportunities in Adopting Blue-Green Infrastructure in Cities of the Developing World" explores challenges and opportunities in adopting BGI in developing world cities, emphasizing its potential as a solution to urbanization issues. It discusses BGI benefits and foundations across environmental, social, and economic aspects. Case studies illustrate successful implementation, showcasing how integrated planning and community involvement can overcome challenges. Collaboration among various stakeholders, including governments, communities, and the private sector, is highlighted as crucial for addressing institutional and financial barriers. Overall, it presents a holistic vision for transforming developing cities into greener, resilient, and inclusive spaces, tailored to their specific contexts.

Chapter "From "Human-Centered" to "System-Oriented": Eco-Cultural Legacy of Feng Shui and Scientific Principles for Establishing Modern Resilient Cities" discusses how cities face ecological challenges due to global warming, such as UHI effects and flood hazards, hindering SDGs. It discusses Feng Shui as a traditional Chinese philosophy offering insights into interacting with nature and addressing environmental changes. By studying the application of Feng Shui in Hong Village, the chapter explores its principles and passive strategies applicable to modern urban systems. The chapter establishes a framework for modern applications of Feng Shui, citing successful projects in South Korea and China that integrate Feng Shui principles into BGI to address urban challenges. It proposes nature-based and soft intervention approaches for constructing and rehabilitating BGIs to enhance urban resilience and achieve SDGs.

Chapter "The Role of Community Engagement in Building Sustainable Blue-Green Infrastructure–Best Practices and Case Studies" discusses the emerging concept of BGI in urban planning, emphasizing its importance for addressing challenges posed by rapid urbanization and climate emergencies. It outlines BGI's multifunctional benefits, such as improving water and air quality, mitigating UHI effects, and enhancing biodiversity. The chapter proposes a framework based on evidence to prioritize community engagement in BGI projects, aiming to guide policymakers, environmentalists, and urban planners toward more holistic and socially inclusive water-sensitive urban planning.

Chapter "Contributions of Vertical Greening to Micro-Climate Control: Exploratory Case Study in a Residential Environment" discusses the benefits of vertical greening or green walls in urban areas, particularly in mitigating heat stress. An exploratory study conducted in Akure, Nigeria, evaluated a vertical greening prototype installed on a residential building. Field measurements showed a reduction in maximum and minimum temperatures compared to a typical building without a green wall. The study suggests that vertical greening contributes to temperature moderation in urban environments. It stresses on policy initiatives to promote the incorporation of vertical greening into green infrastructure in African cities experiencing urban densification. Chapter "Landscape Planning Strategies Within the Scope of Development Plans and Implementation Problems: The Case of Türkiye" discusses development plans initiated by the government since 1963, aimed at addressing various issues such as economy, health, education, and transportation. It examines urbanization problems, climate change, and disasters, particularly within the scope of BGI. The research compares plan decisions with actual implementations and highlights the need for effective execution of planning decisions to build resilient cities in the face of climate change and disasters.

Chapter "Blue-Green Infrastructure for Flood Resilience: Case Study of Indonesia" discusses the increasing risk of urban floods due to climate change, particularly in developing countries like Indonesia, which can lead to significant infrastructure damage. It highlights the importance of BGI in disaster risk reduction and flood management by integrating with urban environmental aspects. The chapter examines issues, gaps, opportunities, and BGI implementation for flood resilience in Indonesian cities such as Bandung, Balikpapan, Samarinda, and Semarang. Case studies illustrate BGI gaps, successes, and threats, identifying policy background, city characteristics, implementation methods, and effectiveness.

Part Three of the book explores scaling and assessment methodologies for urban planning in the context of BGI. Chapters "Modeling Climate Change-Resilient Urban Agriculture in Developing Nations: A Case Study" through "Spatial Accessibility Disparities and Users' Opinion Assessment of Khartoum State Public Green Spaces" present innovative approaches and case studies aimed at modeling climate-resilient urban agriculture, redeveloping open and urban green spaces (UGS), and assessing spatial accessibility disparities in public green spaces. These chapters underscore the importance of incorporating BGI into urban planning frameworks to foster sustainable and equitable development.

Chapter "Modeling Climate Change-Resilient Urban Agriculture in Developing Nations: A Case Study" discusses the importance of implementing sustainable practices in rapidly urbanizing areas, particularly in developing nations vulnerable to climate change. It suggests that BGI can address climate and sustainability challenges in urban environments, including meeting SDGs related to poverty and hunger. Urban agriculture is proposed as a solution to local food security, especially in densely populated cities in Asia, South America, and Africa. The chapter introduces a sunlight-based computational approach for designating agricultural potential on urban surfaces, emphasizing the benefits of building-integrated agriculture (BIA) in minimizing external risks to food production. Hybrid lighting systems are suggested to reduce energy consumption in BIA, thus making it more sustainable for developing nations with limited energy infrastructure. The availability of solar radiation on urban surfaces is highlighted as a crucial factor in selecting BIA locations.

Chapter "Redeveloping Open and Urban Green Spaces for Climate Resilience; A Case Study of Chennai" discusses the importance of open and UGS for urban sustainability, ecosystem services, climate resilience, and human well-being. It presents a GIS-based framework for identifying potential open and UGS areas for climate resilience in Chennai, using satellite imagery. The framework identifies 63 sq. km of land parcels suitable for development as open and UGS, which can help mitigate floods and the UHI effect. The chapter emphasizes the need to incorporate these spaces into urban development master plans and discusses opportunities and barriers, including capacity gaps and governance issues, in Indian cities' efforts to enhance climate resilience through UGS.

Chapter "Spatial Accessibility Disparities and Users' Opinion Assessment of Khartoum State Public Green Spaces" examines the spatial distribution and accessibility of UGSs in Khartoum State, Sudan, given its vulnerability to excessive dry climate conditions due to climate change. Using geospatial and questionnaires, it assesses access to urban parks, community gardens, and city forests for both drivers and pedestrians. Results indicate unequal access, wherein a desire for more well-equipped parks emerges despite the majority requiring entrance fees. The study concludes that UGS accessibility does not align with international quality of life indices.

Finally, Part Four delves into policy issues and sustainable development considerations related to BGI implementation. Chapters "Multifunctional Patterns and Governance Challenges of Blue-Green Infrastructure in Puebla, Mexico" through "Systemic Action Network for Improving Blue-Green Infrastructure Based on the Natural Capital Investigation: The Strategic Plan in Vlorë, Albania" examine multifunctional patterns and governance challenges, the role of BGI in stimulating growth, and systematic action networks for improving BGI based on natural capital investigations. These chapters provide valuable insights into the policy landscape and offer strategic recommendations for advancing BGI agendas at the local, national, and international levels.

Chapter "Multifunctional Patterns and Governance Challenges of Blue-Green Infrastructure in Puebla, Mexico" discusses the multifunctionality of BGI in urban areas, particularly in the City of Puebla and its metropolitan area in central Mexico. It highlights the potential benefits of BGIs, such as open spaces, social interaction areas, climate regulation, and economic well-being. The chapter reviews two factors influencing BGI multifunctionality: size and urban legislation. Using satellite images, it assesses the size and functions provided by BGIs, noting that larger BGIs tend to offer more functions. However, urban legislation related to BGI multifunctionality is found to be lacking across different government levels.

Chapter "Blue-Green Infrastructure as a Stimulant of Growth: Case of Ahmedabad, India" discusses the concept of BGI, which utilizes natural ecosystems and UGS to create sustainable environments. It highlights the role of BGI in enhancing the quality of life, promoting a healthy environment, and stimulating economic and social progress. Taking a case of city of Ahmedabad, it proposes to examine BGI-induced growth through examples of waterfront/riverfront development projects.

Chapter "Systemic Action Network for Improving Blue-Green Infrastructure Based on the Natural Capital Investigation: The Strategic Plan in Vlorë, Albania" explores the significance of natural capital in maintaining human well-being and urban resilience, particularly in the face of climate and Anthropocene challenges. Using the city of Vlorë in Albania as a case study, it examines the practicality of NbS and Integrated Modification Methodology (IMM) in promoting sustainable development. The case study focuses on improving blue-green infrastructure to align with United Nations SDG 11 and regional guidelines. It proposes an innovative urban model for sustainable development, emphasizing the integration of natural capital and NbS principles. The chapter discusses the transferability of IMM research and underscores the importance of multidisciplinary interventions for sustainable urban transformation.

In compiling this volume, we have endeavored to bring together contributions from leading experts and scholars in the field of urban resilience, sustainability, and climate adaptation. Through their collective expertise and interdisciplinary perspectives, we aim to provide a comprehensive resource that will inspire dialogue, inform policy, and catalyze action toward building more resilient and sustainable urban settlements in developing countries. We extend our sincere gratitude to all the contributors for their invaluable insights and scholarly contributions to this volume. We also wish to thank the reviewers for their meticulous feedback and suggestions, which have undoubtedly enhanced the quality and rigor of the chapters.

Lastly, we express our appreciation to the publisher for their support and guidance throughout the publication process. As editors, we hope that this book will serve as a catalyst for continued research, innovation, and collaboration in the field of blue-green infrastructure and sustainable urban development. May it inspire new ideas, initiatives, and partnerships aimed at creating healthier, more resilient, and sustainable cities for generations to come.

New Delhi, India P. K. Joshi
New Delhi, India K. S. Rao
New Delhi, India Rahul Bhadouria
New Delhi, India Sachchidanand Tripathi
Mohali, India Rishikesh Singh

Contents

About the Editors

P. K. Joshi, Ph.D. is Professor in School of Environmental Sciences (SES) and Former Chairperson of Special Centre for Disaster Research (SCDR), Jawaharlal Nehru University (JNU), New Delhi. His research is focused on environmental studies, with specialization in landscape characterization and climate change analysis. He has immensely contributed in understanding geosphere and biosphere interactions. His present research work contributes to assessing impact and vulnerability of climate change, and disaster risk reduction. Now, he has spent more than two decades in actively contributing to a wide range of training, research, and academic programs in the field of geospatial analysis. He possesses a scholastic record coupled with strong leadership skills and networking capabilities. He has made significant contribution as a "Scientific Administrator," and in promoting "Scientific Research" inter-alia through education, research publications, and outreach activities.

K. S. Rao, Ph.D. served as Professor and Head in the Department of Botany and Dean of Faculty of Technology, University of Delhi, India. He also served in different academic and research positions in various reputed organizations/institutes. He has guided 15 Masters (M.Sc.), 3 M.Phil., and 24 Ph.D. students for their degrees in various universities and has published about 25 books and more than 160 research papers in international and national scientific journals. He is member of various academic scientific societies and served as member in several committees constituted by the Government of India to address ecological and environmental challenges. His current area of research includes restoration of degraded lands using agroforestry and land rehabilitation techniques along with urban metabolism and urban ecology.

Rahul Bhadouria, Ph.D. is an Assistant Professor at the Department of Environmental Studies, Delhi College of Arts and Commerce, University of Delhi, New Delhi, India. He obtained his doctoral degree from the Department of Botany, Banaras Hindu University, Varanasi, India. He has published more than 24 papers, 25 book chapters, and 12 edited books in internationally reputed journals/publishers. His current research areas are management of soil C dynamics to mitigate climate change, a perspective on tree seedling survival and growth attributes in tropical dry forests under the realm of climate change, plant community assembly, functional diversity and soil attributes along the forest-savanna-grassland continuum in India, recovery of degraded mountains in central Himalayas and urban ecology.

Sachchidanand Tripathi, Ph.D. is an Associate Professor at the Department of Botany, Deen Dayal Upadhyaya College, University of Delhi, Delhi, India. He obtained his doctoral degree from the Department of Botany, Banaras Hindu University, India. The areas of his interest are plant ecology, soil ecology of various ecosystems, ecophysiology, and urban ecology. He has published more than 60 publications (including research papers, books and book chapters, conference proceedings) with journals and publishers of international repute. He is serving as an editorial board member of several reputed international journals.

Rishikesh Singh, Ph.D. is an Assistant Professor at Amity School of Earth and Environmental Sciences, Amity University Punjab (AUP), Mohali, India. His research interests include various dimensions of ecology and environmental science, viz., herbaceous vegetation analysis, soil carbon dynamics, land-use change and management, management of environmental contaminants, biochar amendment, tree seedling ecology, and urban ecology. He has worked on different ecosystems including dry deciduous forests and savannas, riparian landscapes, and urban and agro-ecosystems of the tropical region. His major contribution to science is in "Biochar and Agro-ecology." He received the Young Scientist Award from NESA, New Delhi in 2022. He has published more than 80 Scopus-indexed publications in various international journals, and has edited 12 books published by Elsevier, Springer Nature, Wiley and CRC Press. He is serving as an Editorial Board member of several reputed international journals, e.g., Tropical Ecology, Biochar, Carbon Research, Frontiers in Forest & Global Change, etc.

Urban Resilience and Sustainability

Strength and Limitation of Nature-Based Solutions Towards Adaptation and Mitigation of Climate Change in Developing Countries

Asma Ben Amor, Hadia Hemmami, Soumeia Zeghoud, and Ilham Ben Amor

Abstract Natural methods provide considerable potential for curtailment cities' susceptibility to climate change and boosting their resilience by promoting the growth of blue and green urban region. Consequently, they may help mitigate the worst consequences of climate change and provide municipalities proactive possibilities for adaptation. We examine the numerous states in which nature-based solutions are pertinent for reducing climate alteration and adapting to it in urban environments, and we develop metrics for gauging their efficacy and corresponding knowledge gaps. We also examine current obstacles and prospective openings for expanding the scope and efficiency of the implementation of nature-based solutions (NbS). In this chapter, we emphasize three key necessarily for future scientific and policy agendas when addressing NbS: (1) create more convincing evidence for natural climate change adaptation and extenuation strategies and increase implementation to increase awareness; (2) Reflexive techniques, which comprise bringing together new networks of society, practitioners, and NbS ambassadors, can be used to adjust for governance issues in putting NbS into practice; (3) When implementing nature-based solutions, take into mind socio-environmental justice and societal cohesion by employing integrated governance strategies that take into account the integrative and transdisciplinary participation of different actors. Taking these requirements into account, NbS may act as tools for climate mitigation and adaptation that also enhance societal well-being, making them excellent choices for investments in sustainable urban development.

A. Ben Amor · H. Hemmami (✉) · S. Zeghoud · I. Ben Amor
Department of Process Engineering and Petrochemical, Faculty of Technology, University of El Oued, 39000 El Oued, Algeria
e-mail: hemmami.h@gmail.com

Renewable Energy Development Unit in Arid Zones (UDERZA), University of El Oued, 39000 El Oued, Algeria

H. Hemmami · S. Zeghoud · I. Ben Amor
Laboratory of Applied Chemistry and Environment, Faculty of Exact Sciences, University of El Oued, P.O. Box 789, 39000 El Oued, Algeria

P. K. Joshi et al. (eds.), *Blue-Green Infrastructure for Sustainable Urban Settlements*,
https://doi.org/10.1007/978-3-031-62293-9_1

Keyword Environmental Implications · Nature-based solutions · Ecosystem services · Urban development

1 Introduction

The traces of climate change are already being touched about the world, with the most notable effects being increased severe weather events, altered rainfall patterns, and rising temperatures (Kabisch et al., 2016). Because of the concentration of infrastructure, property, and population, urban areas are particularly vulnerable. Despite the global north producing over 92% of all GHG emissions (Hickel, 2020), the most affected areas are in the developing world (Change, 2014). According to the most modern Intergovernmental Panel on Climate Change (IPCC) report (Change, 2014), the effects are intensifying as tipping points are being reached, and governments fail to take sufficient action. Even low amounts of warming can have disastrous effects, even if we are facing scenarios of significant temperature increases (Kemp et al., 2022). The sixth mass extinction of life on Earth is a direct result of the biodiversity problem and climate change (Cowie et al., 2022). Nature-based solutions (NbS) has been endorsed more frequently as an alternative to conventional urban infrastructure and for urban adaptation (Arias et al., 2021; Frantzeskaki, 2019; Seddon et al., 2020). The term "Nature-based Solutions" was originally used in a 2008 World Bank study (Sowińska-Świerkosz et al., 2021; Weltbank, 2012) that identified projects addressing the biodiversity and climate problems with a focus on adaptation and vulnerable populations. The International Union for Conservation of Nature (IUCN) defines NbS as initiatives that support biodiversity and human wellbeing in addition to protecting, managing, and restoring natural or modified ecosystems (Cohen-Shacham et al., 2016). The European Commission (EC) offers a more expansive definition of NbS, defining it as cost-effective, nature-inspired, nature-supported solutions that concurrently benefit the society, economy, and environment while promoting resilience (Faivre et al., 2017). Since the impression of NbS is motionless relatively new, there are still several information gaps, including how to promote its applicability to cities (Kabisch et al., 2016). Additionally, there are obstacles to its execution that must be recognized and removed. The goal of this chapter is to fill in any information gaps regarding the usage of NbS for application in an urban setting while also contributing to the assessment of the problem's dimensions. The applied chapter technique focuses on the most relevant published papers with an emphasis on NbS and its adaptation to the urban environment. Several reports from pertinent organizations having a worldwide influence, such as the World Bank, were analyzed in order to comprehend how various governments, agents, and sectors are implementing/supporting these solutions.

2 The Role of NbS in Climate Change

In order to comprehend the significance of NbS within the framework of climate change, our attention is directed toward the following aspects: (a) What are NbS, and how do they differ from urban greening? (b) In addition to addressing the climate crisis, what other challenges can be tackled using these solutions? (c) To what degree can NbS contribute to urban adaptation? (d) How meaningful are these remedies for populations at risk? (e) What challenges and opportunities arise from the transdisciplinary nature of NbS? (f) Lastly, the various typologies of NbS are delineated.

2.1 Basic Ideas and Definitions

NbS refers to strategies that utilize the inherent capabilities of the natural environment to deliver a variety of advantages, encompassing both environmental and societal welfares (Sowińska-Świerkosz & García, 2022). The uncertainty of the notion has been emphasized by several authors (Potschin et al., 2015; Sowińska-Świerkosz & García, 2022), and other organizations offer their own definitions. As per EC, the idea in question is described as an overarching term, incorporating many methodologies such as urban green substructure, ecosystem-based version, blue-green infrastructure, and ecosystem-based disaster danger reduction (Dumitru & Wendling, 2021; Frantzeskaki et al., 2022). NbS encompasses a range of interventions and typologies of land use, spanning natural, rural, and urban settings. These interventions include the establishment and preservation of wetlands, forests, parks, and green belts inside and around metropolitan areas. Additionally, NbS involves the implementation of plants that naturally cleanse wastewater, green walls, green roofs, ecological corridors, and other forms of blue, green, and hybrid substructures. These proposed solutions have the potential to enhance urban pliability, mitigate the hazards associated with urban floods, minimize losses, and promote economic benefits while also providing crucial social and ecological services (Smart, 2021). Significantly, NbS plays a crucial role in fostering sustainable resolutions to enduring environmental predicaments (Eggermont et al., 2015; Bauduceau et al., 2015). The idea of NbS has gained recognition in the discourse surrounding climate change and the biodiversity crisis (Kemp et al., 2022; MacKinnon et al., 2011). While NbS are commonly linked to these issues, it is significant to note that their application is not imperfect to this domain. NbS can also be employed to tackle other challenges, including pest management, food production enhancement, and wastewater treatment (Haines-Young & Potschin, 2014). The capacity of NbS to concurrently accomplish climate adaptation and mitigation goals is one of its main advantages. Numerous scientific evaluations and reports have emphasised this feature, such as the Intergovernmental Panel on the Global Adaptation Commission Report's Climate Change and Land Report (Seddon et al., 2020), Climate Change (Kabisch et al., 2016), and the more recent Sharm El-Sheikh Implementation Plan (Bakošová, 2023). Furthermore, the various advantages they offer are characterized by their relatively affordable nature since they have the potential to yield a cost–benefit ratio exceeding 100:1 (Munang et al., 2013). The adaptability of NbS renders them a very effective method for addressing what may, at times, appear to be an insurmountable task (Arias et al., 2021). It is imperative to acknowledge that they do not possess omnipotent capabilities, and within the framework of mitigation, their influence is significantly less pronounced compared to the substantial decrease in the use of fossil fuels (Anderson et al., 2019).

Differentiating NbS from broader greening initiatives is of utmost significance. In the literature assessment conducted by Sowińska-Świerkosz and García (2022), a set of 11 exclusion criteria for green substructure to be classified as NbS is identified.

1. The absence of operational ecosystems necessitates that NbS encompass natural procedures rather than being just "inspired by nature." This might manifest in several ways, such as the utilization of solar and wind energy or the use of biomimicry.
2. The implementation of NbS necessitates a structured approach, wherein activities are not arbitrary but instead guided by well-defined objectives, intended beneficiaries, and governance mechanisms. It is imperative to reject actions that are devoid of a comprehensive assessment of the prevailing social, economic, and environmental conditions prior to their execution. For instance, creative projects, including plants, would be excluded.
3. The absence of post-implementation goals is problematic, as NbS should be targeted towards addressing pre-existing issues. This would include the exclusion of historical gardens, as one illustrative example.
4. The exclusion of practices such as monoculture and the reduction of "greenwashing" methods would ensure that there is no negative or negligible influence on biodiversity.
5. The benefits of green infrastructure are comparable to those of grey infrastructure. This observation is of special significance as it underscores the necessity for NbS to offer additional advantages compared to alternative approaches. It emphasizes that the extent of human engagement should not be the main determining factor when evaluating the effectiveness of an intervention. Hybrid methods are frequently seen as most appropriate for metropolitan environments (Seddon, 2022).
6. Inequitable allocation of advantages. NbS should provide the concurrent attainment of environmental preservation, human well-being, and economic advantages without prioritizing any one objective above the others.
7. The technique of copy-paste implementation. NbS are contingent upon the unique environment in which they are implemented, and their design necessitates a certain level of knowledge. The implementation of adaptive management and design principles can facilitate the replication of solutions.
8. Adopting a top-down form of governance is indispensable for the successful implementation of NbS, as it necessitates active community engagement.

9. The use of a dynamic management strategy is recommended for NbS, wherein adaptive management and innovation are integral components throughout its duration, grounded in the theory of change. The importance of slide in decision-making cannot be overstated, particularly with regard to the inclusion of public input.
10. It is crucial for NbS to exhibit cost-effectiveness due to the presence of disproportionate financial expenditures in relation to the advantages it provides.
11. Point-scale methods, such as NbS, entail the integration of social and ecological variables at a landscape size, resulting in cumulative consequences. It is significant to note that NbS should not be viewed as separate interventions but rather as interventions that involve interactions across several dimensions.

The United Nations (UN) Organisation is a significant advocate for NbS. At the 2019 United Nations Climate Action Summit, NbS has been designated as one of the nine main action tracks. Following this realization, more than 20 nations and countless organizations have pledged to use nature's resources to combat climate change. These commitments primarily emphasis on the preservation, reforestation, and restoration of ecosystems (Gielen et al., 2019). The enhanced significance of their relevance is seen in the Nationally Determined Contributions (NDCs) that have been presented to the United Nations Framework Convention on Climate Change (UNFCCC). In the year 2015, a mere 66% of the signatories of the Paris Agreement incorporated NbS in their Nationally Determined Contributions (NDCs) (Roemer et al., 2023). However, this figure has significantly increased to 92% in 2021, with NDCs explicitly referencing NbS in relation to both adaptation and mitigation efforts (Castelo et al., 2023).

2.2 *The Crisis Era*

The global community is presently confronted with a plethora of crises that are emerging concurrently (Herman, 2016). The increase in human activities and population growth is placing an unsustainable burden on the natural environment, leading to many crises, such as the deterioration of ecosystems, the destruction of forests, the loss of biodiversity, and the acceleration of climate change (Díaz et al., 2019). The decline of biodiversity is occurring at unprecedented magnitudes within the annals of human history (Hughes et al., 2022), as evidenced by the current endangerment of around one million class (Brondizio et al., 2019). In addition to the aforementioned obstacles, society is confronted with social crises, namely inequality and public health concerns. NbS provides the advantageous capability to effectively tackle these pressing issues as well (Lin et al., 2021).

In 2009, Technical Expert Group 1, the Convention on Biological Diversity, initially acknowledged the interconnections among biodiversity crises and climate change. Subsequently, the IPCC and Intergovernmental Science-Policy Platform on Biodiversity and Ecosystem Services (IPBES) have emphasized the significance of these interlinkages, asserting that addressing them in isolation is not feasible (Pörtner et al., 2021). The recognition of heightened risks arising from interconnected crises is evident across many levels and diverse organizations, including the World Economic Forum (WEF). The annual global dangers report published by WEF serves to identify and classify risks based on five distinct areas, namely: (a) geopolitical risks, (b) environmental risks, (c) technical risks, (d) societal risks, and (e) economic risks. According to a study conducted in 2022, the primary hazards anticipated for the upcoming decade are environmental concerns. These risks include the potential failure to address climate change adequately, the incidence of dangerous weather events, and the ongoing issue surrounding biodiversity (Björnsdottir et al., 2022). Although NbS possess versatility, their primary advantages in urban contexts lie in their ability to mitigate climate effects, enhance urban biodiversity, improve public health, and foster social cohesion and equality (Ulmer et al., 2016). Numerous methods that prioritize biodiversity, such as the implementation of green corridors to facilitate species migration and the expansion of core conservation areas, might also provide advantages for adaptation (Field & Barros, 2014).

When it comes to mitigating climate effects, NbS has the potential to contribute to the regulation of microclimates (Lindén et al., 2016), management of stormwater runoff, and mitigation of the impact of sea level increase (Field & Barros, 2014). NbS have the potential to enhance human welfare (Seddon et al., 2020), including improvements in public health and social advantages. The issue of public health has garnered significant attention, as seen by the World The Health Organization's (McGuire, 2016) attention on the threat presented by climate change. The Lancet study on climate change (Romanello et al., 2021) highlights the impact of climate change on health. Due to rising temperatures, changed rainfall patterns, and an increase in extreme weather events, it has been discovered that climate change has a major negative impact on health. These factors, especially among the world's most vulnerable people, are undoing the centuries-long progress gained in tackling food and water insecurity. According to the study, climate change is exacerbating the ecological factors that facilitate the transmission of various viruses through water, air, food, and vectors. This phenomenon has the potential to meaningfully contribute to the proliferation of illnesses. The established advantages of urban green spaces, irrespective of their classification as NbS, mostly pertain to public health. These benefits include improvements in cardiovascular health, blood pressure regulation, respiratory health, obesity prevention, and diabetes management (Donovan et al., 2013). Additionally, urban trees have a significant role in mitigating air pollution, as shown by existing research (Bhadouria et al., 2022).

Research has demonstrated that being exposed to natural surroundings has positive belongings on mental well-being, including the reduction of stress levels and enhancement of attention abilities (Bratman et al., 2012). There is a correlation between the availability of outdoor green spaces and the overall well-being of children residing in densely populated metropolitan regions (Haaland and Bosch, 2015). Urban green spaces have been found to have a positive influence on public health due to their promotion of physical activity and enjoyment (Haaland and Bosch,

2015). Additionally, at a regional level, NbS can production a crucial part in safeguarding water, sanitation, and hygiene (WASH), which are interconnected with health and have implications across all sectors (Castelo et al., 2023). Additionally, they have the potential to mitigate the occurrence of zoonotic illnesses, like COVID-19, that are frequently associated with the deterioration or loss of ecosystems (Castelo et al., 2023). Recent scientific research has demonstrated that climate change has the potential to augment the incidence of these illnesses since it can lead to the emergence of novel species interactions (Gilbert, 2022). In relation to societal advantages, NbS have been found to mitigate inequality, foster social cohesiveness, decrease instances of violence, and contribute to aesthetic appreciation (Soares et al., 2011). The impact of the aesthetic component on health has been demonstrated to be beneficial, primarily in relative to the visual aspects of the environment (Seresinhe et al., 2015). The initial investigation specifically addressing this matter was published in 1984. Urban green areas have been found to have an impact on the apparent thermal ease, extending beyond their ability to reduce effective temperature (Klemm et al., 2015).

2.3 *Urban Adjustment*

The urgency of climate change, defined as the procedure of adapting to present or anticipated climatic impacts, is widely recognized due to its global occurrence (Field & Barros, 2014). The allocation of climate money has, nonetheless, exhibited a preference for mitigation efforts somewhat than adaptation measures (Halimanjaya & Papyrakis, 2012). Despite the mandate of the Cancun Adaptation Framework to prioritize adaptation on par with mitigation, current statistics indicate a growing disparity between the financing allocated for adaptation and the actual need for adaptation (Neufeldt et al., 2021). According to the Paris Agreement, a yearly transfer of USD 100 billion is mandated, commencing in 2020, from affluent countries to poor nations (UNFCCC, 2015). According to recent studies, it has been determined that the current measures in place would not be enough. This is due to estimations suggesting that the yearly expenses of adaptation in poor nations will vary between USD 155 billion and USD 330 billion by the year 2030 (Neufeldt et al., 2021). In accordance with this particular worry, the Glasgow Climate Pact highlights the need to increase the financial resources allocated by wealthy nations to poor nations, especially for the purpose of adaptation, with the goal of doubling these sums by the year 2025.

Cities are especially susceptible to the effects of climate change (White et al., 2005). According to the UN Secretary-General's remarks at the C40 World Mayors Meeting in 2019, cities play a vital role in the mitigation of climate change due to their significant contribution to carbon emissions. Consequently, the achievement or failure of addressing climate change primarily hinges on the actions taken inside urban areas (Castelo et al., 2023). There will be several impacts arising from the situation, encompassing the loss of infrastructure as well as significant implications for public health. The latter is particularly influenced by weather patterns and various interconnected factors, including a decrease in labor productivity (Smith et al., 2015). The urban heat island (UHI) effect refers to the occurrence of greater temperatures in cities associated to the surrounding natural or rural zones, with a potential temperature difference of up to 8 °C. This is mostly attributed to the high concentration of hard materials inside urban environments (Yan et al., 2020). Moreover, recent studies have revealed that urban areas are experiencing a 29% higher rate of warming compared to rural regions. This observation suggests an uneven amplification of the UHI phenomenon. Urban regions are the primary locations where the mainstream of the global population resides, as evidenced by the fact that 55% of the world's population now inhabits cities. Projections indicate that this figure is anticipated to rise to 68% by the year 2050 (Nations, 2018). The majority of global urban development is projected to occur in the developing world, specifically in the American, Asian, and African continents, accounting for almost 90% of the anticipated expansion. Urban growth has been found to have detrimental effects on the environment, including the reduction of soil permeability and the heightened danger of flooding, in addition to the UHI effect (Skrydstrup et al., 2020). The compounding effect of this phenomenon is further amplified when it is coupled with climate change, interconnected stressors, and the deterioration of natural ecosystems (Kabisch et al., 2016). Hence, the prioritization of urban adaptation in emerging nations is of considerable significance.

Certain regions of the globe are unable to tolerate delays, as exemplified by South Asia, which is projected to witness an increase of around 49 million individuals being thrust into the depths of extreme poverty by the year 2030 (Bank, 2020). The heatwave that occurred in March–April 2022 is considered to be one of the most significantly affected regions globally. It has been determined that climate change has augmented the likelihood of this heatwave by a issue of 30 (Coleman, 2022). The efficacy of NbS in urban adaptation has been extensively validated, resulting in the mitigation of climate-related consequences and the provision of a multitude of public health advantages (Zwierzchowska et al., 2019). NbS holds significant importance in the context of urban adaptation in impoverished countries due to its high cost-effectiveness, ease of implementation, and reduced maintenance costs compared to technical alternatives (Seddon, 2022). The significance of NbS for the Least Developed Countries (LDCs) is particularly pronounced due to their substantial economic reliance on natural resources. The adoption of NbS plays a pivotal role in the protection of natural resources, hence ensuring the well-being of communities. This is of utmost importance since the deterioration of ecosystems intensifies human vulnerability to climate-related consequences and hampers admission to essential resources such as food, water, and clean water. According to the United Nations Framework Convention on Climate Change (UNFCCC), the LDC 2050 Climate Change Vision emphasizes the potential of NbS to enhance the pliability of landscapes and ecosystems in LDCs. The existing information suggests that despite the fact that 45 out of the 46 Nationally Determined Contributions (NDCs) of these countries specifically include nature-related projects, NbS in LDCs are not given significant priority in climate finance (Castelo et al., 2023). According to a recent study, the distribution of funds for climate adaptation projects in LDCs for the period of 2014 to 2018 revealed that nature-related initiatives received less than 10% of the total funding (Roe et al., 2023). In certain regions, such as Bangladesh, there is a tendency to prioritize engineering solutions over NbS in climate adaptation initiatives. This is evident in the fact that 88% of such projects in Bangladesh are engineered solutions, while NbS accounts for just 12%. Despite the results of Narayan et al. (Narayan et al., 2019) in that particular country, it is evident that engineering solutions exhibit a comparatively inferior performance when compared to NbS. One illustrative instance is mangrove forests, which effectively mitigate the adverse impacts of sea level rise, a prominent vulnerability faced by the nation, as opposed to the implementation of seawalls.

2.4 Application to Vulnerable Communities

Low-income areas experience a disproportionate impact from climate change in relation to public health. This is evident via the adverse effects of extreme weather events and increasing temperatures, which have instant ramifications on public health (Beggs et al., 2019). This phenomenon can be attributed not just to a diminished ability to adapt. Low-income people often reside in regions that are prone to flooding and have limited access to green spaces (Deria et al., 2020). These locations not only mitigate UHI impact but also facilitate the absorption of stormwater runoff. The phenomenon of climate change is expected to contribute to a rise in the global population of individuals who are susceptible to adverse impacts. According to the World Bank, if left unattended, climate change has the capacity to cause a rise in poverty levels for a range of 68 million to 132 million individuals by the year 2030 (Jafino et al., 2020). Given the significant social advantages of NbS, such as their capacity to mitigate inequities, they appear to hold special relevance for marginalized groups and developing countries. NbS may be viewed as a sustainable and low-risk strategy that can be implemented in many settings and communities. NbS have the potential to contribute to the mitigation of poverty through the safeguarding of livelihoods (Hou-Jones et al., 2021).

According to the research conducted by Friends of Ecosystem-based Adaptation (FEBA) in 2021, it has been seen that NbS has the potential to mitigate the vulnerability of communities that are at risk of exposure to various hazards. This reduction in vulnerability may be achieved through three distinct mechanisms. There are three key strategies that can be employed to heading the effects of climate change. Firstly, the development of resilience, similar to the principles of agroecology, can enhance both food security and income. Secondly, the frequency of floods and landslides may be successfully reduced by protecting and restoring forests and streams. Last but not least, mangrove growth and restoration in coastal ranges may greatly decrease the susceptibility of towns to tempest flows. Exposed communities have the ability to duplicate the successful solutions offered by NbS, therefore assuming ownership and promoting them (Walz et al., 2021). This process enhances the autonomy and resilience of these communities. The World Bank has highlighted increasing food costs as the primary consequence of climate change that will have the highest influence on severe poverty by the year 2030 (Bank, 2020). Regional differences are expected to occur, with the primary danger in South Asia and Sub-Saharan Africa being related to food costs, whereas in Latin America and East Asia, the Caribbean, and the Pacific, the major risk is associated with health. The development of adaptation programs should prioritize addressing the following threats: urban gardening initiatives in cities can serve as a means to assist low-income groups in mitigating food security problems (Bank, 2020). The involvement of the general public is crucial. Multiple studies have provided empirical evidence supporting the notion that the successful implementation of NbS requires the active involvement of local vulnerable populations, ensuring their full participation in both the long-term and short-term adaptation efforts. It is imperative to safeguard the participation of susceptible communities in climate-related decision-making processes, with a preference for their leadership wherever feasible. The energetic participation of Indigenous groups in the implementation procedure is critical since they possess valuable information about the local natural environment (Seddon, 2022).

2.5 *Dimension of Transdisciplinarity*

The usage of NbS in the setting of urban adaptation is intrinsically characterized by its transdisciplinary and intersectoral nature. Transdisciplinarity is a prominent feature of NbS, as evidenced by scholarly sources (Albert et al., 2019). This characteristic is notably evident in the collaborative processes of cocreation, co-development, and coimplementation of initiatives aimed at addressing various difficulties. The use of natural processes to tackle social difficulties is a longstanding practice well rooted in landscape architecture and planning theory. This approach has been significantly impacted by McHarg's seminal work, "Design with Nature," published in 1969. NbS also derives significant insights from the fields of urban planning, engineering, and environmental sciences (BenDor et al., 2017). In order to effectively apply NbS, it is imperative for urban planning to embrace a fresh and innovative approach towards urban greening. This strategy may be evaluated through the implementation of initiatives aimed at expansion and urban regeneration. The successful execution of NbS necessitates intersectoral cooperation, a task that can prove to be arduous, particularly for municipal administrations in poor nations (Castelo et al., 2023). The inclusion of multi-level governance introduces a heightened level of intricacy to the procedural framework. The oversight of natural resource implementation and management is carried out by a range of varied organizations, leading to the emergence of several co-benefits, including those at the social level (Castelo et al., 2023). Multi-level governance facilitates collaboration and readiness, both of which are crucial for enhancing the ability to tackle climate-related issues. The utilization of these procedures for the purpose of equipping local governments with the ability to effectively address forthcoming crises may be advantageous. However, their current execution may offer difficulties due to the presence of sectoral divisions that commonly occur within their organizational frameworks.

2.6 NbS Typologies

NbS may be classified into several types based on three key factors: (a) the extent of human involvement, (b) the specific problems being treated, and (c) the kind of environment in which they are implemented. The initial typological classification for NbS, which was published in 2015 (Halimanjaya & Papyrakis, 2012), outlined three distinct types based on the level of human interference. Type 1 encompasses interventions that involve negligible or no human interference in ecologies. Type 2 pertains to activities that strive to found ecosystems and sceneries that are both sustainable and multifunctional. Lastly, Type 3 encompasses measures that concentrate on the conception of novel ecosystems or the intensive management of current ones. NbS can also be classified based on the difficulties they aim to tackle. The identification of socioeconomic vulnerability dimensions and the specific climatic impacts being addressed may be categorized into three main strategies: (1) the reduction of exposure, (2) the reduction of sensitivity, and (3) the support of adaptive capacity. IUCN and EC classify NbS based on the specific difficulties they aim to tackle. The International Union for IUCN has identified five key areas that are encompassed by NbS. These categories are human health, food security, water security, disaster menace reduction, and climate change (Frantzeskaki et al., 2022). EC has categorized seven distinct types of NbS based on their respective areas of innovation and investigation. These categories include (a) bolstering coastal resilience; (b) carbon sequestration; (c) regeneration and enhancement of well-being in urban areas; (d) promotion of sustainable utilization of matter and energy; (e) restoration of ecosystems; (f) effective management of watersheds; and (g) insurance of the value provided by ecosystems (Bauduceau et al., 2015). There exists a degree of overlap among these categories. An alternative approach to classifying typologies for NbS is through the categorization of landscape types, as planned by Skrydstrup et al. (2022) in the situation of urban flood management.

1. One example of maintainable urban drainage organizations is the implementation of tiny green spaces, such as accessible rain gardens, green roofs, and swales. The aforementioned systems possess a high degree of flexibility, necessitating little or negligible spatial requirements. The standard spatial scale is smaller than 1 hectare.
2. City parks are expansive green spaces that offer many recreational opportunities and may include natural water features such as lakes. According to the cited study, a city park typically spans an area ranging from 1 to 50 hectares.
3. Nature areas refer to green spaces that provide opportunities for many forms of recreational activities and may include bodies of water, such as wetlands. In contrast to urban parks, peri-urban green spaces exhibit a greater degree of

naturalness, fostering enhanced biodiversity. These areas surpass a minimum size threshold of 50 hectares and are commonly situated in close proximity to urban centers, often taking the form of forests or similar ecosystems.

4. Rivers and streams are predominantly composed of water, hence including riverbanks or floodplains that are accessible to individuals. In general, the focus is on certain segments of the river that accommodate urban recreational activities, often ranging in size from 1 to 60 hectares.

3 Knowledge Gaps

As a nascent discipline, there exists a considerable amount of research that remains to be conducted on NbS. Despite garnering growing attention and serving as a focal point of scholarly investigation (Seddon, 2022), several gaps in knowledge persist and have been duly recognized. In their study, Kabisch et al., (2016) delineate four distinct categories of knowledge gaps pertaining to NbS. These categories include (a) the efficacy of NbS, (b) the social dynamics associated with NbS; (c) the design considerations for NbS; and (d) the application challenges of NbS. The study conducted by FEBA (2021) highlights the importance of data, research, policy coherence, assessment, and the development of instruments that promote the preservation of ecosystems and the implementation of NbS in humanitarian settings. The evaluation of NbS's ability to provide advantages is of significant importance, as it has not yet been comprehensively assessed. In 2022, research was started by Network Nature, a source for NbS financed by the European Commission. The project, created together by ICLEI Europe, IUCN, Oppla, BiodivERsA, and Steinbeis 2i, aimed to identify 142 knowledge gaps across 27 overarching subjects (Calliari et al., 2022). A study conducted by Network Nature (2021) revealed the existence of 29 distinct gaps in knowledge and execution (Calliari et al., 2022). The research paper highlights four distinct categories of knowledge gaps in NbS pertaining to (a) technical designing, (b) assessment, (c) governance, and (d) capability building. These categories are visually shown in Fig. 1.

While there may be variations in the categorization of knowledge gaps among different authors, there is a general consensus on two types: technical design and assessment of efficacy. In addition to the knowledge gaps particular to NbS, there exist some gaps in understanding linked to adaptation despite the significant advancements made in this field over the previous twenty years. Based on the examination of existing scholarly works, this current study delineates four distinct categories of knowledge deficiencies that hold significant relevance in the context of employing NbS for urban adaptation. The topic of discussion pertains to the climate conditions that are anticipated to occur in the future. There exists a degree of ambiguity within the field of climate research as well as in the realm of policy-making, pertaining to the appropriate courses of action to be undertaken by national governments. This phenomenon gives rise to uncertainty and necessitates the continuous revision of estimations. The World Meteorological Organisation (WMO) declared in 2022 that there

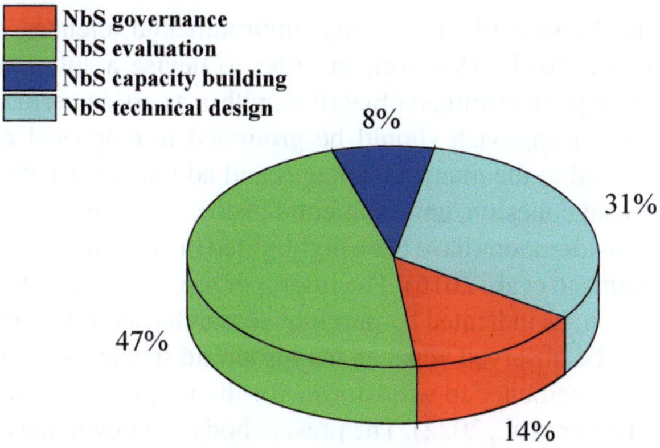

Fig. 1 Knowledge gaps recognized in the Network (2022) desk study

exists a 50% likelihood of attaining the 1.5 °C upsurge above pre-industrial levels, as outlined in the Paris Agreement for the year 2100, during the time frame of 2022 to 2026 (Collier et al., 2022). In the year 2015, the likelihood of this event taking place was nonexistent. An additional illustration of this prevailing ambiguity is demonstrated by a scholarly publication in September 2022, which discloses that the rate of melting seen in Greenland's glaciers seems to surpass prior estimations by a factor of 100 (Schulz et al., 2022). The use of the four representative concentrative pathways (RCPs) enables the examination of potential climatic scenarios in the future; nevertheless, it remains uncertain if any specific RCP, or a variation within their range, will manifest. While many climatic impacts are discernible, their precise magnitude remains to be ascertained via human intervention, unidentified interdependencies within ecosystems, feedback mechanisms, and possibly other factors that are now unknown. Certain regions around the Earth are anticipated to undergo inhabitable conditions as a consequence of the wet-bulb temperature (Suarez-Gutierrez et al., 2020). However, it remains uncertain as to which specific locations will be affected initially and the timeframe within which these impacts will manifest. This information is crucial for the purposes of risk management. In relation to urban areas, cities will witness a significant alteration in their climatic conditions on a global scale. The projected climate conditions by 2050 indicate that cities located nearby the equator, such as Kuala Lumpur, Jakarta, Libreville, Manaus, and Singapore, would likely see significant changes. These changes are expected to be unparalleled in nature (Bastin et al., 2019). Under such circumstances, it is probable that several connected unknown factors may arise. It is vital to comprehend the prospective climate conditions in order to establish objectives and benchmarks pertaining to the execution of NbS in urban variation efforts. Can these solutions sufficiently alleviate the effects of climate change? What is the duration or endpoint of the specified event or timeframe? The use of a time-phased strategy may be deemed most appropriate for effectively aligning adaptation programs with the dynamic nature of continuous developments.

(2) Framework for Technical Design in Theory. The technological know-how required for the usage of NbS for urban adaptation is obtained in large part from

the domains of engineering, environmental sciences, and landscape design (BenDor et al., 2017). However, in order to devise a solution that may effectively accomplish predetermined objectives within a certain period, further study is required. The design approach should be grounded in empirical research, with clear objectives related to the many advantages and additional benefits of NbS, such as biodiversity, social cohesion, environmental justice, and alignment with mitigation efforts. These considerations have been highlighted in preceding research (Frantzeskaki et al., 2022; Kabisch et al., 2016). The improper use of NbS has the potential to result in maladaptation, as indicated by previous research (Seddon, 2022). Instances of maladaptation can be observed when an inappropriate species is selected, such as using species that lack resilience to windstorms for the purpose of providing shelter from windstorms (Turner et al., 2022). The present body of knowledge encompasses the subject matter under consideration. However, instances may arise when this is not the case. It is authoritative for NbS to exhibit resilience in the face of long-term climate change (Turner et al., 2022). With the onset of climate change, urban areas worldwide are seeing shifts in their climatic conditions, rendering certain native or regularly cultivated species less suitable for their respective cities. In instruction to adapt to the anticipated future climate conditions, it will be necessary to introduce new species that possess resilience (Seddon, 2022). Cities whose future climate is predicted to be outside of their existing climatic domain will find it more difficult to adjust to these changes. NbS exhibit a high degree of contextual dependence, despite their foundation in overarching scientific principles (Chausson et al., 2020). Consequently, there are significant prospects for the advancement of research at localized levels. It is imperative to undertake studies that specifically aim to evaluate the elasticity of species in the expression of anticipated future climatic circumstances. The significance of this matter is particularly pronounced in the context of urban trees, as their maturation process is time-consuming, and their ability to provide benefits is contingent upon reaching maturity. With a projected lifespan of 50 years, if these plants were to be cultivated at present, they may conceivably attain maturity amidst an altered environment, so they encountering the potential threat of losing their urban home at that time. The topic of discussion is governance. There are two primary areas of insufficient information pertaining to governance: the identification of optimal preparation and policy frameworks for the implementation of NbS and the establishment of effective mechanisms for stakeholder participation. The integral of NbS has the potential to mitigate knowledge disparities, and it is advisable to promote the amalgamation of policy formulation and scientific inquiry (Seddon, 2022). It is imperative that all citizens have equitable access to the advantages offered by NbS (Kabisch et al., 2016). Several studies suggest that the implementation of NbS in urban zones may have adverse effects on equity, as it potentially raises property costs in the vicinity of these new green initiatives, hence contributing to gentrification (Bockarjova et al., 2020). Engaging with communities is of utmost importance in order to uncover any potential unintended consequences. The topic of assessment will now be discussed. The primary areas of insufficient information about assessments of effectiveness pertain to the evaluation of cost–benefit effectiveness across various temporal and spatial dimensions and within diverse socio-ecological systems. The current approach employed for the mapping of ecology services is not well-suited for the purpose of urban adaptation. This is mostly due to the intricate structure of urban environments, where several types of components cohabit within the urban fabric. Consequently, the complexity involved in this process poses a significant challenge, resulting in a limited number of research that have endeavoured to transfer values associated with urban nature. Establishing goal indicators is a crucial aspect of evaluating performance (Kabisch et al., 2016). These tools will be utilized for the purpose of quantifying, evaluating, and overseeing the effects (Kabisch et al., 2016). Emerging technological advancements in urban greening planning and monitoring, commonly referred to as the Internet of Nature, are expected to overcome some information gaps (Galle et al., 2019). An illustrative instance is an i-tree program, which offers data on the advantages of urban trees across four dimensions: (a) stormwater interception, (b) energy conservation, (c) removal of air pollutants, and (d) reduction of carbon dioxide. The aforementioned advantages are quantified in precise and monetary terms on a yearly basis for each respective city. The inclusion of monitoring is of utmost importance since the monitoring and evaluation of NbS pilot projects can play a significant role in addressing pertinent knowledge deficiencies (Kabisch et al., 2016).

4 Implementation Barriers

Resistance to climate adaptation is commonly seen since urban resilience-building initiatives often use a reactive approach (Amundsen et al., 2010). Kabisch et al., (2016) have recognized many obstacles that hinder the successful implementation of NbS. These barriers include (a) apprehension towards unfamiliar approaches, (b) lack of alignment between long-term and short-term objectives, (c) lack of coherence between immediate activities and overarching plans, (d) compartmentalization within different sectors, and (e) adherence to a growth-oriented paradigm. Furthermore, (Seddon et al., 2020) have categorised three distinct barriers that hinder the successful execution of NbS. These hurdles include: (i) the issue of accurately measuring the impact of NbS, (ii) the difficulty in mobilising sufficient funding for NbS initiative, and (iii) the governance challenges associated with implementing NbS. Based on an examination of the current literature, this study aims to provide a concise overview of five primary obstacles that are unique to the implementation of NbS as an urban adaptation strategy.

(a) Spatial phenomena. Frequently, NbS needs a larger spatial footprint compared to conventional "grey" solutions, especially in the context of water management. Consequently, NbS may encounter competition with alternative spatial allocations for various purposes (Fratini et al., 2012). Urban development initiatives, among other applications, often include several stakeholders with vested interests, hence creating a competitive advantage in acquiring valuable urban property. The participation of the community in decision-making procedures has the potential to rebalance power dynamics, namely by reducing the influence of private entities, such as property developers, and prioritising the interests of the general public, such as the creation and preservation of community green spaces. It is important to acknowledge that the suitability of NbS in urban environments varies based on factors such as terrain, physical structure, and urban density. Consequently, it is crucial to contextualize initiatives to the specific local environment in order to achieve favorable outcomes for both individuals and the environment (Hou-Jones et al., 2021).

(b) The topic of discussion is governance. Municipalities encounter several obstacles when embarking on climate adaptation initiatives. While several affluent countries are actively advocating for NbS adaptation, such as the European Union (EU) (Frantzeskaki & McPhearson, 2022), underdeveloped nations frequently lack the necessary expertise and resources to engage in such efforts. Insufficient knowledge and competence, particularly among municipal council members, pose a significant obstacle to the successful implementation of NbS. The conventional frameworks within municipal administrations are organized according to sectoral divisions, which may impede the adoption of transdisciplinary and intersectoral methods such as NbS (Kabisch et al., 2016). The supremacy of NbS in the context of urban adaptation necessitates a complete approach that encompasses several disciplines and sectors. Additionally, it must be systemic in nature, hence posing a significant challenge to the ability of local administrations, particularly those in developing nations. There is a pressing need for robust backing from international organizations in order to facilitate the design and development process, which should encompass the provision of knowledge tools. The adjustment of urban planning and policies is necessary in order to effectively implement NbS. It has been observed that NbS are seldom put into practice until they are incorporated into planning tools and processes (Nalau et al., 2018). Furthermore, their successful implementation requires their incorporation into viable governance or business implementation models. Frantzeskaki et al. (2019) argue that in order to incorporate NbS into urban planning, it is necessary to establish tools and procedures that adhere to the following principles: (1) planning tools should be adaptable and responsive to living materials; (2) the design method should be grounded in evidence of the advantages and additional benefits derived from the adopted NbS; and (3) a cooperative approach to preparation and design is crucial for the effective implementation of NbS. In order to facilitate effective adaptation efforts, it is imperative for governance institutions to incorporate the participation of local communities. This recognition is evident in the Cancun Adaptation Framework; there is a prevailing worry over the potential failure of NbS to adequately include local perspectives (Seddon, 2022). This risk is particularly pronounced in non-urban adaptation initiatives since urban programs tend to exhibit a higher degree of participatory engagement. Grassroots organizations have exhibited reluctance towards NbS since they see them to be a diversion from more significant forms of activity (Seddon, 2022). The danger arises from the utilization of NbS as a means of greenwashing, as governments find it more convenient to endorse nature-related initiatives rather than directly addressing the fossil fuels sector and implementing a transition towards renewable energy sources. Ensuring complete openness and active involvement of local communities is crucial in addressing these issues and fostering confidence (Seddon, 2022).

(c) Evaluation. Policymakers commonly exhibit a preference for engineering solutions due to their greater ease of calculation and quantification in comparison to NbS (Albert et al., 2019). According to existing research, the successful implementation of NbS relies on the recognition of their transformational potential as being superior to traditional grey infrastructure (Frantzeskaki et al., 2019). The establishment of the advantages of these choices in comparison to other options requires empirical evidence, a task that can be challenging, especially when considering co-benefits. The absence of financial benefits resulting from the adoption of NbS might provide a hindrance to their deployment despite the growing trend of monetizing these solutions (Van Oijstaeijen et al., 2020). In order to tackle this difficulty, it has been necessary to employ specialized valuation methodologies, such as the assessment of indirect public health benefits through the quantification of disease prevention. Despite the ongoing experimentation with novel evaluation and monitoring techniques, research indicates that the current tools for evaluating green infrastructure are not effectively utilized in decision-making procedures when choosing between grey and green substructure options (Van Oijstaeijen et al., 2020).

(d) The subject of finance. Examining nature-related concerns via a financial lens presents considerable difficulties, not limited to governmental entities, since firms encounter knowledge deficiencies in the realm of nature-related finance (Irvine-Broque and Dempsey, 2023). Presently, the primary financier of NbS is the Group of Twenty (G20), allocating an estimated annual sum of USD 120 million. These funds are mostly directed towards domestic initiatives (Ozili, 2022). The primary issue faced by poor nations, who are intended beneficiaries of global climate financing, is the insufficiency of available cash. Presently, the allocation of climate money is providing financial assistance for NbS aimed at adaptation, amounting to an annual range of USD 3.8–8.7 billion. This represents around 0.6–1.4% of the overall climate financing in the year 2018, as reported in sources (Ozili, 2022). Given the pressing nature of urban adaptation and the demonstrated effectiveness of NbS as a dependable and low-risk method, it is recommended that a greater proportion of climate money be directed towards initiatives within this domain. NbS programs have previously received assistance from a number of financing institutions, including the Adaptation Fund, the Global Environmental Facility, and the Green Climate Fund (Morita and Matsumoto, 2021). Nevertheless, the establishment of a dedicated funding mechanism specifically for urban adaptation utilizing NbS has the potential to be transformative by facilitating a more focused approach to implementing actions.

(e) The sociocultural aspect. The failure of humans to demonstrate sufficient respect and recognition of the value inherent in non-human nature has been identified by certain writers as a significant obstacle to the successful implementation of NbS

(Murphy et al., 2021). The prevailing paradigm that has led us to the current era of crises is characterized by a focus on economic expansion, a significant reliance on consumption, and a dependence on inexpensive energy sources. There exists a certain level of acceptance for the necessity of transitioning from a model of continuous expansion to a model of degrowth in the context of development (Köves & Bajmócy, 2022). Degrowth is a scholarly domain and a societal mobilization that critically examines the significance of economic growth, primarily due to its incompatibility with the limitations of limited resources. It serves as an alternative to the prevailing sustainable development framework, which has been deemed ineffective in achieving desired outcomes. The present-day social and cultural individualities and involvements are profoundly intertwined with consumerism and the exploitation of resources to such an extent that achieving a transformation in the existing system appears to be beyond reach. In order to effectuate systemic change, it is imperative to get a comprehensive understanding of the pertinent underlying issues. A problem that warrants attention is the fundamental assumption underlying Western civilisation, which posits a distinct separation between humans and nature. The prevailing cultural paradigm is rooted in the concept of human/nature dualism, an ancient ideology originating from the Western tradition. This ideology posits that humans and their minds belong to a higher category characterized by reason and consciousness, distinct from a lower category encompassing the body, animals, and pre-human entities (Plumwood, 2005). Within the context of hyperseparation, the physical realm might be perceived as a mere means to an end, existing solely to cater to the elevated aware, spiritual, and mindful capacities of human beings. In this perspective, nature is regarded as lacking in individuality, spirit, conscience, and agency, as these attributes are believed to be exclusive to humans. This perspective diverges from the prevailing viewpoint held by the majority of Indigenous communities, wherein they perceive themselves and the natural environment as interconnected entities within a shared ecological framework. This perspective is rooted in the understanding that the sustenance and well-being of human beings are intricately linked to the health and preservation of the natural world (Salmón, 2000). Sustainability has significant importance for Indigenous communities in their relationship with the natural environment, as shown by the Seventh Generation Principle. This principle, established as a law within the Iroquois Confederacy, mandates that the chief is responsible for ensuring sustainable land management practices that can adequately support the needs of future generations up to the seventh generation. Environmental philosophers have underscored the imperative of altering humanity's connection with the natural world as a fundamental first measure to address the various challenges confronting our planet. In accordance with the findings of Irwin et al. (2010), The phenomenon of climate change may be understood as a critical manifestation of humanity's misguided perception of having complete control over the natural world. The anthropocentric perspective prevalent in society can give rise to perilous ideologies, such as the notion of human ecological invincibility. According to Plumwood et al. (2016), the survival of civilization as a species necessitates the development of a new narrative that might alter humanity's perception of the material world. Rather than seeking dominance, this revised narrative would encourage a shift towards negotiation and collaboration with the environment. Given the rapid rate at which life-support systems are deteriorating, it is imperative to acknowledge the genuine potential for society collapse, even if just for the purpose of risk management (Brozović, 2022). According to a new analysis of the renowned publication "The Limits to Growth" from 1972, it has been determined that global society is rapidly approaching a critical juncture when the possibility of societal collapse becomes increasingly imminent. The findings of this assessment suggest that there is a substantial probability of this collapse occurring prior to the year 2040 (Herrington, 2021). The urgency of addressing the underlying causes of the crises necessitates transformative change. In order to ensure the continued existence of humanity, it may be imperative to adopt a philosophical framework that promotes ethical engagement between people and nature (Cajete, 2000).

5 Opportunities

Based on the comprehensive examination of the cited literature, this scholarly article delineates the primary prospects for the use of NbS in the context of urban adaptation. These opportunities encompass (a) the potential for transformative changes in systems, (b) the enhancement of social resilience, (c) the realization of synergistic effects between mitigation and adaptation efforts; and (d) the interconnection between financial mechanisms and knowledge support systems.

(a) The phenomenon of systems undergoing transformation. In order to effectively confront the formidable obstacles posed by climate change and biodiversity crises, it is imperative to implement transformative systems transformation (Wamsler et al., 2022). Systems change is a strategic approach that aims to bring about a fundamental transformation of the existing state of affairs by modifying the foundational structures and supporting processes of a system through a series of targeted interventions (Foster-Fishman et al., 2007). The idea expressed is in line with the notion of transformational adaptation, which refers to adaption strategies that go beyond defensive measures and instead actively contribute to societal transformation in ways that are essential and good. These strategies also aim to concurrently avoid or reduce additional climate harm. NbS are of significance within this particular environment since they operate in harmony with nature rather than in opposition to it. They actively foster resilience and prioritise endeavors towards a green economy as opposed to industrialized activities (Sowińska-Świerkosz & García, 2022). Current study findings are promising, as they indicate a positive relationship between engagement in NbS initiatives and the adoption of sustainable behaviors. The potential of NbS in facilitating transformative changes within systems appears to be encouraging and warrants more investigation (Seddon, 2022).

(b) The concept of social resilience is a significant area of study in academic research. The involvement of local communities and Indigenous populations in the implementation of NbS has the potential to mitigate their vulnerability, encompassing not just economic aspects but also extending to other dimensions. The ability to enhance capabilities and foster creativity is crucial in facilitating adaptation (Devisscher et al., 2021). The inclusion of educational programs is crucial. NbS provide a valuable opportunity to enhance the agency of marginalised communities during a period characterized by significant global disparities. In order to effectively address the issue of inequality and promote sustainable development, it is imperative to shift society's focus away from a resource-extracting paradigm. By embracing renewable resources and striving for equality, genuine solutions may be achieved (Castelo et al., 2023).
(c) Synergies between adaptation and mitigation. NbS plays a crucial role in some regions of the developing world. Significant quantities of irretrievable carbon are found in ecosystems such as peatlands and mangroves, mostly situated in close proximity to the equator. The destruction of these ecosystems would have dire consequences, perhaps triggering uncontrollable climate change. The alleviation of this issue has been recognized as a crucial international concern (Goldstein et al., 2020). Mangrove forests provide the potential to significantly contribute to the adaptation of coastal towns in response to rising sea levels, hence presenting an opportunity to establish robust linkages between adaptation and mitigation efforts. The adoption of this integrated method has the potential to provide significant advantages for several cities, particularly those situated in developing regions and Southeast Asia. This is particularly relevant given that Southeast Asia boasts the most extensive and varied mangrove forests globally, as supported by a study (Siman et al., 2021).
(d) The relationship between funding and knowledge support systems. In order to effectively support NbS for urban adaptation, it is imperative to establish funding streams that are accompanied by a robust knowledge support framework. This will aid in addressing the inherent limits faced by local governments and contribute to the development of institutional capacity. These systems of knowledge support can also facilitate the mapping and resolution of current deficiencies, therefore aiding in the worldwide interchange of knowledge. It is imperative to cultivate synergistic relationships with urban planning agencies and entities, as they inherently possess the capacity to instigate and propel the implementation of NbS (Frantzeskaki & McPhearson, 2022).

6 Futures Prospects

Putting in place "blue-green" infrastructure. (BGI) is of paramount importance in promoting the development of environmentally sustainable metropolitan areas, especially in developing nations that are grappling with the negative consequences of climate change. The following discourse presents an analysis of potential future possibilities and consequences for emerging nations.

- The implementation of climate resilience strategies is crucial for cities in developing nations to effectively mitigate and adapt to the adverse properties of climate change, including but not limited to heightened flood risks, intensified heatwaves, and severe storms. In this regard, BGI (Blue-Green Infrastructure) presents itself as a promising approach to enhance the pliability of these urban areas. Cities may enhance their water management, alleviate heat islands, and address climate-related catastrophes by implementing various strategies such as integrating green roofs, permeable pavements, and urban wetlands.
- Water management is a significant challenge faced by developing nations, characterized by the dual issues of water shortage and inadequate water quality. BGI provides a solution by augmenting water management through the use of strategies such as rainwater gathering, built wetlands, and natural drainage systems. This approach has the potential to offer a dependable and uncontaminated water source for metropolitan communities while alleviating the burden on conventional water resources.
- The promotion of urban biodiversity preservation is a key focus of BGI, which is achieved through the establishment of green areas, the creation of wildlife corridors, and the implementation of native plants. The significance of this matter is particularly pronounced in burgeoning metropolitan areas, where the fast process of urbanization can result in the depletion of natural habitats. The provision of numerous ecosystem services and the enhancement of the general quality of life in urban settings are both significantly improved by the existence of biodiversity within an ecosystem.
- There is a chance that BGI efforts may help developing countries expand economically. Investments in green infrastructure have the potential to generate employment opportunities in several sectors, including building, maintenance, and landscaping. Moreover, the presence of BGI has the potential to draw tourists and enhance property prices, stimulating local economies.
- The implementation of Biophilic Green Infrastructure (BGI) can have a substantial positive impact on public health by promoting the presence of green areas and enhancing air quality. The enhancement of urban inhabitants' well-being may be achieved by the reduction of air pollution and heat-related diseases, as well as the rise in possibilities for physical exercise.
- Community engagement is a crucial aspect of the planning and development of BGI since it allows for the meaningful involvement of local communities in the process of designing their urban environments, therefore empowering them. The utilization of a participatory approach has the potential to cultivate a heightened feeling of ownership and stewardship, resulting in the adoption of more sustainable practices.
- International Collaboration: Developing nations stand to gain advantages by engaging in international alliances and exchanging expertise pertaining to biotechnology and genomics initiatives (BGI). Collaboration between organizations

and governments may facilitate the provision of financial resources, specialized knowledge, and technology transfer, enabling these nations to expedite their transition towards more sustainable urban development practices.
- To improve frameworks for urban planning and policy, it is imperative for developing nations to include concepts of BGI (Blue-Green Infrastructure). Governments have the ability to promote the use of BGI (Blockchain and Distributed Ledger Technology) through many means, such as implementing rules, providing tax incentives, and establishing capacity-building programs for local authorities.
- The use of data and technology, namely GIS (Geographic Information Systems) and remote sensing, can provide assistance in the planning and surveillance of BGI (Blue-Green Infrastructure) initiatives. It is advisable for developing nations to allocate resources towards the establishment of data infrastructure and the use of digital tools in order to enhance the efficiency and efficacy of implementing BGI initiatives.
- The need for education and awareness lies in the need to disseminate knowledge and enlighten individuals regarding the advantages of BGI. The provision of public funding has the potential to stimulate demand for BGI projects and hence contribute to their sustained viability.

In summary, the future outlook for blue-green infrastructure in developing nations is favorable since it has potential advantages encompassing climate resilience, enhanced water management, economic expansion, and public health. Nevertheless, the achievement of desired outcomes necessitates the adoption of a comprehensive strategy encompassing effective governance, active involvement of local communities, collaboration on a global scale, and breakthroughs in technology.

7 Conclusions

NbS involve using natural processes and ecosystems to address environmental challenges. Recognizing nature-based solutions as a sustainable solution for urban adaptation. It is important to realize that although solution-based solutions can play an important role, they may not be the only solution to mitigate climate problems. It often requires a combination of approaches, including technological, policy-based, and community-based initiatives. Incorporating NbS into sustainability efforts is a positive step, as these solutions can contribute to enhancing resilience, biodiversity, and environmental well-being in a positive way general. The execution of this initiative should be conducted within the framework of comprehensive community engagement and empowerment, with the aim of fostering consciousness, adaptability, and ingenuity. In order for NbS to achieve sustained success, it is imperative to make necessary adaptations to urban planning regulations and the overarching approach to urban concerns. This task is naturally complex. The implementation of systems change is imperative, and NbS may serve as a catalyst in fostering nature-centric possibilities, facilitating the transition of societies from a conventional industrialized paradigm to a more harmonious coexistence with the natural environment. Despite the existence of several information gaps and the need for more study, it is imperative to update urban planning regulations promptly since the urgency of urban adaptation necessitates immediate action. In order to safeguard communities, it is imperative to commence pilot projects that address prevailing deficiencies by using insights acquired during the stages of development and evaluation of efficacy. These initiatives play a pivotal role in enhancing capabilities at both global and local scales. In addition to addressing physical adaptation to climate effects, it is imperative to prioritize the development of social resilience and institutional ability. It is imperative to recognize and tackle the issue of limited capacity within local governments, particularly in developing countries. In doing so, it is crucial to consider and investigate potential alternatives, such as the establishment of external special-purpose entities. The involvement of financing organizations in information distribution is an important aspect to consider. This involvement should encompass the provision of knowledge support systems aimed at addressing the limitations of local governments. Ideally, such efforts should be carried out in collaboration with urban planning authorities. There is a pressing need for more engagement in participatory and multidisciplinary research, which aims to provide a comprehensive analysis of the insights and issues associated with the development, implementation, monitoring, and evaluation of pilot projects.

References

Albert, C., Schröter, B., Haase, D., Brillinger, M., Henze, J., Herrmann, S., et al. (2019). Addressing societal challenges through nature-based solutions: How can landscape planning and governance research contribute? *Landscape and Urban Planning, 182*, 12–21.

Amundsen, H., Berglund, F., & Westskog, H. (2010). Overcoming barriers to climate change adaptation—a question of multilevel governance? *Environment and Planning c: Government and Policy, 28*(2), 276–289.

Anderson, C. M., DeFries, R. S., Litterman, R., Matson, P. A., Nepstad, D. C., Pacala, S., et al. (2019). Natural climate solutions are not enough. *Science, 363*(6430), 933–934.

Arias, P., Bellouin, N., Coppola, E., Jones, R., Krinner, G., Marotzke, J., et al. (2021). Climate Change 2021: the physical science basis. *Contribution of Working Group I to the Sixth Assessment Report of the Intergovernmental Panel on Climate Change; technical summary.*

Bakošová, L. (2023). Sharm el-Sheikh Climate Change Conference: Commentary on Selected Issues. *Bratislava Law Review, 7*(1), 115–124.

Bank, W. (2020). *Poverty and shared prosperity 2020: Reversals of fortune.* The World Bank.

Bastin, J.-F., Clark, E., Elliott, T., Hart, S., Van Den Hoogen, J., Hordijk, I., et al. (2019). Understanding climate change from a global analysis of city analogues. *PLoS ONE, 14*(7), e0217592.

Bauduceau, N., Berry, P., Cecchi, C., Elmqvist, T., Fernandez, M., Hartig, T., et al. (2015). *Towards an EU research and innovation policy agenda for nature-based solutions & re-naturing cities: Final report of the horizon 2020 expert group on 'nature-based solutions and re-naturing cities'.*

Beggs, P. J., Zhang, Y., Bambrick, H., Berry, H. L., Linnenluecke, M. K., Trueck, S., et al. (2019). The 2019 report of the MJA–Lancet Countdown on health and climate change: a turbulent year with mixed progress. *Medical Journal of Australia*, 211(11), 490–491. e21.

BenDor, T. K., Spurlock, D., Woodruff, S. C., & Olander, L. (2017). A research agenda for ecosystem services in American environmental and land use planning. *Cities, 60*, 260–271.

Bhadouria, R., Upadhyay, S., Tripathi, S., & Singh, P. (2022). Urban ecology and global climate change. Wiley Online Library.

Björnsdottir, S. H., Jensson, P., Thorsteinsson, S. E., Dokas, I. M., & de Boer, R. J. (2022). Benchmarking ISO risk management systems to assess efficacy and help identify hidden organizational risk. *Sustainability, 14*(9), 4937.

Bockarjova, M., Botzen, W., Van Schie, M., & Koetse, M. (2020). Property price effects of green interventions in cities: A meta-analysis and implications for gentrification. *Environmental Science & Policy, 112*, 293–304.

Bratman, G. N., Hamilton, J. P., & Daily, G. C. (2012). The impacts of nature experience on human cognitive function and mental health. *Annals of the New York Academy of Sciences, 1249*(1), 118–136.

Brondizio, E. S., Settele, J., Diaz, S., & Ngo, H. T. (2019). Global assessment report on biodiversity and ecosystem services of the Intergovernmental Science-Policy Platform on Biodiversity and Ecosystem Services.

Brozović, D. (2022). Societal collapse: A literature review. *Futures*, 103075.

Cajete, G. (2000). Native science: Natural laws of interdependence. *(No Title)*.

Calliari, E., Castellari, S., Davis, M., Linnerooth-Bayer, J., Martin, J., Mysiak, J., et al. (2022). Building climate resilience through nature-based solutions in Europe: A review of enabling knowledge, finance and governance frameworks. *Climate Risk Management*, 100450.

Castelo, S., Amado, M., & Ferreira, F. (2023). Challenges and opportunities in the use of nature-based solutions for urban adaptation. *Sustainability, 15*(9), 7243.

Change, I. C. (2014). Impacts, adaptation and vulnerability. *Part A: global and sectoral aspects. Contribution of working group II to the fifth assessment report of the intergovernmental Panel on Climate Change*, 1132.

Chausson, A., Turner, B., Seddon, D., Chabaneix, N., Girardin, C. A., Kapos, V., et al. (2020). Mapping the effectiveness of nature-based solutions for climate change adaptation. *Global Change Biology, 26*(11), 6134–6155.

Cohen-Shacham, E., Walters, G., Janzen, C., & Maginnis, S. (2016). Nature-based solutions to address global societal challenges. *IUCN: Gland Switzerland, 97*, 2016–2036.

Coleman, J. (2022). Climate change made South Asian heatwave 30 times more likely. *Nature.*

Collier, M. A., O'Kane, T. J., Kitsios, V., & Sandery, P. A. (2022). CSIRO CAFE-60 submissions to the World Meteorological Organization operational decadal forecasts and the international multi-model data exchange. *Journal of Southern Hemisphere Earth Systems Science, 72*(1), 52–57.

Cowie, R. H., Bouchet, P., & Fontaine, B. (2022). The sixth mass extinction: Fact, fiction or speculation? *Biological Reviews, 97*(2), 640–663.

Deria, A., Ghannad, P., & Lee, Y.-C. (2020). Evaluating implications of flood vulnerability factors with respect to income levels for building long-term disaster resilience of low-income communities. *International Journal of Disaster Risk Reduction, 48*, 101608.

Devisscher, T., Spies, J., & Griess, V. C. (2021). Time for change: Learning from community forests to enhance the resilience of multi-value forestry in British Columbia, Canada. *Land Use Policy, 103*, 105317.

Díaz, S., Settele, J., Brondízio, E. S., Ngo, H. T., Agard, J., Arneth, A., et al. (2019). Pervasive human-driven decline of life on Earth points to the need for transformative change. *Science*, 366(6471), eaax3100.

Donovan, G. H., Butry, D. T., Michael, Y. L., Prestemon, J. P., Liebhold, A. M., Gatziolis, D., et al. (2013). The relationship between trees and human health: Evidence from the spread of the emerald ash borer. *American Journal of Preventive Medicine, 44*(2), 139–145.

Dumitru, A., & Wendling, L. (2021). *Evaluating the impact of nature-based solutions: Appendix of methods*. European Commission.

Eggermont, H., Balian, E., Azevedo, J. M. N., Beumer, V., Brodin, T., Claudet, J., et al. (2015). Nature-based solutions: New influence for environmental management and research in Europe. *GAIA-Ecological Perspectives for Science and Society, 24*(4), 243–248.

Faivre, N., Fritz, M., Freitas, T., De Boissezon, B., & Vandewoestijne, S. (2017). Nature-Based Solutions in the EU: Innovating with nature to address social, economic and environmental challenges. *Environmental Research, 159*, 509–518.

Field, C. B., & Barros, V. R. (2014). *Climate change 2014–Impacts, adaptation and vulnerability: Regional aspects*. Cambridge University Press.

Foster-Fishman, P. G., Nowell, B., & Yang, H. (2007). Putting the system back into systems change: A framework for understanding and changing organizational and community systems. *American Journal of Community Psychology, 39*, 197–215.

Frantzeskaki, N. (2019). Seven lessons for planning nature-based solutions in cities. *Environmental Science & Policy, 93*, 101–111.

Frantzeskaki, N., & McPhearson, T. (2022). Mainstream nature-based solutions for urban climate resilience. *BioScience, 72*(2), 113–115.

Frantzeskaki, N., Mahmoud, I. H., & Morello, E. (2022). Nature-based solutions for resilient and thriving cities: Opportunities and challenges for planning future cities. *Nature-based Solutions for Sustainable Urban Planning: Greening Cities, Shaping Cities*, 3–17.

Fratini, C. F., Geldof, G. D., Kluck, J., & Mikkelsen, P. S. (2012). Three Points Approach (3PA) for urban flood risk management: A tool to support climate change adaptation through transdisciplinarity and multifunctionality. *Urban Water Journal, 9*(5), 317–331.

Galle, N. J., Nitoslawski, S. A., & Pilla, F. (2019). The Internet of Nature: How taking nature online can shape urban ecosystems. *The Anthropocene Review, 6*(3), 279–287.

Gielen, D., Castellanos, G., Ruiz, C., Roesch, R., Ratka, S., & Sebastian, T. (2019). Study Navigating the Way to a Renewable Future–Solutions to Decarbonise Shipping: Preliminary findings for the UN Climate Action Summit 2019.

Gilbert, N. (2022). Climate change will force new animal encounters—And boost viral outbreaks. *Nature, 605*(7908), 20–20.

Goldstein, A., Turner, W. R., Spawn, S. A., Anderson-Teixeira, K. J., Cook-Patton, S., Fargione, J., et al. (2020). Protecting irrecoverable carbon in Earth's ecosystems. *Nature Climate Change, 10*(4), 287–295.

Haaland, C., & van Den Bosch, C. K. (2015). Challenges and strategies for urban green-space planning in cities undergoing densification: A review. *Urban Forestry & Urban Greening, 14*(4), 760–771.

Haines-Young, R., & Potschin, M. (2014). Typology/classification of ecosystem services. *Open-NESS Ecosystem Services Reference Book. EC FP7 Grant Agreement*, (308428).

Halimanjaya, A., & Papyrakis, E. (2012). Donor characteristics and the supply of climate change aid. *School of International Development*.

Herman, R. D. K. (2016). Traditional knowledge in a time of crisis: Climate change, culture and communication. *Sustainability Science, 11*, 163–176.

Herrington, G. (2021). Update to limits to growth: Comparing the World3 model with empirical data. *Journal of Industrial Ecology, 25*(3), 614–626.

Hickel, J. (2020). Quantifying national responsibility for climate breakdown: An equality-based attribution approach for carbon dioxide emissions in excess of the planetary boundary. *The Lancet Planetary Health, 4*(9), e399–e404.

Hou-Jones, X., Roe, D., & Holland, E. (2021). *Nature-based Solutions in Action: Lessons from the Frontline*. Bond.

Hughes, A., Shen, X., Corlett, R., Li, L., Luo, M., Woodley, S., et al. (2022). Challenges and possible solutions to creating an achievable and effective Post-2020 Global Biodiversity Framework. *Ecosystem Health and Sustainability, 8*(1), 2124196.

Irvine-Broque, A., & Dempsey, J. (2023). Risky business: Protecting nature, protecting wealth? *Conservation Letters*, e12969.

Irwin, R. (2010). Reflections on modern climate change and finitude. *Climate Change and Philosophy: Transformational Possibilities*, 48–72.
Jafino, B. A., Walsh, B., Rozenberg, J., & Hallegatte, S. (2020). Revised estimates of the impact of climate change on extreme poverty by 2030. The World Bank.
Kabisch, N., Frantzeskaki, N., Pauleit, S., Naumann, S., Davis, M., Artmann, M., et al. (2016). Nature-based solutions to climate change mitigation and adaptation in urban areas: perspectives on indicators, knowledge gaps, barriers, and opportunities for action. *Ecology and Society, 21*(2).
Kemp, L., Xu, C., Depledge, J., Ebi, K. L., Gibbins, G., Kohler, T. A., et al. (2022). Climate Endgame: Exploring catastrophic climate change scenarios. *Proceedings of the National Academy of Sciences, 119*(34), e2108146119.
Klemm, W., Heusinkveld, B. G., Lenzholzer, S., & van Hove, B. (2015). Street greenery and its physical and psychological impact on thermal comfort. *Landscape and Urban Planning, 138*, 87–98.
Köves, A., & Bajmócy, Z. (2022). The end of business-as-usual?–A critical review of the air transport industry's climate strategy for 2050 from the perspectives of Degrowth. *Sustainable Production and Consumption, 29*, 228–238.
Lin, B. B., Ossola, A., Alberti, M., Andersson, E., Bai, X., Dobbs, C., et al. (2021). Integrating solutions to adapt cities for climate change. *The Lancet Planetary Health, 5*(7), e479–e486.
Lindén, J., Fonti, P., & Esper, J. (2016). Temporal variations in microclimate cooling induced by urban trees in Mainz, Germany. *Urban Forestry & Urban Greening, 20*, 198–209.
MacKinnon, K., Dudley, N., & Sandwith, T. (2011). Natural solutions: Protected areas helping people to cope with climate change. *Oryx, 45*(4), 461–462.
McGuire, S. (2016). World cancer report. (2014). Geneva, Switzerland: World Health Organization, international agency for research on cancer, WHO Press, 2015. *Advances in Nutrition, 7*(2), 418–419.
Morita, K., Matsumoto, K., & i. (2021). Governance challenges for implementing nature-based solutions in the Asian region. *Politics and Governance, 9*(4), 102–113.
Munang, R., Thiaw, I., Alverson, K., Mumba, M., Liu, J., & Rivington, M. (2013). Climate change and Ecosystem-based Adaptation: A new pragmatic approach to buffering climate change impacts. *Current Opinion in Environmental Sustainability, 5*(1), 67–71.
Murphy, P., Brereton, P., & O'Brolchain, F. (2021). New materialism, object-oriented ontology and fictive imaginaries: New directions in energy research. *Energy Research & Social Science, 79*, 102146.
Nalau, J., Becken, S., Schliephack, J., Parsons, M., Brown, C., & Mackey, B. (2018). The role of indigenous and traditional knowledge in ecosystem-based adaptation: A review of the literature and case studies from the Pacific Islands. *Weather, Climate, and Society, 10*(4), 851–865.
Narayan, S., Thomas, C., Matthewman, J., Shepard, C. C., Geselbracht, L., Nzerem, K., et al. (2019). Valuing the flood risk reduction benefits of Florida's mangroves. *The Nature Conservancy*, 11.
Nations, U. (2018). Revision of world urbanization prospects. *United Nations: New York, NY, USA*, 799.
Neufeldt, H., Christiansen, L., & Dale, T. W. (2021). Adaptation Gap Report 2021-The Gathering Storm: Adapting to climate change in a post-pandemic world.
Ozili, P. K. (2022). Green finance research around the world: A review of literature. *International Journal of Green Economics, 16*(1), 56–75.
Plumwood, V. (2005). *Environmental culture: The ecological crisis of reason*. Routledge.
Pörtner, H.-O., Scholes, R. J., Agard, J., Archer, E., Arneth, A., Bai, X., et al. (2021). Scientific outcome of the IPBES-IPCC co-sponsored workshop on biodiversity and climate change.
Potschin, M., Kretsch, C., Haines-Young, R., Furman, E., Berry, P., & Baró, F. (2015) 'Nature-based solutions. OpenNESS ecosystem service reference book. OpenNESS synthesis paper'.
Roe, D., Holland, E., Nisi, N., Mitchell, T., & Tasnim, T. (2023). Loss and damage finance should apply to biodiversity loss. *Nature Ecology & Evolution*, 1–3.

Roemer, J., Komulainen, T. M., Scholz, M., & Moatamedi, M. (2023). The review of 102 design support tools for nature-based solutions applications. *International Journal of Hydrology Science and Technology, 15*(4), 327–348.

Romanello, M., McGushin, A., Di Napoli, C., Drummond, P., Hughes, N., Jamart, L., et al. (2021). The 2021 report of the Lancet Countdown on health and climate change: Code red for a healthy future. *The Lancet, 398*(10311), 1619–1662.

Salmón, E. (2000). Kincentric ecology: Indigenous perceptions of the human–nature relationship. *Ecological Applications, 10*(5), 1327–1332.

Schulz, K., Nguyen, A., & Pillar, H. (2022). An Improved and Observationally-Constrained Melt Rate Parameterization for Vertical Ice Fronts of Marine Terminating Glaciers. *Geophysical Research Letters*, 49(18), e2022GL100654.

Seddon, N. (2022). Harnessing the potential of nature-based solutions for mitigating and adapting to climate change. *Science, 376*(6600), 1410–1416.

Seddon, N., Chausson, A., Berry, P., Girardin, C. A., Smith, A., & Turner, B. (2020). Understanding the value and limits of nature-based solutions to climate change and other global challenges. *Philosophical Transactions of the Royal Society B, 375*(1794), 20190120.

Seresinhe, C. I., Preis, T., & Moat, H. S. (2015). Quantifying the impact of scenic environments on health. *Scientific Reports, 5*(1), 16899.

Siman, K., Friess, D. A., Huxham, M., McGowan, S., Drewer, J., Koh, L. P., et al. (2021). Nature-based solutions for climate change mitigation: challenges and opportunities for the ASEAN region.

Skrydstrup, J., Madsen, H. M., Löwe, R., Gregersen, I. B., Pedersen, A. N., & Arnbjerg-Nielsen, K. (2020). Incorporating objectives of stakeholders in strategic planning of urban water management. *Urban Water Journal, 17*(2), 87–99.

Skrydstrup, J., Löwe, R., Gregersen, I. B., Koetse, M., Aerts, J. C., De Ruiter, M., et al. (2022). Assessing the recreational value of small-scale nature-based solutions when planning urban flood adaptation. *Journal of Environmental Management, 320*, 115724.

Smart, U. (2021). Sustainable and Resilient cities: the Power of Nature-based Solutions. *UNEP: Nairobi, Kenya*, 1–32.

Smith, K. R., Chafe, Z., Woodward, A., Campbell-Lendrum, D., Chadee, D., Honda, Y., et al. (2015). Human health: impacts, adaptation, and co-benefits. *Climate Change 2014 Impacts, Adaptation and Vulnerability: Part A: Global and Sectoral Aspects* (pp. 709–754).

Soares, A. L., Rego, F. C., McPherson, E., Simpson, J., Peper, P., & Xiao, Q. (2011). Benefits and costs of street trees in Lisbon, Portugal. *Urban Forestry & Urban Greening, 10*(2), 69–78.

Sowińska-Świerkosz, B., & García, J. (2022). What are Nature-based solutions (NBS)? Setting core ideas for concept clarification. *Nature-Based Solutions, 2*, 100009.

Sowińska-Świerkosz, B., Wójcik-Madej, J., & Michalik-Śnieżek, M. (2021). An assessment of the Ecological Landscape Quality (ELQ) of Nature-Based Solutions (NBS) based on existing elements of Green and Blue Infrastructure (GBI). *Sustainability, 13*(21), 11674.

Suarez-Gutierrez, L., Müller, W. A., Li, C., & Marotzke, J. (2020). Hotspots of extreme heat under global warming. *Climate Dynamics, 55*(3–4), 429–447.

Turner, B., Devisscher, T., Chabaneix, N., Woroniecki, S., Messier, C., & Seddon, N. (2022). The role of nature-based solutions in supporting social-ecological resilience for climate change adaptation. *Annual Review of Environment and Resources, 47*, 123–148.

Ulmer, J. M., Wolf, K. L., Backman, D. R., Tretheway, R. L., Blain, C. J., O'Neil-Dunne, J. P., et al. (2016). Multiple health benefits of urban tree canopy: The mounting evidence for a green prescription. *Health & Place, 42*, 54–62.

UNFCCC, V. (2015). Adoption of the Paris agreement. *Proposal by the President, 282*, 2.

Van Oijstaeijen, W., Van Passel, S., & Cools, J. (2020). Urban green infrastructure: A review on valuation toolkits from an urban planning perspective. *Journal of Environmental Management, 267*, 110603.

Walz, Y., Nick, F., Higuera Roa, O., Nehren, U., & Sebesvari, Z. (2021). Coherence and Alignment among Sustainable Land Management, Ecosystem-based Adaptation, Ecosystem-based Disaster Risk Reduction and Nature-based Solutions.

Wamsler, C., Osberg, G., Panagiotou, A., Smith, B., Stanbridge, P., Osika, W., et al. (2022). Meaning-making in a context of climate change: supporting agency and political engagement. *Climate Policy*, 1–16.

Weltbank. (2012). Biodiversity, Climate Change, and Adaptation: Nature-Based Solutions from the World Bank Portfolio. World Bank.

White, P., Pelling, M., Sen, K., Seddon, D., Russell, S., & Few, R. (2005). *Disaster risk reduction: A development concern.* DfID.

Yan, C., Guo, Q., Li, H., Li, L., & Qiu, G. Y. (2020). Quantifying the cooling effect of urban vegetation by mobile traverse method: A local-scale urban heat island study in a subtropical megacity. *Building and Environment, 169*, 106541.

Zwierzchowska, I., Fagiewicz, K., Poniży, L., Lupa, P., & Mizgajski, A. (2019). Introducing nature-based solutions into urban policy–facts and gaps. Case study of Poznań. *Land Use Policy, 85*, 161–175.

Blue-Green Infrastructure for Urban Resilience and Sustainability in Developing Countries

Kandas Keita and Sory Kourouma

Abstract This chapter comprehensively analyzes the role and importance of blue-green infrastructure in improving urban resilience and sustainability, mainly in developing countries. The Bankruptcy begins by setting the context for urban resilience and sustainability, highlighting the growing importance of cities in the twenty-first century and the challenges posed by urbanization, climate change, economic crises, and pandemics. This highlights the need for a comprehensive strategy to deal with these demanding situations that specialize in adaptability and long-term sustainability. The chapter discusses the capability of the blue-green infrastructure (BGI) method in developing countries. It describes the particularly demanding situations of these countries, including rapid urbanization, limited resources, socioeconomic vulnerabilities, and the impact on adjacent ecosystems. The BGI technique, which integrates sustainable water management with the security of green spaces, is proposed as a response promoting resilience, sustainability, and social equity. This approach benefits all residents, not just the city's elite, by presenting environmental offerings such as improved air and water quality, recreational spaces, and expanded biodiversity. The chapter addresses situations requiring BGI implementation, technical and monetary constraints, governance issues, and the need for cross-sectoral coordination and integrated planning. It highlights the importance of community participation and education to ensure the long-term sustainability of BGI initiatives.

Keywords Climate change · Community participation · Holistic approach · Implementation challenges · Social equity · Sustainable development · Vulnerability

K. Keita (✉) · S. Kourouma
School of Architecture and Urban Planning, Chongqing University, Chongqing 400044, China
e-mail: 11800189@cqu.edu.cn

Higher Institute of Arts of Guinea, Dubreka P.B Box 2421, Republic of Guinea

P. K. Joshi et al. (eds.), *Blue-Green Infrastructure for Sustainable Urban Settlements*,
https://doi.org/10.1007/978-3-031-62293-9_2

1 Introduction

1.1 The Origin and Context of Urban Resilience and Sustainability

The twenty-first century is often called the "century of cities." United Nations projections highlight the growing weight of urban areas in the global demographic fabric (United Nations and Social Affairs, 2014). By announcing that around 70% of the world's population will live in urban zones by 2050, we foresee a planet where cities will dominate (Keita & Kourouma, 2023; Langemeyer & Connolly, 2020). This concentration of people in urban areas will have enormous implications, from increased resource demand to increased infrastructure needs to pressure on local ecosystems (Karabakan & Mert, 2021). A confluence of rising urbanization and global difficulties like climate change, economic crises, and pandemics is sparking concerns about the long-term viability of cities. Urban resilience and sustainability are essential (Baba et al., 2020; Bhadouria et al., 2023).

Urban resilience refers to the aptitude of a city to resist, adapt, and recover quickly from shocks and stresses. It is a holistic approach recognizing that urban challenges are interconnected and often amplified by external factors, such as climate change (Bulkeley et al., 2013). The need for resilience arises from several threats. Natural disasters, for example, can devastate cities, especially those poorly prepared (Acuti et al., 2020). Natural disasters such as hurricanes, floods, earthquakes, and heat waves cost lives and businesses a lot of money. Social tensions, such as rising inequalities, can compromise urban stability (Zeng et al., 2022). Furthermore, in a globalized world, economic and health crises can quickly spread from one region to another, as the COVID-19 pandemic has demonstrated (Bereitschaft & Scheller, 2020; Florida et al., 2021; Teller, 2021; Han et al., 2021; McPhearson et al., 2016).

And when it comes to urban sustainability, it goes beyond simply responding to immediate crises. It focuses on creating cities that can thrive long-term without cooperating with future generations' aptitude to meet their needs (Romero-Lankao et al., 2016b). Think about how your actions in the city might affect the environment, your community, and your wallet. Environmental safeguarding is a cornerstone of sustainability. With increasing population density and consumption, cities significantly impact the environment (Jacobs, 2012). They are responsible for many greenhouse gas emissions, water consumption, and waste production. Therefore, creating sustainable cities means rethinking transport systems, infrastructure, waste management, and energy efficiency.

The idea of urban resilience has its roots in the work of Norwegian psychologist Erik Erikson in the 1970s. Erikson used the term "resilience" to describe a person's ability to overcome moments of crisis and difficulty (Eriksson and Andersson, 2023). This concept was then adapted to urban contexts to reflect the capacity of cities to cope with shocks and disruptions. Over time, urban resilience has been applied to challenges related to natural disasters, economic crises, conflict, and other forms of the disorder.

The city's sustainability is rooted in the 1960s and 1970s environmental movement. A rising consciousness of environmental degradation motivated a reevaluation of municipal design and management techniques. The notion of sustainability was widely disseminated thanks to the 1987 Brundtland Report, "Our Common Future," which emphasized ensuring that future generations also have enough resources to meet their requirements (Schmidt & Guerra, 2018).

Resilience and sustainability are concepts widely used in the social-ecological literature to assess the capacity of social and ecological systems to cope with difficulties (Marchese et al., 2018). Lately, these two ideas have gained importance in the planning literature as they are used to assess the resilience and flexibility of urban systems (Wubneh, 2021). Urban resilience refers to the ability of an urban approach to adapt and function fully to maintain its form, structure, and identity, regardless of challenges. Adaptability allows us to understand cities' historical and cultural transformation during their recovery after a significant historical crisis (Walker, 2020).

Faced with the growing challenges posed by rapid urbanization, undeniable climate change, pressures on resources, and other complex socioeconomic issues, cities have had to fundamentally rethink their approaches to ensure their ability to thrive and adapt in a continually changing world (Fabbricatti et al., 2020). In this dynamic context, urban resilience and sustainability concepts have emerged as essential pillars of urban planning and management (Béné et al., 2018). The origin and context of urban resilience and sustainability are deeply rooted in urban developments, social movements, political changes, and environmental concerns. This analysis looks at the evolution of these concepts and their progressive integration into urban discourses and practices, emphasizing the complexity of the challenges and opportunities that have shaped them (Weichselgartner & Kelman, 2015).

2 Potential in Developing Countries

2.1 *Challenges and Constraints in Developing Countries*

In developing nations, the rapid rate of urbanization combined with limited resources and socioeconomic vulnerabilities poses unique challenges in city and environmental management. Traditional infrastructure struggles to meet the demand for drinking water, sanitation, energy, and housing (Billger et al., 2017). Unregulated urbanization frequently leads to densely populated slums, inequitable distribution of public services, environmental pollution, and deterioration of adjacent ecosystems, constituting prevalent challenges (Nagenborg, 2019). Urban centers undergoing development are inherently more susceptible to the adverse consequences of climate change, including but not limited to occurrences such as flooding, drought, and severe storms. The "Blue-Green Infrastructure" approach has significant potential (Jayasooriya et al., 2017). It provides a pathway to resilience and sustainability and can contribute to social equity by providing environmental services that benefit all residents, not just the urban elite.

Rapid urbanization in developing countries is a phenomenon that is both complex and dynamic, generating a series of significant challenges and constraints in terms of city management and environmental preservation (Van Stigt et al., 2013). Accelerated population growth, limited resources, and socioeconomic vulnerabilities create a particularly demanding context for these nations (Monstadt & Schmidt, 2019). The necessity for a comprehensive and fair strategy for urban development is only one of the many pressing problems that developing countries must address in this article. Managing urban resources is difficult because rapid urbanization frequently coincides with exponential population expansion. Growing demand often overwhelms traditional infrastructure such as drinking water supply, sanitation, energy, and housing (Bibri, 2021). This uncontrolled urbanization also encourages the creation of overcrowded slums. Low-income populations and migrants seeking better economic opportunities often find themselves forced to settle in informal areas, lacking adequate infrastructure and essential services (Mell, 2017). This concentration of vulnerable populations worsens urban inequalities, leaving them without access to quality public services and economic opportunities, which perpetuates the cycle of poverty and marginalization.

Rapid urbanization exerts a tremendous impact on adjacent ecosystems. Deforestation, natural habitat fragmentation, and species extinction result from the increased demand for land for urban growth and inadequate planning. Industrial and transportation-related air and water pollution sources also contribute to declining environmental quality. The health of humans, food supplies, and the natural system are all in jeopardy because of this lack of sustainability (Chelleri et al., 2015; Romero-Lankao et al., 2016a).

2.2 Implementation Challenges

However, despite its benefits, adopting the Blue-Green Infrastructure approach is not without challenges. Local authorities' limited technical and financial capacities in many developing countries can hinder the effective implementation of sustainable infrastructure projects (Bixler et al., 2020). Additionally, the cross-sectoral coordination and integrated planning needed to implement this complex approach can be challenging, especially when immediate development priorities seem more pressing (Weichselgartner & Kelman, 2015). Governance and regulatory issues are also crucial. Insufficient policies, opaque decision-making procedures, and bureaucratic hurdles might hamper the timely completion of "Blue-Green" projects. To address these challenges, there must be close collaboration between municipal governments, federal agencies, civil society groups, and the commercial sector (Liu et al., 2021b).

Given these intricate issues, the "Bleu-Vert Infrastructure" concept is a promising solution. Together, the sustainable management of water resources ("blue" method) and the protection of green areas and ecosystems ("green" approach) form the basis of this integrated strategy (Cristiano, 2018; Meerow & Stults, 2016).

2.3 *Integration of Traditional and Scientific Knowledge*

Integrating traditional and scientific knowledge is essential to address developing countries' complex challenges. Local communities often have a deep understanding of their environment and solutions tailored to their specific contexts. By including them in the planning and implementation process, Blue-Green projects can benefit from holistic and local perspectives (Billger et al., 2017; Karabakan & Mert, 2021). Scientific and technological advances can also be crucial in designing and monitoring "Blue-Green" infrastructure.

2.4 *The "Blue-Green" Infrastructure Concept*

Designing and developing our cities is of utmost importance in a constantly changing world where urban populations continue to grow. Rapid urbanization poses complex challenges regarding natural resource management, climate change resilience, and city dwellers' quality of life (Keita & Kourouma, 2023).In this context, the innovative "Blue-Green" infrastructure concept emerges as a promising solution for shaping more sustainable, greener, and resilient cities. Blue-green infrastructure is based on a simple yet powerful idea: recognizing the close connections between water systems and green ecosystems within urban environments (Kapetas & Fenner, 2020; Shamout et al., 2021). This holistic approach revolves around two fundamental elements, the "blue" elements and the "green" elements, which cooperate to create balanced and harmonious urban spaces, both functional and aesthetic.

The "Blue-Green " infrastructure's " blue " elements encompass sustainable water management in all dimensions. This goes well beyond the simple distribution of drinking water. It is about rethinking the entire water cycle within the city (Nguyen et al., 2019). Stormwater, often considered waste, is treated as a valuable resource. Green roofs, for example, absorb rainwater, reducing the risk of flooding and allowing water to be reused for irrigation and other non-potable needs (Venkataramanan et al., 2019). Underground water storage tanks offer an innovative solution to meet seasonal variations in water demand. Collecting and storing rainwater and treated wastewater, these reservoirs play a vital role in preserving freshwater resources (Liu et al., 2021b). Additionally, environmentally friendly wastewater treatment, including through artificial wetlands, helps purify water before its release into the natural environment.

The "Blue-Green " infrastructure's " green " elements are equally essential. They encompass the creation and preservation of green spaces within urban environments. These green spaces take various forms, from lush parks to rain gardens to artificial wetlands (Austin, 2014). These developments play a specific role in regulating the water cycle and promoting urban biodiversity. Green roofs are not just a rainwater management solution (Kondo et al., 2018). They also provide natural thermal insulation to buildings, reducing dependence on energy-intensive air conditioning systems. Urban parks offer green havens in the heart of the hustle and bustle, allowing residents to connect with nature and enjoy recreational spaces (Nguyen et al., 2019).

However, the "Blue-Green" infrastructure is not limited to environmental and functional benefits. It also positively impacts human well-being (Heinzlef & Serre, 2020). Green spaces provide relaxation and recreation, allowing city dwellers to detach themselves from daily stress and reconnect with nature (de Macedo et al., 2021). Additionally, these spaces play a crucial role in reducing heat stress. By creating islands of coolness within the city, they help to mitigate the effects of increasingly recurrent heat waves in the context of climate change. It requires a collaborative and integrated approach between actors (Song & Yu, 2018). Urban planners play a central role in integrating the principles of "Blue-Green" infrastructure from the design phase of urban plans. Engineers develop innovative technical solutions for water management and the creation of green spaces. Architects integrate these elements into the design of buildings and public spaces. City officials coordinate by implementing policies and incentives to encourage "Blue-Green" practices (Ateş and Erinsel Önder, 2023).

However, the ultimate success of this approach relies on active community participation. Citizens are the ultimate beneficiaries of these urban improvements, and their engagement is essential to ensure the long-term sustainability of the "Blue-Green" infrastructure (Heinzlef & Serre, 2020). Education programs and awareness campaigns must promote awareness of the importance of water management and the conservation of green spaces (Sharifi & Khavarian-Garmsir, 2020).

2.5 *Potential for Resilience and Sustainability*

Rapid urbanization in emerging nations is a complicated phenomenon that calls for innovative approaches to city planning, environmental protection, and preparedness for global threats like climate change (LopezDeAsiain & Díaz-García, 2020). The "Blue-Green Infrastructure" strategy, which combines sustainable water management with the security of green areas, offers a viable answer to these problems (Ahmed et al., 2019). This method has the potential to significantly contribute to making urban areas more robust. Blue-green infrastructure systems can help manage flooding, reduce soil erosion, and increase aquifer recharge, mainly when drainage and irrigation systems are often insufficient (Fenner Andrew, 2017; Drosou et al., 2019).

In this talk, we explore in detail the potential of this approach in building more resilient and sustainable cities in developing countries.

2.6 The Challenges of Urban Resilience

Developing cities often struggle with resilience challenges. Life-threatening weather events, such as floods, storms, and droughts, can have devastating consequences on infrastructure, livelihoods, and the safety of people (Davidson et al., 2019). Additionally, pressure on natural resources due to uncontrolled urbanization can lead to increased environmental degradation, thereby increasing the vulnerability of cities (Reischl et al., 2018).

2.7 The Role of Blue-Green Infrastructure

The Blue-Green Infrastructure approach offers significant potential for strengthening urban resilience in developing countries. Blue-green infrastructure systems, which combine sustainable water management with creating green spaces, can play a crucial role in managing flood risks (Thorne et al., 2018). For example, creating natural water retention ponds in urban parks can assist in mitigating the effects of torrential rains by temporarily capturing excess water (Leng et al., 2020; Ahern, 2011).

Additionally, green spaces integrated into blue-green infrastructure systems can reduce soil erosion. Plant roots hold the soil structure in place, preventing decay during heavy rains. This is particularly important in areas where deforestation and land degradation have left soils vulnerable (Austin, 2014).

2.8 Aquifer Recharge and Water Management

In many developing nations, access to drinking water is a significant concern. Blue-green infrastructure systems can help increase aquifers' recharge and critical groundwater sources. Green spaces can act as natural rainwater catchment areas, replenishing groundwater reserves (Liu & Jensen, 2018). This is particularly important in regions with limited or inefficient water distribution infrastructure.

Additionally, sustainable water management under the "Blue-Green" approach often involves techniques for reusing treated wastewater (Thorne et al., 2018). By collecting and treating wastewater, cities can reduce their dependence on freshwater sources and contribute to long-term water security.

2.9 Green Spaces as Carbon Sinks

Preserving green spaces in blue-green infrastructure systems can also be crucial in combating climate change. Green spaces act as carbon sinks, absorbing carbon dioxide from the atmosphere through photosynthesis. This contributes to lower atmospheric concentrations of greenhouse gases, which in turn helps minimize the severity of climate change (Zeng et al., 2022).

By integrating green spaces into urban areas, developing cities can improve air quality, reduce urban temperatures, and underwrite global efforts to combat climate change.

3 Benefits of the Blue-Green Infrastructure Approach

3.1 Benefits of the "Blue-Green" Approach

The "Blue-Green Infrastructure" approach offers several significant benefits in developing countries. Firstly, it promotes sustainable water management through rainwater harvesting, water reuse treated wastewater, and wetland restoration (O'Donnell and Thorne, 2020; Thorne, 2020). This helps reduce pressure on limited water resources while mitigating the risk of flooding and water shortages.

Second, by preserving and integrating green spaces into urban planning, this approach offers advantages in terms of air quality, thermal regulation, and psychological well-being of citizens (Leng et al., 2020; Liu et al., 2021a). Urban parks, biodiversity corridors, and wetlands act as "green lungs" that improve residents' quality of life and contribute to local biodiversity.

3.2 Social Equity and Community Participation

An essential characteristic of the "Blue-Green" approach is its ability to provide environmental services that benefit the entire community (Liu et al., 2021c). For example, creating parks and wetlands can improve air and water quality, provide recreational opportunities, and increase biodiversity, all benefits accessible to all citizens, regardless of their socio–economic status (Hamann et al., 2020). This helps create social equity, as opposed to solutions that would only benefit a wealthy urban elite. Unlike traditional infrastructure, which can benefit the rich, "Blue-Green Infrastructure" helps the entire population, particularly marginalized groups (Chen, 2008). Community gardens, leisure spaces, and ecological corridors provide areas for meeting and cooperation, thus strengthening the social fabric (Austin, 2014).

In addition, the "Blue-Green" approach encourages community participation in project planning and management. Involving residents in decision-making ensures that the proposed solutions meet the real needs of the population and strengthen their sense of belonging to their urban environment (O'Donnell and Thorne, 2020).

3.3 Integrated Water Resources Management

By integrating water management into the urban ecosystem, the "Blue-Green Infrastructure" approach promotes more efficient and sustainable use of water resources (Özerol et al., 2020). Green spaces can act as natural buffer zones for the absorption of rainwater, thereby reducing the risk of flooding. Natural water storage solutions like wetlands can help alleviate water shortages.

3.4 Improved Quality of Life

Green spaces directly benefit citizens by creating pleasant and healthy environments. Parks and recreation areas promote mental and physical health and strengthen local communities (Mouratidis, 2019). Additionally, increased vegetation can help improve air quality by absorbing air pollutants (Wardekker, 2021; Kim & Lim, 2016).

3.5 Additional Objectives

While resilience focuses on the ability to respond and recover from crises, and sustainability focuses on long-term prosperity, the two concepts are deeply linked. A sustainable city is, by nature, more resilient; for example, green infrastructure, such as green roofs or stormwater management systems, can mitigate the effects of climate change and provide recreational spaces for residents, contributing to mental and physical health (Dhakal & Chevalier, 2017; Shandas et al., 2020).

3.6 Concrete Application Examples

Several developing countries have successfully adopted the Blue-Green Infrastructure approach and learned valuable lessons from it. Several projects around the world illustrate the potential of the Blue-Green Infrastructure approach to strengthen resilience and sustainability in developing countries (Ghofrani et al., 2017).

For example, Singapore has implemented an integrated water management system that combines rainwater harvesting, wastewater recycling, and the creation of green spaces to ensure water security and promote quality of life (Deely et al., 2020). Singapore is a pioneer in integrating water into its urban environment. The city has developed innovative solutions such as "Ne water" (recycled water) to respond to its water while creating water parks and rain gardens to manage stormwater (Liao, 2019).

a. ***The River Reclamation Project in Seoul, South Korea***: The city of Seoul undertook a river restoration project that transformed old roads into green spaces along the waterways. This improved residents' quality of life by providing recreational spaces and helped prevent flooding by creating natural water retention areas (Marin, 2021; Borie et al., 2019)
b. ***The Water Harvesting Program in Chennai, India***: 76 years after its independence, India is a country that faces enormous problems in resource organization and land use planning. In Chennai, the rainwater harvesting initiative started slowly, determined by a group of activists at the height of the water supply crisis. The implementation of rainwater harvesting in Chennai began 18 years ago.

Since 2001, rainwater-harvesting systems have started to be installed in existing buildings and are required in new construction (Jung et al., 2020). The collected rainwater is injected into the subsoil according to various processes, more or less complex, to counterbalance the water withdrawal and thus restore the level of the aquifers. Since adopting this method, there has been a notable increase in water quantities, resulting in a current state of relative stability (Özerol et al., 2020). Therefore, mitigating the presence of saline constituents in aquifers remains a crucial aspect that has not yet been achieved under the implemented plan. Most of the water collected is intended to replenish underground reservoirs, while the remaining portion is designed for various other uses, including industrial and domestic (Liu et al., 2021b).

The depletion of groundwater resources in the Besant Nagari region of Tamil Nadu, India, especially in its capital, Chennai, became visible in the 1990s. This phenomenon can be attributed mainly to the large influx of rural dwellers towards urban areas and rapid urbanization. One aspect to consider is the growing water demand, which affects both the residential and industrial sectors (Bibri, 2020; Bozza et al., 2015). Notably, urban areas require twice as much water per person as rural areas. Additionally, the construction industry plays a vital role in scaling up water extraction from underground reservoirs.

On the other hand, urbanization has led to a decrease in the surface area of unpaved surfaces, preventing the infiltration of precipitation into the ground and consequently hindering the replenishment of underground water reservoirs (Liu et al., 2021b). It is important to note that rainfall in the region occurs in intermittent episodes characterized by short durations and high-intensity levels, which poses challenges for effective absorption. In the early 2000s, a catastrophic scenario happened when around 65% of rainwater flowed into the sea or evaporated before being absorbed by the land.

c. ***Curitiba, Brazil***: Curitiba is often cited as an example of urban planning focused on green spaces and sustainable transportation systems (Acuti et al., 2020). The city has developed a network of parks and recreational areas that act as green lungs while integrating a rapid bus system and cycle paths to reduce car dependence.
d. ***Rotterdam***, in the Netherlands, is often cited as an exemplary city for resilience and sustainability (Fabbricatti et al., 2020). Facing flood risks due to its location below sea level, Rotterdam has invested in impressive infrastructure to manage

water. Rather than simply building barriers, the city created parks that double as retention ponds during floods (Bush & Doyon, 2019).

e. ***Medellín***, Colombia, has taken an innovative approach to addressing social inequality. The city has promoted social and economic integration by investing in transport infrastructure such as cable cars and escalators to connect disadvantaged neighborhoods to financial centers (Reisi et al., 2020).
f. ***In Africa***, A continental effort, the "Great Green Wall" project is an African initiative launched in 2007 by the African Union (AU). This project seeks to plant a buffer zone of trees and plants to prevent further desertification and land degradation in the Sahel area, a semi-arid zone extending from the Atlantic to the Red Sea (Sarker et al., 2020).

The primary goal of this program is to transform a strip of land 15 km wide and 7,775 km long across 20 nations into a patchwork of green and productive regions (McPhillips et al., 2021). Its goals include increasing local population resilience, protecting biodiversity, and providing employment opportunities while restoring 100 million hectares of damaged land by 2030 (Correa Ayram et al., 2016).

3.7 Socioeconomic Impacts

Despite its transformative potential, the Great Green Wall project faces challenges, including financing, coordination between countries, and changing priorities (Zuniga-Teran et al., 2020). However, with growing international support and awareness of its importance, there is renewed hope for realizing this bold vision for Africa. The Great Green Wall is not only a reforestation project but also a holistic initiative that addresses the environmental, social, and economic problems of the Sahel region through reforestation and other measures.

These examples demonstrate the necessity of long-term vision, cross-sector cooperation, and community participation for the success of "Blue-Green" initiatives (Heinzlef & Serre, 2020; Iwaniec et al., 2020).

3.8 Challenges and Perspectives

Although the Blue-Green Infrastructure approach has many benefits, its implementation can encounter challenges. Financial Barriers: One of the significant challenges for implementing blue-green infrastructure lies in allocating and mobilizing sufficient financial resources (Gao et al., 2022). Projects related to this approach may require considerable initial investments. Additionally, the multi-sectoral nature of these projects can complicate the identification of dedicated and stable sources of financing (Deng et al., 2021). The nonexistence of awareness and understanding of the importance and benefits of blue-green infrastructure is another barrier. Effective communication and targeted awareness programs are necessary to gain public and stakeholder support, which is often essential for the success of such projects. Coordinated management among various stakeholders, including governments, private companies, and civil society, is crucial (Van Stigt et al., 2013). This can be particularly complicated due to the variability of each actor's interests, objectives, and application scales. In addition, unbridled urban expansion is also a factor that can hinder the preservation of green spaces and wetlands. In this context, strict urban policies and regulations are necessary to control and guide urbanization in a way that does not compromise environmental resources (United Nations and Social Affairs, 2014).

However, there are many positive prospects. Blue-Green Infrastructure solutions provide short-and long-term benefits regarding sustainability, resilience, and quality of urban life—innovative technologies, such as real-time data monitoring.

4 Disaster and Risk Management

4.1 Natural Disasters and Urban Vulnerability

Natural disasters, such as earthquakes, hurricanes, and floods, have historically affected cities. These events have highlighted the weaknesses of infrastructure, the fragility of urban ecosystems, and social disparities in resilience to shocks (Ribeiro & Gonçalves, 2019). Efforts have been completed to strengthen disaster preparedness and response; reducing the vulnerability of cities to natural disasters requires a comprehensive approach, combining infrastructure improvement, sustainable urban planning, public education, collaboration between the public and private sectors, and the use of technology (LopezDeAsiain & Díaz-García, 2020). By implementing these measures, cities can become more resilient to natural shocks and protect the lives of their residents.

4.2 Improving the Resilience of Urban Infrastructure

Improving urban infrastructure resilience is crucial to mitigate the effects of natural disasters (Mengist et al., 2020). Strengthening buildings and bridges, developing more effective early warning systems, and upgrading water and energy networks to be more resilient to shocks can do this. Cities must invest in upgrading their infrastructure to make it more robust and prepared for disasters (Bruzzone et al., 2021a, 2021b).

a. ***Sustainable urban planning***: Urban planning is crucial in reducing disaster vulnerability. Cities should adopt sustainable planning approaches considering natural hazards (Romero-Lankao et al., 2016a). This includes restricting construction in high-risk areas, preserving raw buffer zones, and creating refuge areas in case of emergency. Urban planning focused on sustainability can help prevent human and economic losses during disasters (Parmesan et al., 2022).

b. ***Public awareness and education***: It is essential to inform and educate residents of urban areas about the risks of natural disasters. Educational campaigns can help people understand how to prepare and respond to an emergency (Reisi et al., 2020).
c. ***Public–Private Sector Collaboration***: Collaboration is crucial to improving urban resilience. Local governments, businesses, and civic organizations must work together to develop disaster preparedness and response strategies (Monstadt & Schmidt, 2019; Mell, 2017).
d. ***Use of technology***: Technological advances such as advanced weather monitoring systems, drones, and geospatial data analysis can be valuable tools for disaster preparedness.

5 Climate Change and Sustainability

Climate change and sustainability are two topics closely linked to challenges that require immediate urgency on a global scale. We will also do our best to provide an in-depth analysis of these two areas and their interconnection.

5.1 Climate Change

"climate change" refers to the global or regional trend of gradually changing average weather conditions over time (Birchall et al., 2022; Anabaraonye et al. 2018). These changes are generally attributed to fossil fuel consumption, deforestation, and industrialization. The repercussions include rising human well-being, rising sea levels, and more severe weather (Mouratidis, 2019; Birchall & Bonnett, 2021).

5.2 Durability

The concept of sustainability refers to the responsible use of natural resources to meet the needs of the current generation without compromising the ability of future generations to meet their own needs. The concept of sustainability, commonly called the "three pillars" of sustainability, is a sustainable society's environmental, economic, and social aspects.

5.3 Interconnection

a. ***Reducing greenhouse gas emissions***: The principal aim of sustainable development is to mitigate greenhouse gas emissions since it plays a crucial role in mitigating the impacts of climate change.
b. ***Resource conservation***: Implementing sustainable practices in managing water, land, and biodiversity resources is paramount in developing climate resilience and reducing the consequences of climate change.
c. ***Circular economy***: Adopting circular economic models can potentially reduce demand for extracted resources, thereby mitigating the climate effect associated with resource extraction.
d. ***Social***: equity refers to the concept that climate change often has unequal effects on marginalized communities. Sustainability concepts emphasize the crucial role of equality and social justice in mitigation and adaptation efforts.
e. ***Adaptation***: The integration of adaptation methods is a common feature of sustainable practices aimed at strengthening the resilience of ecological and social systems in the face of climate change.

6 Strategies and Solutions

Energy Transition: Switching to renewable energy sources to replace fossil fuels.

Sustainable Agriculture: Adopt agricultural practices that consume fewer resources and maintain soil health.

Education and awareness: Educate people about the importance of climate change and sustainability.

Public Policy: Implement policies that encourage sustainable behavior, such as carbon taxes or incentives for renewable energy.

Research and development: Invest in research to develop new technologies and methods that promote sustainability and climate change mitigation.

6.1 The Evolution of Climate Concerns

The recognition of cities as essential actors in the fight against climate change has emerged alongside the advancement of scientific knowledge in this area. Given that a significant share of greenhouse gas emissions can be attributed to urban activities, there is a growing demand for cities to actively mitigate their carbon footprint and respond effectively to the inevitable consequences of climate change, including but not limited to sea level rise, heat waves, and extreme weather events.

6.2 The Transition to Sustainable Urban Centers

A notable rise in the trend toward sustainable urban development marked the decades from the 1990s to the 2000s. Urban strategies and plans have been infused with sustainable development ideas, encompassing preserving natural resources, promoting social equity, and pursuing economic viability (Chelleri & Baravikova, 2021). Adopting initiatives such as Agenda 21 at the Earth Summit in Rio de Janeiro in 1992 encouraged municipal governments to formulate programs to improve their sustainable cities' sustainability (Keita & Kourouma, 2023).

7 The Integration of Urban Resilience and Sustainability

7.1 Convergence of Concepts

Over the past decades, there has been an increasing convergence between urban resilience and sustainability, leading to a blurring of their boundaries (Cariolet et al., 2019; Weichselgartner & Kelman, 2015). This mutual influence has led to a relationship of interdependence between the two concepts. Resilience encompasses the ability to absorb and respond effectively to various disruptions, while sustainability strives to establish sustainable systems that can persist over an extended period (Zeng et al., 2022). Urban areas have recognized the importance of developing resilience to achieve lasting sustainability and vice versa.

7.2 Integrated Approaches

Urban areas increasingly adopt comprehensive strategies considering resilience and sustainability in urban planning and management (Delgado-Ramos & Guibrunet, 2017). This requires the synchronization of multiple sectors, including infrastructure, housing, transport, environment, and risk management (Jung et al., 2020). The advent of initiatives such as intelligent cities has resulted in integrating information and communication technologies into urban areas to improve residents' well-being and mitigate adverse environmental effects (Venkataramanan et al., 2019).

8 Current Challenges and Opportunities

8.1 Social and Economic Inequalities

One of the main obstacles to urban resilience and sustainability is the persistent presence of social and economic disparities within cities (Cook et al., 2021). Natural disasters and economic shocks often affect vulnerable populations, requiring special attention when planning and implementing resilience and sustainability strategies (Hamstead et al., 2021).

8.2 Pressure on Natural Resources

The increase in urban population translates into an increased need for natural resources, including water, electricity, and land (Iwaniec et al., 2021). Urban areas must adopt sustainable strategies to manage their limited resources effectively, promoting energy efficiency, water protection, and the development of renewable energy sources (Cook et al., 2021).

8.3 Climate Change and Extreme Events

Exacerbating climate change poses an ongoing threat to urban areas, with a predicted increase in extreme weather proceedings such as storms, floods, and droughts (Boyd & Juhola, 2015; Mannetti et al., 2021). Urban areas must prioritize implementing adaptation strategies to strengthen their resilience to these risks. This goal can be achieved through improving coastal infrastructure, strengthening stormwater management systems, and promoting urban vegetation to mitigate the impacts of heat waves (Balk et al., 2022; Boyd & Juhola, 2015).

9 Initiatives and Concrete Examples

9.1 Innovative Urban Planning

Urban planners in many cities worldwide have developed new planning approaches to encourage resilience and sustainability. For example, Copenhagen, Denmark, the Danish capital, is committed to becoming the first city to achieve carbon neutrality by 2025 (Ahern et al., 2014; Matovnikov & Matovnikova, 2016). This ambitious goal is being followed through substantial investments in cycling infrastructure, renewable energy sources, and improving the energy efficiency of buildings (Dong et al., 2018). Along the same lines, Singapore has taken several measures, such as implementing rainwater harvesting and utilization, to address the water shortage problem related to drinking water supply (Liao, 2019).

9.2 Urban Renewal and Resilience

In many cities, urban renewal has become a way to strengthen resilience while improving residents' quality of life (Hsu & Chao, 2020; Wei-Hsuan & Chun-Ta, 2014). Medellín, Colombia, is a remarkable example of urban development transformation. Converting former informal settlements into public spaces and parks has resulted in stronger social cohesion and increased resilience to regular flooding (Liu et al., 2023; Ozel et al. 2013).

9.3 Waste Management and Circular Economy

Waste management is an essential element in achieving urban sustainability. Some metropolitan areas have implemented progressive strategies, such as promoting the circular economy, which involves treating waste as a valuable resource (Luttenberger, 2020). San Francisco, situated in the United States, has successfully adopted a comprehensive waste management approach commonly referred to as the zero waste program. This policy aims to advance the concepts of recycling, reusing, and composting, significantly decreasing the amount of garbage disposed of in landfills (Pires & Martinho, 2019; Sharma et al., 2021).

9.4 Citizen Participation and Governance

The effectiveness of urban resilience and sustainability efforts relies mainly on citizen participation and transparent governance (United Nations and Social Affairs, 2014). Examples of cities like Porto Alegre, located in Brazil, have successfully implemented participatory budgeting initiatives, which enable active citizen engagement in the decision-making process regarding budget allocation. This inclusive approach allows individuals to influence investment decisions about urban sustainability programs.

10 With the Aim of Having a Resilient and Sustainable Future

The United Nations created the Sustainable Development Goals (SDG's) to provide a critical framework to guide efforts to achieve sustainability management, focusing on urban environments (Bibri, 2021, Udas-Mankikar & Driver, 2021). Cities worldwide are integrating these goals into their policies and operations to address food insecurity, poverty, healthcare, education, gender inequality, water scarcity, and land degradation (Cohen, 2017).

10.1 Technological and Social Innovation

Innovation plays a crucial role in creating resilient and sustainable cities. Advanced technologies, such as smart sensors, artificial intelligence, and real-time data, facilitate better running of resources and faster response to challenges (Dong et al., 2018). At the same time, social innovation promotes new ways of collaboration between management entities, businesses, non-governmental organizations, and citizens.

10.2 International Collaboration

The issue of urban resilience and sustainability transcends national borders. Exchanges of knowledge and best between cities worldwide play a central role in identifying shared solutions. Efforts such as the C40 Cities Climate Leadership Group aim to unite urban centers dedicated to combating climate change and promoting sustainable practices.

11 Future Perspectives of This Study

Expanding the application of blue-green infrastructure: further study and adoption of BGI in urban design. This involves incorporating smart urban planning that integrates water and green spaces to build resilience to natural calamities, improve environmental conditions and promote human well-being.

Adapting to climate change: tackling the challenges presented by climate change by employing inventive approaches to urban planning and design. This may involve conducting research and implementing innovative technologies and approaches to improve urban resilience, with a specific focus on sustainable resource management.

Integrate technological innovations: use innovations such as smart city applications to improve the effectiveness of urban resilience methods. This may involve using real-time data monitoring, artificial intelligence and other digital tools to manage resources and respond to urban concerns quickly and effectively.

Promote social equity and community participation: ensure that urban resilience efforts are comprehensive and beneficial to all sectors of the urban population, particularly disadvantaged neighborhoods. This may involve involving local people in the planning and decision-making process to ensure that resilience measures align with the diverse requirements of urban residents.

Multidisciplinary and collaborative approaches: In this segment, the need to adopt a multidisciplinary method of urban resilience is highlighted. This approach requires the cooperation of a very large number of stakeholders, such as urban planners, engineers, policymakers and network participants. Collaboration between these organizations would be very important to achieve the objective of developing solutions that are both broad and effective in terms of resilience.

12 Conclusion

This conclusion discusses blue-green infrastructure (BGI) and its crucial importance for urban resilience and sustainability in developing countries, it emphasizes the giant role that blue-green infrastructure (BGI) plays and the United Nations Sustainable Development Goals (SDGs) in providing a framework for monitoring sustainability, particularly in the context of developing countries. It emphasizes BGI's ability to address various urban issues, including climate change, rapid urbanization and socioeconomic inequality, food insecurity, poverty, healthcare, education, inequality between women, water shortages and land degradation.

We also highlight that successful BGI implementation requires integrated planning, collaboration across sectors, community engagement, and schooling, which are all important additives to successful BGI implementation. In the latest installment of the bankruptcy, BGI's relevance is reaffirmed. BGI is a comprehensive approach that not only provides environmental benefits but also promotes social equity and economic viability. The focus is on how BGI can help create more resilient and adaptive cities able to cope with the impacts of climate change and growing urban pressures. As a result, it is an important method for sustainable urban development in developing countries.

References

Acuti, D., Bellucci, M., & Manetti, G. (2020). Company disclosures concerning the resilience of cities from the Sustainable Development Goals (SDGs) perspective. *Cities, 99*, 102608.

Ahern, J., Cilliers, S., & Niemelä, J. (2014). The concept of ecosystem services in adaptive urban planning and design: A framework for supporting innovation. *Landscape and Urban Planning, 125*, 254–259.

Ahern, J. (2011). From fail-safe to safe-to-fail: Sustainability and resilience in the new urban world. *Landscape and Urban Planning, 100*, 341–343.

Ahmed, S., Meenar, M., & Alam, A. (2019). Designing a Blue-Green Infrastructure (BGI) network: Toward water-sensitive urban growth planning in Dhaka, Bangladesh. *Land, 8*, 138.

Anabaraonye, B., Okon, O. E., Dibia, S. I., Onwuzurike, U., Olisah, N. C. & Ewa, B. O. Climate change education and multi-stakeholder partnership projects to scale up climate resilience in Africa. *environment (Barrios et al., 2006; Coulibaly et al., 2020; Rigaud et al., 2018)*, 11, 27.

Ateş, M. & Erinsel Önder, D. (2023). A local smart city approach in the context of smart environment and urban resilience. *International Journal of Disaster Resilience in the Built Environment, 14*, 266–284.

Austin, G. 2014. *Green infrastructure for landscape planning: integrating human and natural systems*, Routledge.

Baba, K., Nagata, Y., Kawakubo, S. & Tanaka, M. 2020. A framework and indicators of resilience. *Resilient Policies in Asian Cities: Adaptation to Climate Change and Natural Disasters*, 3–45.

Balk, D., Tagtachian, D., Jiang, L., Marcotullio, P., Cook, E. M., Jones, B., Mustafa, A., & McPhearson, T. (2022). Frameworks to envision equitable urban futures in a changing climate: A multi-level, multidisciplinary case study of New York City. *Frontiers in Built Environment, 8*, 949433.

Béné, C., Mehta, L., McGranahan, G., Cannon, T., Gupte, J., & Tanner, T. (2018). Resilience as a policy narrative: Potentials and limits in the context of urban planning. *Climate and Development, 10*, 116–133.

Bereitschaft, B., & Scheller, D. (2020). How might the COVID-19 pandemic affect 21st century urban design, planning, and development? *Urban Science, 4*, 56.

Bhadouria, R., Tripathi, S., Singh, P., Joshi, P.K. and Singh, R., 2023. Urban Metabolism and Global Climate Change: An Overview. *Urban Metabolism and Climate Change: Perspective for Sustainable Cities*, pp.3–22.

Bibri, S. E. 2020. *Advances in the leading paradigms of urbanism and their amalgamation: compact cities, eco–cities, and data–driven smart cities*, Springer.

Bibri, S. E. (2021). Data-driven smart sustainable cities of the future: An evidence synthesis approach to a comprehensive state-of-the-art literature review. *Sustainable Futures, 3*, 100047.

Billger, M., Thuvander, L., & Wästberg, B. S. (2017). In search of visualization challenges: The development and implementation of visualization tools for supporting dialogue in urban planning processes. *Environment and Planning b: Urban Analytics and City Science, 44*, 1012–1035.

Birchall, S. J., & Bonnett, N. (2021). Climate change adaptation policy and practice: The role of agents, institutions and systems. *Cities, 108*, 103001.

Birchall, S. J., Macdonald, S., & Baran, N. N. (2022). An assessment of systems, agents, and institutions in building community resilience to climate change: A case study of Charlottetown, Canada. *Urban Climate, 41*, 101062.

Bixler, R. P., Lieberknecht, K., Atshan, S., Zutz, C. P., Richter, S. M., & Belaire, J. A. (2020). Reframing urban governance for resilience implementation: The role of network closure and other insights from a network approach. *Cities, 103*, 102726.

Borie, M., Pelling, M., Ziervogel, G., & Hyams, K. (2019). Mapping narratives of urban resilience in the global south. *Global Environmental Change, 54*, 203–213.

Boyd, E., & Juhola, S. (2015). Adaptive climate change governance for urban resilience. *Urban Studies, 52*, 1234–1264.

Bozza, A., Asprone, D., & Manfredi, G. (2015). Developing an integrated framework to quantify resilience of urban systems against disasters. *Natural Hazards, 78*, 1729–1748.

Bruzzone, M., Dameri, R. P., & Demartini, P. (2021b). Resilience reporting for sustainable development in cities. *Sustainability, 13*, 7824.

Bruzzone, M., Dameri, R. & Demartini, P. 2021a. Resilience Reporting for Sustainable Development in Cities. Sustainability 2021, 13, 7824. s Note: MDPI stays neutral with regard to jurisdictional claims in published ....

Bulkeley, H., Carmin, J., Broto, V. C., Edwards, G. A., & Fuller, S. (2013). Climate justice and global cities: Mapping the emerging discourses. *Global Environmental Change, 23*, 914–925.

Bush, J., & Doyon, A. (2019). Building urban resilience with nature-based solutions: How can urban planning contribute? *Cities, 95*, 102483.

Cariolet, J.-M., Vuillet, M., & Diab, Y. (2019). Mapping urban resilience to disasters–A review. *Sustainable Cities and Society, 51*, 101746.

Chelleri, L., & Baravikova, A. (2021). Understandings of urban resilience meanings and principles across Europe. *Cities, 108*, 102985.

Chelleri, L., Waters, J. J., Olazabal, M., & Minucci, G. (2015). Resilience trade-offs: Addressing multiple scales and temporal aspects of urban resilience. *Environment and Urbanization, 27*, 181–198.

Chen, Y.-S. (2008). The driver of green innovation and green image–green core competence. *Journal of Business Ethics, 81*, 531–543.

Cohen, M. (2017). A systematic review of urban sustainability assessment literature. *Sustainability, 9*, 2048.

Cook, E. M., Berbés-Blázquez, M., Mannetti, L. M., Grimm, N. B., Iwaniec, D. M. & Muñoz-Erickson, T. A. 2021. Setting the stage for co-production. *Resilient urban futures*, 99–111.

Correa Ayram, C. A., Mendoza, M. E., Etter, A. & Salicrup, D. R. P. (2016). Habitat connectivity in biodiversity conservation: A review of recent studies and applications. *Progress in Physical Geography, 40*, 7–37.

Cristiano, S. 2018. Systemic thoughts on ecology, society, and labour. *Through the Working Class*, 9.

Davidson, K., Nguyen, T. M. P., Beilin, R., & Briggs, J. (2019). The emerging addition of resilience as a component of sustainability in urban policy. *Cities, 92*, 1–9.

de Macedo, L. S. V., Picavet, M. E. B., de Oliveira, J. A. P., & Shih, W.-Y. (2021). Urban green and blue infrastructure: A critical analysis of research on developing countries. *Journal of Cleaner Production, 313*, 127898.

Deely, J., Hynes, S., Barquín, J., Burgess, D., Finney, G., Silió, A., Álvarez-Martínez, J. M., Bailly, D., & Ballé-Béganton, J. (2020). Barrier identification framework for the implementation of blue and green infrastructures. *Land Use Policy, 99*, 105108.

Delgado-Ramos, G. C., & Guibrunet, L. (2017). Assessing the ecological dimension of urban resilience and sustainability. *International Journal of Urban Sustainable Development, 9*, 151–169.

Deng, H., Zhang, K., Wang, F., & Dang, A. (2021). Compact or disperse? Evolution patterns and coupling of urban land expansion and population distribution evolution of major cities in China, 1998–2018. *Habitat International, 108*, 102324.

Dhakal, K. P., & Chevalier, L. R. (2017). Managing urban stormwater for urban sustainability: Barriers and policy solutions for green infrastructure application. *Journal of Environmental Management, 203*, 171–181.

Dong, L., Wang, Y., Scipioni, A., Park, H.-S., & Ren, J. (2018). Recent progress on innovative urban infrastructures system towards sustainable resource management. *Resources, Conservation and Recycling, 128*, 355–359.

Drosou, N., Soetanto, R., Hermawan, F., Chmutina, K., Bosher, L., & Hatmoko, J. U. D. (2019). Key factors influencing wider adoption of blue–green infrastructure in developing cities. *Water, 11*, 1234.

Eriksson, E. & Andersson, T. 2023. The ‘service turn’in a new public management context: a street-level bureaucrat perspective. *Public Management Review*, 1–25.

Fabbricatti, K., Boissenin, L., & Citoni, M. (2020). Heritage Community Resilience: Towards new approaches for urban resilience and sustainability. *City, Territory and Architecture, 7*, 17.

Fenner Andrew, R. (2017). Spatial evaluation of multiple benefits to encourage multi-functional design of sustainable drainage in blue-green cities. *Water, 9*, 953.

Florida, R., Rodríguez-Pose, A. & Storper, M. (2021). Cities in a post-COVID world. *Urban studies.*

Gao, Z., Zhang, Q., Xie, Y., Wang, Q., Dzakpasu, M., Xiong, J., & Wang, X. (2022). A novel multi-objective optimization framework for urban green-gray infrastructure implementation under impacts of climate change. *Science of the Total Environment, 825*, 153954.

Ghofrani, Z., Sposito, V. & Faggian, R. (2017). A comprehensive review of blue-green infrastructure concepts. *International Journal of Environment and Sustainability,* 6.

Hamann, F., Blecken, G.-T., Ashley, R. M., & Viklander, M. (2020). Valuing the multiple benefits of blue-green infrastructure for a Swedish case study: Contrasting the economic assessment tools B£ ST and TEEB. *Journal of Sustainable Water in the Built Environment, 6*, 05020003.

Hamstead, Z. A., Iwaniec, D. M., Mcphearson, T., Berbés-Blázquez, M., Cook, E. M. & Muñoz-Erickson, T. A. (2021). *Resilient urban futures,* Springer Nature.

Han, S., Sim, J., & Kwon, Y. (2021). Recognition changes of the concept of urban resilience: Moderating effects of COVID-19 pandemic. *Land, 10*, 1099.

Heinzlef, C., & Serre, D. (2020). Urban resilience: From a limited urban engineering vision to a more global comprehensive and long-term implementation. *Water Security, 11*, 100075.

Hsu, K.-W., & Chao, J.-C. (2020). Economic valuation of green infrastructure investments in urban renewal: The case of the station district in Taichung, Taiwan. *Environments, 7*, 56.

Iwaniec, D. M., Cook, E. M., Davidson, M. J., Berbés-Blázquez, M., Georgescu, M., Krayenhoff, E. S., Middel, A., Sampson, D. A., & Grimm, N. B. (2020). The co-production of sustainable future scenarios. *Landscape and Urban Planning, 197*, 103744.

Iwaniec, D. M., Berbés-Blázquez, M., Cook, E. M., Grimm, N. B., Mannetti, L. M., Mcphearson, T. & Muñoz-Erickson, T. A. (2021). Positive Futures. *Resilient urban futures*, 85–97.

JACOBS, M. 2012. *Green growth: Economic theory and political discourse*, Grantham Research Institute on Climate Change and the Environment London.

Jayasooriya, V., Ng, A., Muthukumaran, S., & Perera, B. (2017). Green infrastructure practices for improvement of urban air quality. *Urban Forestry & Urban Greening, 21*, 34–47.

Jung, H., Yun, S., Kim, T., & Kim, S. (2020). Estimating values of urban green space in seoul: Application of hierarchical linear regression model on hedonic housing price analysis. *Journal of Environmental Policy and Administration, 28*, 213–242.

Kapetas, L., & Fenner, R. (2020). Integrating blue-green and grey infrastructure through an adaptation pathways approach to surface water flooding. *Philosophical Transactions of the Royal Society A, 378*, 20190204.

Karabakan, B., & Mert, Y. (2021). Measuring the Green Infrastructure Resilience in Turkey. *Chinese Journal of Urban and Environmental Studies, 9*, 2150014.

Keita, K., & Kourouma, S. (2023). Assessment of Policy and Legal Frameworks of Urban Green Infrastructure Development: Republic of Guinea. *Buildings, 13*, 1945.

Kim, D., & Lim, U. (2016). Urban resilience in climate change adaptation: A conceptual framework. *Sustainability, 8*, 405.

Kondo, M. C., Fluehr, J. M., McKeon, T., & Branas, C. C. (2018). Urban green space and its impact on human health. *International Journal of Environmental Research and Public Health, 15*, 445.

Langemeyer, J., & Connolly, J. J. (2020). Weaving notions of justice into urban ecosystem services research and practice. *Environmental Science & Policy, 109*, 1–14.

Leng, L., Mao, X., Jia, H., Xu, T., Chen, A. S., Yin, D., & Fu, G. (2020). Performance assessment of coupled green-grey-blue systems for Sponge City construction. *Science of the Total Environment, 728*, 138608.

LIAO, K.-H. (2019). The socio-ecological practice of building blue-green infrastructure in high-density cities: What does the ABC Waters Program in Singapore tell us? *Socio-Ecological Practice Research, 1*, 67–81.

Liu, L., & Jensen, M. B. (2018). Green infrastructure for sustainable urban water management: Practices of five forerunner cities. *Cities, 74*, 126–133.

Liu, H.-Y., Jay, M., & Chen, X. (2021a). The role of nature-based solutions for improving environmental quality, health and well-being. *Sustainability, 13*, 10950.

Liu, S.-C., Peng, F.-L., Qiao, Y.-K., & Zhang, J.-B. (2021b). Evaluating disaster prevention benefits of underground space from the perspective of urban resilience. *International Journal of Disaster Risk Reduction, 58*, 102206.

Liu, Z., Cheng, W., Jim, C. Y., Morakinyo, T. E., Shi, Y., & Ng, E. (2021c). Heat mitigation benefits of urban green and blue infrastructures: A systematic review of modeling techniques, validation and scenario simulation in ENVI-met V4. *Building and Environment, 200*, 107939.

Liu, X., Wang, N., Li, Z., Jia, R. & Qiao, Z. (2023). Research on Time Series and Spatial Gradient of Urban Heat Island Expansion from the Perspective of Urban Renewal. *IEEE Journal of Selected Topics in Applied Earth Observations and Remote Sensing.*

Lopezdeasiain, M., & Díaz-García, V. (2020). The importance of the participatory dimension in urban resilience improvement processes. *Sustainability, 12*, 7305.

Luttenberger, L. R. (2020). Waste management challenges in transition to circular economy–case of Croatia. *Journal of Cleaner Production, 256*, 120495.

Mannetti, L., Berbés-Blázquez, M., Cook, E., Iwaniec, D. M., Grimm, N. B., Lloyd, R., Mcphearson, T. & Muñoz-Erickson, T. A. (2021). The UREx guide to scenarios.

Marchese, D., Reynolds, E., Bates, M. E., Morgan, H., Clark, S. S., & Linkov, I. (2018). Resilience and sustainability: Similarities and differences in environmental management applications. *Science of the Total Environment, 613*, 1275–1283.

MARIN, J. 2021. Global resilience models and territories of the South. A critical review. *International Journal of Disaster Risk Reduction,* 66, 102541.

Matovnikov, S., & Matovnikova, N. (2016). Innovative urban planning methods for the urban landscape design in the Volgograd agglomeration. *Procedia Engineering, 150*, 1966–1971.

McPhearson, T., Pickett, S. T., Grimm, N. B., Niemelä, J., Alberti, M., Elmqvist, T., Weber, C., Haase, D., Breuste, J., & Qureshi, S. (2016). Advancing urban ecology toward a science of cities. *BioScience, 66*, 198–212.

McPhillips, L. E., Matsler, M., Rosenzweig, B. R., & Kim, Y. (2021). What is the role of green stormwater infrastructure in managing extreme precipitation events? *Sustainable and Resilient Infrastructure, 6*, 133–142.

Meerow, S., & Stults, M. (2016). Comparing conceptualizations of urban climate resilience in theory and practice. *Sustainability, 8*, 701.

Mell, I. C. 2017. Green infrastructure: reflections on past, present and future praxis. Taylor & Francis.

Mengist, W., Soromessa, T., & Legese, G. (2020). Method for conducting systematic literature review and meta-analysis for environmental science research. *MethodsX, 7*, 100777.

Monstadt, J., & Schmidt, M. (2019). Urban resilience in the making? The governance of critical infrastructures in German cities. *Urban Studies, 56*, 2353–2371.

Mouratidis, K. (2019). Compact city, urban sprawl, and subjective well-being. *Cities, 92*, 261–272.

Nagenborg, M. (2019). Urban resilience and distributive justice. *Sustainable and Resilient Infrastructure, 4*, 103–111.

Nguyen, T. T., Ngo, H. H., Guo, W., Wang, X. C., Ren, N., Li, G., Ding, J., & Liang, H. (2019). Implementation of a specific urban water management-Sponge City. *Science of the Total Environment, 652*, 147–162.

O'donnell, E. & Thorne, C. 2020. Urban flood risk management: the blue–green advantage. *Blue–green cities: Integrating urban flood risk management with green infrastructure.* Ice Publishing.

Ozel, B., Mecca, S. & Dipasquale, L. (2013). Architectural Heritage, Resilience and Urban Renewal.

Özerol, G., Dolman, N., Bormann, H., Bressers, H., Lulofs, K., & Böge, M. (2020). Urban water management and climate change adaptation: A self-assessment study by seven midsize cities in the North Sea Region. *Sustainable Cities and Society, 55*, 102066.

PARMESAN, C., MORECROFT, M. D. & TRISURAT, Y. 2022. *Climate change 2022: Impacts, adaptation and vulnerability.* GIEC.

Pires, A., & Martinho, G. (2019). Waste hierarchy index for circular economy in waste management. *Waste Management, 95*, 298–305.
Reischl, C., Rauter, R., & Posch, A. (2018). Urban vulnerability and adaptation to heatwaves: A case study of Graz (Austria). *Climate Policy, 18*, 63–75.
Reisi, M., Sabri, S., Agunbiade, M., Rajabifard, A., Chen, Y., Kalantari, M., Keshtiarast, A., & Li, Y. (2020). Transport sustainability indicators for an enhanced urban analytics data infrastructure. *Sustainable Cities and Society, 59*, 102095.
Ribeiro, P. J. G., & Gonçalves, L. A. P. J. (2019). Urban resilience: A conceptual framework. *Sustainable Cities and Society, 50*, 101625.
Romero-Lankao, P., Gnatz, D. M., & Sperling, J. B. (2016a). Examining urban inequality and vulnerability to enhance resilience: Insights from Mumbai, India. *Climatic Change, 139*, 351–365.
Romero-Lankao, P., Gnatz, D. M., Wilhelmi, O., & Hayden, M. (2016b). Urban sustainability and resilience: From theory to practice. *Sustainability, 8*, 1224.
Sarker, M. N. I., Khatun, M. N., Alam, G. M. & Islam, M. S. Big data driven smart city: way to smart city governance. 2020 International Conference on Computing and Information Technology (ICCIT-1441), 2020. IEEE, 1–8.
Schmidt, L. & Guerra, J. 2018. Sustainability: dynamics, pitfalls and transitions. *Changing societies: legacies and challenges. Vol. 3. The diverse worlds of sustainability*, 27–53.
Shamout, S., Boarin, P., & Wilkinson, S. (2021). The shift from sustainability to resilience as a driver for policy change: A policy analysis for more resilient and sustainable cities in Jordan. *Sustainable Production and Consumption, 25*, 285–298.
Shandas, V., Matsler, A. M., Caughman, L., & Harris, A. (2020). Towards the implementation of green stormwater infrastructure: Perspectives from municipal managers in the Pacific Northwest. *Journal of Environmental Planning and Management, 63*, 959–980.
Sharifi, A., & Khavarian-Garmsir, A. R. (2020). The COVID-19 pandemic: Impacts on cities and major lessons for urban planning, design, and management. *Science of the Total Environment, 749*, 142391.
Sharma, H. B., Vanapalli, K. R., Samal, B., Cheela, V. S., Dubey, B. K., & Bhattacharya, J. (2021). Circular economy approach in solid waste management system to achieve UN-SDGs: Solutions for post-COVID recovery. *Science of the Total Environment, 800*, 149605.
Song, W., & Yu, H. (2018). Green innovation strategy and green innovation: The roles of green creativity and green organizational identity. *Corporate Social Responsibility and Environmental Management, 25*, 135–150.
TELLER, J. 2021. Urban density and Covid-19: towards an adaptive approach. *Buildings & Cities*, 2.
Thorne, C. R., Lawson, E. C., Ozawa, C., Hamlin, S. L., & Smith, L. A. (2018). Overcoming uncertainty and barriers to adoption of Blue-Green Infrastructure for urban flood risk management. *Journal of Flood Risk Management, 11*, S960–S972.
Thorne, C. R. 2020. *Blue–Green Cities: Integrating urban flood risk management with green infrastructure*, ICE Publishing.
UDAS-MANKIKAR, S. & DRIVER, B. 2021. Blue-green infrastructure: An opportunity for Indian cities. *ORF Occasional Paper*.
UNITED NATIONS, D. O. E. & SOCIAL AFFAIRS, P. D. 2014. World Urbanization Prospects: The 2018 Revision, Custom Data Acquired via Website.
van Stigt, R., Driessen, P. P., & Spit, T. J. (2013). Compact City Development and the Challenge of Environmental Policy Integration: A Multi-Level Governance Perspective. *Environmental Policy and Governance, 23*, 221–233.
Venkataramanan, V., Packman, A. I., Peters, D. R., Lopez, D., McCuskey, D. J., McDonald, R. I., Miller, W. M., & Young, S. L. (2019). A systematic review of the human health and social well-being outcomes of green infrastructure for stormwater and flood management. *Journal of Environmental Management, 246*, 868–880.
Walker, B. (2020). Resilience: what it is and is not. *Ecology and Society*, 25.

Wardekker, A. (2021). Contrasting the framing of urban climate resilience. *Sustainable Cities and Society, 75*, 103258.

Weichselgartner, J., & Kelman, I. (2015). Geographies of resilience: Challenges and opportunities of a descriptive concept. *Progress in Human Geography, 39*, 249–267.

WEI-HSUAN, S. & CHUN-TA, T. 2014. A Study on Vulnerability of Urban for Identifying Urban Renewal Area– A Case Study of Taipei, Kaohsiung City. *The World SB14*, 1–7.

WUBNEH, M. (2021). Urban resilience and sustainability of the city of Gondar (Ethiopia) in the face of adverse historical changes. *Planning Perspectives, 36*, 363–391.

Zeng, X., Yu, Y., Yang, S., Lv, Y., & Sarker, M. N. I. (2022). Urban resilience for urban sustainability: Concepts, dimensions, and perspectives. *Sustainability, 14*, 2481.

Zuniga-Teran, A. A., Staddon, C., de Vito, L., Gerlak, A. K., Ward, S., Schoeman, Y., Hart, A., & Booth, G. (2020). Challenges of mainstreaming green infrastructure in built environment professions. *Journal of Environmental Planning and Management, 63*, 710–732.

Integrating Blue-Green as Next-Generation Urban Infrastructure in Developing Countries

Meenakshi Chaurasia, Kajal Patel, Ranjana Singh, Rahul Bhadouria, and K. S. Rao

Abstract The climate emergency and rapid population growth present challenges to the sustainability of urban design in cities worldwide. Nature-based solutions, such as reviving blue-green infrastructure (BGI) are gaining global traction for sustainable urban planning and effective urban space management. Developing BGI is a new concept that integrates water and vegetation management for ecosystem and human wellbeing. Several studies have documented multifunctional aspects of BGI including climate change adaptation, heat mitigation, land restoration and improvement of life quality. However, pertinent design strategies for this approach are remain unfledged in developing countries. The present chapter reviews the benefits and application of different BG solutions and also discusses the scientific information in the planning, revitalization, and maintenance of BGI as an imperative tool for achieving sustainable development in developing countries. This would aid in comprehending the priorities, differences and similarities in the development of BGI as a next-generation urban infrastructure.

Keywords Blue-green infrastructure · Climate change · Climate resilience · Nature-based solutions

Abbreviations

UN-Habitat	United Nations Human Settlements Programme
WWAP	World Water Assessment Programme
BGI	Blue-green infrastructure

M. Chaurasia (✉) · K. Patel · R. Bhadouria · K. S. Rao
Department of Botany, University of Delhi, New Delhi 110007, India
e-mail: meenakshichaurasia2@gmail.com

R. Singh
Governmental Model Degree College, Arniya, Bulandshahr, UP 203131, India

R. Bhadouria
Delhi College of Arts and Commerce, University of Delhi, New Delhi 110023, India

P. K. Joshi et al. (eds.), *Blue-Green Infrastructure for Sustainable Urban Settlements*,
https://doi.org/10.1007/978-3-031-62293-9_3

IPCC	Intergovernmental Panel on Climate Change
IPBES	Intergovernmental Science-Policy Platform on Biodiversity and Ecosystem Services
EC	European Commission
UHI	Urban Heat Island
CO_2	Carbon Dioxide
PM2.5	Particulate Matter of $\leq$2.5 μm size

1 Introduction

The impacts of urbanisation and climate change are interacting in perilous ways, endangering the world's environmental and socioeconomic stability. Currently, more than 55% of the total world population lives in urban areas and consume 78% of the world's energy and produces more than 60% of greenhouse gas emissions (UN Habitat, 2023). The current scenario for global climate change and demographic trends demonstrates the considerable contribution of urban regions to climate change at the local or regional scale, while also emphasising the potentially catastrophic impact of climate change on urban populations such as more intense rainfall and heat stress (Stephenson et al., 2010). Cities in developing nations, lacking sufficient infrastructure to tackle emerging environmental challenges, are most affected by these climatic changes (Bazrkar et al., 2015; Bhadouria et al., 2022; DePaul, 2012). Sustainability and resilience are significant concepts that centre on comprehending the key aspects of urban life in order to address problems and ensure viable urban futures (Romero-Lankao et al., 2016). This calls for a paradigm change in existing urban infrastructure; it must be reinforced and made resilient to the projected population expansion and able to endure the outcomes of impending climate change (Rozos et al., 2013). Addressing the repercussions of urbanization and climate change all while keeping up with economic and socially equitable growth presents a considerable challenge; however, it also offers an incentive to reimagine our approach to achieving sustainable urban development. Many of the most important development challenges including efficient energy and water utilization, biodiversity preservation, disaster risk management and climate change adaptation can be dealt with in parallel through prioritising sustainably developed urban areas under a wider development framework. (Kabisch et al., 2016). A pertinent method to developing sustainable cities is the incorporation of nature based solutions into mainstream infrastructure systems, which can help in successfully managing the water, energy, and climate interrelationship while providing more resilient ecosystem services at lower costs (De Silva et al., 2022; Seddon et al., 2020). UN World Water Development stressed how nature-based solutions (including green infrastructure) might aid in the achievement of the 2030 SDGs (WWAP, 2018). One of the strategies of overarching nature-based solutions for cities is the blue-green infrastructure (BGI) approach which strives to safeguard ecosystem services and human well-being and provide resilience to ecosystems and society against climate risk (Plieninger et al., 2022).

BGI outlines a systematically and methodically designed network of natural and semi natural urban features including spaces and water bodies designed to provide a wide range of environmental benefits, enhance the life quality of its residents and ensure an equitable and biodiverse future (Henderson et al., 2022). The BGI approach aims to the habitability, safety, and resilience centres to make them more liveable and sustainable (Brown & Mijic, 2019; Lafortezza et al., 2013; Puchol-Salort et al., 2021; Sánchez & Govindarajulu, 2023). BGI include elements like green roofs, street greens, permeable paving, engineered stormwater detention or storage basins, green façades, parks and open spaces, swales, rain gardens and retention basins and many others (Ahmed et al., 2019; Almaaitah et al., 2021; Kabisch et al., 2017; Klompmaker et al., 2023). BGI elements can be categorized based on their position, such as on-ground elements (e.g., bioretention areas, swales, permeable pavements,, parks, and other vegetated areas), above-ground modules (e.g., green roofs and walls, rain downspouts and barrels, green), or below-ground elements (e.g., stormwater storage and retention basins) and other under public spaces and structures (Brears, 2018; Kimic & Ostrysz, 2021). The common BGI elements and their benefits are described in Table 1.

More recently the concept of BGI has been enhanced and broadened by recognising the role that urban blue and green spaces, such as parks, greenway paths, forests, and lakes can play in holistically managing the escalating environmental issues and has been widely acknowledged by national and international bodies and agendas (IPCC, 2014; Brink et al., 2016; IPBES, 2019; EC, 2019). Nonetheless, most research on BGI has been undertaken mostly in developed cities of North American and European countries. Hence, little is known about the application of BGI solutions on the ground level in the planning of developing countries (Almaaitah et al., 2021; Brink et al., 2016; Liao et al., 2017). Cities in emerging nations experiencing rapid rise in population, poverty, and unemployment rates are significantly distinct from those in the developed countries in terms of socioeconomic state, environmental circumstances, and resource dynamics. Furthermore, the many rapidly expanding metropolitan regions of the Global South are vulnerable to number of environmental issues owing to decennial exponential expansion of urban regions and lax implementation of urban planning and regulatory measures (Drosou et al., 2019). Hence, requires context-specific BGI solutions that consider the unique social and cultural issues of developing cities to tackle urban challenges at the local level (Dhyani et al., 2020, 2022). In addition to serving regulatory and cultural purposes, BGI also supports the subsistence of marginalized urban hence, BGI planning and implementation in such emerging metropolitan centres necessitate a particular strategy. Also, there is a lack of critical formal guidance for spatial planning for BGI despite several pieces of research in the area of landscape planning and management proposing BGI as next-generation infrastructure, indicating that it remains a challenge (Romero-Duque et al., 2020).

Although, landscape architectures, engineers and scientific community are becoming progressively more interested in BGI, its widespread application remains

Table 1 Different types of blue-green spaces and their benefits

	Blue-green solutions	Description	Benefits
1	Bioretention areas (strips/ponds)	Engineered soils and vegetation used in shallow landscaped depressions aimed to reduce run-off rates and volumes simultaneously remediating pollution	1. Appealing landscape features that are capable of self-irrigate and fertilise 2. Habitat and biodiversity 3. Evapotranspiration-induced cooling of the local microenvironment
2	Buffer strip and hedgerow	Buffer strips are made up of grassland, marsh, scrub, or trees for protecting and improving natural components like streams, ponds, ditches, and woodlands	1. Slow down water flow 2. Reduce soil run-off 3. Provide water bank stabilisation and habitat
3	Channels and rills	Water channels with open surfaces and hard edges designed at varying widths to suit the urban landscape	1. Water purification 2. Contribute to Urban Drainage Systems 3. Act as pre-treatment to remove silt
4	Dune reinforcement and strengthening	Sand dunes reinforced with vegetation plantation or by installation of internal structures within dune systems	1. Preventing erosion 2. Create an effective flood barrier 3. Increases reflected energy
5	Filtration strips	Strips of land that is frequently integrated into adjacent land uses such as public open spaces or road verges in order to transit water running off a site before entering a watercourse	1. Vegetative filter strip removes contaminants or polluting solids from runoff 2. Positive impact on biodiversity
6	Green roof/green wall	A multi-layered system that uses vegetation and/or a drainage layer to cover different surfaces of a building	1. Improve the urban environment and purify air 2. Improve the thermal performance of a building by providing shading
7	Green screens	Composed of welded wire trellises that form a robust and lightweight modular panel for supporting vegetation (climbing and hedging plants) in outdoor applications	1. Save energy by building cooling requirements 2. Mitigate the urban heat island effect 3. Create comfortable outdoor spaces

(continued)

Table 1 (continued)

	Blue-green solutions	Description	Benefits
8	Greenway	Ecologically significant corridors/strips of undeveloped land near an urban area, set aside for recreational use or environmental protection	1. Protect natural systems: rivers, coastal areas and ridgelines 2. Maintain bio-diversity 3. Provide recreational opportunities 4. Historical heritage and cultural values
9	Parks (local–regional)	natural, semi-natural or artificially designed open public areas for recreational and aesthetic purposes	1. Boost the biodiversity and habitat value 2. Improve the physical and mental well-being of inhabitants 3. Mitigate air and noise pollution
10	Permeable pavements/ surfaces	Comprises various forms of pavement, pavers, and other devices that offer stormwater infiltration while also serving as a structural surface	1. Sustainable water management 2. Reduce heat at ground level
11	Urban ponds	Ponds are water bodies that act as detention basins for storing surface water run-off	1. Alleviate flood risk 2. Ameliorate ~~of~~ climatic fluctuations 3. Improve the quality of water runoff 4. Enhance biodiversity and amenity values
12	Rain gardens	Shallow depression with free-draining soil and vegetation cover that act as absorbent and can withstand infrequent transient flooding	1. Reduce rainwater runoff volume 2. Purify rainwater by filtering pollutants 3. Enhance biodiversity
13	Rainwater harvesting	Rainwater is the collection and storage of on-site rain for reuse	1. Reduce runoff and floods 2. Reduce water demand, especially in drought conditions
14	Sediment capture pond	Engineered ponds placed in drainage networks to help slow down the water Runoff and the carrying of large quantities of sediments from upstream to downstream	1. Managing the effects of ditch construction and maintenance and road work 2. Temporary measure for preserving water quality at construction sites 3. Capturing sediment in 4. Agricultural runoff

(continued)

Table 1 (continued)

	Blue-green solutions	Description	Benefits
15	Urban Forestry	Areas for practising cultivation and processing, and distribution of food in cities	1. Provision for food, fibre and fuel 2. Local climate regulation 3. Carbon sequestration
16	Vegetated Swales and Ditches	A vegetated swale is a broad, shallow, open, heavily planted with a dense and diverse type of trees, shrubs, and/or grasses designed for conveying water	1. Infiltration and treatment of stormwater 2. Prevent soil erosion 3. Improve water quality
17	Temporary Flood Water Storage/Detention basin	Open area constructed to temporarily or permanently retain floodwater during rain or flood events	1. Water quality improvement via solids/trap sediment, adsorption or biochemical degradation of pollutants

restricted. This could be because of the requirements of their pragmatic multidisciplinary approach, limited pertinent information and experience among experts, and ambiguities related to their ecological, economic and cultural functioning. (Drosou et al., 2019; Thorne et al., 2018). Optimization of BGI layout according to concurrent multiple demands and their efficiency has received less attention (Evans et al., 2019; Zhou & Wu, 2020). The present chapter aims to highlight the multiple benefits of integration of BGI solutions as next generation infrastructure in urban ecosystem and its contribution to the improvement of urban ecosystems. It also provides overview of the numerous fields in which potential BGI solutions could be applied and lays out the fundamental framework needed for their successful implementation in the built environment. This chapter also addresses the different challenges and barriers that affect the adoption of BGI, which are crucial to consider when putting BGI into practice in rapidly expanding cities of developing countries facing similar environmental issues.

2 Benefits of BGI as Next-Generation Infrastructure

The term "blue-green infrastructure" has gained popularity in recent years as an efficient approach to enhancing living conditions including improved water quality, protection from extreme climatic events and enhanced public amenities (Kimic & Ostrysz, 2021). There is a increasing number of research studies supporting the numerous advantages of urban BGI solutions to city dwellers (Liao, 2019; Sánchez & Govindarajulu, 2023) and its scope in different types of settings is also rapidly expanding (Kimic & Ostrysz, 2021; Van Oijstaeijen et al., 2020). Urban blue-green spaces have been acknowledged to exhibit notable potential to improve health and welfare of urban population while providing a wide array of socioeconomic, ecological and health co-benefits such as mitigation and adaptation to climate extremes, urban heat island (UHI) effects, increased noise, air pollution, water related issues (Ghofrani et al., 2017; Well & Ludwig, 2021). Apart from mitigating the impact of extremities, BGI helps to continuously enhance urban environmental conditions (Sánchez & Govindarajulu, 2023). Blue-green spaces promote environmental well-being by improving hydrological issues like stormwater runoff, poor water quality and water shortage, preserving biodiversity, creating recreational and aesthetic value and fostering a greener and sustainable environment (Brears, 2018; Well & Ludwig, 2021).

As the benefits delivered by BGI are linked to the variety of highly interrelated ecosystem services provisioned by blue-green elements for example water quality improvement, mitigation of climate extremes (Dhyani et al., 2022; Liu & Jensen, 2018), biodiversity conservation and provision of cultural, psychological and health facilities (Kati & Jari, 2016; Völker & Kistemann, 2015), cost reduction (Naumann et al., 2011; Hsu & Chao, 2021) and adaptation to risk posed by climate changes (Kazmierczak & Carter, 2010; Almaatiah et al., 2021). These services fall into four categories: cultural (recreational and aesthetic values), regulating (air quality regulation, water management, climatic resilience, and carbon storage), provisioning (habitat, food, and energy), and supporting (soil functioning and biogeochemical cycling). These services contribute to the improvement of the social, cultural, economic, psychological, and spiritual well-being of individuals.

BGI offers many environmental benefits such as reducing flood risk, stormwater pollutant reductions, reducing sewer overflow events, improving quality of potable water, increasing groundwater recharge, alleviating UHI effects, improving air quality, etc. (Antoszewski et al., 2020; Badach et al., 2022; Ghofrani et al., 2017; Hamel & Tan, 2022). Additional environmental benefits of urban blue and green spaces include flood risk reduction (Alves et al., 2019); stormwater management (Liao et al., 2017); improved biodiversity (Barbosa et al., 2019; Donati et al., 2022); carbon sequestration (Fenner, 2017; Hamann et al., 2020); and habitat provision for wildlife (Nguyen et al., 2021). Economic benefits include the replacement of costly grey infrastructure with less capital-intensive infrastructure, reduced water treatment costs, reduced cost of maintenance, increased natural capital, the addition of and many more (Ahmed et al., 2019; Brears, 2018; Kapoor et al., 2020; Raymond et al., 2017).

These environmental and economic values extend to numerous social benefits. For example, BGI can also improve levels of physical and pyschological health by providing clean air, enhancing the quality of indoor spaces, ensuring water safety, increasing resilience to climate change, and other important factors that are essential for a healthier environment. Improved transport networks, increased socialization among citizens, reduced noise pollution, and the cooling effect provided by green corridors have been shown clearly to contribute to improving people's health (Drosou et al., 2019; Well & Ludwig, 2021). Urban blue-green spaces can foster positive social interaction within communities and provide social support mechanisms (Lamond &

Everett, 2019; Thorne et al., 2018). This can enhance notion of affiliation and contentment with their neighbourhood in residents, benefitting their psychological health (Brears & Brears, 2018).

3 Application of Integrated Blue-Green Infrastructure

Urban areas encounter an extensive array of environmental problems that entail substantial threats to their future social and economic well-being. These challenges include risks to food and water security, increased energy and carbon footprints, extreme weather events such as such as storms, flooding, cyclones, coastal erosion, and threat to natural habitats (Hobbie & Grimm, 2020). In terms of offering mitigation and resilience to climate change risk, the multi-scalar and multi-functional urban BGI infrastructure can be implemented for the following main benefits and services that represent the outlook for BGI benefits globally.

Mitigating climatic extremes: Being responsible for a large proportion of global greenhouse gas emissions, urban centres have contributed significantly to climate change (Wamsler et al., 2013). Extreme weather events, such as heavy precipitation, floods, heatwaves, forest fires, and droughts, are becoming more frequent and intense, causing great concerns about food and watersecurity, destruction of natural ecosystems, damage to infrastructures, and other consequences (Weiskopf et al., 2020). These events are putting immense strain on the intricate critical infrastructure that supports cities and its population (Gasper et al., 2011). Thus, initiatives aimed at reducing the impact of extreme climatic events on urban centres must be weighed towards the necessity to improve the climate resilience of cities. Urban BGI development is one such strategy that could enhance to the capacity of urban centres to cope with the present and forthcoming risk posed by climate change (Almaaitah et al., 2021). BGI is an important means of dealing with extreme hydrological events (i.e., flooding). The interconnected network of water reservoirs, wetlands, and associated open spaces developed along river bodies, serve a number of interrelated purposes such as water storage, regulation of the river system to prevent flooding conditions, nature conservation by providing habitat for plant and animal wildlife, water quality improvement of polluted water using decentralised blue-green spaces, and creation of area to foster suitable recreational activities (Ahmed et al., 2019; Ghofrani et al., 2017; Hamel & Tan, 2022; Kazmierczak & Carter, 2010; Liu et al., 2019). Incorporation of BGI in urban planning have been shown to lower the number of fatalities from severe heat waves (Gunawardena et al., 2017). Moreover, several case studies have shown a considerable improvement in planning and designing of urban drainage (Davis & Naumann, 2017; Wijayanti et al., 2023). Perpetual change in climatic conditions over time has prompted sustainable engineering solutions and corresponding anthropogenic environmental modifications such as the widespread integration of BGI solutions in urban infrastructure planning (Matos et al., 2022). It is important to note that implementing BGI is essential to enhancing urban stormwater management and mitigate other the adverse impacts of climate change. A deeper comprehension of benefits of incorporation of BGI for ecosystem provisioning can be applied to shape the resilience of the urban centres (Kimic & Ostrysz, 2021).

Water quality improvement: BGI's efficiency to improve water quality has been evidenced by several case studies (Berndtsson, 2010; Hamel & Tan, 2022; Stagge et al., 2012). To achieve desirable results in water quality improvement, most researchers highlight the importance of soil media composition and regular maintenance. In addition to mitigating the water runoff and flooding, BGI components such as swales, detention basins, rain gardens, ditches, green roofs have been demonstrated to significantly improve the water quality by eliminating suspended solid particles, heavy metals, hydrocarbons, and other nutrients (Davis et al., 2012; Liu et al., 2019). These BGI methods often improve water quality by increasing infiltration and storage, delaying runoff peaks, lowering runoff rates and volumes, and controlling the movement of contaminants (Chen et al., 2019). Among the investigated measures, the establishment of grass strips was reported to be the most effective in improving water quality (Hamel & Tan, 2022; Stagge et al., 2012). Most researches emphasise the relevance of soil media composition and frequent management in achieving optimal results in water quality enhancement.

Temperature reductions: Increase in impervious surfaces, reduction in vegetation and water cover increase the air and surface temperature. Higher temperatures resulting from UHI modifies the urban microclimate, augment air pollution, increase energy consumption and rate of morbidity and mortality posing threats to urban residents (Gunawardena et al., 2017; Kousis et al., 2021). Strategic inclusion of urban blue and green forms in urban infrastructure can modify the urban microclimate and lower the UHI effect. Urban green spaces provide cooling by improving evapotranspiration, filtering pollutants and by providing shading, that further reduces temperatures, whilst waterbodies create cooling effect through evaporating cooling (Gunawardena et al., 2017). Several studies have demonstrated the cooling impact of blue green spaces on city temperature in various urban development scenarios (Antoszewski et al., 2020; Pérez et al., 2014).

Energy saving: Increasing demand for energy for economic development has necessitated measures to reduce the demand and consumption of energy in developing countries. Research have shown that introducing green elements such as, parks, greenways, green roofs, and walls in urban neighbourhood can profoundly reduce the energy demands of individual buildings making them more energy-efficient (Almaatiah et al., 2021; Lamond & Everett, 2023). Determining the overall benefits incurred by BGIs is difficult due to the heterogeneity in analysing the multiple benefits and energy expenses of the BGI put at multiple scales. Numerous studies, however, imply that BGI development can lower the energy consumption and production cost by reducing heating and cooling costs while also solving other environmental issues (Lamond & Everett, 2023; Siehr et al., 2022).

Air quality improvement: Air pollution has emerged as a significant problem that endangers the health of metropolitan populations (Kampa & Castanas, 2008). For example, long-term exposure to ambient air pollution is associated with greater susceptibility to cardiovascular and respiratory diseases, with some evidence suggesting that this may have a greater impact on lower-income populations (Makri &

Stilianakis, 2008; Mannucci & Franchini, 2017). Additionally, there is growing evidence of a shift in the distribution of fine particulate matter ($PM_{2.5}$) concentration in the atmosphere across the socioeconomic gradient, with many urban settings in developing countries experiencing increases in $PM_{2.5}$ concentrations compared to cities in high-income countries (van Donkelaar et al., 2015). Numerous studies have revealed that nature-based strategies, including urban BGI, can reduce the exposure to elevated levels of ambient particulate matter and other types of air pollutants (Lin et al., 2011; Kabisch & Van Den Bosch, 2017; Yu et al., 2020). Urban ventilation corridors can alleviate the poor air quality and UHI effect in polluted city cores by improving air circulation between the cities and the adjacent areas (Badach et al., 2022; Han et al., 2022).

Carbon sequestration: Terrestrial carbon sequestration is one of the valuable advantages of BGI. Carbon storage may be less immediate, but it becomes apparent over time (Brears, 2018; Fenner, 2017; Ghofrani et al., 2020; O'Donnell et al., 2021). Urban vegetation can be a critical tool for reducing carbon footprint of cities, by trapping and storing atmospheric carbon in the form of vegetation and soil biomass (Niemelä et al., 2010; Vargas-Hernández et al., 2023). Numerous research studies have evidenced the efficiency of the vegetation component of BGI in reducing the CO_2 levels in the atmosphere and their potential role as a net CO_2 sink (Zhang et al., 2014; Strohbach et al., 2012; Sun et al., 2019, Li & Wang, 2021). BGI benefits for carbon storage may be less immediate, but they become apparent over longer time periods at the regional, national, and international levels.

Habitat improvement: The development of green corridors as a part of BGI can provide ecological services such as habitat improvement, increase mobility networks, social interaction among residents, and noise reduction (Ghofrani et al., 2017). Habitat improvement will provide residence and sustenance to migratory birds, insects, and other microscopic species. Further, it serves as a nursery for certain species that require specific habit to flourish in their initial stage of development (Donati et al., 2022). BGI also provides an intriguing service that reduces noise and enhances comfort. Blue-green spaces, such as permeable pavements and green roofs are known to significantly lower noise pollution (Connelly & Hodgson, 2008).

Recreation and aesthetics: Introduction of BGI in urban forms promotes leisure pursuits like exercise, picnics, hiking, and nature observation. Urban settings that have well-planned and structured blue and green space enhance aesthetic experience of residents and have more monetary value, desirable to investors (O'Donnell et al., 2021). Another important factor in determining how well-received BGI solutions are by society is their aesthetic appreciation in urban settings The link between urban greening and property value is a well-established indicator that reflect positive aesthetic impact of blue-green infrastructure projects. Numerous studies have also looked into how green infrastructure affects property values (Tate et al., 2022).

BGI concepts are most widely implemented in developed nations. The sustainable urban drainage systems in the UK, water-sensitive urban design in Australia, the low-impact development and green stormwater infrastructure in the USA, the 'sponge city' in China and the Active, Beautiful, Clean Waters Programme in Singapore are some of the evidences of successful implementation of BGI (Ghofrani et al.,

2017). However, in economically emerging countries like India, the idea of BGI and its real-world applications in urban planning are still at a very rudimentary stage (Kaur & Gupta, 2022). Although BGI is relatively a new concept in India it is gaining momentum. Blue-green components are being included in plans in Indian cities such as Delhi, Bengaluru, Bhopal, Guwahati and Madurai to reduce pollution, increase resilience to severe climatic events, improve health, assure efficient resource utilization, and encourage economic development (Dhyani et al., 2022). The case studies in Chennai and Kochi demonstrate the scientific community's involvement, deliberation of social and environmental requirements and active participation to mitigate flooding, improve water quality, and manage solid waste (Sánchez & Govindarajulu, 2023). Similarly, the adoption of the water corridor in Dhaka evidences the development of BGI solutions for flood control and stormwater management. Policymaker interviews and site visits in Indonesia have revealed the incorporation of small-scale BGI measures such as biopores and rainwater storage for bioretention (Drosou et al., 2019). In Singapore, engineered measures such as raingardens, greenroofs, and built wetlands have been installed to improve stormwater quality (Hamel & Tan, 2022). The Mekong Delta region of populated Vietnam city has implemented blue-green infrastructure initiatives, such sponge parks, to manage flooding (Toledo-Gallegos et al., 2022). Although the availability of research articles and documentation of city projects suggest comprehensive BGI-integrated planning, they are overlooked as low impact strategy for solving urban issues in developing countries (Ahmed et al., 2019; Prescott et al., 2021; Hamel & Tan, 2022). Realization of BGI concept as a sustainability strategy to address both urban development and global environmental concerns require appropriate framework to deploy, monitor, and evaluate integrated urban BGI solutions (de Macedo et al., 2021).

4 Framework for Integration of BGI in Urban System

The BGI promotes a healthy environment for its living beings by offering habitable space and opportunities for active transport, recreation, and culture which require inclusive community involvement and stewardship. Urban infrastructures should not only be designed sustainably to facilitate the implementation of various environmental solutions but also should be accommodating and receptive to the demands of all of their users. A single discipline strategy is insufficient to address all domains of concern (Well & Ludwig, 2021). Effective implementation of BGI stipulate an innovative approach that addresses three basic interrelated requisites i.e., inclusive community participation; proactive engagement of stakeholders at multiple scales; and pertinent implementation of BGI solutions that is compatible with the local conditions (Drosou et al., 2019). In the local context of BGI adoption, social and cultural factors can help or hinder these requirements. An enhanced conception of socio-economic and cultural concerns enables the creation of consolidated and disseminated policies that promote comprehensive, dynamic, and suitable BGI solutions that can be executed and embraced by local communities at the level of the bottom line (Drosou et al., 2019). This can be achieved by including liveability and environmental sustainability in its goal alongside economic targets (Hersperger et al., 2019; Pamukcu-Albers et al., 2021).

To accomplish this, stakeholder engagement at multiple levels is constantly sought and factored into the decision-making processes of BGI integration into the urban system. Planning, development, and maintenance of BGI projects require a multi-stakeholder approach to proactively adopt a collaborative mindset and a readiness to act.

The integration of a multiscalar, multidimensional strategy for BGI implementation would benefit at the systems, project and user levels (Fig. 1). Numerous examples from urban areas demonstrate that a methodical, collaborative involvement of local governmental bodies, social community organisations, and private sector corporations is required to facilitate successful adaptation and mitigation to climate change (Forino et al., 2015; Fünfgeld, 2010). Participation of stakeholder participation and firm cooperation amongst several groups including planners, developers, politicians and scientists can aid in widespread application of BGI and the subsequent functioning of the system (Bhadouria et al., 2023; Nesshöver et al., 2017; Zuniga-Teran et al., 2020). Planning elements include leaders' involvement, public participation, stakeholder engagement, and finance from sources other than local budgets are all viewed as important requisites for planning (Hassenforder et al., 2020; Hughes, 2015).

BGI integration cannot be solely the domain of governments, politicians, and other stakeholders; for it to be sustainable, local communities must be included in new stewardship techniques. Although BGI is becoming more common in urban planning literature, there is still obscurity regarding practical approaches for assuring community understanding, awareness, and approval. Significant community participation in conjunction government, lawmakers, developers, and other political officials will raise awareness about blue-green projects among citizens and encourage participation in the conception, creation, implementation, and oversight of such activities.

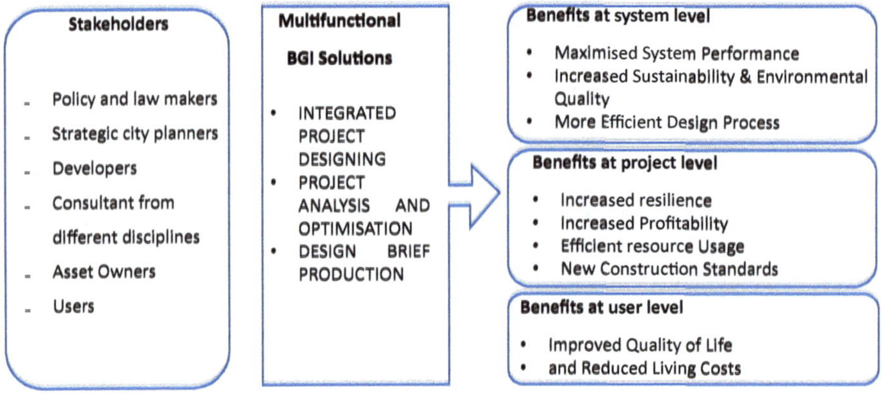

Fig. 1 Benefits of the BGI approach for stakeholders at multilevel

Citizen participation is required to effect behavioural changes, and this is frequently accomplished by conveying the possible advantages to persuade communities to realise and value the BGI (Charlesworth & Warwick, 2011). Urban BGI initiates must take into account the requirements and beliefs of the residents in regard to the benefits of BGI in order to achieve acceptable adaptation, change in attitude, and public acceptance (Derkzen et al., 2017).

Each city faces distinct environmental challenges socioeconomic status, resource availability, climatic condition (Pampel, 2014). Thus, cities worldwide require context-specific solutions within an appropriate strategic framework that is subjective to spatial and temporal urban changes. While planning and developing urban public spaces, it is essential to recognise appropriate BGI solutions in connection to spatial and functional, environmental, and social factors at the local level. Deployment of BGI as next generation infrastructure for developing sustainable and resilient cities is more pragmatic in economically emerging developing nations lacking robust infrastructure network, unlike developed nations (Fonseca et al., 2022). They may integrate sustainable practices into their current and future economic development and overall expansion. It is more sustainable to invest in green energy, water reuse, resource recovery, and environment conservation as opposed to rushing to develop through traditional, degrading industrial practices (Jordan, 2019). Cities must adhere to a standardized design approach to develop contextual technical protocol based on available performance data and residents's demands and beliefs. Design framework that conforms with regional environmental and social circumstances and responsive to their specific set of challenges and concerns (Hammel & Tan, 2022; Bhadouria et al., 2023; Kimic & Ostrysz, 2021; O'Donnell et al., 2017).

Environmental aspect: Considering the many environmental advantages, including reduced risk of disasters, adaptation to and mitigation of climate change, and management of other environmental challenges, BGI solutions can significantly influence the natural capital of cities. BGI systems have turned into a highly useful instrument for building and reinstating sustainable living conditions by improving soil, water and air quality, protecting habitats, and improving resistance and adaptability to climate extremes (Ghofrani et al., 2017; Lamond & Everett, 2019; O'Donnell et al., 2017). However, the impact of BGI on the environment can vary depending on the unique characteristics of each BGI solution, particularly in terms of its integration with local hydrological conditions and its ecological performance. Inappropriate application of BGI for biodiversity preservation might provide the circumstances for invasion of non-native species.

Spatial and functional aspects: It is imperative to take into account the design obstacles associated with the spatial execution of different BGI solutions in conjunction with the pre-existing spatial components within an urban setting. Cities have different kinds of public spaces, including roadways and parking lots, public squares, parks, surrounding structures, etc. They take up spaces of varied sizes and serve a variety of purposes. Unfortunately, BGI's physical infrastructure such as dams and dikes limits space for recreation and greenery and hinders biodiversity restoration, etc. (Kimic & Ostrysz, 2021). Furthermore, one aspect that hinders and reduces the efficiency of the applications, operation, and advancement of environmentally viable solutions is the shortage of space for the deployment of new BGI elements, even those that interfere less with the structures of the city (Liu & Jensen, 2018).

Social aspect: The contribution of BGI to the mitigation of multiple environmental issues effectively contributes to better positively impacts liveability of city as well as the health and well-being of its inhabitants (Kabisch et al., 2017; Keeler et al., 2019). In turn, the success of BGI functioning is decided by the perception of the residents. Therefore, for the successful integration of BGI, an interdisciplinary framework for adaptive planning and design is essential, to confirm that measures executed are socially viable and widely accepted in community (Raymond et al., 2017; Sánchez & Govindarajulu, 2023). For BGI efforts to be successful, it is critical to include the local community in the designing and operative stages (Everett et al., 2018). Implementation of solutions that are little differentiated, ineffective, and aesthetically unappealing and tend to be poorly maintained and executed and may evoke negative perceptions.

The development of envisaged BGI in urban planning requires an integrative designing process to identify and analyse issues and potential solutions via systematic stakeholder involvement, to enhance resource identification and integration to reduce life-cycle costs and increase operating and resource efficiency and sustainability and to develop resilience to climate change and weather extremes. A framework for BGI design includes a collection of concept solutions that have been optimised together with the specific criteria from all the stakeholders in charge of infrastructure planning, financing, design, delivery, and maintenance. The design brief helps the developer make well-informed, compliant and robust project decisions to enhance the efficiency of BGI solutions Not only will the project be more environmentally friendly, but it will also result in substantial capital savings, particularly in terms of operational expenditures (Bozovic et al., 2017).

The BGI Design Brief is created in three primary steps. The first step entails establishing project objectives based on established needs, performance benchmarks and indicators. Collecting configuring project specifications and data concerning design (e.g. prevailing issues, local climatic conditions, planning opportunities and availability of natural resources). The second stage is project analysis and optimisation. It involves the conceptual conceptualization of measures for target indicators, assessment of possible synergies between various urban features in order to propose prospective solutions and thorough comparative evaluation for the development of optimised integrated solutions. This phase covers project concerns such as feasibility, funding and finance. The third stage involves the production and approval of BGI design comprising the project's specifications (depending on inputs from participant) and implementation, management and monitoring of optimised BGI solutions (Bozovic et al., 2017).

The need of implementing BGI cannot be overstated in terms of enhancing urban environmental management and mitigating the adverse impacts of climate change. There exist numerous approaches to surmount the obstacles that impede the complete operation of BGI, and the understanding of ecosystem services can be leveraged to design resilient urban centre. However, a deeper comprehension of the potential opportunities and constraints associated with deploying BGI solutions in urban areas as a substitute for conventional systems, as well as the creation of sustainable urban environment, is required. The following section will throw some light on some of these barriers and challenges that restrain the execution and functioning of BGI.

5 Barriers and Challenges to BGI Integration

Urban zones often face several problems in adoption BGI initiatives, one of which is the establishment of a well-structured BGI. Integrating BGI into the built environment is a complicated and uncertain process in and of itself (Ghofrani et al., 2017; O'Donnell et al., 2017). For instance, highly varied land uses, populations, and utilities in the current urban context, would create a variety of demands for BGI development distributed in various patterns with varying intensities (Meerow & Newell, 2017). The development of integrated BGI applications is hampered by several technical, social, legal, economic, and institutional hurdles. Recent research has provided vital insights into these constraints, which varied greatly among nations with varying legislation, economic levels, and industrial growth (Almaaitah et al., 2021; Drosou et al., 2019). In addition to these impediments, biophysical barriers such as uncertainty in physical science/engineering, a lack of sufficient space, and future land use and climate had been also identified (Almaaitah et al., 2021; Drosou et al., 2019).

The discourse surrounding the socio-economic-environmental aspect of technological BGI innovations and ambiguities and risk involved in their implementation are included in technical barriers. These challenges are often manifested in the form of environmental and societal doubts about the long-term functioning of engineered BGI and the assumption related to expenses of BGI maintenance and difficult to deliver (Thorne et al., 2018). findings revealed that the key technical issues are uncertainty in hydrologic performance and maintenance problems (Thorne et al., 2018). The reluctance of locals to the risks associated with BGI further creates obstructions to the uptake of these engineered solutions in metropolitan areas.

Institutional and social impediments typically emerge as a result of mismanagement and a lack of conscious policy, such as tardy adoption of sustainable solutions or a lack of a long-term plan for their operation and supervision (O'Donnell et al., 2017). The framework for BGI integration generally involves long ranged strategic planning, execution and maintenance procedures, which may have to be modified by administrations' immediate actions and decision-making cycles due to administrative changes in the city that cause a decoupling between immediate actions and prospective goals. In some circumstances, the project's maintenance obligations are still unclear, which puts the long-term delivery of the expected socioeconomic and environmental advantages at risk. Scientifically validated options and information are frequently lacking, even in places wherein futuristic policy proposals go through adaptive monitoring for adopting novel inventive measures. These variables contribute to a mismatch between policy acceptance and ready-to-apply scientific findings and conceptions (Frantzeskaki & Tilie, 2014; Hansen et al., 2015). Concurrently, changes in administrative policies concentrating on execution specifics only focus on short-term goals and lack long-term ecological projects and their maintenance. Another major barrier to action is the trapping of knowledge in "sectoral silos" which refers to situations in which the different actors involved in a BGI project operate on independently or with objectives that are incompatible with those of other actors (Hansen et al., 2015; Naumann et al., 2011). Sectoral silos leads to organisational disintegration that hinders the progress and success of BGI interventions (Redondo Bermúdez et al., 2022).

These concerns, which are unique to BGI, are mostly circumscribed by broader concerns like the need for socially equitable benefits of climate change risk management and the challenge of engaging the community in intricate technical, operational, and planning matters related to infrastructure projects (Thorne et al., 2018; Zuniga-Teran et al., 2020). Community participation and individual citizens' engagement play a crucial role in improving existing living conditions in settlements. Lack of community and/or end-user participation in the planning and system design phases of any BGI facility may have detrimental effects on both input and results. Certainly, the previously described difficulties are exacerbated by combination sets of BGI because of the more complicated hydro-ecological functioning of individual BGI measures and their expected greater maintenance cost. Climate change-oriented BGI must take into account the importance of hydrologic and ecologic interconnection between each BGI measures BGI programmes support one another when they are properly connected (Voskamp & Van de Ven, 2015).

Budget constraints are identified as a crucial challenge for BGI integration in higher-income countries as well as in low-income countries such as India (Flores et al., 2021; O'Donnell et al., 2021; Sánchez & Govindarajulu, 2023). The financial challenges placed on the implementation of BGI in low-income nations are justified by their restricted financial capacity. However, the primary cause of financial barriers in higher-income nations are the absence of financial sources and the incapacity to obtain capital for both original investments and operation. Budgetary constraints, scheduling constraints, workforce cuts, and a lack of experience might also hinder such initiatives from using available funding sources. Moreover, financing of BGI projects may cause social injustice, addressing these inequalities is critical for expansion of BGI adoption (Amaral et al., 2021). It is widely recognised that ongoing maintenance procedures and the absence of systematic monitoring present both operational and budgetary obstacles to BGI integration (Almaatitah et al., 2021; Suleiman, 2021). Andenaes et al., (2021) discuss lack of risk management as one of the technological issues. The development of framework for risk reduction and processual understanding could be effective ways to handle the difficulty. The space available for BGI installation is strictly limited due to the lack of vacant land in high-density locations. This necessitates the use of a layout model that allows for great efficiency and adaptability in BGI development (Zhou & Wu, 2020). The problem of how to optimise BGI layout and simultaneously and effectively address multiple service demands has received less attention, despite the fact that numerous prior studies have examined layout organisation strategies corresponding to a single BGI service demand. This issue is becoming more pressing in contemporary planning process (Evans et al., 2019).

Apart from creating the necessary infrastructure for the environment and welfare, people's aspirations, perseverance, and hope for a brighter future are just as crucial as their capacity to establish novel economic ecologies. As part of a sustainable development project, it is equally important for citizens to reconsider their daily lives in order for the community to evolve (Knierbein & Viderman, 2018). Evidence outside Southeast Asia suggests that challenges to BGI implementation in informal settlements are similar to those in formal areas in cities (Sinharoy et al., 2019). Particularly factors such as inadequate ability and experience, financial obstacles, ineffective governance, a lack of preliminary information, a consensus of disservices, a shortage of space that contends with other land uses, and a low capacity for climate adaptation have been identified as significant contributors to the limited understanding and lack of pertinent valuation data on BGI (Hamel & Tan, 2022; Mulligan et al., 2020). Scholars regard the BGI technique as a catapulting solution in which embrace more modern hydrological management measures that challenges encountered by wealthy economies (e.g., centralised drainage systems to diminish water conservation services) (Rogers et al., 2019). Conducting further investigation in this field and considering the local governance scenario will be helpful in comprehending the hydrological performance, socioeconomic advantages, and exploring implementation opportunities and problems (Hamel & Tan, 2022; Mulligan et al., 2020; Sinharoy et al., 2019).

Due to a lack of adequate financing and human resources, many Indian cities struggle with knowledge gaps and institutional barriers that restrict them from responding to climate risks through resilience measures, as well as from recognising and seizing opportunities for BGI-related actions, as previously reported by Kabisch et al. (2016). In order for BGI to actually improve city resilience, it is imperative that it comprehend these obstacles and the intricate factors that serve to strengthen them. Planning for BGI must be forward-looking and goal-oriented in order to address these issues (Pamukcu-Albers et al., 2021).

6 Conclusion and Future Perspectives

When planned and executed in harmony with their environment, Blue Green (BG) solutions can significantly improve the host city's resilience, resource efficiency, quality of life, and sustainability as a whole. This chapter highlights the social, environmental and economic benefits of BGI initiative and its application in various fields. The chapter also discussed the technical, institutional and social barriers to the implementation of BGI in urban centres. Confluence of social concerns of urban dwellers and policy makers serves to enhance comprehension of the local context and fosters a more well-informed and comprehensive process for developing and executing successful BGI design. Understanding various emerging BGI solutions and challenges related to its implementation in developing countries can can serve as a model other rapidly urbanising urban centres in middle and lower income developing countries like India experiencing various socio-ecological issues. As these countries are set to experience unprecedented population growth and urban development in the near future and require long-term sustainable solutions to address the housing, employment, and amenity needs of their new inhabitants. Past and ongoing initiatives related to BGI have yielded a wealth of valuable practices and insights that can shape future endeavours in this direction. It is more critical than ever to advocate for the advantages and economic viability of ecosystem centred initiatives such as BGI. To gain political support across all levels, obtain financing and commercial sector involvement, enhance decisiveness, and eventually promote BGI adoption, we must strengthen the scientific foundation and evidence base for BGI. A scenario of triple-win outcomes, encompassing economic advantages, environmental preservation, and social empowerment, can be achieved through effectively constructed BGI strategies aimed at bolstering resilience against impact of climate change. Following are the specific recommendations.

- Piloting a local context-based environmental risk assessment to guide the planning and design of BGI projects.
- Utilizing multisectoral (science, technology, and policy-making) revolutions to optimize the most suitable BGI options.
- Coordinating efforts through a comprehensive, government-wide and society-wide approach when implementing BGI initiatives.
- Involving pertinent parties in the conception, execution, and evaluation of BGI programs
- Prioritizing adaptation strategy that emphasizes investments in nature based solutions such as BGI, to fortify city resiliency to climate and disaster.
- Implementing available knowledge in management managerial strategies to record and assemble best practices and field lessons gained.
- Employing a straightforward quantitative evaluation method based on expert knowledge to increase the appeal of BGI solutions.

References

Ahmed, S., Meenar, M., & Alam, A. (2019). Designing a Blue-Green Infrastructure (BGI) network: Toward water-sensitive urban growth planning in Dhaka, Bangladesh. *Land, 8*(9), 138. https://doi.org/10.3390/land8090138

Almaaitah, T., Appleby, M., Rosenblat, H., Drake, J., & Joksimovic, D. (2021). The potential of Blue-Green infrastructure as a climate change adaptation strategy: A systematic literature review. *Blue-Green Systems, 3*(1), 223–248. https://doi.org/10.2166/bgs.2021.016

Alves, A., Gersonius, B., Kapelan, Z., Vojinovic, Z., & Sanchez, A. (2019). Assessing the Co-Benefits of green-blue-grey infrastructure for sustainable urban flood risk management. *Journal of Environmental Management, 239*, 244–254. https://doi.org/10.1016/j.jenvman.2019.03.036

Amaral, M. H., Benites-Lazaro, L. L., de Almeida Sinisgalli, P. A., da Fonseca Alves, H. P., & Giatti, L. L. (2021). Environmental injustices on green and blue infrastructure: Urban nexus in a macrometropolitan territory. *Journal of Cleaner Production, 289*, 125829. https://doi.org/10.1016/j.jclepro.2021.125829

Andenæs, E., Time, B., Muthanna, T., Asphaug, S., & Kvande, T. (2021). Risk reduction framework for blue-green roofs. *Buildings, 11*(5), 185. https://doi.org/10.3390/buildings11050185

Antoszewski, P., Świerk, D., & Krzyżaniak, M. (2020). Statistical review of quality parameters of blue-green infrastructure elements important in mitigating the effect of the urban heat island in the temperate climate (C) zone. *International Journal of Environmental Research and Public Health, 17*(19), 7093. https://doi.org/10.3390/ijerph17197093

Badach, J., Szczepański, J., Bonenberg, W., Gębicki, J., & Nyka, L. (2022). Developing the urban blue-green infrastructure as a tool for urban air quality management. *Sustainability, 14*(15), 9688. https://doi.org/10.3390/su14159688

Barbosa, A., Martín, B., Hermoso, V., Arévalo-Torres, J., Barbière, J., Martínez-López, J., Domisch, S., Langhans, S. D., Balbi, S., Villa, F., & Delacámara, G. (2019). Cost-effective restoration and conservation planning in Green and Blue Infrastructure designs. A case study on the Intercontinental Biosphere Reserve of the Mediterranean: Andalusia (Spain)–Morocco. *Science of the Total Environment, 652*, 1463–1473. https://doi.org/10.1016/j.scitotenv.2018.10.416

Bazrkar, M. H., Zamani, N., Eslamian, S., Eslamian, A. & Dehghan, Z. (2015). Urbanization and Climate Change. In: Leal Filho, W. (eds) *Handbook of Climate Change Adaptation*. Springer, Berlin, Heidelberg. https://doi.org/10.1007/978-3-642-38670-1_90

Berndtsson, J. C. (2010). Green roof performance towards management of runoff water quantity and quality: A review. *Ecological Engineering, 36*(4), 351–360. https://doi.org/10.1016/j.ecoleng.2009.12.014

Bhadouria, R., Upadhyay, S., Tripathi, S., & Singh, P. (Eds.). (2022). *Urban ecology and global climate change*. John Wiley & Sons.

Bhadouria, R., Tripathi, S., Singh, P., Joshi, P.K. and Singh, R. (2023). Urban Metabolism and Global Climate Change: An Overview. *Urban Metabolism and Climate Change: Perspective for Sustainable Cities*, pp. 3–22.

Bozovic, R., Maksimovic, C., Mijic, A., Smith, K. M., Suter, I. & Van Reeuwijk, M. (2017). Blue green solutions, a systems approach to sustainable, resilient and cost-efficient urban development. London: imperial college. Accessed on 14.06.2023. https://www.climate-kic.org/wp-content/uploads/2017/10/BGD-Guide.compressed.pdf

Brears, R. C. & Brears, R. C. (2018). From traditional grey infrastructure to blue-green infrastructure. In: *Blue and Green Cities*. pp. 1–41. Palgrave Macmillan, London. https://doi.org/10.1057/978-1-137-59258-3_1

Brears, R. C. (2018). Blue-Green Infrastructure in Managing Urban Water Resources. In: *Blue and Green Cities*. pp 43–61. Palgrave Macmillan, London. https://doi.org/10.1057/978-1-137-59258-3_2

Brink, E., Aalders, T., Ádám, D., Feller, R., Henselek, Y., Hoffmann, A., Ibe, K., Matthey-Doret, A., Meyer, M., Negrut, N. L., & Rau, A. L. (2016). Cascades of green: A review of ecosystem-based adaptation in urban areas. *Global Environmental Change, 36*, 111–123. https://doi.org/10.1016/j.gloenvcha.2015.11.003

Brown, K., & Mijic, A. (2019). Integrating green and blue spaces into our cities: Making it happen. *Grantham Institute, 30*, 1–10.

Charlesworth, S. M. & Warwick, F. (2011). 15 Adapting to and Mitigating Floods Using Sustainable Urban Drainage Systems. Flood Hazards: Impacts and Responses for the Built Environment, p.207.

Chen, J., Liu, Y., Gitau, M. W., Engel, B. A., Flanagan, D. C., & Harbor, J. M. (2019). Evaluation of the effectiveness of green infrastructure on hydrology and water quality in a combined sewer overflow community. *Science of the Total Environment, 665*, 69–79. https://doi.org/10.1016/j.scitotenv.2019.01.416

Connelly, M., & Hodgson, M. (2008). Sound transmission loss of extensive green roofs-Field test results. *Canadian Acoustics, 36*(3), 74–75.

Davis, A. P., Stagge, J. H., Jamil, E., & Kim, H. (2012). Hydraulic performance of grass swales for managing highway runoff. *Water Research, 46*(20), 6775–6786. https://doi.org/10.1016/j.watres.2011.10.017

Davis, M. & Naumann, S. (2017). Making the Case for Sustainable Urban Drainage Systems as a Nature-Based Solution to Urban Flooding. In: Kabisch, N., Korn, H., Stadler, J., Bonn, A. (eds) Nature-Based Solutions to Climate Change Adaptation in Urban Areas. *Theory and Practice of Urban Sustainability Transitions*. Springer, Cham. https://doi.org/10.1007/978-3-319-56091-5_8

de Macedo, L. S. V., Picavet, M. E. B., de Oliveira, J. A. P., & Shih, W. Y. (2021). Urban green and blue infrastructure: A critical analysis of research on developing countries. *Journal of Cleaner Production, 313*, 127898.

De Silva, A., Amaratunga, D., & Haigh, R. (2022). Green and blue infrastructure as nature-based better preparedness solutions for disaster risk reduction: key policy aspects. *Sustainability, 14*(23), 16155. https://doi.org/10.3390/su142316155

DePaul, M. (2012). Climate change, migration, and megacities: Addressing the dual stresses of mass urbanization and climate vulnerability. *Paterson Review of International Affairs, 12*, 145–162.

Derkzen, M. L., Van Teeffelen, A. J., & Verburg, P. H. (2017). Green infrastructure for urban climate adaptation: How do residents' views on climate impacts and green infrastructure shape adaptation preferences? *Landscape and Urban Planning, 157*, 106–130. https://doi.org/10.1016/j.landurbplan.2016.05.027

Dhyani, S., Karki, M. & Gupta, A. K. (2020). Opportunities and advances to mainstream nature-based solutions in disaster risk management and climate strategy. In: *Nature-based solutions for resilient ecosystems and societies*, pp. 1–26. https://doi.org/10.1007/978-981-15-4712-6_1

Dhyani, S., Singh, S., Basu, M., Dasgupta, R. & Santhanam, H. (2022). Blue-Green Infrastructure for Addressing Urban Resilience and Sustainability in the Warming World. In: *Blue-Green Infrastructure Across Asian Countries: Improving Urban Resilience and Sustainability* pp. 1–22. Singapore: Springer Singapore. https://doi.org/10.1007/978-981-16-7128-9_1

Donati, G. F., Bolliger, J., Psomas, A., Maurer, M., & Bach, P. M. (2022). Reconciling cities with nature: Identifying local Blue-Green Infrastructure interventions for regional biodiversity enhancement. *Journal of Environmental Management, 316*, 115254. https://doi.org/10.1016/j.jenvman.2022.115254

Van Donkelaar, A., Martin, R. V., Brauer, M., Boys, B. L. (2015). Use of satellite observations for long-term exposure assessment of global concentrations of fine particulate matter. *Environmental Health Perspectives, 123*(2):135–143. https://doi.org/10.1289/ehp.14086 46

Drosou, N., Soetanto, R., Hermawan, F., Chmutina, K., Bosher, L., & Hatmoko, J. U. D. (2019). Key factors influencing wider adoption of blue–green infrastructure in developing cities. *Water, 11*(6), 1234. https://doi.org/10.3390/w11061234

EC (2019) Additional information on the review of implementation of the green infrastructure strategy, SWD (2019) 184 final. https://eur-lex.europa.eu/legal-content/EN/TXT/PDF/?uri=CELEX:52019SC0184&rid=1 Accessed on 21.08.23

Evans, A. J., Firth, L. B., Hawkins, S. J., Hall, A. E., Ironside, J. E., Thompson, R. C., & Moore, P. J. (2019). From ocean sprawl to blue-green infrastructure–A UK perspective on an issue of global significance. *Environmental Science and Policy, 91*, 60–69. https://doi.org/10.1016/j.envsci.2018.09.008

Everett, G., Lamond, J. E., Morzillo, A. T., Matsler, A. M., & Chan, F. K. S. (2018). Delivering green streets: An exploration of changing perceptions and behaviours over time around bioswales in Portland, Oregon. *Journal of Flood Risk Management, 11*, S973–S985. https://doi.org/10.1111/jfr3.12225

Fenner, R. (2017). Spatial evaluation of multiple benefits to encourage multi-functional design of sustainable drainage in blue-green cities. *Water, 9*(12), 953. https://doi.org/10.3390/w9120953

Flores, C. C., Vikolainen, V. & Crompvoets, J. (2021). Governance assessment of a blue-green infrastructure project in a small size city in Belgium. The potential of Herentals for a leapfrog to water sensitive. *Cities, 117*, 103331. https://doi.org/10.1016/j.cities.2021.103331

Fonseca, K., Espitia, E., Breuer, L., & Correa, A. (2022). Using fuzzy cognitive maps to promote nature-based solutions for water quality improvement in developing-country communities. *Journal of Cleaner Production, 377*, 134246. https://doi.org/10.1016/j.jclepro.2022.134246

Forino, G., von Meding, J., & Brewer, G. J. (2015). A conceptual governance framework for climate change adaptation and disaster risk reduction integration. *International Journal of Disaster Risk Science, 6*, 372–384.

Frantzeskaki, N., & Tilie, N. (2014). The dynamics of urban ecosystem governance in Rotterdam, The Netherlands. *Ambio, 43*, 542–555. https://doi.org/10.1007/s13280-014-0512-0

Fünfgeld, H. (2010). Institutional challenges to climate risk management in cities. *Current Opinion in Environmental Sustainability, 2*(3), 156–160. https://doi.org/10.1016/j.cosust.2010.07.001

Gasper, R., Blohm, A., & Ruth, M. (2011). Social and economic impacts of climate change on the urban environment. *Current Opinion in Environmental Sustainability, 3*(3), 150–157. https://doi.org/10.1016/j.cosust.2010.12.009

Ghofrani, Z., Sposito, V. & Faggian, R. (2017). A comprehensive review of blue-green infrastructure concepts. *International Journal of Environment and Sustainability*, 6(1).

Ghofrani, Z., Sposito, V. & Faggian, R. (2020). Maximising the value of natural capital in a changing climate through the integration of blue-green infrastructure. Journal of Sustainable Development of Energy, Water and Environment Systems, 8(1), pp. 213–234. https://doi.org/10.13044/j.sdewes.d7.0279

Gunawardena, K. R., Wells, M. J., & Kershaw, T. (2017). Utilising green and blue space to mitigate urban heat island intensity. *Science of the Total Environment, 584*, 1040–1055. https://doi.org/10.1016/j.scitotenv.2017.01.158

Hamann, F., Blecken, G. T., Ashley, R. M., & Viklander, M. (2020). Valuing the multiple benefits of blue-green infrastructure for a Swedish case study: Contrasting the economic assessment tools B£ ST and TEEB. *Journal of Sustainable Water in the Built Environment, 6*(4), 05020003. https://doi.org/10.1061/JSWBAY.000091

Hamel, P., & Tan, L. (2022). Blue–green infrastructure for flood and water quality management in Southeast Asia: Evidence and knowledge gaps. *Environmental Management, 69*(4), 699–718. https://doi.org/10.1007/s00267-021-01467-w

Han, L., Zhao, J., Zhang, T., & Zhang, J. (2022). Urban ventilation corridors exacerbate air pollution in central urban areas: Evidence from a Chinese city. *Sustainable Cities and Society, 87*, 104129. https://doi.org/10.1016/j.scs.2022.104129

Hansen, R., Frantzeskaki, N., McPhearson, T., Rall, E., Kabisch, N., Kaczorowska, A., Kain, J. H., Artmann, M., & Pauleit, S. (2015). The uptake of the ecosystem services concept in planning discourses of European and American cities. *Ecosystem Services, 12*, 228–246. https://doi.org/10.1016/j.ecoser.2014.11.013

Hassenforder, E., Barreteau, O., Daniell, K. A., Ferrand, N., Kabaseke, C., Muhumuza, M., & Tibasiima, T. (2020). The effects of public participation on multi-level water governance, lessons from Uganda. *Environmental Management, 66*, 770–784. https://doi.org/10.1007/s00267-020-01348-8

Henderson, H., Bush, J., Kozak, D. (2022). Mainstreaming Blue Green Infrastructure in Cities: Barriers, Blind Spots, and Facilitators. In: Brears, R.C. (eds) *The Palgrave Encyclopedia of Urban and Regional Futures*. Palgrave Macmillan, Cham. https://doi.org/10.1007/978-3-030-87745-3_270

Hersperger, A. M., Grădinaru, S., Oliveira, E., Pagliarin, S., & Palka, G. (2019). Understanding strategic spatial planning to effectively guide development of urban regions. *Cities, 94*, 96–105. https://doi.org/10.1016/j.cities.2019.05.032

Hobbie, S. E., & Grimm, N. B. (2020). Nature-based approaches to managing climate change impacts in cities. *Philosophical Transactions of the Royal Society, 375*, 20190124. https://doi.org/10.1098/rstb.2019.0124

Hsu, K. W., & Chao, J. C. (2021). Study on the value model of urban green infrastructure development—A case study of the central district of Taichung city. *Sustainability, 13*(13), 7402. https://doi.org/10.3390/su13137402

Hughes, S. (2015). A meta-analysis of urban climate change adaptation planning in the US. *Urban Climate, 14*, 17–29. https://doi.org/10.1016/j.uclim.2015.06.003
IPBES, W. (2019). Intergovernmental science-policy platform on biodiversity and ecosystem services. Summary for Policy Makers of the Global Assessment Report on Biodiversity and Ecosystem Services of the Intergovernmental Science-Policy Platform on Biodiversity and Ecosystem Services. IPBES Secretariat, Bonn, Germany.
Intergovernmental Panel on Climate Change (IPCC). (2014). Summary for Policymakers. In Climate Change 2013—The Physical Science Basis: Working Group I Contribution to the Fifth Assessment Report of the Intergovernmental Panel on Climate Change (pp. 1–30). Cambridge University Press. https://doi.org/10.1017/CBO9781107415324.004
Jordan, E. (2019). Nature-based Solutions: The advantage of being a developing country. https://iwa-network.org/nature-based-solutions-the-advantage-of-being-a-developing-country/. Accessed on 13.08.2023
Kabisch, N. & Van Den Bosch, M. A. (2017). Urban green spaces and the potential for health improvement and environmental justice in a changing climate. Nature-Based Solutions to Climate Change Adaptation in Urban Areas: Linkages between Science, Policy and Practice, pp.207–220.
Kabisch, N., Frantzeskaki, N., Pauleit, S., Naumann, S., Davis, M., Artmann, M., Haase, D., Knapp, S., Korn, H., Stadler, J. & Zaunberger, K. (2016). Nature-based solutions to climate change mitigation and adaptation in urban areas: perspectives on indicators, knowledge gaps, barriers, and opportunities for action. *Ecology and Society*, *21*(2). https://doi.org/10.5751/ES-08373-210239
Kabisch, N., Stadler, J., Korn, H. & Bonn, A. (2017). Nature-Based solutions for societal goals under climate change in urban areas–Synthesis and Ways Forward. In: Kabisch, N., Korn, H., Stadler, J., Bonn, A. (eds) Nature-Based Solutions to Climate Change Adaptation in Urban Areas. Theory and Practice of Urban Sustainability Transitions. Springer. https://doi.org/10.1007/978-3-319-56091-5_19
Kampa, M., & Castanas, E. (2008). Human health effects of air pollution. *Environmental Pollution, 151*(2), 362–367. https://doi.org/10.1016/j.envpol.2007.06.012
Kapoor, V., Tripathi, S., Devi, R.S., Srivastava, P. and Bhadouria, R., 2020. Ecological economics of an urban settlement: an overview. *Urban Ecology*, pp.91–110.
Kati, V., & Jari, N. (2016). Bottom-up thinking—Identifying socio-cultural values of ecosystem services in local blue–green infrastructure planning in Helsinki, Finland. *Land Use Policy, 50*, 537–547. https://doi.org/10.1016/j.landusepol.2015.09.031
Kaur, R., & Gupta, K. (2022). Blue-Green Infrastructure (BGI) network in urban areas for sustainable storm water management: A geospatial approach. *City and Environment Interactions, 16*, 100087. https://doi.org/10.1016/j.cacint.2022.100087
Kazmierczak, A. & Carter, J. (2010). Adaptation to climate change using green and blue infrastructure. A database of case studies. https://orca.cardiff.ac.uk/id/eprint/64906/1/Database_Final_no_hyperlinks.pdf
Keeler, B. L., Hamel, P., McPhearson, T., Hamann, M. H., Donahue, M. L., Meza Prado, K. A., Arkema, K. K., Bratman, G. N., Brauman, K. A., Finlay, J. C., & Guerry, A. D. (2019). Social-ecological and technological factors moderate the value of urban nature. *Nature Sustainability, 2*(1), 29–38. https://doi.org/10.1038/s41893-018-0202-1
Kimic, K., & Ostrysz, K. (2021). Assessment of blue and green infrastructure solutions in shaping urban public spaces—spatial and functional, environmental, and social aspects. *Sustainability, 13*(19), 11041. https://doi.org/10.3390/su131911041
Klompmaker, J. O., Hart, J. E., Bailey, C. R., Browning, M. H., Casey, J. A., Hanley, J. R., Minson, C. T., Ogletree, S. S., Rigolon, A., Laden, F., & James, P. (2023). Racial, ethnic, and socioeconomic disparities in multiple measures of blue and green spaces in the United States. *Environmental Health Perspectives, 131*(1), 017007. https://doi.org/10.1289/EHP11164
Knierbein, S. & Viderman, T. eds. (2018). *Public space unbound: urban emancipation and the post-political condition.* Routledge.

Kousis, I., Pigliautile, I., & Pisello, A. L. (2021). Intra-urban microclimate investigation in urban heat island through a novel mobile monitoring system. *Scientific Reports, 11*(1), 9732. https://doi.org/10.1038/s41598-021-88344-y

Lafortezza, R., Davies, C., Sanesi, G., & Konijnendijk, C. C. (2013). Green Infrastructure as a tool to support spatial planning in European urban regions. *Forest-Biogeosciences and Forestry, 6*(3), 102. https://doi.org/10.3832/ifor0723-006

Lamond, J., & Everett, G. (2019). Sustainable Blue-Green Infrastructure: A social practice approach to understanding community preferences and stewardship. *Landscape and Urban Planning, 191*, 103639. https://doi.org/10.1016/j.landurbplan.2019.103639

Lamond, J., & Everett, G. (2023). Willing to have, willing to help, or ready to own—Determinants of variants of stewardship social practices around Blue-Green Infrastructure in dense urban communities. *Frontiers in Water, 5*, 1. https://doi.org/10.3389/frwa.2023.1048494

Li, P., & Wang, Z. H. (2021). Environmental co-benefits of urban greening for mitigating heat and carbon emissions. *Journal of Environmental Management, 293*, 112963. https://doi.org/10.1016/j.jenvman.2021.112963

Liao, K. H. (2019). The socio-ecological practice of building blue-green infrastructure in high-density cities: What does the ABC Waters Program in Singapore tell us? *Socio-Ecological Practice Research, 1*(1), 67–81. https://doi.org/10.1007/s42532-019-00009-3

Liao, K. H., Deng, S. & Tan, P. Y. (2017). Blue-green infrastructure: new frontier for sustainable urban stormwater management. In: *Greening cities: Forms and functions* (pp. 203–226). Singapore: Springer Singapore. https://doi.org/10.1007/978-981-10-4113-6_10

Lin, W., Wu, T., Zhang, C., & Yu, T. (2011). Carbon savings resulting from the cooling effect of green areas: A case study in Beijing. *Environmental Pollution, 159*(8–9), 2148–2154. https://doi.org/10.1016/j.envpol.2011.02.035

Liu, L., & Jensen, M. B. (2018). Green infrastructure for sustainable urban water management: Practices of five forerunner cities. *Cities, 74*, 126–133. https://doi.org/10.1016/j.cities.2017.11.013

Liu, L., Fryd, O., & Zhang, S. (2019). Blue-green infrastructure for sustainable urban stormwater management—lessons from six municipality-led pilot projects in Beijing and Copenhagen. *Water, 11*(10), 2024. https://doi.org/10.3390/w11102024

Makri, A., & Stilianakis, N. I. (2008). Vulnerability to air pollution health effects. *International Journal of Hygiene and Environmental Health, 211*(3–4), 326–336. https://doi.org/10.3390/w11102024

Mannucci, P. M., & Franchini, M. (2017). Health effects of ambient air pollution in developing countries. *International Journal of Environmental Research and Public Health, 14*(9), 1048. https://doi.org/10.3390/ijerph14091048

Matos, S., Viardot, E., Sovacool, B.K., Geels, F.W. & Xiong, Y. (2022). Innovation and climate change: A review and introduction to the special issue. *Technovation*, 102612. https://doi.org/10.1016/j.technovation.2022.102612

Meerow, S., & Newell, J. P. (2017). Spatial planning for multifunctional green infrastructure: Growing resilience in Detroit. *Landscape and Urban Planning, 159*, 62–75. https://doi.org/10.1016/j.landurbplan.2016.10.005

Mulligan, J., Bukachi, V., Clause, J. C., Jewell, R., Kirimi, F., & Odbert, C. (2020). Hybrid infrastructures, hybrid governance: New evidence from Nairobi (Kenya) on green-blue-grey infrastructure in informal settlements. *Anthropocene, 29*, 100227. https://doi.org/10.1016/j.ancene.2019.100227

Naumann, S., Davis, M., Kaphengst, T., Pieterse, M. & Rayment, M. (2011). Design, implementation and cost elements of Green Infrastructure projects. Final report, European Commission, Brussels, 138.

Nesshöver, C., Assmuth, T., Irvine, K. N., Rusch, G. M., Waylen, K. A., Delbaere, B., Haase, D., Jones-Walters, L., Keune, H., Kovacs, E., & Krauze, K. (2017). The science, policy and practice of nature-based solutions: An interdisciplinary perspective. *Science of the Total Environment, 579*, 1215–1227. https://doi.org/10.1016/j.scitotenv.2016.11.106

Nguyen, T. T., Meurk, C., Benavidez, R., Jackson, B., & Pahlow, M. (2021). The effect of blue-green infrastructure on habitat connectivity and biodiversity: A case study in the Ōtākaro/Avon River catchment in Christchurch, New Zealand. *Sustainability, 13*(12), 6732. https://doi.org/10.3390/su13126732

Niemelä, J., Saarela, S. R., Söderman, T., Kopperoinen, L., Yli-Pelkonen, V., Väre, S., & Kotze, D. J. (2010). Using the ecosystem services approach for better planning and conservation of urban green spaces: A Finland case study. *Biodiversity and Conservation, 19*, 3225–3243.

O'Donnell, E. C., Lamond, J. E., & Thorne, C. R. (2017). Recognising barriers to implementation of Blue-Green Infrastructure: A Newcastle case study. *Urban Water Journal, 14*(9), 964–971. https://doi.org/10.1080/1573062X.2017.1279190

O'Donnell, E. C., Netusil, N. R., Chan, F. K., Dolman, N. J., & Gosling, S. N. (2021). International perceptions of urban blue-green infrastructure: A comparison across four cities. *Water, 13*(4), 544. https://doi.org/10.3390/w13040544

Pampel, F. C. (2014). The varied influence of SES on environmental concern. *Social Science Quarterly, 95*(1), 57–75. https://doi.org/10.1111/ssqu.12045

Pamukcu-Albers, P., Ugolini, F., La Rosa, D., Grădinaru, S. R., Azevedo, J. C., & Wu, J. (2021). Building green infrastructure to enhance urban resilience to climate change and pandemics. *Landscape Ecology, 36*(3), 665–673. https://doi.org/10.1007/s10980-021-01212-y

Pérez, G., Coma, J., Martorell, I., & Cabeza, L. F. (2014). Vertical Greenery Systems (VGS) for energy saving in buildings: A review. *Renewable and Sustainable Energy Reviews, 39*, 139–165. https://doi.org/10.1016/j.rser.2014.07.055

Plieninger, T., Thapa, P., Bhaskar, D., Nagendra, H., Torralba, M., & Zoderer, B. M. (2022). Disentangling ecosystem services perceptions from blue infrastructure around a rapidly expanding megacity. *Landscape and Urban Planning, 222*, 104399. https://doi.org/10.1016/j.landurbplan.2022.104399

Prescott, M. F., Dobbie, M. F., & Ramirez-Lovering, D. (2021). Green infrastructure for sanitation in settlements in the global south: A narrative review of socio-technical systems. *Sustainability, 13*(4), 2071. https://doi.org/10.3390/su13042071

Puchol-Salort, P., O'Keeffe, J., van Reeuwijk, M., & Mijic, A. (2021). An urban planning sustainability framework: Systems approach to blue green urban design. *Sustainable Cities and Society, 66*, 102677. https://doi.org/10.1016/j.scs.2020.102677

Raymond, C. M., Frantzeskaki, N., Kabisch, N., Berry, P., Breil, M., Nita, M. R., Geneletti, D., & Calfapietra, C. (2017). A framework for assessing and implementing the co-benefits of nature-based solutions in urban areas. *Environmental Science and Policy, 77*, 15–24. https://doi.org/10.1016/j.envsci.2017.07.008

Redondo Bermúdez, M. D. C., Kanai, J. M., Astbury, J., Fabio, V., & Jorgensen, A. (2022). Green Fences for Buenos Aires: Implementing Green Infrastructure for (More than) Air Quality. *Sustainability, 14*(7), 4129. https://doi.org/10.3390/su14074129

Rogers, B., Ramirez, D., Marthanty, D. R., Arifin, H. S., Farrelly, M., Fowdar, H., Gunn, A., Holden, J., Kaswanto, R. L., Marino Zamudio, R., McCarthy, D., Novalia, W., Payne, E., Suwarso, R., Syaukat, Y., Urich, C., Wright, A. & Yuliantoro, D. (2019). Leapfrogging pathways for a water sensitive Bogor. Australia-Indonesia Centre (AIC). https://research.monash.edu/en/publications/leapfrogging-pathways-for-a-water-sensitive-bogor

Romero-Duque, L. P., Trilleras, J. M. , Castellarini, F., & Quijas, S. (2020). Ecosystem services in urban ecological infrastructure of Latin America and the Caribbean: How do they contribute to urban planning? *Science of The Total Environment, 728*, 138780. https://doi.org/10.1016/j.scitotenv.2020.138780

Romero-Lankao, P., Gnatz, D. M., Wilhelmi, O., & Hayden, M. (2016). Urban sustainability and resilience: From theory to practice. *Sustainability, 8*(12), 1224. https://doi.org/10.3390/su8121224

Rozos, E., Makropoulos, C., & Maksimović, Č. (2013). Rethinking urban areas: An example of an integrated blue-green approach. *Water Science and Technology: Water Supply, 13*(6), 1534–1542. https://doi.org/10.2166/ws.2013.140

Sánchez, F. G., & Govindarajulu, D. (2023). Integrating blue-green infrastructure in urban planning for climate adaptation: Lessons from Chennai and Kochi, India. *Land Use Policy, 124*, 106455. https://doi.org/10.1016/j.landusepol.2022.106455

Seddon, N., Chausson, A., Berry, P., Girardin, C. A., Smith, A., & Turner, B. (2020). Understanding the value and limits of nature-based solutions to climate change and other global challenges. *Philosophical Transactions of the Royal Society B, 375*(1794), 20190120. https://doi.org/10.1098/rstb.2019.0120

Siehr, S. A., Sun, M., & Aranda Nucamendi, J. L. (2022). Blue-green infrastructure for climate resilience and urban multifunctionality in Chinese cities. *Wiley Interdisciplinary Reviews: Energy and Environment, 11*(5), e447.

Sinharoy, S. S., & Pittlu ck, R. & Clasen, T. (2019). Review of drivers and barriers of water and sanitation policies for urban informal settlements in low-income and middle-income countries. *Utilities Policy, 60*, 100957. https://doi.org/10.1016/j.jup.2019.100957

Stagge, J. H., Davis, A. P., Jamil, E., & Kim, H. (2012). Performance of grass swales for improving water quality from highway runoff. *Water Research, 46*(20), 6731–6742. https://doi.org/10.1016/j.watres.2012.02.037

Stephenson, J., Newman, K., & Mayhew, S. (2010). Population dynamics and climate change: What are the links? *Journal of Public Health, 32*(2), 150–156.

Strohbach, M. W., Arnold, E., & Haase, D. (2012). The carbon footprint of urban green space—A life cycle approach. *Landscape and Urban Planning, 104*(2), 220–229. https://doi.org/10.1016/j.landurbplan.2011.10.013

Suleiman, L. (2021). Blue green infrastructure, from niche to mainstream: Challenges and opportunities for planning in Stockholm. *Technological Forecasting and Social Change, 166*, 120528. https://doi.org/10.1016/j.techfore.2020.120528

Sun, Y., Xie, S., & Zhao, S. (2019). Valuing urban green spaces in mitigating climate change: A city-wide estimate of aboveground carbon stored in urban green spaces of China's Capital. *Global Change Biology, 25*(5), 1717–1732. https://doi.org/10.1111/gcb.14566

Tate, C., Hunter, R., Wang, R., Akaraci, S., Burns, K., Garcia, L. & Clarke, M. (2022). The Contribution of Urban Green and Blue Spaces to the United Nation's Sustainable Development Goals: An Evidence Gap Map. Available at SSRN 4264138. https://doi.org/10.2139/ssrn.4264138

Thorne, C. R., Lawson, E. C., Ozawa, C., Hamlin, S. L., & Smith, L. A. (2018). Overcoming uncertainty and barriers to adoption of Blue-Green Infrastructure for urban flood risk management. *Journal of Flood Risk Management, 11*, S960–S972. https://doi.org/10.1111/jfr3.12218

Toledo-Gallegos, V. M., My, N. H., Tuan, T. H., & Börger, T. (2022). Valuing ecosystem services and disservices of blue/green infrastructure. Evidence from a choice experiment in Vietnam. *Economic Analysis and Policy, 75*, 114–128. https://doi.org/10.1016/j.eap.2022.04.015

United Nations- Habitat (2023). Urban energy. https://unhabitat.org/topic/urban-energy. Accessed on 12.07.2023

Van Oijstaeijen, W., Van Passel, S., & Cools, J. (2020). Urban green infrastructure: A review on valuation toolkits from an urban planning perspective. *Journal of Environmental Management, 267*, 110603. https://doi.org/10.1016/j.jenvman.2020.110603

Vargas-Hernández, J. G., Pallagst, K., & Zdunek-Wielgołaska, J. (2023). Urban green spaces as a component of an ecosystem. *Sustainable Development and Environmental Stewardship: Global Initiatives Towards Engaged Sustainability* (pp. 165–198). Springer International Publishing.

Völker, S., & Kistemann, T. (2015). Developing the urban blue: Comparative health responses to blue and green urban open spaces in Germany. *Health and Place, 35*, 196–205. https://doi.org/10.1016/j.healthplace.2014.10.015

Voskamp, I. M., & Van de Ven, F. H. (2015). Planning support system for climate adaptation: Composing effective sets of blue-green measures to reduce urban vulnerability to extreme weather events. *Building and Environment, 83*, 159–167. https://doi.org/10.1016/j.buildenv.2014.07.018

Wamsler, C., Brink, E., & Rivera, C. (2013). Planning for climate change in urban areas: From theory to practice. *Journal of Cleaner Production, 50*, 68–81. https://doi.org/10.1016/j.jclepro.2012.12.008

Weiskopf, S. R., Rubenstein, M. A., Crozier, L. G., Gaichas, S., Griffis, R., Halofsky, J. E., Hyde, K. J., Morelli, T. L., Morisette, J. T., Muñoz, R. C., & Pershing, A. J. (2020). Climate change effects on biodiversity, ecosystems, ecosystem services, and natural resource management in the United States. *Science of the Total Environment, 733*, 137782. https://doi.org/10.1016/j.scitotenv.2020.137782

Well, F., & Ludwig, F. (2021). Development of an integrated design strategy for blue-green architecture. *Sustainability, 13*(14), 7944. https://doi.org/10.3390/su13147944

Wijayanti, Y., Julianto, A., Cahyono, C., & Setyandito, O. (2023). Analysis of flooding and nature-based solutions in the improvement of urban drainage system. *In IOP Conference Series: Earth and Environmental Science, 1218*, 012010. https://doi.org/10.1088/1755-1315/1218/1/012010

WWAP- United Nations World Water Assessment Programme (UN Water). (2018). *The United World Water Development Report 2018: Nature based Solutions for water*. UNESCO.

Yu, Z., Yang, G., Zuo, S., Jørgensen, G., Koga, M., & Vejre, H. (2020). Critical review on the cooling effect of urban blue-green space: A threshold-size perspective. *Urban Forestry and Urban Greening, 49*, 126630. https://doi.org/10.1016/j.ufug.2020.126630

Zhang, C., Tian, H., Pan, S., Lockaby, G., & Chappelka, A. (2014). Multi-factor controls on terrestrial carbon dynamics in urbanized areas. *Biogeosciences, 11*(24), 7107–7124. https://doi.org/10.5194/bg-11-7107-2014

Zhou, C., & Wu, Y. (2020). A planning support tool for layout integral optimization of urban blue–Green infrastructure. *Sustainability, 12*(4), 1613. https://doi.org/10.3390/su12041613

Zuniga-Teran, A. A., Staddon, C., de Vito, L., Gerlak, A. K., Ward, S., Schoeman, Y., Hart, A., & Booth, G. (2020). Challenges of mainstreaming green infrastructure in built environment professions. *Journal of Environmental Planning and Management, 63*(4), 710–732. https://doi.org/10.1080/09640568.2019.1605890

Understanding the Role of Blue-Green Infrastructure in Abatement of Urban Heat Island Effect

Kshama Gupta, Bhoomika Ghale, R. Sarath, Ravnish Kaur, and Arijit Roy

Abstract Blue-Green Infrastructure (BGI) is identified as one of the most effective solutions for alleviating the adverse impacts of Urban Heat Island (UHI) as it is found to create 2–4 °C cooler environment than the adjacent built-up areas. The rapid urbanization, which often comes at a loss of BGI, going to worsen the UHI phenomenon looking at the current and future trends of climate change. Hence, a temporal and spatial analysis is conducted in the city of Singapore to comprehend the correlation between improvement in BGI and their influence on temperature reduction by employing thermal remote sensing data. It is observed that a nearly 8% increase in BGI is able to reduce temperature by 1.21 °C in Singapore. The size of BGI is also found to significantly impact effective cooling distance (distance to which cooling effect of BGI is observed), which increases with an increase in the size of BGI. The findings of this study have considerable implications for countries of developing regions also (the study was conducted in Singapore, a developed country due to the primary stage of BGI planning in developing regions) as it not only analyses the positive impact of BGI on abatement of UHI but also emphasizes the need for distributed network of BGI for sustainable built environment.

Keywords Climate change · Thermal remote sensing · Landsat 8 · Effective cooling distance · Sustainable environment · Urban microclimate

1 Introduction

Blue-Green Infrastructure (BGI) is an interconnected network of blue and green spaces, providing nature-based solutions. Empirical derivations by Robitu et al. (2006) revealed that the presence of vegetation and water ponds within urban area cities alters the energy balance and can reduce the air temperature by 40.8 K for 96.5% solar radiation. Hence, there is a need to incorporate urban design practices

K. Gupta (✉) · B. Ghale · R. Sarath · R. Kaur · A. Roy
Indian Institute of Remote Sensing, 4, Kalidas Road, Dehradun, Uttarakhand 248001, India
e-mail: kshama@iirs.gov.in

P. K. Joshi et al. (eds.), *Blue-Green Infrastructure for Sustainable Urban Settlements*,
https://doi.org/10.1007/978-3-031-62293-9_4

that prioritize BGI, as enhancing the amount of vegetation and water sites is considered as highly effective strategy for mitigating the Urban Heat Island (UHI) effect. An UHI phenomenon, widely recognized, especially in large urban conglomerates, is a localized phenomenon of heat accumulation, leading to elevated temperature in urban areas, surpassing those in the neighboring rural areas. The phrase "heat island" refers to how cities behave as heat-absorbing islands within a broader region (Thompson, 1981; Yang et al., 2016). UHI phenomenon was first observed and quantified by Luke Howard where he has found that air temperatures in London downtown area were usually higher than those in the surrounding countryside (Howard, 1833). In the last century, globally numerous studies have found evidence of a noticeable temperature increase over urban areas due to a rapid increase in urban population (Kim, 1992; Lee et al., 2019). Lee et al. (2019) studied several Asian megacities and observed that the dynamics of urbanization aggravated temperature rise and climate patterns between 1992 and 2012. Increased daytime temperatures and decreased nighttime cooling due to UHI phenomenon contribute to health issues and fatalities, including general discomfort and respiratory difficulties, heat cramps and exhaustion, and non-fatal heat stroke (Voogt, 2004). UHI raise the requirement for air conditioning to keep buildings cool, which increases the overall energy demand. It has been found that summertime energy demand may increase by 2–4% for every 10 °C temperature increase (Akbari et al., 2001). This rise in the use of air conditioners for cooling and other anthropogenic activities exacerbates emissions of Green House Gases (GHGs) which further adds to the phenomenon of global warming. Rising global temperatures as a consequence of climate change can further amplify temperature rise in urban areas. Hence, this ongoing interaction between climate change, the UHI effect, and increased energy demand create a vicious cycle of feedback loop where increased energy use, combined with reliance on fossil fuel-based sources, results in increased GHG emissions, aggravate climate change and climate change-induced UHI effect. It will result in hotter urban microclimates that may worsen heat stress on urban populations which consist of nearly more than 55% of the global population. It also has a severe impact on air quality, surface and ground water quality, local weather and climate, enhanced precipitation rates, etc. in urban areas.

Green infrastructure (GI), such as rooftop gardens, vertical greening, and urban agriculture, can be used in addition to increasing urban green cover through the planting of trees, urban forests, and other greenspaces. The blue infrastructure such as natural or man-made water bodies within urban areas provides evaporative cooling effects. Moreover, these waterbodies aid in natural ventilation, allow precipitation to infiltrate into the ground and replenishing groundwater (Thompson, 2018; Wang et al., 2022). Rainwater harvesting, rain gardens, bioswales, and retention ponds are examples of effective water management practices that can collect and store rainwater, reduce runoff and boost vegetation development. Hence, development of BGI can be adapted for decreasing the UHI impact, reducing energy consumption, and as a sustainable planning practices to break this feedback loop (Gunawardena et al., 2019; Wang, 2022). It is well understood that implementation of BGI can significantly modulate the urban environment and hence, understanding the role of BGI in adapting to and mitigating climate change by regulating temperature and thus minimizing the UHI effect needs to be studied in detail for sustainable management of urban areas and formulation of adaptation and mitigation strategies. Therefore, this chapter focuses on the significance of BGI and its relationship with UHI. The chapter will further explore various approaches to analyze BGI and UHI, with a primary emphasis on geospatial technologies. Finally, the chapter will conclude by presenting a case study of Singapore, aiming to conduct spatio-temporal analysis to examine the impact of BGI on mitigating UHI effects.

2 Blue-Green Infrastructure

2.1 Concept and Definition

In 1986, E. O. Wilson proposed that humans are innately connected with nature, also known as biophilia (Gaekwad et al., 2022). Natural infrastructure i.e., the blue-green spaces ranging from urban parks, gardens, lakes, and ponds have always been an intrinsic part of urban areas. However, these open spaces in the cities are generally perceived as aesthetical and recreational in nature and are often overlooked as an important component of an urban ecosystem for a range of environmental benefits they provide. Aquatic and terrestrial ecosystems such as rivers, lakes, seas, wetlands, forests, grasslands, and deserts provide all the energy and materials that are essential for life on Earth. These resources delivered by nature are also called ecosystem services which are further categorized in provisioning services, such as food, water and natural resources; regulating services which involve the regulation of UHI, flood, micro climate and other natural hazards, purification of water etc.; supporting services such as shelter; and cultural services such as recreation and tourism (Semeraro et al., 2017). Historically, cities have been located near such ecosystems that directly or indirectly benefit human well-being and quality of life. However, due to rapid urbanization, there has been a radical shift from how we used to plan our cities. The natural layer underneath the highly constructed environment were increasingly spoiled through the direct and indirect impacts of urbanization. The loss of blue and green spaces in and around cities causes significant losses in vital biotic (flora and fauna) and abiotic (soil) components which negatively impacts the inherent capacity of natural environment to provide the goods and services, including clean air, food, water, and energy, necessary for human existence.

Realizing the significance of conserving blue and green spaces for achieving a resilient and sustainable city, Benedict and McMahon (2002) pioneered the concept of reintegrating nature into city landscape by introducing Green Infrastructure (GI) in city planning. GI was defined by Mark Benedict, USA as "an interconnected network of natural areas and other open spaces that conserves natural ecosystem values and functions, sustains clean air and water, and provides a wide array of benefits to people and wildlife" (Benedict & McMahon, 2002). However, in recent times, there has been shift from GI to BGI which aims at restoring the natural hydrological systems in the city and linking them with the vegetation systems to bring back highly urbanized environments to its natural state (Dreiseitl & Wanschura, 2006). BGI is a well-defined network of natural (trees, woodlands, lakes, wetlands) and man-made (green roofs, rain gardens, bio-swales, etc.) systems in high-density urban environments which provides multiple water-related, biodiversity-related, and climate-related ecosystem services to urbanites (Ghofrani et al., 2016). In simpler terms, BGI, an extension of the concept of GI, mimics natural processes in man-made environments, thus creating an environment resilient to UHI, floods, heat waves, droughts, and other extreme weather-related events (Depietri & McPhearson, 2017; Hamel & Tan, 2021).

2.2 *Importance of Blue-Green Infrastructure*

BGI, as understood now, is an ecologically based design approach which place emphasis on an interconnected network of natural and semi-natural landscapes rather than the conservation of few selected sites of recreation or relaxation that makes the city accessible to wildlife and people alike (Benedict & McMahon, 2012; Bartesaghi et al. 2017). This network of blue-green spaces are multifunctional landscape as compared to its grey alternatives which attempt to provide a diverse range of benefits, emcompassing benefits related to water, biodiversity, and climate (Korkou et al., 2023; Semeraro et al., 2017). Water-related benefits by BGI aims at reinstating the natural hydrological processes by managing storm water runoff through slowing down, decentralization and spreading, enhancing soaking of run-off below ground, evaporation and release of water into the natural environment and thus providing services like groundwater recharge, reduced storm water discharge, prevention of soil erosion and increased quality of storm water flow (Dreiseitl & Wanschura, 2006).

A study conducted by Singh and Mishra explores the relationship between deforestation in Sanjay Gandhi National Park in Mumbai and its effect on drinking water supply in the region and concluded that due to deforestation, water turbidity increased by 8.41% and hence the cost incurred on water treatment increased by 1.58% (Singh & Mishra, 2014). Many scientific studies in the recent past have also reported the enhanced effects of BGI on biodiversity in the city (Boussema et al., 2022; Kong et al., 2010; Zhang et al., 2019). For instance, the restoration project of the false creek in Vancouver, British Columbia has restored the creek life to what it used to be and as a result of the naturalization process, the creek is also witnessing the return of salmon along with a variety of dragon flies and bugs. BGI also helps ensure healthy urban living conditions for its citizens by mitigating climate changes such as reducing UHI effects which will be further explored in this chapter (Marando et al., 2022). Furthermore, BGI offers numerous additional benefits, such as social benefits and economic benefits. The ability of BGI to provide social, economic and environmental services simultaneously makes it a viable solution for creating sustainable, resilient cities (Wang et al., 2021).

2.3 Challenges in Implementing Blue-Green Infrastructure

There are several examples demonstrating the successful implementation of BGI to mitigate the impacts of climate change, exemplified by Oslo, Norway (Wilbers et al., 2022); Stuttgart, Germany (Winker et al., 2022); New York, USA (CNY RPDB 2012), but most of them are centered in the developed region. While the concept of BGI was developed worldwide in a common response to the current and future climate change impacts, the interventions will vary from region to region adapting to the natural, social-economic, cultural, and political environment of each specific area. Any BGI study and its practical application in the study region requires an interdisciplinary approach and the availability of data from various departments such as the underlying geology, topography, soil cover, groundwater status, vegetation cover, land use map, demographics, type of flora and fauna, climate parameters, surface water flows, etc. The current BGI studies in developed countries have a vast resource of similar open-source databases, easily available to implement. However, due to the non-availability of data in developing nations, there is a lack of research in the development of BGI solutions to specifically address various environmental challenges in its urban areas (Hamel & Tan, 2021; Raška et al., 2022). Another major challenge affecting the application of BGI in cities is more economical in nature. In spite of many ecological and societal benefits that BGI provide, most of them are not quantifiable in nature (Drosou et al., 2019). The economical aspect of BGI adaption is very important in the decision-making process as the investment in BGI should be efficient as well. Furthermore, the implementation of BGI planning is often lacking due to the lack of availability of open public space. In a dense urban conurbation of developing region where land is highly expensive, a change in land use, particularly in highly commercialised zones is an issue and a hindrance. As emphasized in a study by Drousou et al. (2019), societal perception is integral in the successful implementation of BGI in the cities. In most of the developing nations, BGI applications are still viewed as low-impact development strategies and as planning obligations and not a system as a whole and hence most of the developing nations are still far behind in implementing integrated BGI approaches for harnessing the number of benefits provided by BGI. Moreover, these countries in the global south are also struggling with more pertinent socio-economic issues which take precedence such as poverty, food security, provision to clean water, sanitation, and education to all. In spite of all the issues, the developing countries now share a keen interest in the implementation of BGI measures as it is expected that this region is going to witness a huge spurt in the growth of urban populations in the near future. That calls for planning the urban areas in a much more sustainable manner much before the situation goes out of control.

3 Urban Heat Island

The solar radiation reaching both the city and surrounding countryside is equivalent, the temperature difference between the two is related to the concentration of human activity and urban construction, as well as the effectiveness of heat absorption and retention on respective surfaces (Singh et al., 2017). The surface in rural areas is pervious and often covered with vegetation. They absorb soil water through their roots and utilize it for metabolism. Excess water is stored in their stems and leaves, eventually making its way to microscopic holes on the underside of leaves. It then travels through evapotranspiration, which acts as nature's air conditioner (Paschalis et al., 2021). On the other hand, extensive removal of natural vegetation and water-bodies in urban areas occurs to fulfill the demands of the urban population (Nuruzzaman, 2015). These surfaces are replaced by concrete, asphalt, and other materials with high thermal mass. These materials having low albedo absorb all wavelengths of light energy and convert them to heat, resulting in a warm surface (Bouyer et al., 2011). Moreover, the impermeable properties of these materials lead to heat trapping in cities, as they do not allow water to pass through them. As a result, they lack the beneficial cycle of flowing and evaporating water that helps to cool the surface and surrounding environment. Furthermore, the heat reflected from a building is trapped by surrounding structures in urban settings, forming urban canopy, and this exacerbates the UHI effect (Masson, 2006). Tall buildings and narrow streets can hamper air circulation and impede natural ventilation, which would normally cool a region. Blockage of winds due to these structures leads to the warming of the air in urban areas (Etheridge & Ford, 2008). Other anthropogenic activities, including industrial processes, car emissions and energy usage emit heat and pollutants into the environment, intensifying the temperature disparity between urban and rural areas i.e., creating UHIs.

UHI has far-reaching consequences on the health of the urban populace and the environmental quality of urban areas (Voogt, 2004). Other than heat-related illnesses and deaths, use of air conditioners as an adaptation measure to high temperatures in urban areas and use of thermal power plants to meet rising energy demand, releases air pollutants such as ground-level ozone, nitrogen oxides, carbon dioxide, etc. These pollutants deteriorate air quality and create an envelope of air pollutants over the urban areas, which further enhances the UHI effect. UHI also affects the surface and ground water quality, as water flowing over pavement and rooftop surfaces are heated due to the high temperatures of these surfaces. High water temperature has an impact on the metabolism and reproduction of aquatic organisms (Bhargava et al., 2017). It has also been observed that local weather and climate is also impacted by UHI and lead to changes in local wind and humidity patterns, formation of fog and cloud, and increased intensity of precipitation. The heat induced by UHI contributes to the stronger upward wind movement, creating a local low-pressure zone, which attracts cool air from nearby regions, stimulating cloud formation and precipitation process. As a result, with today's fast-paced urbanization and industrialization, UHI is considered as one of the major threats to human civilization.

4 Blue-Green Infrastructure and Urban Heat Island

BGI can assist considerably alleviate the negative impacts of UHI in urban areas by reducing heat with vegetation and water using multifarious ways such as green roofs, green facades, urban forests, gardens, street trees and waterbodies such as ponds, canals, lakes, wetlands and rivers with riparian edges (Oliveira et al., 2011; Syafii et al., 2016). The combination of blue (water) and green infrastructure (vegetation) creates a microclimate that provides shade, reduces solar radiation, and promotes cooling through evapotranspiration, thereby lowering the local temperatures (Oliveira et al., 2011). As the sun heats the water surface, the water molecules gain energy and transform into water vapor, which increases the humidity levels. This process absorbs heat from the surrounding air, resulting in evaporative cooling. As cool air moves over the water surface, it absorbs moisture and cools, resulting in a localized cooling effect that significantly influence the microclimate in the area. As a result, the air temperature near water bodies is often lower compared to surrounding areas (Tominaga et al., 2015). Moreover, these waterbodies aid in natural ventilation, allow precipitation to infiltrate into the ground, replenishing groundwater and reducing the amount of heat-absorbing runoff. By maintaining moisture in the soil, BGI helps in regulating soil and air temperatures, minimizing overheating during summers. When water bodies are combined with GI, such as vegetation and trees, the cooling effect can be enhanced. Vegetation near waterbodies benefit from increased humidity, which promotes evapotranspiration as explained above, thus cooling the air and surfaces through the release of moisture. This vegetation provides shade for buildings, streets, and outdoor spaces, thus limiting direct sunlight exposure and reducing heat absorption by buildings and pavements (Abdulateef & Al-Alwan, 2022). When considering the cooling and ventilation impacts in urban locations, it is critical to consider elements such as BGI size, prevailing wind patterns, and the surrounding built environment. However, when properly built and integrated, BGI can act as natural air conditioners, facilitating ventilation and lowering the ambient temperature in urban areas (Almaaitah et al., 2021). As mentioned earlier, BGI combines blue and green elements and a well-integrated network comprise of BGI elements ranging from building scale (green roof, green façade, vegetated balconies) to neighborhood scale (green parking, urban trees, urban parks, vegetated waterbodies) can effectively assist in mitigating UHI and in long term can be an effective measure for mitigating impacts of climate change.

5 Approaches to Study Blue-Green Infrastructure and Urban Heat Island

Several major cities worldwide have adopted revised urban planning and design approaches that integrate blue elements such as rivers, lakes, wetlands, and water infrastructure in conjunction with green elements like trees, parks, gardens, and woods (Udas-Mankikar & Driver, 2021). Assessing the effectiveness of BGI in mitigating the UHI effect requires a multidisciplinary approach that involves field measurements, temperature monitoring, remote sensing, thermal imaging, vegetation analysis, and more. Surveys are conducted to investigate the perceptions of BGI to improve thermal comfort, air quality, and overall quality of life (Donnell et al., 2021; Klemm et al., 2015). BGI is known to naturally cool the surrounding area, however, one must comprehend the connection between land cover and Land Surface Temperature (LST). The topographical, geometrical, and structural factors of BGI such as distance from the nearest BGI feature, shape and size, edge density, etc., affect the effectiveness of UHI mitigation (Antoszewski et al., 2020). Assessing the impact of BGI on UHI mitigation typically involves deploying a network of temperature and solar radiation sensors across the urban area. These sensors are strategically placed in areas with BGI elements and in those without them. Data collected from these sensors are then used to analyze diurnal and seasonal temperature variations. By comparing temperatures in BGI-enhanced areas to those in non-BGI areas, researchers can evaluate the cooling effect (Shafique & Kim, 2017). However, implementing this method can be challenging in practice due to equipment constraints and the associated costs.

Urban climate models that are computationally efficient are one more approach to study cooling functionality of BGI. Du et al. (2019) suggested an optimized design method for waterbody and greenspaces to maximize the urban cool island (UCI) and improve the urban thermal environment using computational fluid dynamics (CFD) simulation. The Air Temperature Response to Green/Blue-Infrastructure (TARGET) model developed by Broadbent et al. (2018) assesses the surface and street-level temperature in cities where BGI elements are used to reduce heat. Similarly, other computational models such as ENVI-met, Temperature of Urban Facets in 3-D (TUF-3D), Vegetated Derivative (VTUD-3D), Canyon Air Temperature (CAT) model are used for understanding and simulating the cooling impact of BGI (Abdulateef & Al-Alwan, 2022). However, CFD models are computationally intensive and thus has been applied till now to study BGI and urban systems at micro scales covering smaller extents.

Geospatial technologies offer a cost-effective and practical solution to overcome these challenges. It includes technologies such as Remote Sensing, Geographic Information Systems, Global Navigation Satellite Systems, Web Mapping, Mobile mapping and many more. It has been extensively utilized for urban planning and research, including urban monitoring (Gupta, 2013; Jat et al., 2008), urban suitability analysis (Bodhankar et al., 2022), quantification and assessment of urban green spaces (Gupta et al., 2012, 2016, 2020; Loi et al., 2015), site selection studies, multi-criteria analysis, study and modeling of urban growth (Chowdhury & Maithani, 2014; Kumar et al., 2021), 3D modeling of urban areas (Gupta et al., 2015, 2017a), urban canopy parameters (Gupta, 2022; Gupta et al., 2022), urban hazard (Rao et al., 2021), urban climate research and modeling (Voogt & Oke, 2003; Wong 2005; Gupta et al., 2017b; Bhavana et al., 2018; Gupta et al., 2021; Theethai Jacob et al., 2023) etc. There is no substitute of these technologies for understanding and analyzing urban issues due to their synoptic, repetitive and geolocation characteristics. It is necessary to integrate multiscale geospatial technologies in a seamless manner, with near-real-time digital maps that host spatially referenced data along with corresponding metadata (Blaschke et al., 2011). The GIS technologies when combined with the remote sensing technologies give more opportunities in editing, analysing, extracting and modeling geospatial data. For example, one can easily integrate census data along with building footprints extracted with remote sensing imageries and further use it for taxation purposes (Ali et al., 2020). An added advantage of geospatial technologies in urban scenarios is that apart from providing a bird eye view, modern remote sensing technologies can acquire data in third dimension and various analysis like modelling will contribute to a more comprehensive understanding of the existing spatial scenarios.

5.1 Geospatial Technologies for Assessment of Blue-Green Infrastructure

Geospatial technologies help in mapping, analysing, modeling and interpreting BGI in city areas as green spaces and blue spaces. Collectively the BGI are a network of vegetation and waterbodies within a city. Indices like Normalized Difference Vegetation Index (NDVI), Normalized Difference Water Index (NDWI), image classification helps in extracting green space and blue space from multispectral satellite imageries. Classification utilizing high-spatial resolution images with Object-Based Image Analysis (OBIA) technique can yield better results for mapping of BGI (Shekhar & Aryal, 2019). A series of such imageries of different time period can be used for temporal analysis to understand the change in BGI. The method of digitizing features is time consuming but are still used to extract vegetation and waterbodies. Once the features are extracted, further analysis can be performed using GIS tools and models. It has been demonstrated that geospatial data can be effectively used for analysing urban green spaces quantitatively and qualitatively such as quantification of percentage, type, proximity, multi-criteria analysis and stress monitoring (Gupta et al., 2020). Accessibility to green or blue spaces can be taken into consideration (Gupta et al., 2016) and regions with low accessibility can be chosen as sites to develop BGI. It can be used efficiently for analysing spatial distribution, its allocation wrt planning guidelines and identification of potential BGI networks (Kaur & Gupta, 2022). Hyperspectral remote sensing extends the scope of remote sensing by evaluating the condition of vegetation and waterbodies in the region. Geospatial technology thus plays a significant role in analysing, interpreting, planning and mapping BGI in the urban domain.

5.2 *Geospatial Technologies for Analysis of Urban Heat Island*

Urban climate research have predominantly employed data on LST and emissivity data derived from satellite thermal infrared (TIR) imagery for analysing LST patterns and its correlation with surface features. This approach is utilized to identify the UHI and establish connections between LSTs and surface energy fluxes to characterize landscape patterns and processes (Blaschke et al., 2011; Gupta et al., 2017b; Bhatt et al., 2023). Several satellites provide LST at coarser resolutions of 1 km like Moderate Resolution Imaging Spectroradiometer (MODIS), Geostationary Operational Environmental Satellites (GOES), Visible Infrared Imaging Radiometer Suite (VIIRS), whereas, Advanced Spaceborne Thermal Emission and Reflection radiometer (ASTER) and Landsat series acquire thermal data at 100 m and Landsat gives data at 30 m resampled resolution. LST can be derived from Landsat by converting from DN to Radiance value of TIR band and calculating emissivity and applying various algorithms like single channel method, split window method, radiative transfer model and many more (Yu et al., 2014). The LST thus calculated can be used to derive temperature profiles, detecting temperature anomalies, identifying UHI and Urban Cool Islands. Estimating the UHI involves calculating the disparity between the temperature of the urban area and surrounding rural areas. Research indicates a stong positive correlation between LST and built-up density, while a negative correlation is observed with vegetation and water bodies (Halder et al., 2021; Kafy et al., 2021; Lu & Weng, 2006), which reflects effect of LULC changes on LST (Abdulateef & Al-Alwan, 2022; Kafy et al., 2021). TIR data can also enhance the precision of generating land cover and impervious surface maps as compared to relying solely on optical data. Distinguishing impervious surface cover from pervious cover is facilitated by leveraging the distinctive thermal responses using Landsat TIR data. This approach aids in minimizing errors associated with underestimation of densely built-up areas and overestimation of less developed ones (Lu & Weng, 2006).

5.3 *Geospatial Technologies for Integrated study of Blue-Green Infrastructure and Urban Heat Island*

Remote sensing, including satellite and aerial imagery, TIR sensors, and multispectral data, provide a comprehensive understanding of temperature variations over larger urban areas. These data can be used to assess temperature variations, identify UHI hotspots, and evaluate the cooling impact of BGI. Moreover, the progress in remote sensing techniques and the availability of open-access satellite data have significantly increased accessibility for researchers and urban planners. This provides a valuable alternative for evaluating the efficacy of BGI in alleviating the UHI without an extensive sensor network. This assessment can be achieved using Landsat 8 and field data. Yusof et al. (2019) analyzed the potential of heritage trees in mitigating

UHI and found that having more trees can lower the LST in urban areas by roughly 5.3 °C. To determine the degree to which BGI can reduce UHI, various indices such as NDVI, Universal Thermal Climate Index (UTCI), and Thermal Sensation Vote (TSV) (Fung & Jim, 2020; Lehnert et al., 2021) are correlated with LST. Wu et al. (2019) utilized the NDWI to assess the maximum cooling distance from a lake in Wuhan. The findings indicated a significant association between the maximum cooling distance and the cooling impact of BGI elements in urban areas. With technological developments, Unmanned Aerial Vehicles (UAVs) equipped with optical and TIR cameras can capture high-resolution aerial imagery. This imagery has the capability to offer intricate details regarding LULC and the distribution of BGI elements. In recent study by Qin et al. (2022), an enhanced methodology termed "UAV + IRs + 3 T" was introduced. This approach integrates UAV data, TIRS, and a three-temperature model (3 T), to estimate urban evapotranspiration. Furthermore, it was able to show significant differences in mean intra-day evapotranspiration rates among different plant species. The multispectral and hyperspectral sensors on UAVs can capture data about vegetation health which is crucial for understanding the condition of green spaces within BGI systems (Lee et al., 2021; Nasi et al., 2018).

6 Case Study: Spatio-temporal analysis to study impact of Blue-Green Infrastructure on Urban Heat Island mitigation

Many research studies have proven high potential of geospatial technologies for quantitative and qualitative analysis of BGI as well as UHI (Yusof et al., 2019). However, most of the studies has focussed on the study of either BGI or UHI separately. Very few studies have analysed BGI in conjunction with UHI phenomenon (Jia & Wang, 2022). Besides, most of the studies on integrated assessment of BGI and UHI focuses more on the impact of decreasing BGI on UHI as implementation of BGI is limited to only a handful of cities. Therefore, studies on the positive impact of BGI enhancement on the mitigation of UHI is also very limited. Hence, this study seeks to analyse the extent to which improvement in BGI percentage in any urban area reduces UHI and temperature. The selection of study area is critical for this study as the city which has successfully implemented BGI needs to be selected for the purpose of this study. After careful examination of the literature, Singapore was selected as study area as it has successfully implemented BGI starting from year 2006 and completed in the year 2015–16.

6.1 Study Area Profile

Singapore is a sovereign island nation in Southeast Asia, positioned at the southern extremity of the Malay Peninsula. Geographically, it is situated at 1°17′N and 103°50′E, encompassing Singapore Island and 60 smaller islets. The overall land area spans 734 square kilometers, and it boasts a coastline stretching 193 km (refer to Fig. 1). The topography of the island is predominantly hilly, with nearly two-thirds of its landscape lying below an elevation of 15 m above sea level.

6.1.1 Climate

Singapore is located entirely between the first and second parallels, i.e., 1.5 degrees north of the equator. The Köppen climatic classification for Singapore is Af (tropical rainforest climate) and displays not much variation in temperature and pressure annually. The region experiences high humidity levels and copious amounts of rainfall as a result of its geographic location and coastal exposure, resulting in typically warm and wet climate. The northeast monsoon, which lasts from November to March, is the wettest and windiest time of year, with rainfall averaging more than 250 mm each month on average in December. On the other hand, the southwest monsoon (May–September) is the time of year with the least amount of rain and the lightest winds, with July seeing the lowest monthly rainfall total of less than 7 inches. Intense

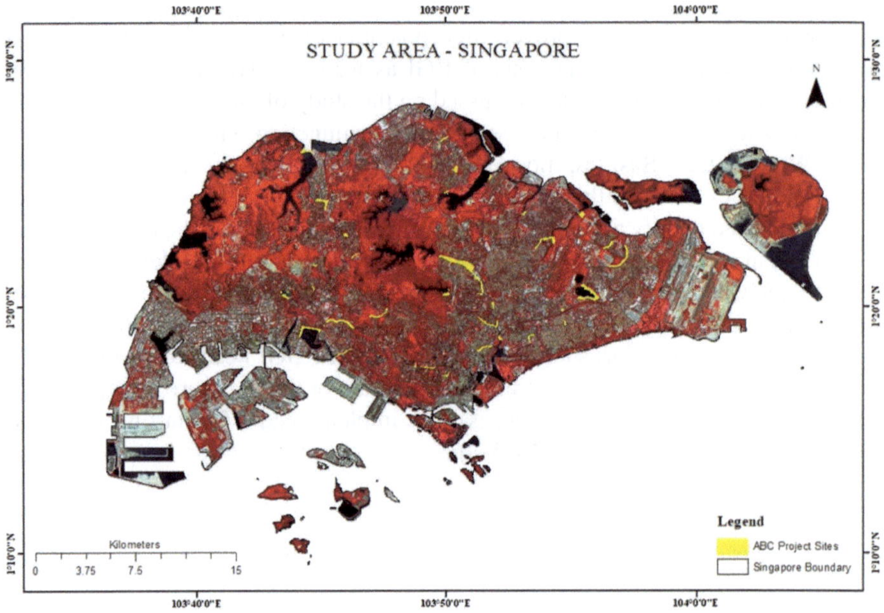

Fig. 1 Study area of Singapore representing Active Blue and Clean (ABC) Waters Project Sites

afternoon rains and thunderstorms characterize inter monsoonal months of April and October with slow air movements. Every day of the year, rain falls somewhere on the island of Singapore, which receives an average of 95 inches of precipitation annually.

6.1.2 Demography

Singapore is highly urbanized and has a dense population of 5.457 million in 2021 (Census, 2020) with lowest rates of population growth and birth in entire Southeast Asian region. Because of its high standards of hygiene and access to a better healthcare system, Singapore has a low infant mortality rate and a high average life expectancy. The majority of the people who live there are Malays and Indians, making up three-fourths of the population.

6.1.3 Blue-Green Infrastructure in Singapore

Active Beautiful Clean (ABC) Waters Programme was launched in Singapore in 2006, forming an integral component of the nation's stormwater management policy. The ABC Waters Programme has successfully completed over 60 projects and was able to address the issue of urban flooding (Lim & Lu, 2016). It is a manifestation of Singapore's approach of Low-Impact Development (LID) concepts and practices (Lim & Lu, 2016). The primary objective of the ABC Waters initiative was to enhance the overall quality and aesthetic appeal of streams, rivers, and lakes. This was achieved through a comprehensive approach that involved seamless integration of reservoirs, drains, and canals with their surrounding environments. Efforts were made to collect, treat and store stormwater, involving the creation of catchments on roads, pedestrian walkways, and building features. Storage solutions were implemented using water bodies and waterways. Additionally, the project placed a strong emphasis on treating stormwater through various means, including constructing wetlands, sedimentation basins, and vegetated and bio-retention swales (Udas-Mankikar and Driver 2021).

6.2 Datasets Used

The administrative boundary of Singapore is retrieved from the Stanford university website (https://earthworks.stanford.edu/catalog.html?f%5Bdc_format_s%5D%5B%5D=Shapefile&f%5Bdct_spatial_sm%5D%5B%5D=Singapore+%28City%29&per_page=10&sort=solr_year_i+desc%2C+dc_title_sort+asc). The ABC Water sites where the BGI implementations is done in Singapore is retrieved from government data portal of Singapore (https://beta.data.gov.sg/datasets?query=ABC%20Waters%20Sites). For a preliminary assessment, MODIS Day Time Land Surface Data at 1 km spatial resolution is used. The annual mean composite for each year,

Table 1 Details of the various data used in the study

S. No	Data	Spatial resolution (m)	Temporal resolution (day)	Source	Date
1	MODIS	1000	4 passes per day	Google Earth Engine	Annual median images from 2014 to 2021
2	LANDSAT 8 OLI & TIR	30	16	Google Earth Engine	
3	Singapore Administrative Boundary	–	–	Stanford University website	2015
4	Singapore ABC Waters project site	–	–	Singapore Government data portal	2023

i.e., from 2014 to 2021, is calculated for Singapore considering the fact that the BGI implementation was completed by 2016. Landsat 8 data was used to study temporal as well as spatial assessment of the temperature profile in detail. The Landsat 8 OLI bands offer a spatial resolution of 30 m and TIRS gives thermal remote sensing data at 30 m which is acquired at 100 m spatial resolution. The OLI bands of Blue, Green, Red, NIR and SWIR is used for LULC classification and emissivity calculation. For LST calculation, TIR band (Band 10) at 10.60–11.19 μm wavelength is used. The various data used in the study is given in Table 1.

6.3 *Methodology*

6.3.1 Generation of BGI Maps

The locations of BGI were required to do temperature trend analysis in BGI areas and also for spatial analysis. The locations of BGI i.e., the ABC Waters project site was retrieved as kml files from the official site of ABC waters. The administrative boundary was obtained from Stanford University. The kml files obtained were converted to shapefiles and were used for further analysis.

6.3.2 Temporal Trend Analysis Using MODIS LST

The Day time LST products of MODIS were obtained from Google Earth Engine (GEE) Collections. The annual median imagery was obtained for the time period of seven years from 2014 to 2021 for Singapore. Further, average temperature was calculated within its administrative boundary of study area. The average temperature was plotted and the trend is studied. The trend was further analysed by performing Mann Kendall Test in python environment.

6.3.3 Estimation of LST from Landsat 8

The LST was estimated from the Landsat 8 for both the study areas using the Single Channel Method, where only one thermal band is used, in this case, band 10 (10.60–12.51 μm) of Landsat 8. The proportional vegetation was calculated from NDVI to estimate emissivity from OLI sensor of Landsat 8 of the corresponding dates. B10 of Landsat 8 was used since the B11 TIRS was having stray light problem and using single channel method performs better than split window algorithms (Cristóbal et al., 2018). The Band 10 thermal data was first atmospherically corrected and at sensor radiance temperature was calculated using Eq. 1.

$$\text{Radiance}, L_\lambda = \frac{L_{max-L_{min}}}{QCal_{max-QCal_{min}}} * \left(Qcal - QCal_{min}\right) + L_{min} \tag{1}$$

L_{max} and L_{min} represents the minimum and maximum value and $QCal_{max}$ and $QCal_{min}$ represents maximum and minimum DN values (0–255). QCal is the DN value of raster.

The emissivity (Eq. 4) was calculated using proportional vegetation (Pv) (Eq. 3), which was calculated using NDVI image (Eq. 2) extracted from band 4 and band 5 of OLI sensor of Landsat data. Landsat TIRS Band 10 DN number is converted into at sensor radiance values to estimate LST (Eq. 1). Then the LST was calculated using Single Channel Method (Eq. 5).

$$\text{Normalised Difference Vegetation Index}, NDVI = \frac{NIR - R}{NIR + R} \tag{2}$$

where, NIR is Near Infrared band and R is red band.

$$\text{Proportional vegetation}, Pv = \left(\frac{NDVI - NDVI_{min}}{NDVI_{max} - NDVI_{min}}\right) \tag{3}$$

where, $NDVI_{max}$ and $NDVI_{min}$ are maximum and minimum values of NDVI

$$\text{Emissivity}, \varepsilon = 0.004 * Pv + 0.986 \tag{4}$$

$$\text{Land Surface Temperature}, LST = \left(\frac{k_2}{(ln\frac{\varepsilon k_1}{L_\lambda} + 1)}\right) - 273 \tag{5}$$

where k_1 and k_2 are constants which are 774.8853 and 1321.0789 respectively, ε is emissivity, L_λ is radiance value.

6.3.4 Development of Code in Google Earth Engine

GEE platform was used for calculating LST and image classification, as the study involves multi temporal datasets. Median images were generated annually for each year for Singapore from 2014 to 2021 as cloud cover was high due to its tropical climate and persistent rainfall. The NDVI, Proportional vegetation, Emissivity, LST were further calculated using expressions in GEE and the code was able to calculate LST for each image for the time period provided. LULC Classification using Classification and Regression Tree (CART) algorithm (Shetty, 2019) was also performed by developing small codes in GEE platform.

6.3.5 Temporal Trend Analysis with Landsat 8

The LST maps obtained using Landsat 8 data from GEE were further analysed by averaging temperature of the study area. The average values thus obtained was plotted and trendline was plotted. Mann Kendall test was performed in python environment to validate the trend of annual temperature change. Average temperature of each BGI were also analysed using the BGI shapefiles for the study areas for all the years during the time period and their trend was studied to determine the extent of the decrease in temperature due to BGI implementation. Further, Mann–Kendall test was performed to detect and quantify trends in temperature data collected over a period of time. Temperature trend analysis using the Mann–Kendall test helps in understanding the long-term behavior of temperature patterns, detecting potential climate change signals, and providing insights for environmental monitoring, planning, and decision-making. By applying this test, researchers can determine whether there is a statistically significant upward or downward trend in temperature measurements.

6.3.6 Land Use Land Cover Classification and Change Analysis

The LULC of a particular area are known to have an impact on its temperature (Abdulateef & Al-Alwan, 2022; Kafy et al., 2021). LULC maps were generated for the entire study area in order to study change in BGI percentage over the duration of study i.e. 2014 -2021. The classification is carried out in GEE using the blue, green, red, NIR and SWIR bands of OLI sensor of Landsat 8 from 2014 to 2021. Classification and Regression Tree (CART), a machine learning classifier available on GEE, were employed to classify the entire study area in broad LULC classes viz. vegetation, water bodies and barren/built up/impervious areas. The areas under BGI (Vegetation and water) were calculated and the percentage of change was analysed. The percentage change is further analysed with temporal change in temperature to quantify the impact of enhancing BGI on LST.

6.3.7 Spatial Analysis Through Temperature Profile Across BGI

Spatial analysis was performed to quantify the distance to which the BGI reduce temperature and also the reduction in magnitude of temperature as a function of BGI size and distance. The centroids of the BGI were created and using the 3D analyst tool in Arc Map, the transects were plotted in all the four cardinal directions and using the LST layer of the latest image of 2021, temperature profile was plotted. To analyse the impact of BGI on LST, the distance to which the temperature is gradually decreasing is noted as the effective cooling distance. To assess the impact of BGI, the temperature range was determined by calculating the temperature difference between the BGI centroid and effective cooling distance. This analysis was conducted to understand the magnitude of temperature reduction resulting from the presence of BGI.

6.4 Results and Discussions

6.4.1 Temperature Trend Analysis

The average annual day time LST of MODIS Terra shows a decreasing trend in temperature. The LST computed at 30 m resolution using Landsat 8 data (Fig. 2a), also shows decreasing trend of temperature in Singapore. The average temperature which was 26.17 °C in 2018 got reduced to 24.87 °C in 2021, a decrease of 1.21 °C. For the purpose of this analysis, the average temperature of year 2020 was considered as an outlier as low temperatures were recorded worldwide during this year due to COVID 19 pandemic situation (Fig. 2b). It is to be noted that although BGI implementation was completed by 2016 in Singapore but the temperature was rising till 2018 and then it started decreasing gradually. Mann Kendall score of -29.0 further confirms the decreasing trend in LST, which can be attributed to the increase in BGI percentage which is further evaluated in the LULC change analysis.

The LULC classification was performed using CART classification to classify three classes, viz., vegetation, water, and built-up areas (Fig. 3a). Implementation of BGI in the study area leads to an augmentation in the percentage of vegetation and water bodies which is evident from LULC change analysis and shows an increase of 8.08% during the time period 2014–2021 in Singapore (Fig. 3b). The implementation of BGI in Singapore was done by ABC waters project and there was a significant increase in BGI. It is also observed that whenever the BGI has increased temperature shows a declining trend.

6.4.2 Effect of Blue-Green Infrastructure on Cooling Distance

The temperature variation from the blue and green space is studied by plotting temperature profiles along the four directions East, North, West and South (Fig. 4) to analyse

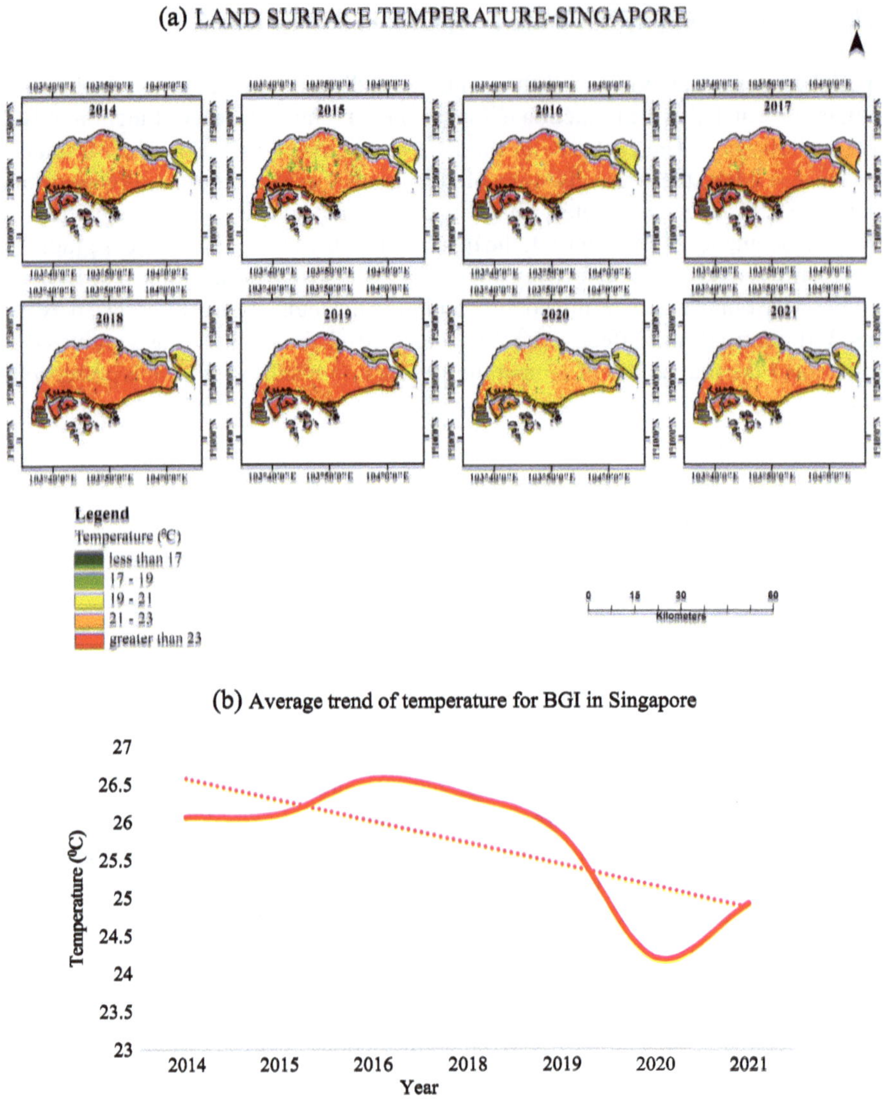

Fig. 2 **a** Land Surface Temperature map for Singapore; **b** Average trend of temperature for BGI in Singapore

the impact of BGI size on effective cooling distance. It is found that LST increases as you move away from BGI. It is found that size of BGI has a significant impact on effective cooling distance. Effective cooling distance is found to be 200 m for BGI (water bodies and vegetation) of areas less than 50,000 m^2, whereas it is found to be 249 m and 512 m respectively if size of BGI increases from 50,000 to 100,000 m^2 and greater than 100,000 m^2 (Table 2).

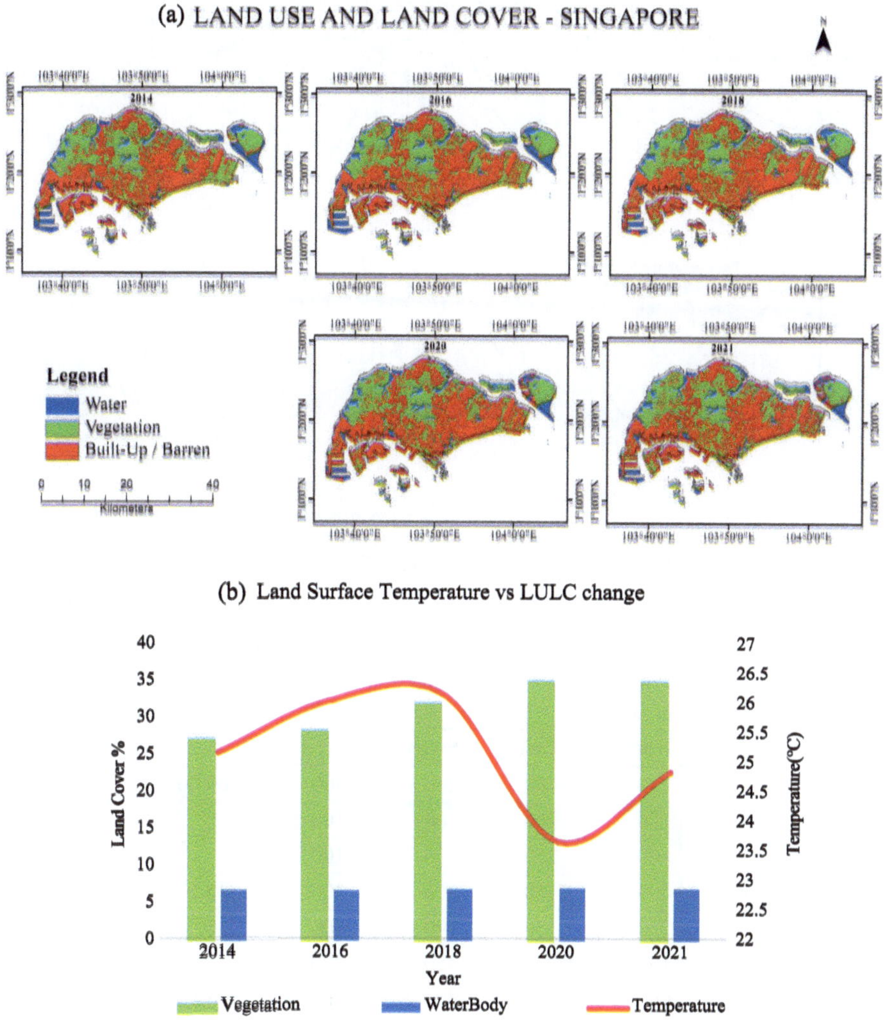

Fig. 3 **a** Land use land cover for Singapore; **b** Land surface temperature and Land use land cover change

The study has successfully analyzed the temporal and spatial impacts of BGI on the temperature pattern in an urban area. Temporal analysis of LST displayed a decreasing trend after the implementation of BGI, however it is to be noted that temperature started to reduce after two years of successful BGI implementation. Therefore, it can be concluded that the effcts of BGI implementation on the urban environment is not instant and will take some time to be observed.

The research also demonstrated the potential of open-source remote sensing data and geospatial technologies for conducting temporal and spatial trend analysis of the influence of BGI on LST variations and the UHI effect within urban environment.

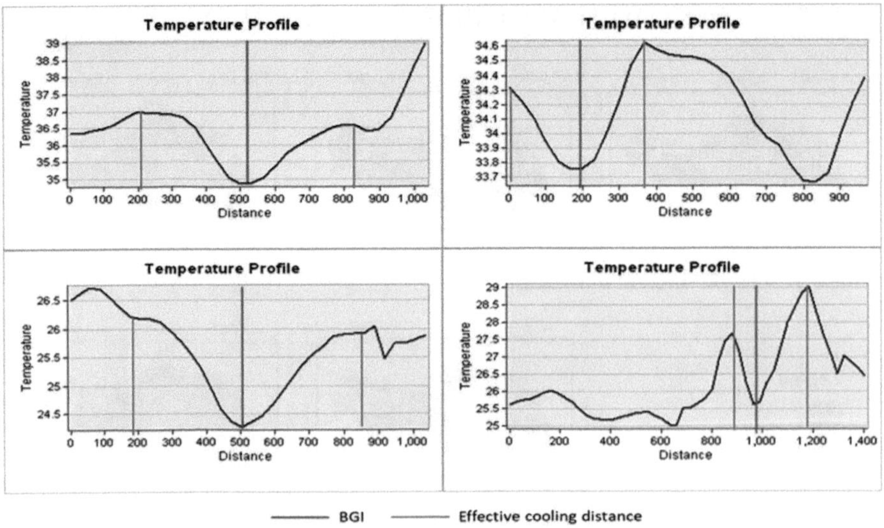

Fig. 4 Effective distance of cooling from Blue-Green Infrastructure

Table 2 Effective distance of cooling from centroid of Blue-Green Infrastructure

Area (m^2)	Effective distance of cooling from centroid of BGI				Average distance (m)
	North	East	South	West	
<50,000	250	400	200	400	200.00
	100	200	100	100	
	100	250	200	100	
50,000–100,000	400	160	300	300	249.17
	250	350	100	400	
	150	230	150	200	
>100,000	600	600	400	700	512.50
	500	700	700	600	
	600	100	300	350	

Findings from the temporal trend analysis, utilizing both MODIS LST and LST derived from Landsat, indicate a positive impact of BGI on temperatures in urban areas. The temperature was steadily decreasing in Singapore. The temperature which was at 26.17 °C in 2018 was brought down to 24.87 °C in 2021. The quantification of the magnitude to which the temperature can go down with a significant amount of increase in BGI was possible with the help of LULC change analysis and temperature change analysis from remote sensing data during the time period considered for the study. For an increase of 8.08% in BGI in the study area, 1.21°C of temperature is reduced in Singapore. The results obtained corroborates with the earlier studies which has also shown that increase in BGI percentage has significant impact on the reduction of temperatures in urban areas and UHI phenomenon (Abdulateef & Al-Alwan, 2022; Kemarau & Eboy, 2020; Wu et al., 2018). Annual median images were used for obtaining LST for temporal and spatial trend analysis in this study as study area falls in tropical wet climate and has high amount of cloud cover and persistent rainfall due to which not a single image was available as cloud free images.

Spatial analysis of BGI with LST highlighted the need for well-distributed network of BGI as effective cooling distance (distance upto which cooling effect of BGI is observed) is found to have a significant relation with size of BGI. Bigger the size of BGI, effective cooling distance will also increase. However, it does not imply that spatial distribution of BGI can be overlooked in order to have larger size BGI sites. Rather it is found that if BGI is implemented in the form of well distributed network it may have more impact on urban macro and micro environments.

The study employed freely available coarse to medium resolution remote sensing data, specifically MODIS and Landsat for conducting a city-level analysis. This choice was driven by the limited availability of high-resolution thermal remote sensing data. However, in future the availability of high-resolution thermal data, either from upcoming satellite missions or through UAV platforms, could facilitate more detailed site-level analysis of BGI and its impact on LST and the UHI effect.

6.5 Conclusion

Blue-Green Infrastructure approaches are nature-oriented for the development of a sustainable city in its meteorological, economic, and social aspects. The study area selected for this study (Singapore) falls in developed region, as there are very few examples of BGI implementation in developing region and that also is in initial stage, it might not be able to provide a conclusive result. Nonetheless, the significance of results obtained is still valid for developing regions. From the cases of developed countries where BGI implementation was successful, it is evident that BGI helps in reducing the temperature. The well distributed network of BGI has a great potential to reduce temperatures in urban areas and significant impact on mitigating UHI. It would be better to incorporate BGI in planning in developing countries so that development occurs in a sustainable way. Many countries in the world especially in developing region is now promoting BGI development through government schemes and policies for example Government of India has given significant importance to development of urban green spaces and rejuvenation of urban water bodies through its national level projects of Atal Mission for Rejuvenation and Urban Transformation (AMRUT) and smart city program. The quantification of extent to which the BGI can reduce the temperature can be practically applied on formulating planning guidelines and norms so that blue and green space allocation is done efficiently and effectively. The study can be furthermore incorporated with very high resolution thermal data such as from UAV imageries and surface temperature dynamics in urban areas can be studied in more depth. The study also opens scope for incorporating factors such as terrain characteristics and urban geometry to assess the impact of BGI on UHI.

References

Abdulateef, M. F., & Al-Alwan, H. A. (2022). The effectiveness of urban green infrastructure in reducing surface urban heat island. *Ain Shams Engineering Journal, 13*(1), 101526. https://doi.org/10.1016/j.asej.2021.06.012

Akbari, H., Pomerantz, M., & Taha, H. (2001). Cool surfaces and shade trees to reduce energy use and improve air quality in urban areas. *Solar Energy, 70*(3), 295–310. https://doi.org/10.1016/S0038-092X(00)00089-X

Ali, D. A., Deininger, K., & Wild, M. (2020). Using satellite imagery to create tax maps and enhance local revenue collection. *Applied Economics, 52*(4), 415–429. https://doi.org/10.1080/00036846.2019.1646408

Almaaitah, T., Appleby, M., Rosenblat, H., Drake, J., & Joksimovic, D. (2021). The potential of Blue-Green infrastructure as a climate change adaptation strategy: A systematic literature review. *Blue-Green Systems, 3*(1), 223–248. https://doi.org/10.2166/bgs.2021.016

Antoszewski, P., Świerk, D., & Krzyżaniak, M. (2020). Statistical review of quality parameters of blue-green infrastructure elements important in mitigating the effect of the urban heat island in the temperate climate (C) zone. *International Journal of Environmental Research and Public Health, 17*(19), 7093. https://doi.org/10.3390/ijerph17197093

Bartesaghi Koc, C., Osmond, P., & Peters, A. (2017). Towards a comprehensive green infrastructure typology: A systematic review of approaches, methods and typologies. *Urban Ecosystems, 20*, 15–35. https://doi.org/10.1007/s11252-016-0578-5

Benedict, M. A., & McMahon, E. T. (2002). Green infrastructure: Smart conservation for the 21st century. *Renewable Resources Journal, 20*(3), 12–17.

Benedict, M.A. and McMahon, E.T. (2012). Green infrastructure: linking landscapes and communities. *Island press.*

Bhargava, A., Lakmini, S., & Bhargava, S. (2017). Urban Heat Island Effect: It's relevance in urban planning. *Journal of Biodiversity & Endangered Species, 5*(187), 2020. https://doi.org/10.4172/2332-2543.1000187

Bhatt, M.M., Gupta, K., Danodia, A., Chakroborty, S.D. and Patel, N.R. (2023). Detailed urban roughness parametrization for anthropogenic heat flux estimation using earth observation data. *Heliyon, 9*(7). https://doi.org/10.1016/j.heliyon.2023.e18361.

Bhavana, M., Gupta, K., Pal, P. K., Kumar, A. S., & Gummapu, J. (2018). Evaluation of High Resolution Urban LULC for Seasonal Forecasts of Urban Climate using WRF Model. *ISPRS Annals of the Photogrammetry, Remote Sensing and Spatial Information Sciences, 4*, 303–310. https://doi.org/10.5194/isprs-annals-IV-5-303-2018

Blaschke, T., Hay, G. J., Weng, Q., & Resch, B. (2011). Collective sensing: Integrating geospatial technologies to understand urban systems—An overview. *Remote Sensing, 3*(8), 1743–1776. https://doi.org/10.3390/rs3081743

Bodhankar, S., Gupta, K., Kumar, P., & Srivastava, S. K. (2022). GIS based multi-objective urban land allocation approach for optimal allocation of urban land uses. *Journal of the Indian Society of Remote Sensing, 50*(4), 763–774. https://doi.org/10.1007/s12524-022-01492-0

Boussema, S.B.F., Cohen, M. and Allouche, F.K., 2022. Green and blue infrastructure design in a semi-arid region. *Frontiers in Environmental Science, 10*. https://doi.org/10.3389/fenvs.2022.1061256.

Bouyer, J., Musy, M., Huang, Y. and Athamena, K. (2011). Mitigating urban heat island effect by urban design: forms and materials. *Cities and Climate Change: Responding to an Urgent Agenda*, pp.164–181.

Broadbent, A. M., Coutts, A. M., Tapper, N. J., & Demuzere, M. (2018). The cooling effect of irrigation on urban microclimate during heatwave conditions. *Urban Climate, 23*, 309–329. https://doi.org/10.1016/j.uclim.2017.05.002

Census, 2020. Census of Population 2020. Department of Statistics, Singapore. https://www.singstat.gov.sg/census2020/census-2020.

Chowdhury, P. R., & Maithani, S. (2014). Modelling urban growth in the Indo-Gangetic plain using nighttime OLS data and cellular automata. *International Journal of Applied Earth Observation and Geoinformation, 33*, 155–165. https://doi.org/10.1016/j.jag.2014.04.009

CNY RPDB, Central New York Regional Planning & Development Board. (2012). Green Infrastructure Planning for Improved Stormwater Management in Central New York. Retrieved from https://www.cnyrpdb.org/docs/reports/GI_PlanningReport_2011.pdf

Cristóbal, J., Jiménez-Muñoz, J. C., Prakash, A., Mattar, C., Skoković, D., & Sobrino, J. A. (2018). An improved single-channel method to retrieve land surface temperature from the Landsat-8 thermal band. *Remote Sensing, 10*(3), 431. https://doi.org/10.3390/rs10030431

Depietri, Y. and McPhearson, T. (2017). Integrating the grey, green, and blue in cities: nature-based solutions for climate change adaptation and risk reduction. Nature-based solutions to climate change adaptation in urban areas: linkages between science, policy and practice, pp. 91–109. https://doi.org/10.1007/978-3-319-56091-5_6.

Donnell, E. C., Netusil, N. R., Chan, F. K. S., Dolman, N. J., & Gosling, S. N. (2021). International perceptions of urban blue-green infrastructure: A comparison across four cities. *Water, 13*(4), 544. https://doi.org/10.3390/w13040544

Dreiseitl, H. and Wanschura, B. (2006). Strengthening blue-green infrastructure in our cities: enhancing blue-green infrastructure & social performance in high density urban environments. *Ramboll Foundation*. Retrieved from http://www.ramboll.com/LCL.

Drosou, N., Soetanto, R., Hermawan, F., Chmutina, K., Bosher, L., & Hatmoko, J. U. D. (2019). Key factors influencing wider adoption of blue-green infrastructure in developing cities. *Water, 11*(6), 1234. https://doi.org/10.3390/w11061234

Du, H., Cai, Y., Zhou, F., Jiang, H., Jiang, W., & Xu, Y. (2019). Urban blue-green space planning based on thermal environment simulation: A case study of Shanghai. *China. Ecological Indicators, 106*, 105501. https://doi.org/10.1016/j.ecolind.2019.105501

Etheridge, D., & Ford, B. (2008). Natural ventilation of tall buildings–options and limitations. *CTBUH 8th world congress* (pp. 3–5). School of the Built Environment, University of Nottingham.

Fung, C. K. W., & Jim, C. Y. (2020). Influence of blue infrastructure on lawn thermal microclimate in a subtropical green space. *Sustainable Cities and Society, 52*, 101858. https://doi.org/10.1016/j.scs.2019.101858

Gaekwad, J.S., Sal Moslehian, A., Roös, P.B., Walker, A. (2022). A meta-analysis of emotional evidence for the biophilia hypothesis and implications for biophilic design. *Frontiers in Psychology, 13*, pp.2476. https://www.frontiersin.org/articles/https://doi.org/10.3389/fpsyg.2022.750245.

Ghofrani, Z., Sposito, V., & Faggian, R. (2016). Designing resilient regions by applying blue-green infrastructure concepts. *WIT Transactions on Ecology and the Environment, 204*, 493–505. https://doi.org/10.2495/SC160421

Gunawardena, K., Kershaw, T., & Steemers, K. (2019). Simulation pathway for estimating heat island influence on urban/suburban building space-conditioning loads and response to facade material changes. *Building and Environment, 150*, 195–205.

Gupta, K. (2013). Unprecedented growth of Dehradun urban area: A spatio-temporal analysis. *International Journal of Advancement in Remote Sensing, GIS and Geography, 1*(2), 47–56.

Gupta, K. (2022). Retrieval of Urban Canopy Parameters from Very High Resolution Optical Stereo and Multi-Spectral Data. Copyright no. L-113996/2022. New Delhi, India: Copyright Office, Govt. of India.

Gupta, K., Bhardwaj, A., Kumar, P., & Pushpalata, P. (2015). Procedural rule-based 3D city modeling and visualization using high resolution satellite data. *International Journal of Advancement in Remote Sensing, GIS and Geography, 3*(2), 16–25.

Gupta, K., Garg, P., Gupta, P. K., Debnath, A., Roy, A., & Shukla, Y. (2022). An innovative approach for retrieval of gridded urban canopy parameters using very high resolution optical satellite stereo. *International Journal of Remote Sensing, 43*(12), 4378–4409. https://doi.org/10.1080/01431161.2022.2112108

Gupta, K., Kumar, P., Pathan, S. K., & Sharma, K. P. (2012). Urban Neighborhood Green Index—A measure of green spaces in urban areas. *Landscape and Urban Planning, 105*(3), 325–335. https://doi.org/10.1016/j.landurbplan.2012.01.003

Gupta, K., Puntambekar, K., Roy, A., Pandey, K. and Kumar, P. (2020). Smart Environment Through Smart Tools and Technologies for Urban Green Spaces. *In Smart Environment for Smart Cities*, T M Vinod Kumar (Ed.), pp.149–194. Springer, Singapore.

Gupta, K., Pushpalata, P., John, S., Bhardwaj, A., Kumar, P., Kumar, A. S. (2017a). Comparative evaluation of Pleiades, Cartosat-2 and Kompsat-3 Stereo data for DSM and 3D model generation. *Asian Conference on Remote Sensing*, pp. 23–27.

Gupta, K., Pushpalata, P., Rajwal, T., Nekkla, S., Kumar, P. and Kumar, A. S., 2017b. Investigating the relationship of urban form and function with surface temperature patterns: a case study of Chandigarh. *Asian Conference on Remote Sensing*, pp. 23–27, October, 2017, Delhi.

Gupta, K., Pushplata, L., & A., Ghosh Dastidar, P., Malleswara Rao, J., Thakur, P., Gummapu, J.S. and Senthil Kumar, A.,. (2021). Modeling seasonal variation in urban weather in sub-tropical region of Delhi. *Journal of the Indian Society of Remote Sensing, 49*, 193–213.

Gupta, K., Roy, A., Luthra, K., & Maithani, S. (2016). GIS based analysis for assessing the accessibility at hierarchical levels of urban green spaces. *Urban Forestry & Urban Greening, 18*, 198–211. https://doi.org/10.1016/j.ufug.2016.06.005

Halder, B., Bandyopadhyay, J., & Banik, P. (2021). Evaluation of the climate change impact on urban heat island based on land surface temperature and geospatial indicators. *International Journal of Environmental Research, 15*, 819–835. https://doi.org/10.1007/s41742-021-00356-8

Hamel, P., & Tan, L. (2021). Blue-green infrastructure for flood and water quality management in southeast asia: evidence and knowledge gaps. *Environmental Management, 69*(4), 699–718. https://doi.org/10.1007/s00267-021-01467-w

Howard, L. (1833). *The Climate of London*. Vol.-I-III.

Jat, M. K., Garg, P. K., & Khare, D. (2008). Monitoring and modelling of urban sprawl using remote sensing and GIS techniques. *International Journal of Applied Earth Observation and Geoinformation, 10*(1), 26–43.

Jia, S., & Wang, Y. (2022). Comparison of different blue-green infrastructure strategies in mitigating urban heat island effects and improving thermal comfort. *Construction Research Congress, 2022*, 357–366.

Kafy, A. A., Al Rakib, A., Akter, K. S., Rahaman, Z. A., Mallik, S., Nasher, N. R., Hossain, M. I., & Ali, M. Y. (2021). Monitoring the effects of vegetation cover losses on land surface temperature dynamics using geospatial approach in Rajshahi City, Bangladesh. *Environmental Challenges, 4*, 100187. https://doi.org/10.1016/j.envc.2021.100187

Kaur, R., & Gupta, K. (2022). Blue-Green Infrastructure (BGI) network in urban areas for sustainable storm water management: A geospatial approach. *City and Environment Interactions., 16*, 100087. https://doi.org/10.1016/j.cacint.2022.100087

Kemarau, R. A., & Eboy, O. V. (2020). Analyses Water Bodies Effect in Mitigation of Urban Heat Effect: Case Study Small Size Cities Kuching, Sarawak. In *IOP Conference Series: Earth and Environmental Science, 540*(1), 012010. https://doi.org/10.1088/1755-1315/540/1/012010

Kim, H. H. (1992). Urban heat island. *International Journal of Remote Sensing, 13*(12), 2319–2336.

Klemm, W., Heusinkveld, B. G., Lenzholzer, S., Jacobs, M. H., & Van Hove, B. (2015). Psychological and physical impact of urban green spaces on outdoor thermal comfort during summertime in The Netherlands. *Building and Environment, 83*, 120–128. https://doi.org/10.1016/j.buildenv.2014.05.013

Kong, F., Yin, H., Nakagoshi, N., & Zong, Y. (2010). Urban green space network development for biodiversity conservation: identification based on graph theory and gravity modeling. *Landscape and Urban Planning, 95*, 16–27. https://doi.org/10.1016/j.landurbplan.2009.11.001

Korkou, M., Tarigan, A. K. M., & Hanslin, H. M. (2023). The multifunctionality concept in urban green infrastructure planning: A systematic literature review. *Urban Forestry & Urban Greening, 85*, 127975. https://doi.org/10.1016/j.ufug.2023.127975

Kumar, V., Singh, V. K., Gupta, K., & Jha, A. K. (2021). Integrating cellular automata and agent-based modeling for predicting urban growth: A case of Dehradun City. *Journal of the Indian Society of Remote Sensing, 49*(11), 2779–2795. https://doi.org/10.1007/s12524-021-01418-2

Lee, G., Hwang, J., & Cho, S. (2021). A novel index to detect vegetation in urban areas using UAV-based multispectral images. *Applied Sciences, 11*(8), 3472. https://doi.org/10.3390/app11083472

Lee, K., Kim, Y., Sung, H. C., Ryu, J., & Jeon, S. W. (2019). Trend analysis of urban heat island intensity according to urban area change in Asian mega cities. *Sustainability, 12*(1), 112. https://doi.org/10.3390/su12010112

Lehnert, M., Brabec, M., Jurek, M., Tokar, V., & Geletic, J. (2021). The role of blue and green infrastructure in thermal sensation in public urban areas: A case study of summer days in four Czech cities. *Sustainable Cities and Society, 66*, 102683. https://doi.org/10.1016/j.scs.2020.102683

Lim, H. S., & Lu, X. X. (2016). Sustainable urban stormwater management in the tropics: An evaluation of Singapore's ABC Waters Program. *Journal of Hydrology, 538*, 842–862. https://doi.org/10.1016/j.jhydrol.2016.04.063

Loi, D. T., Tuan, P. A., & Gupta, K. (2015). Development of an index for assessment of urban green Spacesat city level. *International Journal of Remote Sensing Applications, 5*(1), 78. https://doi.org/10.14355/ijrsa/2015.05.009

Lu, D., & Weng, Q. (2006). Use of impervious surface in urban land-use classification. *Remote Sensing of Environment, 102*(1–2), 146–160. https://doi.org/10.1016/j.rse.2006.02.010

Marando, F., Heris, M. P., Zulian, G., Udías, A., Mentaschi, L., Chrysoulakis, N., Parastatidis, D., & Maes, J. (2022). Urban heat island mitigation by green infrastructure in European Functional Urban Areas. *Sustainable Cities and Society, 77*, 103564. https://doi.org/10.1016/j.scs.2021.103564

Masson, V. (2006). Urban surface modeling and the meso-scale impact of cities. *Theoretical and Applied Climatology, 84*, 35–45. https://doi.org/10.1007/s00704-005-0142-3

Nasi, R., Honkavaara, E., Blomqvist, M., Lyytikäinen-Saarenmaa, P., Hakala, T., Viljanen, N., Kantola, T., & Holopainen, M. (2018). Remote sensing of bark beetle damage in urban forests at individual tree level using a novel hyperspectral camera from UAV and aircraft. *Urban Forestry & Urban Greening, 30*, 72–83. https://doi.org/10.1016/j.ufug.2018.01.010

Nuruzzaman, M., 2015. Urban heat island: causes, effects and mitigation measures-a review. *International Journal of Environmental Monitoring and Analysis, 3*(2), pp. 67–73. https://doi.org/10.11648/j.ijema.20150302.15.

Oliveira, S., Andrade, H., & Vaz, T. (2011). The cooling effect of green spaces as a contribution to the mitigation of urban heat: A case study in Lisbon. *Building and Environment, 46*(11), 186–2194. https://doi.org/10.1016/j.buildenv.2011.04.034

Paschalis, A., Chakraborty, T.C., Fatichi, S., Meili, N. and Manoli, G. (2021). Urban forests as main regulator of the evaporative cooling effect in cities. *AGU Advances, 2*(2), p.e2020AV000303.

Qin, L., Yan, C., Yu, L., Chai, M., Wang, B., Hayat, M., Shi, Z., Gao, H., Jiang, X., Xiong, B., & Mao, P. (2022). High-resolution spatio-temporal characteristics of urban evapotranspiration measured by unmanned aerial vehicle and infrared remote sensing. *Building and Environment, 222*, 109389. https://doi.org/10.1016/j.buildenv.2022.109389

Rao, P., Gupta, K., Roy, A., & Balan, R. (2021). Spatio-temporal analysis of land surface temperature for identification of heat wave risk and vulnerability hotspots in Indo-Gangetic Plains of India. *Theoretical and Applied Climatology, 146*(1), 567–582. https://doi.org/10.1007/s00704-021-03756-0

Raška, P., Bezak, N., Ferreira, C., Kalantari, Z., Banasik, K., Bertola, M., Bourke, M., Cerdà, A., Davids, P., Brito, M., Evans, R., Finger, D., Halbac-Cotoara-Zamfir, R., Housh, M., Hysa, A., Jakubínský, J., Kapovic, S. M., Kaufmann, M., Keesstra, S., & Hartmann, T. (2022). Identifying barriers for nature-based solutions in flood risk management: An interdisciplinary overview using expert community approach. *Journal of Environmental Management, 310*, 114725. https://doi.org/10.1016/j.jenvman.2022.114725

Robitu, M., Musy, M., Inard, C., & Groleau, D. (2006). Modeling the influence of vegetation and water pond on urban microclimate. *Solar Energy, 80*(4), 435–447. https://doi.org/10.1016/j.solener.2005.06.015

Semeraro, T., Aretano, R. and Pomes, A., 2017. Green infrastructure to improve ecosystem services in the landscape urban regeneration. In *IOP conference series: materials Science and engineering, 245*(8), pp. 082044. https://doi.org/10.1088/1757-899X/245/8/082044.

Shafique, M., & Kim, R. (2017). Application of Green blue roof to mitigate heat island phenomena and resilient to climate change in urban areas: A case study from Seoul, Korea. *Journal of Water and Land Development, 33*(1), 165–170. https://doi.org/10.1515/jwld-2017-0032

Shekhar, S., & Aryal, J. (2019). Role of geospatial technology in understanding urban green space of Kalaburagi city for sustainable planning. *Urban Forestry & Urban Greening, 46*, 126450. https://doi.org/10.1016/j.ufug.2019.126450

Shetty, S. (2019). Analysis of machine learning classifiers for LULC classification on Google Earth Engine. University of Twente.

Singh, P., Kikon, N., & Verma, P. (2017). Impact of land use change and urbanization on urban heat island in Lucknow city, Central India. A remote sensing-based estimate. *Sustainable Cities and Society, 32*, 100–114. https://doi.org/10.1016/j.scs.2017.02.018

Singh, S., & Mishra, A. (2014). Deforestation-induced costs on the drinking water supplies of the Mumbai metropolitan, India. *Global Environmental Change, 27*(1), 73–83. https://doi.org/10.1016/j.gloenvcha.2014.04.020

Syafii, N. I., Ichinose, M., Wong, N. H., Kumakura, E., Jusuf, S. K., & Chigusa, K. (2016). Experimental study on the influence of urban water body on thermal environment at outdoor scale model. *Procedia Engineering, 169*, 191–198. https://doi.org/10.1016/j.proeng.2016.10.023

Theethai Jacob, A., Jayakumar, A., Gupta, K., Mohandas, S., Hendry, M. A., Smith, D. K., Francis, T., Bhati, S., Parde, A. N., Mohan, M., & Mitra, A. K. (2023). Implementation of the urban parameterization scheme in the Delhi model with an improved urban morphology. *Quarterly Journal of the Royal Meteorological Society, 149*(750), 40–60. https://doi.org/10.1002/qj.4382

Thompson, K. (1981). The question of climatic stability in America before 1900. *Climatic Change, 3*(3), 227–241. https://doi.org/10.1007/BF02423216

Thompson, R. (2018). *Evaluating the Benefits, Sustainability, and Resilience of Green Infrastructure on a Sustainable Residential Home*. University of Maryland.

Tominaga, Y., Sato, Y., & Sadohara, S. (2015). CFD simulations of the effect of evaporative cooling from water bodies in a micro-scale urban environment: Validation and application studies. *Sustainable Cities and Society, 19*, 259–270. https://doi.org/10.1016/j.scs.2015.03.011

Udas-Mankikar, S. and Driver, B. (2021). Blue-green infrastructure: An opportunity for Indian cities. *ORF Occasional Paper, 317*.

URL for ABC sites https://beta.data.gov.sg/datasets?query=ABC%20Waters%20Sites

URL of boundary https://earthworks.stanford.edu/catalog.html?f%5Bdc_format_s%5D%5B%5D=Shapefile&f%5Bdct_spatial_sm%5D%5B%5D=Singapore+%28City%29&per_page=10&sort=solr_year_i+desc%2C+dc_title_sort+asc

Voogt, J. A., & Oke, T. R. (2003). Thermal remote sensing of urban climates. *Remote Sensing of Environment, 86*(3), 370–384.

Voogt, J.A. (2004). Urban heat islands: hotter cities. American Institute of Biological Sciences. An ActionBioscience.org original article. Retrieved from http://www.actionbioscience.org/environment/voogt.html.

Wang, W., Wang, D., Chen, H., Wang, B., & Chen, X. (2022). Identifying urban ventilation corridors through quantitative analysis of ventilation potential and wind characteristics. *Building and Environment, 214*, 108943.

Wang, Y., Chang, Q., & Fan, P. (2021). A framework to integrate multifunctionality analyses into green infrastructure planning. *Landscape Ecology, 36*, 1951–1969. https://doi.org/10.1007/s10980-020-01058-w

Wang, Z. H. (2022). Reconceptualizing urban heat island: Beyond the urban-rural dichotomy. *Sustainable Cities and Society, 77*, 103581.

Wilbers, G. J., Bruin, K. D., Dähnn, I. S., Lekkerkerk, W., Li, H., & Ballinas, M. B. P. (2022). Investing in urban blue-green infrastructure—assessing the costs and benefits of stormwater management in a peri-urban catchment in Oslo, Norway. *Sustainability, 14*, 1934. https://doi.org/10.3390/su14031934

Winker, M., Jutta, D., Michaela, R., Engelbert, S., & Melina, S. (2022). Enhancing blue-green infrastructure in German cities with the involvement of urban society: Insights from Frankfurt/Main and Stuttgart. *Blue-Green Systems, 4*(2), 230–246. https://doi.org/10.2166/bgs.2022.017

Wong, E., 2005. Green roofs and the US Environmental Protection Agency's heat island reduction initiative.

Wu, C., Li, J., Wang, C., Song, C., Chen, Y., Finka, M., & La Rosa, D. (2019). Understanding the relationship between urban blue infrastructure and land surface temperature. *Science of Total Environment, 694*, 133742. https://doi.org/10.1016/j.scitotenv.2019.133742

Wu, D., Wang, Y., Fan, C., & Xia, B. (2018). Thermal environment effects and interactions of reservoirs and forests as urban blue-green infrastructures. *Ecological Indicators, 91*, 657–663. https://doi.org/10.1016/j.ecolind.2018.04.054

Yang, L., Qian, F., Song, D. X., & Zheng, K. J. (2016). Research on urban heat-island effect. *Procedia Engineering, 169*, 11–18. https://doi.org/10.1016/j.proeng.2016.10.002

Yu, X., Guo, X., & Wu, Z. (2014). Land surface temperature retrieval from Landsat 8 TIRS—Comparison between radiative transfer equation-based method, split window algorithm and single channel method. *Remote Sensing, 6*(10), 9829–9852. https://doi.org/10.3390/rs6109829

Yusof, N. S., Hussain, N. H. M., & Rusli, N. (2019). Analysing street heritage trees surface temperature for UHI mitigation using remote sensing and GIS application. *Malaysian Journal of Sustainable Environment, 6*(2), 77–95.

Zhang, Z., Meerow, S., Newell, J. P., & Lindquist, M. (2019). Enhancing landscape connectivity in detroit through multifunctional green corridor modeling and design. *Urban Forestry & Urban Greening, 38*, 305–317. https://doi.org/10.1016/j.ufug.2018.10.014

Role of Blue-Green Infrastructure in Achieving Sustainable Development Goals in Urban Centres of Developing countries

Kajal Patel, Meenakshi Chaurasia, Ranjana Singh, Rahul Bhadouria, and K. S. Rao

Abstract The central idea for establishing the 2030 agenda for sustainable development is achieving well-being for all while protecting the environment and welfare for future generations with 17 interlinked Sustainable development goals (SDGs). These goals are no poverty, zero hunger, good health and well-being, quality education, gender equality, clean water and sanitation, affordable and clean energy, decent work and economic growth, industry, innovation and infrastructure, reduced inequalities, sustainable cities and communities, responsible consumption and production, climate action, life below water, life on land, peace, justice, and strong institutions, and partnerships for the goals. To achieve these goals blue-green infrastructure (BGI) can play a vital role as it promotes sustainable and resource-efficient living. Utilizing urban planning, ecological restoration, and infrastructure to foster a sense of community with the natural world, the BGI network offers the "ingredients" for resolving urban and climatic concerns. Under BGI the water bodies like rivers and tanks are indicated by the color "blue." and for trees, parks, and gardens, use "green.". BGI has numerous advantages and produces effects that cut across the economic, social, and environmental sectors. It contributes to the control of regional ecosystems as well as the reduction of pollutants and urban temperature. These in turn have a significant economic impact, for instance, terrace gardens encourage lower building surface temperatures, which reduces the need for cooling and, consequently, the demand for energy and power. Reduced consumption of vital raw materials and resources,

The original version of the chapter has been revised: A correction to this chapter can be found at https://doi.org/10.1007/978-3-031-62293-9_18

K. Patel · M. Chaurasia · R. Bhadouria · K. S. Rao
Department of Botany, University of Delhi, New Delhi 110007, India

R. Singh (✉)
Government Model Degree College, Arniya, Bulandshahr, UP 203131, India
e-mail: vermaranjana09@gmail.com

R. Bhadouria
Delhi College of Arts and Commerce, University of Delhi, New Delhi 110023, India

P. K. Joshi et al. (eds.), *Blue-Green Infrastructure for Sustainable Urban Settlements*,
https://doi.org/10.1007/978-3-031-62293-9_5

the prevention of pollution, and a decrease in carbon emissions are further financial and economic advantages. This chapter reviews the applicability of BGIs in achieving SDGs at the ground level and further highlights the implications, benefits, and challenges of their implementation. It is known that an effective and efficient infrastructure like BGI allows multipurpose uses and several advantages that directly or indirectly help to meet the targets in order to achieve the SDGs.

Keywords Blue-green infrastructure · Climate change · Climate resilience · Nature-based solutions

1 Introduction

It is predicted that 68% of people on the planet will live in cities by 2050. Elbeltagi et al. (2021) found that the growing urban population has put a strain on our natural resources and disturbed the urban ecosystems. Urbanization has led to the construction of roads, businesses and industrial spaces, residences, office buildings, parking lots, and other structures on top of meadows, woodlands, and agricultural areas. According to Paul et al. (2018), The microclimate and land usage have also changed as a result of urbanization. The conversion of rural areas into urban areas and significant changes in land use can lead to the occurrence of heat waves, droughts, and floods (Mohtar et al., 2020). These changes can also have devastating effects on urban ecosystems. Blue-green infrastructure (BGI) has been promoted as a long-term, sustainable approach to lessen the detrimental effects of urbanization on the ecology. A network of natural features incorporated into built-up regions to create an urban landscape is known as urban blue-green infrastructure (BGI). These characteristics are derived from either flora (green), water (blue), or both. These types of designs can be seen in parks, rivers, ponds, swales (shallow channels or drains), grassy areas, and rain gardens. So are green roofs and walls (Oberndorfer et al., 2007). Blue-green infrastructure is essential for reducing greenhouse gas emissions and preparing for climate change, in addition to offering a host of other benefits to humans and wildlife (Brown & Mijic, 2019).

Due to its significant contribution to sustainable development (potential social, environmental and health benefits)—Both on a micro- and in-situ level, the application of BGI has gained momentum in recent years. The services provided by BGI implementation have been assessed in several studies. To contribute to the regulation of services, this chapter primarily focuses on comprehending and assessing the performance of various BGIs (Puchol-Salort et al., 2021). It also gives a general review of BGI's numerous advantages and how it helps to improve urban ecosystems. In this chapter, a variety of professional perspectives on various BGI technologies, their value, services, and effective BGI methods are discussed. From the viewpoint of several nations around the world, this chapter illustrates how BGI can be used in practice. Though the global blue-green interventions that are currently in place may vary in terms of density, scale, problem nature, and goal, Indian cities and others will be able to utilize the lessons learned from these to create blue-green infrastructure frameworks to address local issues and challenges.

2 Importance of SDGs

According to Salvia et al. (2019), throughout the last thirty years, the notion of sustainable development has garnered significant interest; nonetheless, the means of achieving this objective remain mostly unclear. In this direction, the framework of Sustainable development goals (SDG) mainly comprises 169 targets that aim to achieve the goal of a developed world by preserving the environment and future generations' needs. The 17 goals hit various targets like poverty, no hunger, climate change, life on land and underwater, and so on (Henderson & Loreau, 2023). The ultimate results can help us to overcome big challenges such as pollution, global warming, poverty, hunger, etc. Despite being interconnected, neither sustainability nor the goals can be defined since they do not specify the trade-offs and synergies that relate to the total well-being of humans and the environment (Bhadouria et al., 2023; Costanza et al., 2016). According to Fuso Nerini et al. (2018), there is a close relationship between the UN's SDGs and energy availability, agricultural production, medical facilities, water treatment, urban ecosystems, sustainable consumption and infrastructure development, climate change policies and renewable energy.

A collapse of non-renewable resources, agricultural land, and the ability to absorb pollutants was anticipated by the theory of Limits to Growth in the 1970s as continuous growth would approach its limit in the twenty-first century (Gómez-Baggethun & Naredo, 2015). The model concluded that the pursuit of growth at the expense of the environment would result in an irreversible decline in population and industrial capacity. It also investigated interconnections among the population, industrial production, food supply, pollution, and non-renewable resources. It was also concluded by the authors that economic and ecological stability could be attained by cutting down consumption and redistributing wealth. Following this scenario, almost a decade later the idea of sustainability appeared in the report of Brundtland (1987). This idea as we know it today emphasizes the importance of preserving both present and future generations' access to resources. Since then, various discussions have been going on over more coupled human–environment models using various socioeconomic indicators such as inequality, well-being, social learning, food trade, rebound effects, and investment in sustainable development. Since the announcement of the UN's framework of SDGs, numerous models have been developed to analyze the overlaps and trade-offs between goals, as well as to measure indicators or quantify goal achievement (Costanza et al., 2016; De la Poza et al., 2021; Kapoor et al., 2020; Spaiser et al., 2017).

3 BGIs as a Nature-Based Solution

The term "blue-green infrastructure" refers to a broad category of nature-based solutions that comprise two components i.e., green (land) and blue (water). Water bodies such as rivers, canals, ponds, wetlands, floodplains, and water treatment plants are referred to as "Blue" infrastructure; on the other hand, "Green" infrastructure includes fields, forests, parks, trees, and hedgerows. BGI directly addresses climate change, urban resilience and health and wellness. Several countries defined BGI as "a strategically planned network of natural and semi-natural areas with environmental features that are designed and managed to deliver a wide range of ecosystem services which can enhance the quality of life". The idea entails urban development where water bodies and land are interdependent and flourish together while providing environmental and social benefits (Udas-Mankikar & Driver, 2021). Water purification, air quality, recreational space, and climate adaptation and mitigation are just a few of the numerous ecosystem services that help to advance biodiversity, foster employment possibilities, and support a green economy.

As mentioned in the earlier section, infrastructure shows a close synergy with socioeconomic development and sustainability. Even though nature-based solutions are now fundamental to sustainable urban and landscape planning schemes, cross-scale integration sharpens the focus on single- and small-scale solutions, which in turn impacts flows of ecosystem services, raising the possibility that prior gains in green–blue infrastructure planning will be reversed (Langemeyer & Baró, 2021).

BGI is a two-component approach which are interconnected but both of them have their benefits separately. Therefore, to understand the importance of BGI, firstly it's important to understand their roles. The interconnected networks of multipurpose green space that offer many advantages and can support sustainable development are frequently used to define green infrastructure. Parks, open spaces, street and private trees, playgrounds, woods, private gardens, allotments, and green roofs and walls are some of these characteristics. These "green" infrastructures show benefits (Brown & Mijic, 2019; Dhyani et al., 2022; Ghofrani et al., 2017; Hamann et al., 2020):

1. Trees in landscaping increase a property's value by 5% to 15% and help it sell more quickly.
2. The cost of winter heating can be reduced by up to 25% by strategically placing trees as windbreaks around buildings.
3. One mature tree has a 21.6 kg annual capacity to absorb carbon.
4. Customers are willing to pay 9% to 12% more for goods and services in a shopping center with large, well-maintained trees.
5. Street trees contribute to a 15–20% reduction in particulate matter concentration, and increased canopy cover helps to reduce air pollution.
6. Carbon dioxide emissions are lowered by 3–7% for every 10% increase in urban tree canopy.

On the other hand, ponds, lakes, streams, rivers, and stormwater management systems are examples of urban water infrastructure that is commonly referred to as "blue infrastructure." Green infrastructure is sometimes used to refer to sustainable drainage systems, which are typically classified under this category. The benefits of the blue infrastructure are (Dhyani et al., 2022; Kopp et al., 2021):

1. A mature tree needs up to 200 L of water each day for evapotranspiration, a process that helps to clean and cool the air.
2. Various water streams help to flush out the pollutants and help to decontaminate the environment.
3. Without harming tree health, the potential attenuation area in specialized soil can hold 22–35% more water by volume of soil.
4. The amount of runoff from storms is decreased by 2% for every 5% of tree canopy cover.
5. Healthy urban trees may store up to 70% of the water that falls during the first hour of a rain event on their canopy and branch structure.
6. When both the components merge to plan a proper urban structure, its benefits also get multiplied in various aspects (Brown & Mijic, 2019; Dhyani et al., 2022; Ghofrani et al., 2017):
1. Environmental benefits: BGI can help to improve ecosystem health if utilized in sectors such as transportation, water and housing. Their incorporation in urban infrastructure planning not only benefits humans but also nature.
2. Social benefits: The design and aesthetics of the landscape can help define the character of the urban city. The aesthetic and moral values are enhanced by green streetscapes and landscapes. BGI can boost outdoor activities, lower urban temperatures and provide shelter in public areas, all of which encourage social interaction.
3. Economic benefits: Implementing blue-green initiatives in the city may benefit the local population financially. Low surface temperatures reduce the need for cooling, which lowers energy requirements because there is less demand for cooling. Green infrastructure will protect the buildings from high temperatures thus increasing the life of the building and lowering the maintenance costs also.

Implementing them can also lessen noise, improve resource efficiency, biodiversity, and amenity value, as well as manage temperature, improve air and water quality, improve flood control, and remove carbon dioxide. These all contribute to bettering urban ecosystems and human well-being. Particularly against rising temperatures and flooding, two of the main climate risks, these characteristics increase resilience to the consequences of climate change. Provisioning services, regulating services, and socio-cultural services are the three classes into which all benefits of BGIs can be divided (Kati & Jari, 2016) (Fig. 1).

Overall, it can be concluded that integration of BGI in urban infrastructure during city planning is a game-changer and win–win approach that could be helpful to save water, energy, and the environment at the same time thus promoting to achievement the SDGs.

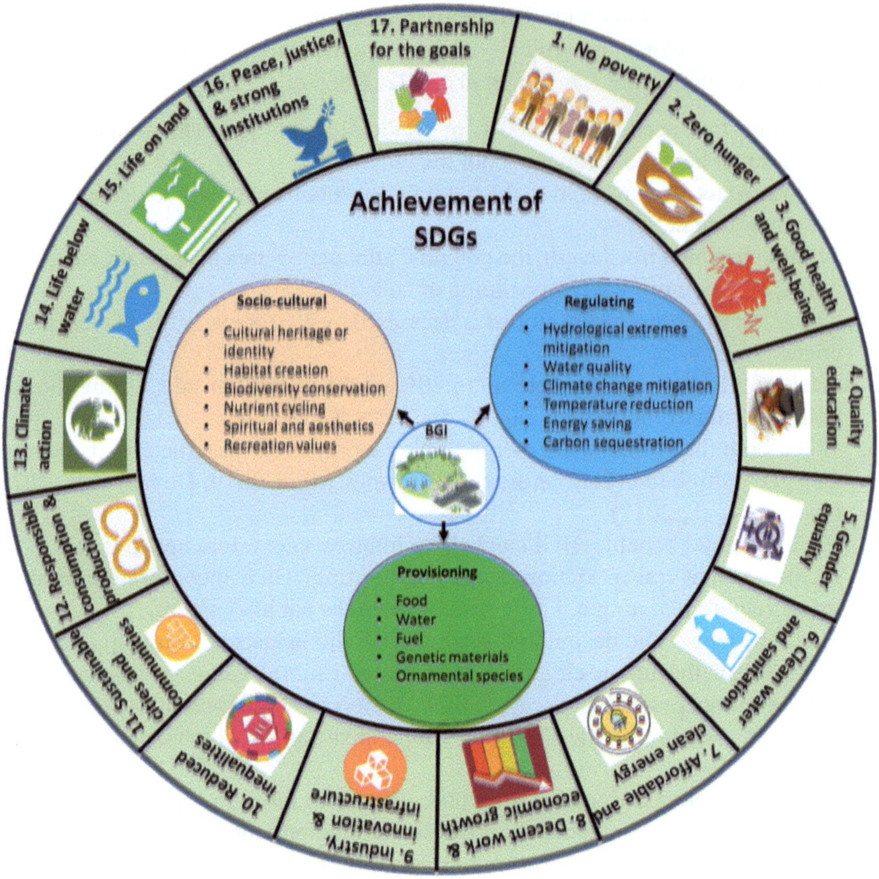

Fig. 1 Illustration of the role played by BGI in the achievement of various sustainable development goals (SDGs)(Modified). *Source* https://www.isglobal.org/en/-/sdgs-and-global-health

4 Achieving Various SDGs Targets Using BGIs

It is clear from the mentioned section that BGIs can be extremely important in achieving sustainable development. Numerous studies have been conducted recently to support this claim, some of which are summarised here and demonstrate the importance of BGIs in providing comprehensive services to humanity (Udas-Mankikar & Driver, 2021).

4.1 Mitigation of Hydrological Extremities

Climate change is rising as the most concerning issue because of its extreme results in the form of drought due to rising temperatures and flooding due to rising sea levels. The inclusion of various BGIs can help to fight these extreme conditions. These infrastructures could be categorized into three different components i.e., surface, above-surface, and sub-surface and accomplish their functioning through three processes i.e., infiltration retention, detention, and storage retention. Various infrastructures under these three components are described below which perform differently in three processes (Marks et al., 2022):

1. Surface component: Regional agriculture, bioretention cells, parks and forests, stormwater trees, permeable pavement, stormwater flow-through planters, and bioretention swales are examples of infrastructure are surface BGIs that can regulate hydrological conditions by using infiltration retention process. Further, surface detention ponds and water squares utilize the detention process while rainwater harvesting, retention storage basins, regional wetlands, and seasonal storage utilize the storage retention process.
2. Above-surface component: Similar infrastructures like green facades, green roofs, and trees are found above the surface to perform infiltration retention. While blue roofs perform detention and rainwater performs the storage retention process.
3. Sub-surface: Infrastructures like subsurface storage, subsurface storage tanks, and surface storage perform infiltration retention, detention and storage retention respectively.

4.2 Improvement of Water Quality

Many studies reported the potential and effectiveness of various BGIs in improving water quality. Implementation of BGIs removes various pollutants and excessive minerals from the water. In a study, 12 Best management practices were analyzed under 16 scenarios where they found a reduction of runoff volume decreased by 0–26.5%, total suspended particles by 0.3–53.6%, total nitrogen by 0.3–34.2%, and total phosphorus by 0.3–47.4% (Liu et al., 2016). Among different methods, it was observed that the most successful case scenario involved implementing grass strips throughout 25% of the area. Davis et al. (2010) also found that urban BGIs enhance the physiological characteristics of urban runoff along with reducing peak runoff by removing heavy metals, nutrients, hydrocarbons, and suspended solids. Odefey (2009) reported that total runoff volume and peak flow reduction are interrelated. Further, studies reported a decline in overall pollutant loading, and reduced impacts of acid rain through the utilization of green roofs (Gregoire & Clausen, 2011), which depend on the pollutant, type of vegetation, soil resources, fertilizers, and local environment. Contamination of stormwater by various natural and anthropogenic activities is also well-reported and BGIs like permeable pavements (PP) comprised plants help to remove the contamination and overall run-off volume to drainage systems. Furthermore, PP also reduces individual pollutants like total nitrogen, phosphorus, suspended solids, and various heavy metals like zinc, copper, iron, lead, cadmium, etc. by at least 50% (Fassman & Blackbourn, 2010; Zachary Bean et al., 2007). Another BGI named bioretention cells which are landscape depressions that capture and infiltrate stormwater runoff also help to reduce water pollutants and run-off volumes (Costello et al., 2020; Wardynski & Hunt, 2012).

4.3 Climate Change Mitigation

The Intergovernmental Panel on Climate Change (IPCC) (Change, 2014), states that greenhouse gas emissions and global warming are changing the earth's climate. BGIs are reported to greatly decline the heat-island effect and other specific impacts of climate change on urban areas (Bhat et al., 2017). Compared to non-urban locations, the air and surface temperatures are often warmer in metropolitan areas. Roads, paved areas, and buildings retain heat during the day, which is released at night and causes temperatures to rise (RICS, 2011). The growth of BGI boosts the amount of vegetation, which increases the amount of incident radiation that is reflected and diffused into the atmosphere, resulting in a general cooling effect (Oberndorfer et al., 2007). Furthermore, scenarios involving climate change are expected to have an impact on mean sea level rise, storm severity and frequency, and extreme weather. Adoption of BGIs can significantly aid in reducing temperature rise and its subsequent effects (Malik et al., 2018). Thus, urban planning techniques may prove to be an essential tool for city planning when the consequences of climate change are taken into account where BGI could play a vital role hence, their assessment is also important (Matthews et al., 2015).

4.4 Reduction in Temperature

There is no doubt that urban microclimates have greater temperatures, higher wind speeds, and lower rainfall amounts than either natural landscapes or rural locations (Alexandri E and Jones, 2008). Thus, the improvement of urban areas' microclimate might be seen as one of the main environmental advantages of the BGI, especially about urban forests and trees (Santamouris, 2013). Liu and Niyogi (2019) reported that urban microclimate gets affected by high temperatures caused due to urban heat island effects which result in climate changes, global warming, and severities in rainfall events. Multiple studies have indicated that green infrastructure reduces the urban heat island effect through natural processes such as evapotranspiration, humidification, and the cooling effect of plants. According to Hamada and Ohta (2010), an increase in the average amount of green cover can result in a daytime temperature reduction of around 2.3 °C. Some studies reported that in dense biological areas, in-situ cooling can reach up to 7 °C, which results in a significant drop in daytime temperatures (Onishi et al., 2010). According to Alexandri and Jones (2008), green walls and roofs can chill a space by up to 8 °C. Thus, we can conclude that extreme heat and stress-related mortality can be decreased by urban planning and BGI. The detrimental effects of heat waves on human health have been documented by several researchers. An investigation carried out in London found that for every degree above 21.5 °C, the death rate increases by as much as 3%. Increasing a city's green space can therefore have a big impact on the city's average daily temperature (Gill et al., 2007).

4.5 *Saving of Energy*

All nations are looking at strategies for lowering energy use and demand because the availability of energy is the foundation for a nation's economic development (Oberndorfer et al., 2007). In cities with a moderate climate, the contribution of green infrastructures like, green roofs, roadside trees, urban green spaces, and others has come to be seen as a viable and efficient way to lower energy costs by cooling the buildings (Bilgili & Gökyer, 2012). Urban BGI significantly contributes to climate change adaptation by reducing air and surface temperatures. Lower energy use and more thermal comfort result from the temperature drop. As evidenced by observable metrics like room temperature, energy savings, and turbulent flows, the application of BGI reduces energy usage. According to BGIs, in urban residential areas, the cost of cooling and heating has decreased by 20–50% and 10–15%, respectively (Heisler, 1986; Nowak, 2023). Similarly, a study conducted in Chicago, US also revealed the reduced total energy use for cooling by 5–10% and by 10% for heating (Wang & Chen, 2014).

4.6 *Carbon Sequestration Potential of BGIs*

The organic matter found in soil and long-lived plants that serves as a long-term fixation and storage of atmospheric carbon dioxide (CO_2) also known as carbon sequestration, is one of the beneficial ecosystem services provided by BGI (Davies et al., 2011). According to Baes et al. (1977), vegetation-rich soil is excellent at absorbing and storing atmospheric carbon and hence, they are crucial tools for reducing a person's direct or indirect carbon footprint. During the day, the urban BGI absorbs CO_2 from the atmosphere through photosynthesis. Additionally, CO_2 is eliminated, which refers to CO_2 that is trapped during the development of biomass and in subterranean soils, both of which are eventually very beneficial in lessening the consequences of climate change (Velasco & Roth, 2010). The net CO_2 sequestration in the urban environment will depend on the strength and efficiency of the source and sink. In a study carried out in Leicester City (United Kingdom), According to

Davies et al. (2011), home gardens store an average of 7.6 tC ha^{-1} in carbon, with above-ground vegetation storing 31.6 tC ha^{-1}.

4.7 Habitat Improvement, Recreation, and Aesthetics

A greater amount of vegetation is present as a result of BGI development in metropolitan areas, supporting a wide range of plants and wildlife along with the provision of comfort. Green space growth and improved habitat are directly related. Native species put in BGI components like rain gardens can improve performance even further. There are two ways that habitat creation in parks and other natural areas might benefit the environment. It first offers migratory birds, insects, and other small creatures a place to live and nourishment. Its habitat nursery role, which is to provide space for species that need it while they grow and establish, is its second purpose. An economic analysis of habitats is carried out using the market worth of the products harvested there (Balany et al., 2020; Perrotti & Iuorio, 2019).

Cities that have been thoughtfully developed and have a lot of green space are more attractive to visitors, residents, and investors. The aesthetic value of neighborhoods rises when there is more greenery in urban settings. The well-documented correlation between urban greening and property value can be used as evidence for the beneficial effects of green infrastructure measures on aesthetics. An increase in green space enhances metropolitan areas' aesthetic value greatly. Urban BGI initiatives have a favorable impact on aesthetics, which is reflected in the well-established relationship between urban greening and property value. People prefer to live in regions with more green spaces, thus a property buyer would be willing to pay more for an environmentally friendly and visually beautiful home. Due to this, properties next to urban woods are valued higher.

Numerous studies reveal an increase in property value in regions encircled by trees and other greenery. According to a study, adding a tree to the front of a property can improve its value by 2–10% (Wachter & Wong, 2008). Another study done in Portland, Oregon, found that street trees decrease the time a residential property spends on the market by 1.7 days and raise the selling price of the property by US $8870 (Donovan & Butry, 2009). A thorough investigation into the advantages of green infrastructure in Philadelphia also examines how it affects property values. While property prices have increased significantly in BGI and green space areas, it is challenging to quantify the aesthetic gains and prevent double counting of advantages such as enhanced air and water quality, energy efficiency, and flood management (Ghods et al., 2023; Pandit & Laband, 2010).

5 Existing Global and Indian Practices and Projects

Numerous nations throughout the world have adopted and studied urban BGI techniques. Based on the scale of the implementation these BGI projects can be categorized into three classes i.e., microscale, mesoscale and macroscale (Haghighatafshar et al., 2018). The application of BGI at various scales in various nations is discussed in this section. Here we will briefly explain the BGI implementation at mesoscale using case studies. The projects were chosen to show a wide range of strategies and methods from various angles. These studies demonstrate the advantages and disadvantages of BGI planning and implementation while illuminating the present planning methodology. The examination is concentrated on the BGI's green and blue parts working in concert. In order to do this, the project's current vegetation, water flow patterns, and surrounding environment were investigated to spot potential interfaces and optimization opportunities.

For instance, as an illustration of an urban-scale BGI project, Potsdamer Platz in Berlin and the High Line in New York City were both suggested, and both initiatives address the conditions of their respective enormous urban areas (see for detail Well and Ludwig, 2020). But there are various problems raised by these undertakings. On one hand, the High Line project concentrates on the BGI adoption's aesthetic benefits as an example of green infrastructure. On the other hand, in the Potsdamer Platz project, In the crowded inner-city area, decentralized drainage has been made possible by the deployment of a rainwater management system (Fig. 2). Further, along with the implementation of High Line Project using BGI techniques the pre-existing trees were replanted to turn the grey cemented infrastructure into green. Therefore, plants that could grow well in thin soil with a meager supply of nutrients and minerals and a high tolerance for drought and natural conditions of the US East Coast were chosen for the implementation in this project (Nyman, 2019). According to Friends of the High Line (2018), the design demonstrates a new blurring of the lines between architecture, environment, and infrastructure. It was observed that the porous designs of the walkways were able to absorb about 90% of the rainwater, making enough water accessible to irrigate plant beds (Nyman, 2019). From the aforementioned justifications, it can be concluded that the urban BGI strategy served as the foundation for the design and implementation of the High Line project. Yet the water management idea was only a supporting element in the project design, which was primarily concerned with producing a lovely and inviting green space. In the absence of such, adjacent buildings might have served as water sources (gray water, rainfall). Therefore, the whole design was not built upon the creation of a networked notion.

Thus, the improvement of aesthetic quality and recreational value were the key design strategies. The project was created to regulate the demand for sustainable development to some level while keeping the derivation of water benefits as low as feasible. The concept's resource-saving strategy might be acceptable up to a point, however, a low water requirement lowers evapotranspiration, which reduces its cooling effect on metropolitan systems.

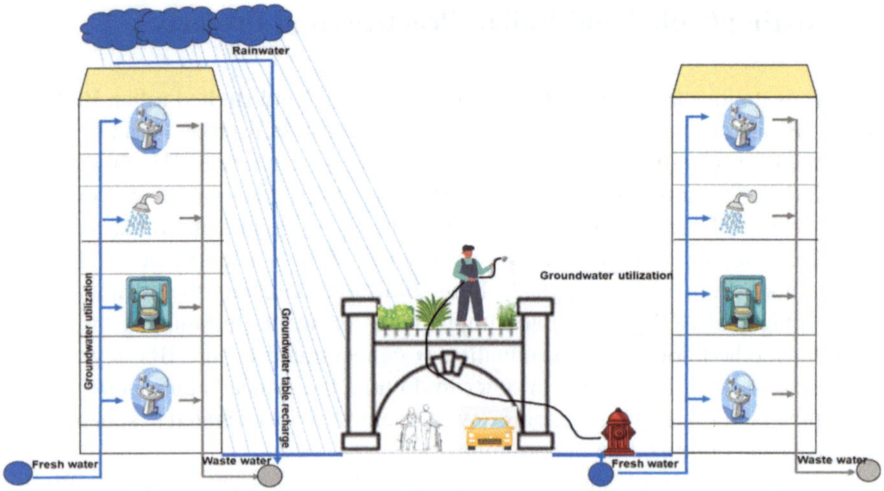

Fig. 2 High Line project in New York City: The vegetation concept has a low water requirement (*Source* Modified Well & Ludwig, 2020)

Now let's discuss about the Potsdamer Platz Square project in Berlin, Germany which was redesigned in the early 1990s (see for detail Well and Ludwig, 2020). According to Grant et al. (2012), water was incorporated to achieve zero discharge from rains and for this combination of a variety of retention strategies and various runoff water retention technologies were needed (Fig. 3). It included the green roofs with a surface area of 12,000 m^2 which causes immediate evapotranspiration of more than 50% of the rainwater. Then, the collected runoff was placed in five cisterns with a combined 2600 m^3 water holding capacity. The newly developed artificial water zones offer an additional 1300 m^3 of water storage capacity because of the fluctuating water levels (Dreiseitl & Grau, 2006). The improvement of water quality is a further advantage that these reservoirs may offer. The planted biotopes built around water bodies can filter debris and improve ecological harmony (Well & Ludwig, 2020). Only three times every 10 years is it possible to discharge huge amounts of extra water into the neighboring Landwehr Canal. According to Hiltrud and Pierre (2011), this amount is equivalent to the value of an unsealed site. Thus, nearly 85% of the total amount of precipitation that is received each year is either evaporated, retained, or flows off. The remaining 15% of the collected rainwater can be used for a variety of tasks, such as flushing toilets in offices or watering plants (Well & Ludwig, 2020).

Despite BGI's many advantages and growing popularity in smart cities, the potential for synergistic links has not yet been investigated. As a result, the gray water is not included in the water notion. The requirement of high construction costs is the key drawback for the adoption of BGI solutions for flood protection and water management. The required network of vast water storage facilities and related connections has not yet become the norm. Hence, a combination of storage volume reduction and the benefits of a microclimate can therefore be used in these circumstances.

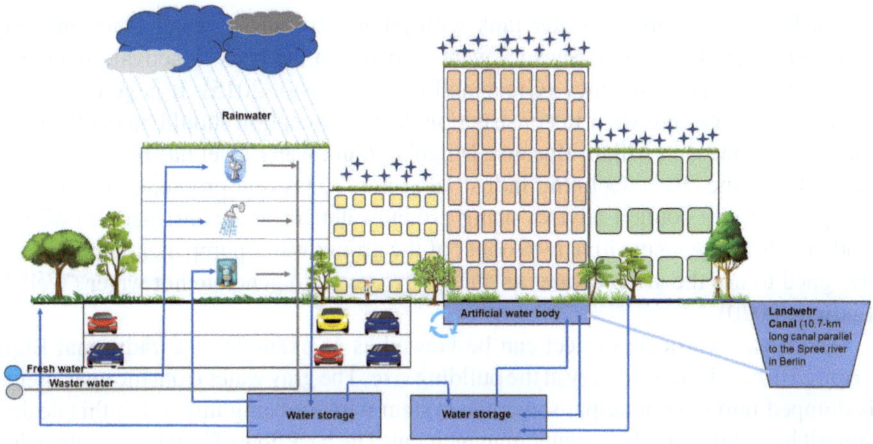

Fig. 3 Potsdamer Platz square project in Berlin, Germany: By integrating several water retention technologies, the rainwater management system achieves 0% runoff. Where, precipitation is gathered and evaporated by man-made blue and green infrastructures. (*Source* Modified Well & Ludwig 2020)

Another case study is based in the Italian city of Milan, where the Bosco Verticale project was finished in 2015 (see for detail Well and Ludwig, 2020). According to Ishween (2021), the project aims to provide sustainable housing, reforest, and naturalize the city. To produce a dense canopy for the two towers constructed for these purposes, some 20,000 plants—including 700 trees—were planted throughout the project (Giacomello & Valagussa, 2015). For this particular project, the selection of the trees for the plantation was done based on endurance to temperature changes of range −1 and 31 °C and also based on a certain standard for heights to withstand wind loads. To ensure their successful relocation to the new site (Housing Towers), young tree saplings were pre-gowned (Ishween, 2021; Tokuç & İnan, 2017). The plan sought to offer adequate green cover and the best possible shading. Therefore, maximum shading and a suitable amount of green cover were the goals of the design. The project's sole focus was increasing the amount of vegetation, and to that end, the idea of water management was also modified (Fig. 4). There are various advantages to vertical gardening and plantations, according to research. The project had several advantages, but it also helped keep the buildings' energy balance positive. The premise, however, was unable to be independently verified. Rather than using experimental observations, Giacomello and Valagussa (2015) reported data based on assumptions and models. Solar energy from the plantation is balanced to the annual average using a façade simulation. Thus, the energy savings apply equally to both flat façades and to buildings with unplanted balconies alone (Giacomello & Valagussa, 2015). These computations were used to estimate the required water volume, and the central system's groundwater was automatically pumped for irrigation. The notion of using gray water to irrigate plants was also entertained, although it was eventually abandoned (Well & Ludwig, 2020). The unique water distribution system for each tower includes a water tank with pumps to transfer water from one part of the structure to another. Sensors were placed into the beds to activate irrigation to maximize water efficiency (Giacomello & Valagussa, 2015). It is generally not advised to use groundwater in construction. But an exception should be made due to the circumstances in Milan. Because Milan's groundwater level has been rising for years due to the city's declining industrial sector. To prevent problems like flooded garages, the city council advises using groundwater for new construction (Well & Ludwig, 2020). As a positive byproduct of the groundwater pump, heat pumps were designed to use the groundwater as an energy source to generate hot water (Well & Ludwig, 2020).

The Bosco Verticale project can be viewed as an example of a traditional BGI-inspired urban design strategy at the building size. The gray water from the apartments is dumped into the municipal sewerage system without being utilized in this design since it lacks integrated rainwater management. The backflow of irrigation water adds

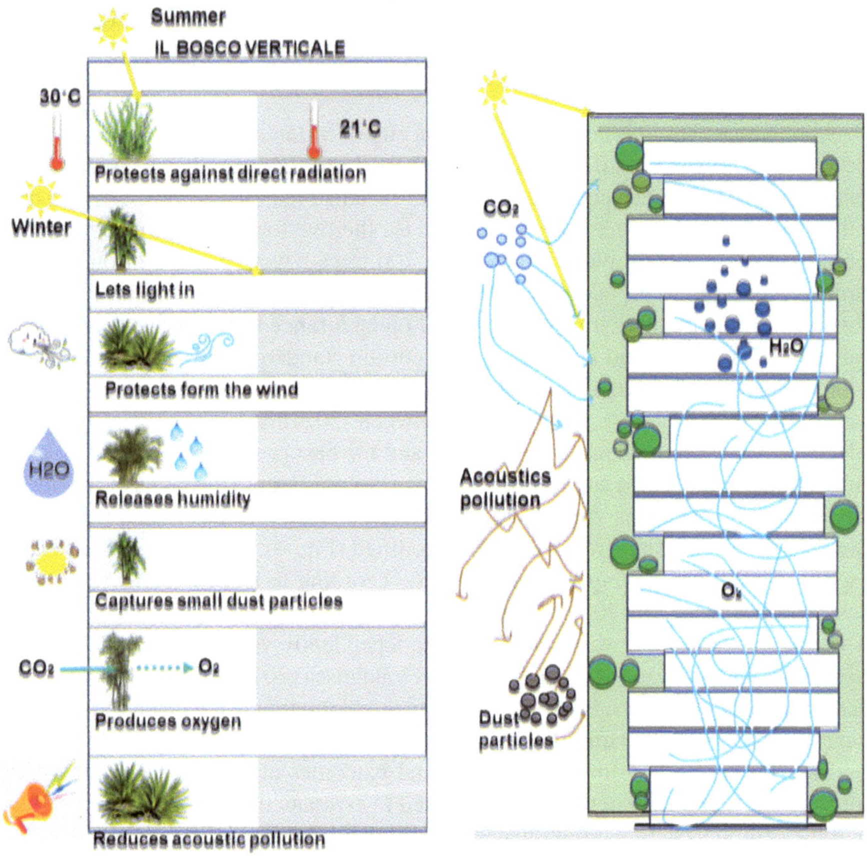

Fig. 4 Irrigation in Bosco Verticale and the life cycle concept of Bosco Verticale (*Source Modified from Stefano Boeri Architetti official website and* Ishween, 2021)

to the volume of wastewater and puts a strain on the city's sewerage infrastructure. This argument contends that rather than reducing the burden on the central system, the buildings generate a large amount of wastewater.

Our next case study is from China, whose expanding population and resource scarcity have increased the urgency of the need for sustainable urban development. When preparing for urban development projects, keeping the usage of water and other resources in balance is crucial (Sarkar et al., 2018). A cutting-edge approach for managing rainwater is the sponge cities of China (Nie & Jia, 2018). In central, southern, and eastern China, there are now 26 provinces with sponge cities (Zhou, 2015). Water harvesting structures, green rooftop areas, the use of water-intensive vegetation, porous roads and pavements, and the creation of basins like artificial lakes and wetlands to catch excess runoff are all features of sponge city designs. These elements support the entire urban water management system by regulating the runoff peak, total runoff, and pollutants through infiltration, interception, storage, and filtration processes (Eckart et al., 2017). Sponge cities replicate and create natural hydrological conditions by maximizing infiltration capacity, thereby bringing the urban hydrological cycle as close to nature as is practical. Sponge Cities struggle with several issues, according to Zhang et al. (2019), including ineffective management, a dearth of studies on cooperative urban management, exorbitant management costs, and the feasibility of local urban cooperation. The control of sources of waste from green space and rainfall, as well as technology from best management practices (BMPs), are the main ways that low-impact development design is accomplished. Improved green rainwater infrastructure is essential for building better rain cities, rain villages, and water-cycle communities. China's Sponge City (Ma et al., 2020) eloquently illustrates how nature absorbs, drains, and reuses rainwater. Sponge City aims to enhance urban planning and management through the complete utilization of green spaces, water systems, roads, buildings, and ecosystems. The collected rainwater is then released gradually to facilitate natural accumulation, infiltration, improvement of water quality, and stormwater runoff (Fig. 5) (Hopkins et al., 2018). Stormwater management initiatives are implemented in China according to a top-down strategy. When implementing various BGI projects, cooperation between organized civic societies and the many implementing authorities is lacking. Even more obstacles stand in the way of these initiatives' public participation. Less frequently, and usually at a later stage, the public is involved. Furthermore, since choices are made top-down, moving from the central administration to the municipal, then the local government (Zhou, 2015), the general public's viewpoint has little influence on the decision-making process. This implies that earlier in the project's development, higher public participation can be easily attained.

Now we'll discuss the case study from India, according to Parikh et al. (2016), Indian cities are expanding in terms of both area and population, and they are also increasingly susceptible to climate extremes like floods, heat waves, and cyclones. India's cities experienced a 30% population growth rate between 2001 and 2011. Low standards for service delivery and infrastructure development characterize the peri-urban area that is in the process of moving from a rural to an urban environment. According to the World Bank (2013), India's cities, towns, and municipalities

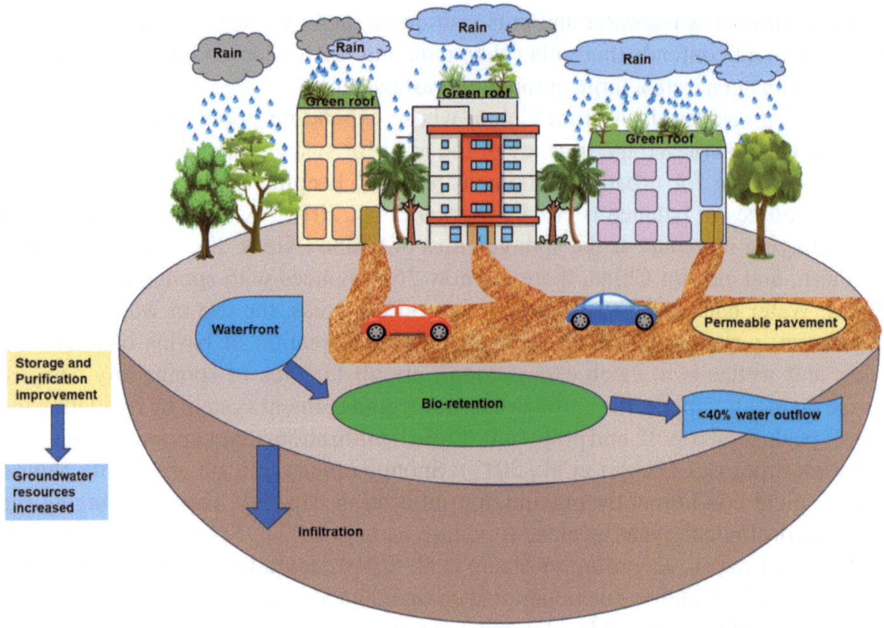

Fig. 5 Illustration of the Sponge city concept (*Source* Modified from Nei & Jia, 2018)

all have significant flaws in their land use planning and building bylaws. Urban India has a significant BGI adoption rate, but ecology and land use are not given the proper consideration they deserve while developing urban infrastructure, even though they might be extremely important in reducing the risks associated with natural disasters (Govindarajulu, 2014, Bhadouria et al. 2022). Thus, engineered or gray infrastructure, which contributes to problems like flash floods and poor water quality in metropolitan areas, has received the majority of attention during India's infrastructure development. Gray infrastructure is presently under scrutiny because of the ineffective transformation of natural capital into financial capital, which worsens a country's social capital and damages the environment. According to recent global assessments, green and blue infrastructure, such as forests and lakes, as well as sustainable infrastructure (SI), that are continuously being depleted or ignored are the best solutions to deal with both economic and environmental issues, are SI. The world's leading intergovernmental agency, UNEP, views the SI as a path towards sustainable development and provides methods for monitoring the shift to a green economy. Therefore, to reduce future losses from major climatic and natural catastrophes, enough money must be invested in risk-reduction measures. Adequate investments are also justified in light of the elevated severity and regularity of such extreme occurrences predicted by future climatic projections (Field & Barros, 2014). In India, building urban resilience has to be a top priority in order to promote sustainable urban development and lessen financial losses brought on by catastrophic weather events. A significant paradigm change has occurred with the adoption of the SDG by the

United Nations and the World Bank's switch from infrastructure action plans to the Sustainable Infrastructure Action Plan (SIAP) starting in 2009. By reducing reliance on energy-intensive gray infrastructure, sustainable infrastructure can assist address global problems like rising sea levels and high atmospheric carbon concentrations, among others. There aren't many studies that focus primarily on urban resilience that have been done in India, especially when it comes to improving green infrastructures, dealing with land use issues, getting ready for potential disasters caused by climate change (Kumar et al., 2016), and having strong institutional response mechanisms like early warning systems and disaster management plans. Indian urban governance has advanced significantly since 1994 and the Urban Local Bodies (ULB) now have more control over municipal budgets and governance. Urban governance changes gave ULBs new authority over a variety of tasks, including controlling land use, urban planning, delivering essential services, and enhancing green infrastructure to increase urban resilience (Govindarajulu, 2020). The ULBs have the authority to impose taxes as well, and the money they raise may be used to fund infrastructure and service provision in cities. Therefore, there is a large window of opportunity for the adoption and application of BGI approaches, if the local government is supportive and the technology is made widely known among the populace.

6 Limitations and Recommendations for BGIs

The limitations faced by BGIs over their entire existence vary in severity. A successful BGI deployment requires consideration of advice from experts, users, and other target groups. To deal with the many difficulties related to BGI implementation, certain nations (including the USA and India) have established certification schemes. The implementation of a certification program that provides practitioners with continual training enhances urban resilience. To get over regulatory roadblocks, the implementing agencies ought to set up regulatory frameworks that highlight BGI's complexity. To successfully execute BGI, which is extremely complex and challenging, there must be collaboration across various levels of government. The answer to this problem may lie in the establishment of new, well-funded regulatory bodies that lack complicated organizational structures.

Successful implementation necessitates a people-centered strategy based on prior research (Webler & Tuler, 2006). Research on climate change and sustainable development can address the issue of socioeconomic challenges by showing that political will is a necessary but distant prerequisite for tackling all five concerns coherently and thoroughly (Oulahen et al., 2018).

According to Campbell et al. (2016), an inclusive process including stakeholders at the community and organizational levels may be necessary for the successful implementation of BGI. Compared to gray infrastructure projects, BGI efforts might be less expensive. however, this does not increase their popularity or economic impact. These include the challenge of calculating the environmental advantages, the prohibitive maintenance costs for low-income families, rules set by the government, and public resistance to BGI adoption. Depending on the country of implementation, different obstacles are encountered in the development and adoption of BGI. Administrative, political, economic, and societal obstacles could exist. In Table 1, there is a list of the challenges involved in the creation, growth, and upkeep of BGI. Before starting a BGI project, it is important to identify any potential obstacles that can arise so that decision-makers can make the necessary plans. Decision-makers can evaluate projects according to how flexible and applicable they are throughout their whole life cycle by using the system to identify impediments. Experts should determine the possibility of encountering obstacles and their level of severity. It would be adequate to compare the projects using guidance from engineers, legislators, and local management. Involving all pertinent stakeholders at the outset of the BGI project is typically advised to avoid and identify any potential bottlenecks as early as possible.

7 Conclusion and Future Perspectives

In order to help planners to become more aware of potential obstacles and difficulties encountered during BGI implementation. This chapter reviewed a variety of studies and important information about BGI technologies. The act of identifying the barriers, however, is challenging in and of itself because it depends on numerous socioeconomic and environmental contexts. In the stages from conception to distribution to application, with a focus on various challenges, the tactics and procedures of BGI require additional research and improvement. It is important to learn more about the factors that influence the performance and uptake of BGIs. Once the underlying causes or factors have been thoroughly understood, potential issues and flaws in BGI can be rectified. It might be possible to introduce BGI to the public by designing it with the upcoming issues brought on by climate change in mind. It is important to emphasize studies on the use of BGI in real-world settings while also advancing the science. Although the benefits of BGI in urban water management (in flood hazard prevention and water quality improvement) are not new, further research must be done to assess the advantages of BGI as environmental management for urban biodiversity and heat reduction instead of potential future hazards from climate change in addition to the cultural benefits. Building integrated solutions and identifying drivers and associated impediments in the context of BGIs could significantly increase urban resilience and sustainability. Consequently, a better understanding of BGI services would lead to more efficient designs for BGI systems, which would subsequently assist in their promotion through the appropriate use of evidence. Based on the above-mentioned points, the future perspectives of BGI are as follows:

1. Implementation of BGI in urban areas should also include the analysis of the case studies demonstrating the significance of scientific data and local community involvement in resilient transformation processes.
2. Planning for land use, urban infrastructure, and the ability of the local government to reduce the risk of disasters were all lacking therefore, these gaps can be filled.

Table 1 List of some of the barriers for the development of blue-green infrastructure (*Source* Modified from Deely et al., 2020)

1. Negative past experience	15. Funding and costs	29. No long-term vision, strategy, monitoring and/or evaluation	43. Uncertainty over performance and cost
2. Low priority and/or competing priorities	16. Who pays	30. Including climate change and environmental science in policy/ design standards	44. Impacts of natural hazards arising from
3. Political leadership and champions	17. Lack of knowledge, education and/or awareness	31. Keeping BGI on the political agenda	45. Retrofitting grey infrastructure
4. Securing effective leadership/effective future	18. Lack of awareness of BGI multi-functionality	32. Equity issue i.e., who are winners and losers	46. Urban morphology
5. Legislation, regulations and governance	19. Technocratic path dependence	33. Understanding potential stakeholders	47. Impact on existing ecosystems
6. How to change, need to change policy to support	20. Uncertainty over the impacts of climate change	34. Lack of human resources	48. Construction length, Length of time to benefits
7. Responsibilities and ownership	21. which are the appropriate responses to climate change	35. Lack of knowledge around the impacts of grey Infrastructure	49. Lack of identifiable examples in a geographically similar area
8. Institutional capacity and expertise	22. Public preferences, Willingness to pay for BGI	36. lack of maintenance standards/ difficulties maintaining BGI	50. Identifying and quantifying, monetising the multiple benefits
9. Behaviours and culture	23. Lack of Community buy-in, public engagement	37. Grey vested policies	51. Monitoring and Control
10. Interagency fragmentation/ Issues with partnership Working	24. Lack of community empowerment	38. Lack of available data	52. Lack of financial Incentives e.g. government subsidies
11. Promoting interagency work	25. Future land use	39. Physical characteristics of the site/sites	53. Transaction Cost

(continued)

Table 1 (continued)

12. Poor communication and knowledge sharing	26. Lack of landowner buy-in	40. Lack of available space	54. Price Setting
13. Using climate science in policymaking	27. Lack of developmental and procedural guidance	41. Physical science, engineering uncertainties	55. Lack of Additionality
14. Unclear roles	28. Lack of political will	42. Uncertainty over future infrastructure and maintenance requirements	56. Free rider problem

3. This chapter gives the complete idea regarding benefits, challenges, and structural framework along with case studies regarding BGI and it is required to reassess the planning tools in light of a climate-resilient future.
4. To facilitate the incorporation of adaptation into urban planning, it is necessary to enhance methods, expand knowledge regarding novel community engagement models, and cooperate with municipal technical staff.

References

Alexandri, E., & Jones, P. (2008). Temperature decreases in an urban canyon due to green walls and green roofs in diverse climates. *Building and Environment, 43*(4), 480–493. https://doi.org/10.1016/j.buildenv.2006.10.055

Baes, C. F., Goeller, H. E., Olson, J. S., & Rotty, R. M. (1977). Carbon dioxide and climate: the uncontrolled experiment: Possibly severe consequences of growing CO_2 release from fossil fuels require a much better understanding of the carbon cycle, climate change, and the resulting impacts on the atmosphere. *American Scientist, 65*(3), 310–320. https://doi.org/10.2172/7145150

Balany, F., Ng, A. W., Muttil, N., Muthukumaran, S., & Wong, M. S. (2020). Green infrastructure as an urban heat island mitigation strategy—a review. *Water, 12*(12), 3577. https://doi.org/10.3390/w12123577

Bhadouria, R., Upadhyay, S., Tripathi, S., & Singh, P. (Eds.). (2022). *Urban ecology and global climate change*. John Wiley & Sons, Incorporated.

Bhadouria, R., Tripathi, S., Singh, P., Joshi, P.K., & Singh, R. (2023). Urban Metabolism and Global Climate Change: An Overview. *Urban Metabolism and Climate Change: Perspective for Sustainable Cities*, pp. 3–22. https://doi.org/10.1007/978-3-031-29422-8_1

Bhat, S.A., Pandit, B.A., Khan, J.N., Kumar, R., & Jan, R. (2017). Water requirements and irrigation scheduling of maize crop using CROPWAT model. *International Journal of Current Microbiology and Applied Sciences, 6*(11), 1662–1670. https://doi.org/10.20546/ijcmas.2017.611.199

Bilgili, B. C., & Gökyer, E. (2012). Urban green space system planning. *Landscape Planning, 360*, 107–123. https://doi.org/10.5772/45877

Brown, K., & Mijic, A. (2019). Integrating green and blue spaces into our cities: Making it happen. *Grantham Institute, 30*, 1–10. chrome-extension://efaidnbmnnnibpcajpcglclefindmkaj/https://www.imperial.ac.uk/media/imperial-college/grantham-institute/public/publications/briefing-papers/Integrating-green-and-blue-spaces-into-our-cities---Making-it-happen-.pdf

Brundtland, G.H., (1987). What is sustainable development? *Our Common Future, 8*(9), 1–40. Chrome-extension://efaidnbmnnnibpcajpcglclefindmkaj/https://www.latrobe.edu.au/__data/assets/pdf_file/0005/554927/Sustainability-Plan-2013-2017.pdf

Campbell, L. K., Svendsen, E. S., & Roman, L. A. (2016). Knowledge co-production at the research–practice interface: Embedded case studies from urban forestry. *Environmental Management, 57*, 1262–1280. https://doi.org/10.1007/s00267-016-0680-8

Change, I.P.O.C. (2014). Ipcc. *Climate change*.

Costanza, R., Daly, L., Fioramonti, L., Giovannini, E., Kubiszewski, I., Mortensen, L. F., Pickett, K. E., Ragnarsdottir, K. V., De Vogli, R., & Wilkinson, R. (2016). Modelling and measuring sustainable wellbeing in connection with the UN Sustainable Development Goals. *Ecological Economics, 130*, 350–355. https://doi.org/10.1016/j.ecolecon.2016.07.009

Costello, D. M., Hartung, E. W., Stoll, J. T., & Jefferson, A. J. (2020). Bioretention cell age and construction style influence stormwater pollutant dynamics. *Science of the Total Environment, 712*, 135597. https://doi.org/10.1016/j.scitotenv.2019.135597

Davies, Z. G., Edmondson, J. L., Heinemeyer, A., Leake, J. R., & Gaston, K. J. (2011). Mapping an urban ecosystem service: Quantifying above-ground carbon storage at a city-wide scale. *Journal of Applied Ecology, 48*(5), 1125–1134. https://doi.org/10.1111/j.1365-2664.2011.02021.x

Davis, A. P., Traver, R. G., & Hunt, W. F. (2010). Improving urban stormwater quality: Applying fundamental principles. *Journal of Contemporary Water Research and Education, 146*(1), 3–10. https://doi.org/10.1111/j.1936-704x.2010.00387.x

De la Poza, E., Merello, P., Barberá, A., & Celani, A. (2021). Universities' reporting on SDGs: Using the impact rankings to model and measure their contribution to sustainability. *Sustainability, 13*(4), 2038. https://doi.org/10.3390/su13042038

Deely, J., Hynes, S., Barquín, J., Burgess, D., Finney, G., Silió, A., Álvarez-Martínez, J. M., Bailly, D., & Ballé-Béganton, J. (2020). Barrier identification framework for the implementation of blue and green infrastructures. *Land Use Policy, 99*, 105108. https://doi.org/10.1016/j.landusepol.2020.105108

Dhyani, S., Basu, M., Santhanam, H., & Dasgupta, R. (2022). Blue-green infrastructure across Asian countries: Improving urban resilience and sustainability. *Springer*. https://doi.org/10.1007/978-981-16-7128-9

Donovan, G. H., & Butry, D. T. (2009). The value of shade: Estimating the effect of urban trees on summertime electricity use. *Energy and Buildings, 41*(6), 662–668. https://doi.org/10.1016/j.enbuild.2009.01.002

Dreiseitl, H., & Grau, D. (2006). *Wasserlandschaften: Planen*. De Gruyter.

Eckart, K., McPhee, Z., & Bolisetti, T. (2017). Performance and implementation of low impact development–A review. *Science of the Total Environment, 607*, 413–432. https://doi.org/10.1016/j.scitotenv.2017.06.254

Elbeltagi, A., Azad, N., Arshad, A., Mohammed, S., Mokhtar, A., Pande, C., Etedali, H. R., Bhat, S. A., Islam, A. R. M. T., & Deng, J. (2021). Applications of Gaussian process regression for predicting blue water footprint: Case study in Ad Daqahliyah, Egypt. *Agricultural Water Management, 255*, 107052. https://doi.org/10.1016/j.agwat.2021.107052

Fassman, E. A., & Blackbourn, S. (2010). Urban runoff mitigation by a permeable pavement system over impermeable soils. *Journal of Hydrologic Engineering, 15*(6), 475–485. https://doi.org/10.1061/(asce)he.1943-5584.0000238

Field, C. B., & Barros, V. R. (2014). Climate change 2014–Impacts, adaptation and vulnerability: Regional aspects. *Cambridge University Press*, https://doi.org/10.1017/cbo9781107415379

Friends of the High Line, (2018). *The High Line; Sustainable Practices*. Retrieved on 18 09 2023 from https://www.thehighline.org/sustainable-practices/

Fuso Nerini, F., Tomei, J., To, L. S., Bisaga, I., Parikh, P., Black, M., Borrion, A., Spataru, C., Castán Broto, V., Anandarajah, G. & Milligan, B. (2018). Mapping synergies and trade-offs between energy and the sustainable development goals. *Nature Energy, 3*(1), 10–15.

Ghods, M., Aghamohammadi Zanjirabad, H., Vafaeinejad, A., Behzadi, S., & Gharagozlo, A. (2023). Optimizing the shade coverage of trees on a block of residential buildings using GIS and ACO (case study: Semnan, Iran). *International Journal of Environmental Science and Technology, 20*(1), 489–502. https://doi.org/10.1007/s13762-022-04537-6

Ghofrani, Z., Sposito, V., & Faggian, R. (2017). A comprehensive review of blue-green infrastructure concepts. *International Journal of Environment and Sustainability, 6*(1). https://doi.org/10.24102/ijes.v6i1.728

Giacomello, E., & Valagussa, M. (2015). *Vertical greenery: Evaluation the High-Rise Vegetation of the Bosco Verticale.* Council of Tall Buildings and Urban Habitat.

Gill, S. E., Handley, J. F., Ennos, A. R., & Pauleit, S. (2007). Adapting cities for climate change: The role of the green infrastructure. *Built Environment, 33*(1), 115–133. https://doi.org/10.2148/benv.33.1.115

Gómez-Baggethun, E., & Naredo, J. M. (2015). In search of lost time: The rise and fall of limits to growth in international sustainability policy. *Sustainability Science, 10*, 385–395. https://doi.org/10.1007/s11625-015-0308-6

Govindarajulu, D. (2014). Urban green space planning for climate adaptation in Indian cities. *Urban Climate, 10*, 35–41. https://doi.org/10.1016/j.uclim.2014.09.006

Govindarajulu, D. (2020). Strengthening institutional and financial mechanisms for building urban resilience in India. *International Journal of Disaster Risk Reduction, 47*, 101549. https://doi.org/10.1016/j.ijdrr.2020.101549

Grant, S. B., Saphores, J. D., Feldman, D. L., Hamilton, A. J., Fletcher, T. D., Cook, P. L., Stewardson, M., Sanders, B. F., Levin, L. A., Ambrose, R. F., & Deletic, A. (2012). Taking the "waste" out of "wastewater" for human water security and ecosystem sustainability. *Science, 337*(6095), 681–686. https://doi.org/10.1126/science.1216852

Gregoire, B. G., & Clausen, J. C. (2011). Effect of a modular extensive green roof on stormwater runoff and water quality. *Ecological Engineering, 37*(6), 963–969. https://doi.org/10.1016/j.ecoleng.2011.02.004

Haghighatafshar, S., Nordlöf, B., Roldin, M., Gustafsson, L. G., la Cour Jansen, J., & Jönsson, K. (2018). Efficiency of blue-green stormwater retrofits for flood mitigation–Conclusions drawn from a case study in Malmö, Sweden. *Journal of Environmental Management, 207*, 60–69. https://doi.org/10.1016/j.jenvman.2017.11.018

Hamada, S., & Ohta, T. (2010). Seasonal variations in the cooling effect of urban green areas on surrounding urban areas. *Urban Forestry & Urban Greening, 9*(1), 15–24. https://doi.org/10.1016/j.ufug.2009.10.002

Hamann, F., Blecken, G. T., Ashley, R. M., & Viklander, M. (2020). Valuing the multiple benefits of blue-green infrastructure for a Swedish case study: Contrasting the economic assessment tools B£ ST and TEEB. *Journal of Sustainable Water in the Built Environment, 6*(4), 05020003. https://doi.org/10.1061/jswbay.0000919

Heisler, G.M. (1986). Energy savings with trees. *Journal of arboriculture (USA).* https://doi.org/10.48044/jauf.1986.026

Henderson, K., & Loreau, M. (2023). A model of Sustainable Development Goals: Challenges and opportunities in promoting human well-being and environmental sustainability. *Ecological Modelling, 475*, 110164. https://doi.org/10.1016/j.ecolmodel.2022.110164

Hiltrud, P., & Pierre, B. (2011). Urban Green-blue Grids for Sustainable and Dynamic Cities. *Coop for Life, Delft.*

Hopkins, K. G., Grimm, N. B., & York, A. M. (2018). Influence of governance structure on green stormwater infrastructure investment. *Environmental Science & Policy, 84*, 124–133. https://doi.org/10.1016/j.envsci.2018.03.008

Ishween, A. (2021). Grays to greens: A place where humans and nature coexist A case study of Bosco Verticale, Milan, India. *Descriptio, 3*(1). https://journalsacfa.apeejay.edu/index.php/descriptio/article/view/5

Kapoor, V., Tripathi, S., Devi, R.S., Srivastava, P. and Bhadouria, R., 2020. Ecological economics of an urban settlement: an overview. Urban Ecology, pp.91–110.

Kati, V., & Jari, N. (2016). Bottom-up thinking—Identifying socio-cultural values of ecosystem services in local blue–green infrastructure planning in Helsinki, Finland. *Land Use Policy, 50*, 537–547. https://doi.org/10.1016/j.landusepol.2015.09.031

Kopp, J., Frajer, J., Lehnert, M., Kohout, M., & Ježek, J. (2021). Integrating concepts of blue-green infrastructure to support multidisciplinary planning of sustainable cities. *Problemy Ekorozwoju, 16*(2), 137–146. https://doi.org/10.35784/pe.2021.2.14

Kumar, P., Geneletti, D., & Nagendra, H. (2016). Spatial assessment of climate change vulnerability at city scale: A study in Bangalore, India. *Land Use Policy, 58*, 514–532. https://doi.org/10.1016/j.landusepol.2016.08.018

Langemeyer, J., & Baró, F. (2021). Nature-based solutions as nodes of green-blue infrastructure networks: A cross-scale, co-creation approach. *Nature-Based Solutions, 1*, 100006. https://doi.org/10.1016/j.nbsj.2021.100006

Liu, J., & Niyogi, D. (2019). Meta-analysis of urbanization impact on rainfall modification. *Scientific Reports, 9*(1), 7301. https://doi.org/10.1038/s41598-019-42494-2

Liu, W., Chen, W., Feng, Q., Peng, C., & Kang, P. (2016). Cost-benefit analysis of green infrastructures on community stormwater reduction and utilization: A case of Beijing. *China Environ. Manag., 58*, 1015–1026. https://doi.org/10.1007/s00267-016-0765-4

Ma, Y., Jiang, Y., & Swallow, S. (2020). China's sponge city development for urban water resilience and sustainability: A policy discussion. *Science of the Total Environment, 729*, 139078. https://doi.org/10.1016/j.scitotenv.2020.139078

Malik, S., Bano, H., Rather, R.A., & Ahmad, S. (2018). Cloud seeding; its prospects and concerns in the modern world-A review. *Int. J. Pure App. Biosci, 6*(5), 791–796. https://doi.org/10.18782/2320-7051.6824

Marks, N. K., Hosseiny, H., Bill, V. P., Ahn, K. L., Crimmins, M. C., Kremer, P., & Smith, V. B. (2022). Spatial integration of urban runoff modeling, heat, and social vulnerability for blue-green infrastructure planning and management. *Journal of Water Resources Planning and Management, 148*(11), 05022007. https://doi.org/10.1061/(asce)wr.1943-5452.0001593

Matthews, T., Lo, A. Y., & Byrne, J. A. (2015). Reconceptualizing green infrastructure for climate change adaptation: Barriers to adoption and drivers for uptake by spatial planners. *Landscape and Urban Planning, 138*, 155–163. https://doi.org/10.1016/j.landurbplan.2015.02.010

Mohtar, W. H. M. W., Abdullah, J., Abdul Maulud, K. N., & Muhammad, N. S. (2020). Urban flash flood index based on historical rainfall events. *Sustainable Cities and Society., 56*, 102088. https://doi.org/10.1016/j.scs.2020.102088

Nie, L. M., & Jia, H. F. (2018). China's sponge city construction: Ambition and challenges. *Vann, 2*, 159–167. https://doi.org/10.1016/j.jenvman.2017.11.018

Nowak, D.J. (2023). Improved Air Quality and Other Services from Urban Trees and Forests. In *Engineering and Ecosystems: Seeking Synergies Toward a Nature-Positive World* (pp. 215–245). Cham: Springer International Publishing. https://doi.org/10.1007/978-3-031-35692-6_10

Nyman, M. (2019). Food, meaning-making and ontological uncertainty: Exploring 'urban foraging' and productive landscapes in London. *Geoforum, 99*, 170–180. https://doi.org/10.1016/j.geoforum.2018.10.009

Oberndorfer, E., Lundholm, J., Bass, B., Coffman, R. R., Doshi, H., Dunnett, N., Gaffin, S., Köhler, M., Liu, K. K., & Rowe, B. (2007). Green roofs as urban ecosystems: Ecological structures, functions, and services. *BioScience, 57*(10), 823–833. https://doi.org/10.1641/b571005

Odefey, C. (2009). U.S. Patent No. 7,547,554. U.S. Patent and Trademark Office, Washington, DC.

Onishi, A., Cao, X., Ito, T., Shi, F., & Imura, H. (2010). Evaluating the potential for urban heat-island mitigation by greening parking lots. *Urban Forestry & Urban Greening, 9*(4), 323–332. https://doi.org/10.1016/j.ufug.2010.06.002

Oulahen, G., Klein, Y., Mortsch, L., O'Connell, E., & Harford, D. (2018). Barriers and drivers of planning for climate change adaptation across three levels of government in Canada. *Planning Theory & Practice, 19*(3), 405–421. https://doi.org/10.1080/14649357.2018.1481993

Pandit, R., & Laband, D. N. (2010). Energy savings from tree shade. *Ecological Economics, 69*(6), 1324–1329. https://doi.org/10.1016/j.ecolecon.2010.01.009

Parikh, J. K., Jindal, P., & Sandal, G. (2016). Climate Resilient Urban Development: Vulnerability Profile of Indian Cities. *Cambridge University Press.* https://doi.org/10.1007/978-81-322-2541-6_16

Paul, S., Ghosh, S., Mathew, M., Devanand, A., Karmakar, S., & Niyogi, D. (2018). Increased spatial variability and intensification of extreme monsoon rainfall due to urbanization. *Scientific Reports,* 8, 3918. https://doi.org/10.1038/s41598-018-22322-9. https://doi.org/10.1038/s41598-018-22322-9

Perrotti, D., & Iuorio, O., (2019). Green infrastructure in the space of flows: an urban metabolism approach to bridge environmental performance and user's wellbeing. *Planning Cities with Nature: Theories, Strategies and Methods*, pp. 265–277. https://doi.org/10.1007/978-3-030-01866-5_18

Puchol-Salort, P., O'Keeffe, J., van Reeuwijk, M., & Mijic, A. (2021). An urban planning sustainability framework: Systems approach to blue green urban design. *Sustainable Cities and Society, 66*, 102677. https://doi.org/10.1016/j.scs.2020.102677

RICS. (2011). Green Infrastructure in Urban Areas, RICS Information Paper, first ed. www.ricsbooks.com.

Salvia, A. L., Leal Filho, W., Brandli, L. L., & Griebeler, J. S. (2019). Assessing research trends related to Sustainable Development Goals: Local and global issues. *Journal of Cleaner Production, 208*, 841–849. https://doi.org/10.1016/j.jclepro.2018.09.242

Santamouris, M. (2013). On the built environment—the urban influence. In *Energy and Climate in the Urban Built Environment* (pp. 3–18). Routledge London, UK.

Sarkar, S., Butcher, J. B., Johnson, T. E., & Clark, C. M. (2018). Simulated sensitivity of urban green infrastructure practices to climate change. *Earth Interactions, 22*(13), 1–37. https://doi.org/10.1175/ei-d-17-0015.1

Spaiser, V., Ranganathan, S., Swain, R. B., & Sumpter, D. J. (2017). The sustainable development oxymoron: Quantifying and modelling the incompatibility of sustainable development goals. *International Journal of Sustainable Development & World Ecology, 24*(6), 457–470. https://doi.org/10.1080/13504509.2016.1235624

The Stefano Boeri Architetti website, (online), Available: https://www.stefanoboeriarchitetti.net/en/vertical-foresting/?location=Milano%2C+MI%2C+Italia. Accessed on 18–09–2023.

Tokuç, A., & İnan, T. (2017). October. A Green Outlook to Tall Building Facades via Milan's "Vertical Forest". In *Conference paper: International Advanced Technologies Symposium (IATS).* chrome-extension://efaidnbmnnnibpcajpcglclefindmkaj/https://www.researchgate.net/profile/Ayca-Tokuc-2/publication/320556587_A_Green_Outlook_to_Tall_Building_Facades_via_Milan's_Vertical_Forest/links/5b3760c40f7e9b0df5dcf97e/A-Green-Outlook-to-Tall-Building-Facades-via-Milans-Vertical-Forest.pdf

Udas-Mankikar, S., & Driver, B. (2021). Blue-green infrastructure: An opportunity for Indian cities. *ORF Occasional Paper*, (317). chrome-extension://efaidnbmnnnibpcajpcglclefindmkaj/https://www.researchgate.net/profile/Berjis-Driver-2/publication/353616196_Blue-Green_Infrastructure_An_Opportunity_for_Indian_Cities/links/61063b7a0c2bfa282a12cc35/Blue-Green-Infrastructure-An-Opportunity-for-Indian-Cities.pdf

Velasco, E., & Roth, M. (2010). Cities as net sources of CO2: Review of atmospheric CO2 exchange in urban environments measured by eddy covariance technique. *Geography Compass, 4*(9), 1238–1259. https://doi.org/10.1111/j.1749-8198.2010.00384.x

Wachter, S. M., & Wong, G. (2008). What is a tree worth? Green-city strategies, signaling and housing prices. *Real Estate Economics, 36*(2), 213–239. https://doi.org/10.1111/j.1540-6229.2008.00212.x

Wang, H., & Chen, Q. (2014). Impact of climate change heating and cooling energy use in buildings in the United States. *Energy and Buildings, 82*, 428–436. https://doi.org/10.1016/j.enbuild.2014.07.034

Wardynski, B. J., & Hunt, W. F., III. (2012). Are bioretention cells being installed per design standards in North Carolina? A field study. *Journal of Environmental Engineering, 138*(12), 1210–1217. https://doi.org/10.1061/(asce)ee.1943-7870.0000575

Webler, T., & Tuler, S. (2006). Four perspectives on public participation process in environmental assessment and decision making: Combined results from 10 case studies. *Policy Studies Journal, 34*(4), 699–722. https://doi.org/10.1111/j.1541-0072.2006.00198.x

Well, F., & Ludwig, F. (2020). Blue–green architecture: A case study analysis considering the synergetic effects of water and vegetation. *Frontiers of Architectural Research, 9*(1), 191–202. https://doi.org/10.1016/j.foar.2019.11.001

World Bank. (2013). *Urbanization beyond municipal boundaries: Nurturing metropolitan economies and connecting peri-urban areas in India.* The World Bank. https://documents1.worldbank.org/curated/en/373731468268485378/pdf/757340PUB0EPI0001300pubdate02021013.pdf. Accessed on 18–09–2023.

Zachary Bean, E., Frederick Hunt, W., & Alan Bidelspach, D. (2007). Evaluation of four permeable pavement sites in eastern North Carolina for runoff reduction and water quality impacts. *Journal of Irrigation and Drainage Engineering, 133*(6), 583–592. https://doi.org/10.1061/(asce)0733-9437(2007)133:6(583)

Zhang, L., Sun, X., & Xue, H. (2019). Identifying critical risks in Sponge City PPP projects using DEMATEL method: A case study of China. *Journal of Cleaner Production, 226*, 949–958. https://doi.org/10.1016/j.jclepro.2019.04.067

Zhou, Y. (2015). State power and environmental initiatives in China: Analyzing China's green building program through an ecological modernization perspective. *Geoforum, 61*, 1–12. https://doi.org/10.1016/j.geoforum.2015.02.002

Management and Development of Blue-Green Infrastructure

Challenges and Opportunities in Adopting Blue-Green Infrastructure in Cities of the Developing World

Kandas Keita and Sory Kourouma

Abstract Challenges and opportunities in adopting blue-green infrastructure in cities across the developing world focus on the impact of urbanization in developing countries and the potential of blue-green infrastructure as a solution to urban challenges. It discusses the benefits and foundations of blue-green infrastructure, including the environmental, social and economic benefits of cities. The paper also discusses the challenges and opportunities of implementing this approach in developing cities, such as financing, urban planning and public attitudes, with case studies illustrating successful examples. These examples demonstrate how integrated planning and community participation can overcome these challenges, leading to more sustainable and climate-responsive cities. The paper highlights the importance of an integrated approach to the adoption of blue-green infrastructure. It highlights the need for collaboration between different urban actors, including governments, local communities and the private sector. This cooperation is crucial to overcoming institutional and financial obstacles. The document also emphasizes the importance of education and awareness to foster public acceptance and large infrastructure. It offers a holistic vision for transforming developing cities into greener, resilient and inclusive spaces while taking into account the challenges specific to these contexts.

Keywords Climate change · Green spaces · Pollution · Rainwater management · Stormwater management · Sustainable water management · Water quality

1 Introduction

Urbanization is a dynamic and complex process that shapes modern society. The movement of people from rural areas to urban centers, an accelerating trend, manifests it (Ribeiro & Gonçalves, 2019; Romero-Lankao et al., 2016b). Urbanization

K. Keita (✉) · S. Kourouma
School of Architecture and Urban Planning, Chongqing University, Chongqing 400044, China
e-mail: 11800189@cqu.edu.cn

Higher Institute of Arts of Guinea, Dubreka P.B Box 2421, Republic of Guinea

P. K. Joshi et al. (eds.), *Blue-Green Infrastructure for Sustainable Urban Settlements*,
https://doi.org/10.1007/978-3-031-62293-9_6

rates in developing countries have shown a large gap compared to their counterparts in industrialized nations during the past two centuries. This disparity has been caused by industrialization. Sub-Saharan Africa and South Asia are experiencing unpreceenerdented urbanization. This phenomenon significantly affects both regions' economies, cultures, and ecosystems. Many interconnected factors cause the change; Urbanization's impact on agricultural practices and urban residents' climate change responses are two of these components. When well-managed, urbanization offers many opportunities (Ruggerio, 2021). Cities are engines of economic growth, centers of innovation and creativity. They facilitate scale economies in service delivery, increase social and cultural exchange chances, and are often at the vanguard of social and political transformation. The spread of cities can raise people's quality of life by expanding their access to public goods like schools and hospitals. The challenges to urban resilience and sustainability posed by rapid urbanization in emerging nations are complicated. Problems in urban areas range from social issues to ecological disasters (Sharifi & Yamagata, 2016). Pressure on water, energy, and food resources is increasing with population growth and climate change. Sustainable urban development necessitates creative responses to socioeconomic disparities and the impacts of climate change (Mouratidis, 2019, Hughes, 2013, Bhadouria et al., 2023). Additionally, the effects of climate change, such as floods, heat waves, and storms, endanger the resilience of infrastructure and urban communities.

However, rapid urbanization has drawbacks. The population and infrastructure of cities make them susceptible to natural and artificial calamities. Urban resilience, or a city's capacity to absorb and recover from shocks, is crucial, especially in light of climate change. Pollution, waste management, water and food security, and biodiversity protection threaten city sustainability (Kumar et al., 2022; Maurya et al., 2023). Many towns have growing social, economic, and spatial inequities, which strain urban government (Chan et al., 2018).

The "blue-green infrastructure" approach is a promising strategy to address these challenges. It combines water resources management (blue) with developing and maintaining green spaces (green) within an integrated urban planning and design framework (Pereira et al., 2020). This multidisciplinary approach draws on engineering, environmental science, landscape architecture, and urban planning principles to create sustainable urban ecosystems. In developing countries, where pressures on resources are extreme, adopting the Blue-Green Infrastructure approach can pave the way for more sustainable and resilient cities (Liu et al., 2021b). Water management systems aim to optimize water availability, quality, and distribution for multiple uses. It involves water collection, treatment, storage, and reuse of water. Examples of blue-green strategies in infrastructure design include rain gardens, green roofs, and rainwater collecting systems. They serve as "green lungs" for cities, contribute to biodiversity, provide recreational spaces, and can even serve as natural barriers against flooding and erosion (Molavi, 2018).

Because of the increasing complexity of environmental and social challenges, cities, especially developing countries, must build resilience and sustainability to protect the environment and provide better living conditions for their citizens (Drosou et al., 2019). The effects of global warming, rapid urbanization, and rising populations pose severe threats to these cities. Adopting blue-green infrastructure is often proposed as an effective solution to improve urban resilience and sustainability (Bellezoni et al., 2021). Blue-green infrastructure refers to integrated water and green space management systems, which seek to combine hydraulic elements, such as lakes, rivers, and retention ponds, with green spaces, such as parks, gardens, and green roofs, to create a more harmonious and sustainable urban environment which have the potential to solve several problems including the management of runoff, pollution, and loss of biodiversity (de Macedo et al., 2021).

2 Importance of Its Adoption in Developing Cities

Implementing blue-green infrastructure is important beyond simple aesthetic or environmental factors. The proposed approach encompasses a holistic strategy aimed at achieving sustainable urbanization. This method offers many benefits that strengthen cities' resilience, improve their inhabitants' well-being, and promote economic growth (Leitner et al., 2018). Implementing blue-green infrastructure in developing cities is essential for various reasons, viewed from a multidisciplinary perspective. Benefits encompass different environmental, social, and economic dimensions (Boyd et al., 2022). Each of these axes is important in sustainable and resilient urban development (de Macedo et al., 2021). The main objective of this new strategy is to effectively integrate sustainable water management practices and biodiversity conservation in urban environments.

Here is a brief overview of the importance of Blue-Green infrastructure in this context:

A. Rainwater Management:

Bleu-Vert infrastructure is designed to effectively manage stormwater, reducing the risk of flooding in urban areas. Green spaces, artificial wetlands, retention basins, and green roofs help slow the flow of rainwater, filter it, and temporarily store it. Stormwater management helps prevent flooding, protect urban infrastructure, and improve water quality (Dhakal & Chevalier, 2017).

B. Improvement of Air Quality and Urban Life:

Elements of Blue-Green infrastructure, such as parks, green spaces, and trees, promote the filtration of urban air by capturing fine particles and providing oxygen (Jayasooriya et al., 2017). In addition, these green spaces offer leisure and relaxation areas for citizens, thus contributing to a better quality of life in urban areas.

C. Urban Biodiversity:

Blue-Green infrastructure creates habitats for urban flora and fauna. Green spaces and artificial bodies of water attract various species, promoting biodiversity in urban areas. This also helps educate citizens on the importance of nature conservation in urban areas.

D. Resilience to Climate Change:

In the context of climate change, Blue-Green infrastructure contributes to strengthening the resilience of cities. It helps mitigate the effects of heat waves by creating shaded areas and reducing urban heat islands (Marchese et al., 2018).In addition, sustainable rainwater management makes it possible to face the challenges linked to increasingly intense precipitation and droughts.

E. Integration of Nature in the City:

Blue-Green infrastructure promotes the integration of nature into urban environments, creating more pleasant, sustainable, and resilient cities (Acuti et al., 2020). It also encourages citizen awareness of the importance of nature and responsible management of natural resources. Blue-Green infrastructure is essential to creating more sustainable, resilient, liveable cities. It offers considerable benefits regarding rainwater management, improving air quality and urban life, preserving biodiversity, and strengthening resilience to climate change, making it an essential element of modern urban planning focused on sustainability.

3 The Foundations of Blue-Green Infrastructure

The foundations of Blue-Green infrastructure are based on a holistic approach to sustainable water management and environmental preservation within urban areas (Al-Azzawi & Abaas, 2023; Dhakal & Chevalier, 2017; McPhearson et al., 2016). This innovative approach integrates natural elements, such as water, vegetation, and biodiversity, into urban planning to create more resilient, pleasant, and sustainable urban spaces (Correa Ayram et al., 2016; Liu et al., 2021a).

A. Presentation of the Environmental, Social, and Economic Advantages of This Approach.

The Blue-Green infrastructure approach offers a range of significant environmental, social, and economic benefits in the urban context (Kapetas & Fenner, 2020; Yamagata & Sharifi, 2018). Here is a detailed presentation of these advantages.

B. Environmental Benefits:

Sustainable water management and Bleu-Vert infrastructure reduce stormwater runoff into sewers, which helps prevent flooding (Liu et al., 2021a, Fenner Andrew, 2017). By filtering and slowing the flow of rainwater, it also promotes the recharge of groundwater, which is essential for the sustainable management of water resources. This approach reduces water pollution by limiting contaminants by filtering rainwater through plants and natural soils. (Nguyen et al., 2019; Venkataramanan et al., 2019). It thus contributes to the preservation of watercourses and aquatic ecosystems. Bleu-Vert infrastructure creates habitats for urban fauna and flora, promoting environmental biodiversity (Liu et al., 2021b). Bleu-Vert infrastructure's green spaces and green roofs provide shade and reduce local temperatures, helping to mitigate the effects of urban heat islands (Boyd & Juhola, 2015; Hamann et al., 2020). Promoting vegetation and reducing dependence on air conditioning helps reduce greenhouse gas emissions in urban areas (Dar et al., 2021; Yamagata & Sharifi, 2018).

C. Social Benefits:

Green spaces and parks created by Blue-Green infrastructure improve the overall quality of life by providing residents with opportunities for recreation, relaxation, and social interaction, thereby contributing positively to their well-being (Venkataramanan et al., 2019). This technique improves individuals' mental and physical well-being by facilitating their participation in physical activities and relaxation outdoors in natural environments (Acuti et al., 2020). Blue-green infrastructure projects have the potential to function as educational instruments, effectively raising public awareness regarding environmental concerns and the importance of water and biodiversity governance.

D. Economic Benefits:

Reduced maintenance expenses, green spaces often require less maintenance than grey infrastructure, such as underground pipes. For cities, this could mean long-term savings (Liu et al., 2021b). Green areas and well-designed blue-green infrastructure make a community more desirable to investors and potential residents, which can boost property values (Baba et al., 2020). Planning, designing, and implementing blue-green infrastructure projects creates urban planning, construction, and environmental management jobs. Sustainable and resilient cities are increasingly attracting businesses. Sustainability-focused companies can benefit from blue-green infrastructure (Austin, 2014).

Implementing the Blue-Green infrastructure strategy presents various environmental, social, and economic benefits. The process under study is very beneficial for sustainable urban development in the twenty-first century due to its promotion of responsible water management, preservation of biodiversity, improvement of residents' quality of life, and creation of economic prospects (Cariolet et al., 2019, Fenner Andrew, 2017).

4 Reference to Success Stories from Developed World Cities Adopting Bleu-Green Infrastructure (BGI)

Blue-green infrastructure, also known as green infrastructure or stormwater infrastructure, encompasses incorporating natural elements such as green spaces, trees, wetlands, and permeable surfaces into the planning of cities to effectively manage stormwater, improve water quality, and cultivate more aesthetically pleasing urban environments. Here are examples of metropolitan areas within developing countries that have effectively executed blue-green infrastructure initiatives:

A. Green City, Clean Waters in Philadelphia:

The primary objective of this program is to oversee stormwater management effectively by employing a comprehensive strategy integrating green infrastructure and urban landscaping (see Fig. 1a). With an annual precipitation rate exceeding one million gallons per acre, Philadelphia has allocated approximately $2.4 billion in public funds to establish an "urban mosaic of green stormwater infrastructure." (see Fig. 1b). This project has successfully created over a thousand acres of "green space," capable of absorbing significant rainfall and mitigating pollution annually (see Fig. 1c). Additionally, these areas serve as recreational spaces accessible to the public (see Fig. 1d).

B. Copenhagen, Denmark-Cloudburst Masterplan:

Copenhagen is frequently cited as a striking illustration of the effective implementation of blue-green infrastructure. Following an extreme rainfall event in 2011, Copenhagen formulated a comprehensive strategy to address and mitigate the impacts of such events (see Fig. 2a,b,c). This plan provided for implementing several measures, including installing green roofs, developing pedestrian streets, and creating designated places for rainwater collection. These initiatives have been crucial in reducing vulnerability to flooding, improving the availability of accessible green space, and increasing the overall water quality in the city's canal network (Heinzlef & Serre, 2020; Reischl et al., 2018; Romero-Lankao et al., 2016a) (Fig. 2d). The municipality implemented a comprehensive six-step procedure to conduct in-depth data collection and analysis to develop strategy and implement blue-green infrastructure initiatives (see Fig. 2e). These projects aim to effectively respond to the predicted increase in extreme precipitation, which is expected to occur more frequently in the coming years. The project encompassed a comprehensive process of planning and

Fig. 1 **a** Green City, Clean Waters in Philadelphia. *Source* https://waterbucket.ca/gi/2018/03/27/green-infrastructure-a-tale-of-two-american-cities-philadelphia-and-chicago/. **b** The Philadelphia Water Department Green City. *Source* https://gridphilly.com/blog-home/2011/11/01/go-with-the-flow-the-philadelphia-water-departments-green-ci-html/. **c** Philadelphia Water Department, Green City Clean. **d** Philadelphia Water Department, Green City Clean Waters Program. **c** and **d** *Source* https://fr.slideshare.net/njfuture/philadelphia-water-department-green-city-clean-waters-program

Fig. 1 (continued)

Fig. 1 (continued)

executing blue-green infrastructure (BGI) solutions to manage rainwater effectively (Du & Li, 2019; Fenner Andrew, 2017; Van Oijstaeijen et al., 2020).

C. Green Your Laneway in Melbourne, Australia:

In 2017, Melbourne launched four pilot projects to transform its central thoroughfares into green spaces to mitigate rising temperatures, improving rainwater filtration while providing shade and improving air quality (see Fig. 3a). This project includes installing rainwater tanks, green roofs, green walls, and rain gardens, transforming

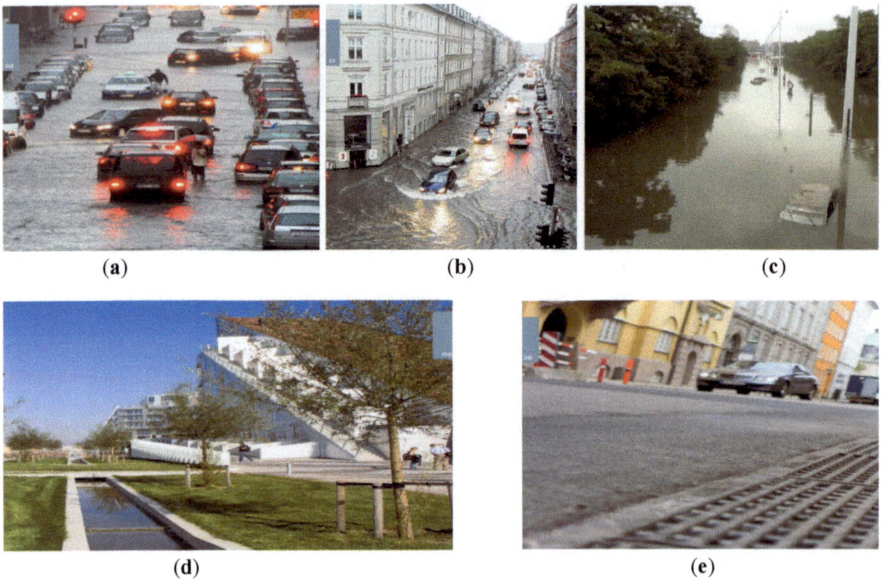

Fig. 2 (**a–e**) *Source* CiTy of Copenhagen Cloudburst ManageMent Plan 2012

these lanes into green sanctuaries within the urban core (Acuti et al., 2020; Ahern et al., 2014) (see Fig. 3b and c).

Fig. 3 **a** *Source* Greening laneways - City of Melbourne. *Source* https://www.melbourne.vic.gov.au/community/greening-the-city/green-infrastructure/Pages/greening-laneways.aspx. **b** Melbourne laneways to be transformed into green spaces. *Source* https://www.australiandesignreview.com/news/melbourne-laneways-to-be-transformed-into-green-spaces/. **c** Shows Green Your Laneway in Melbourne. Source: https://www.greenerspacesbetterplaces.co

Each of these projects illustrates the potential of integrating natural elements into urban environments to effectively address environmental issues such as stormwater management and adaptation to climate change (Jalkanen et al., 2020; Kondo et al., 2018). These efforts also illustrate the importance of collaboration between the public and private sectors and community engagement to ensure the effectiveness of blue-green infrastructure in developed cities.

5 Challenges in Adoption of Blue-Green Infrastructure in Cities of Developing Countries

Implementing blue-green infrastructure, sometimes called integrated stormwater management infrastructure (ISMI), in urban areas of developing countries could face many obstacles arising from various socio-economic, environmental, and institutional issues.

Here are some of these challenges:

5.1 Limited Funding:

- ***Budgetary constraints of developing cities:*** Developing countries often have financial resources when allocating funds for costly infrastructure initiatives, such as implementing large Integrated Stormwater Management Systems Infrastructure (ISMI) (Venkataramanan et al., 2019). Obtaining sufficient funds for developing, executing, and maintaining this blue-green infrastructure can be a significant obstacle.
- ***The need for significant investment:*** Blue-green infrastructure systems, including retention ponds, artificial wetlands, and landscaping, require a considerable initial capital outlay, becoming a potential obstacle for local governments (Ribeiro & Gonçalves, 2019; Sachs et al., 2022).

5.2 Demographic Pressure:

- ***Rapid urban population growth:*** Cities in developing countries often face rapid population expansion, leading to unanticipated urbanization and increased infrastructure needs. Integrating blue-green infrastructure into an already established urban fabric can make it difficult.
- ***Unplanned use of urban space:*** The exponential expansion of urban areas can lead to sub-optimal use of urban space, characterized by the construction of buildings and the provision of infrastructure that fails to integrate sustainable stormwater management practices.

- ***Land use conflicts:*** Urban areas in developing countries often have high population density and face intense rivalry for access to land. Implementing integrated rainwater management could pose challenges where spaces are limited, and there is a demand for urban expansion, including residential and commercial sectors.
- ***Land ownership and land access:*** Land ownership, illegal occupation, and urban density could make it difficult to acquire land for retention ponds, constructed wetlands, and other aspects of blue-green infrastructure.
- ***Multiple pollution:*** Pollution problems from numerous sources pose a significant challenge to stormwater management in urban regions of emerging countries. Stormwater can harbor many contaminants from roads, solid waste, and untreated sewage.
- ***Climate Challenges:*** Developing countries often exhibit increased sensitivity to the impacts of climate change, encompassing erratic precipitation patterns and severe weather events. The integrated stormwater management infrastructure design must be adapted to address the abovementioned issues.

5.3 Lack of Awareness:

- ***Need for increased awareness among residents and decision-makers***: Successful implementation of blue-green infrastructure requires attention and education of the local population and decision-makers, ensuring their understanding of the benefits associated with this approach (Billger et al., 2017; Bixler et al., 2020). A lack of awareness of the benefits of ISMI can lead to reluctance to accept change and hinder the progress of community efforts. However, overcoming these obstacles can be a formidable task due to cultural and linguistic differences.
- ***Difficulties in communicating the benefits of the Blue-Green approach***: To promote the use of blue-green infrastructure, it is imperative to effectively communicate its benefits, which include flood mitigation, improved quality of water, and the creation of green spaces (Jayasooriya et al., 2017, Mell, 2017).
- ***Climatic and environmental factors:*** The design of an integrated stormwater management infrastructure (ISMI) should incorporate adaptability to consider the diversity of local ecological variables prevalent in developing countries (Miles & Ebrey, 2017; Nguyen et al., 2019). Containing ISMI solutions can introduce complexities in terms of design and functionality.

5.4 Complex Management:

To solve these problems, urban cities in developing countries often need to explore innovative financing methods, strengthen their institutional capacities, encourage community participation, and cultivate local expertise in integrated infrastructure of rainwater management (Heinzlef & Serre, 2020; Keita & Kourouma, 2023; Sachs et al., 2022; Zeng et al., 2022). Collaboration between local governments, international organizations, non-governmental organizations (NGOs), and private sector entities is of paramount importance to effectively mobilize financial resources, raise awareness among stakeholders, formulate supportive policies, and provide assistance technical training (Shandas et al., 2020; Deely et al., 2020; Gao et al., 2022).

6 Opportunities and Solutions Related to the Adoption of Blue-Green Infrastructure

a. **Improved quality of life:**
- *Positive impact on air quality:* Incorporating green spaces and artificial wetlands within the blue-green infrastructure facilitates air purification through the pollutant absorption of suspended contaminants in the air and the subsequent release of oxygen (Thorne et al., 2018).
- *Promoting biodiversity and creating recreational spaces:* Blue-green zones play a crucial role in supporting the local ecosystem by providing habitat for various species, promoting urban biodiversity, and providing recreational areas to residents, thereby improving their overall quality of life (Weichselgartner & Kelman, 2015).

b. **Resilience to climate change:**

Mitigating the effects of floods, storms, and heatwaves: Implementing blue-green infrastructure, such as retention ponds and wetlands, is crucial in mitigating flood risks through storage and effective rainwater management. In addition, the presence of green spaces helps to reduce heat waves by providing shade and subsequent cooling of the environment (Gao et al., 2022).

Strengthening urban resilience: Blue-green infrastructure enhances the strength of urban areas against climate shocks, thus contributing to the protection of citizens and the preservation of critical urban infrastructure (Liu & Jensen, 2018).

c. **Job creation:**
- *Local employment opportunities in design and maintenance:* The implementation, construction, and management of blue-green infrastructure create employment opportunities at the local level, encompassing several sectors such as landscape architecture, rainwater management, and maintenance of green spaces (Verma & Raghubanshi, 2018).

- *Economic stimulus for the city*: Investments in blue-green infrastructure can stimulate the local economy by generating associated economic activities, such as manufacturing environmentally friendly building materials (Weichselgartner & Kelman, 2015; Liao, 2019; Bulkeley et al., 2013).

d. **Long-term economic benefits:**

- *Reducing health costs from air pollution*: Using vegetation and stormwater management strategies could lead to better air quality, potentially reducing health care costs associated with respiratory and cardiovascular problems (O'Donnell and Thorne, 2020; McPhearson et al., 2016).
- *Savings on maintenance costs of traditional infrastructure:* Implementing blue-green infrastructure has the potential to extend the lifespan of conventional infrastructure, which would lead to a decrease in expenses associated with maintenance and repairs in the long term (Heinzlef & Serre, 2020).

e. **Technological innovation**:

- *Promotion of research and development of green technologies:* Adopting blue-green infrastructure promotes the study and advancement of sustainable, environmentally friendly technologies, exemplified by developing advanced stormwater management systems (McPhearson et al., 2016).
- *Encouraging innovation in the field of sustainability:* Implementing new strategies for blue-green infrastructure can function as a paradigm and inspire further progress in urban sustainability.

Adopting blue-green infrastructure offers many opportunities to improve the quality of life, strengthen urban resilience, stimulate the local economy, achieve long-term savings, and promote technical progress. To achieve these benefits, it is imperative to allocate resources to effectively strategize and execute these systems while actively involving local stakeholders (Sachs et al., 2022).

a. Exploring unique opportunities in developing country cities to integrate Blue-Green infrastructure, including designing sustainable cities.

The integration of blue-green infrastructure in urban areas of emerging countries presents particular prospects, encompassing the potential for building sustainable cities from the start.

Holistic urban planning: Cities in developing countries have the potential to establish comprehensive urban planning strategies that include blue-green infrastructure elements from the design phase. This involves integrating sustainable rainwater management, creating green spaces, and promoting biodiversity in the design of neighbourhoods and infrastructure (Thorne, 2020).

Rational use of space: In town planning and urban development, integrating blue-green zones, such as parks, retention ponds, and rain gardens, offers the potential for optimization of space use (Dar et al., 2021). This approach achieves the dual objective of properly managing rainwater while simultaneously providing recreational areas for residents.

Integrated Stormwater Management: Growing and developing urban areas can adopt comprehensive stormwater management systems that effectively collect, store, and treat stormwater in an environmentally sustainable manner (Dhakal & Chevalier, 2017; O'Donnell and Thorne, 2020). This intervention helps mitigate flooding, improve water quality, and reduce adverse impacts on nearby water resources.

Promotion of urban biodiversity and innovation: Blue-green spaces have the potential to serve as habitats for local flora and fauna, thus making a valuable contribution to urban biodiversity conservation. Implementing ecological corridors and conservation areas within urban environments can establish a symbiotic relationship between human populations and the natural world (Heinzlef & Serre, 2020; Shandas et al., 2020). Early integration of blue-green infrastructure into urban design promotes sustainability innovation by stimulating the emergence of new technologies and optimal approaches to stormwater management (Venkataramanan et al., 2019).

Climate resilience: Cities in developing countries generally face increased vulnerability to the impacts of climate change. Integrating blue-green infrastructure from the start improves the urban area's ability to withstand and recover from flooding, heat waves, and other extreme weather events (Birchall & Bonnett, 2021).

Long-term savings: Implementing blue-green infrastructure in the early stages of urban development can generate substantial long-term benefits, primarily in terms of reduced stormwater management costs, improved public health (by improving air and water quality), and maintenance needs for infrastructure (Deely et al., 2020; Gao et al., 2022).

Community Participation: Active citizen participation in the early stages of the planning and design process can cultivate a sense of ownership and responsibility for blue-green spaces within the city.

Cities in developing countries have a distinct potential to strategically plan sustainable urban landscapes by integrating blue-green infrastructure from the outset (Matovnikov & Matovnikova, 2016). This initiative can potentially improve the general well-being of residents, strengthen their capacity to adapt to the impacts of climate change, support the preservation of diverse urban ecosystems, and stimulate innovative practices while reducing long-term sustainable costs. Effective use of these opportunities requires implementing strategic urban planning and establishing collaborative partnerships between stakeholders (Gao et al., 2022).

b. A presentation of adaption measures, including the recycling of rainwater, the development of urban parks, and the enhancement of waste management.

Implementing adaptation techniques, such as using rainwater, creating urban parks, and improving waste management, can substantially contribute to developing sustainable and resilient cities in developing countries. Development refers to the process of growth and progress in various aspects of society.

a. **Reuse of rainwater:**

- ***Collection and storage of rainwater:*** Rainwater harvesting systems can be implemented to capture rainwater that precipitates on roofs and impermeable surfaces. Subsequently, the water can be retained in reservoirs or basins (Heinzlef & Serre, 2020).

- ***Rainwater treatment:*** Collected rainwater can be treated to remove toxins and impurities. After treatment, water can be used for many purposes, such as irrigating green spaces, flushing toilets, washing automobiles, and even replenishing groundwater levels (Thorne et al., 2018).
- ***Reduction in demand for drinking water:*** Rainwater reuse helps alleviate the need for drinking water, thereby facilitating the conservation of water resources and mitigating water treatment expenses.

b. **Creation of urban parks:**

- ***Maximization of green spaces:*** Urban parks provide many benefits, such as creating recreational areas for residents, regulating temperature, mitigating air pollution, and promoting biodiversity.
- ***Rainwater management:*** Urban parks have the potential to incorporate blue-green elements, such as retention ponds and wetlands, into their design, which can effectively manage stormwater, mitigate flood risks, and improve water quality (Shandas et al., 2020, O'Donnell and Thorne, 2020).
- ***Strengthening social cohesion:*** Urban parks often function as gathering spaces for the local community, thus promoting social cohesion and improving the well-being of residents (Venkataramanan et al., 2019).

c. ***Improved waste management:***

- ***Selective waste collection:*** Implementing selective programs can promote recycling and waste reduction practices, thereby contributing to environmental conservation (Heinzlef & Serre, 2020).
- ***Recovery of organic waste:*** Managing organic waste, including kitchen and green waste, can use various methods, such as composting or anaerobic digestion, to generate fertilizer or energy resources.
- ***Raising awareness about waste reduction:*** Awareness campaigns can encourage residents to reduce waste production and adopt more sustainable practices.

Integrating these adaptation measures into urban planning can effectively maximize their benefits. One potential approach is to create metropolitan parks equipped with stormwater management systems, which can function as reservoirs to reuse stormwater (Dhakal & Chevalier, 2017; Venkataramanan et al., 2019). This strategy addresses the need for stormwater management and provides additional benefits, such as providing residents with green space and recreational facilities (McPhillips et al., 2021; O'Donnell and Thorne, 2020).

7 C. We Are Highlighting Innovative Financing Models and Public–private Partnerships.

Successful implementation of blue-green infrastructure projects in developing countries' urban areas may require inventive financial frameworks and public–private partnership collaborations. These measures are crucial to overcoming budgetary limitations and effectively mobilizing the necessary resources (Sharifi, 2019).

A. Crowdfunding:

Cities can encourage community and citizen participation by launching crowdfunding campaigns for specific blue-green infrastructure projects (Ihlebæk et al., 2021). For example, owners can contribute financially to constructing an urban park with stormwater management infrastructure.

B. Green Bonds:

Municipalities can use green bonds to obtain financing for blue-green infrastructure initiatives. These bonds are designed to invest in environmentally friendly projects and appeal to investors who prioritize environmental sustainability (Tuofu et al., 2021).

C. Sustainable Infrastructure Investment Funds:

Governments, development banks, and financial institutions can create investment funds allocated explicitly to blue-green infrastructure initiatives within deprived nations. These funds can attract official and private investors (Jacobs, 2012).

D. Public–Private Partnerships (PPPs):

Public–private partnerships (PPPs) have the potential to simplify the design, funding, construction, and operation of blue-green infrastructure projects. Private companies can offer financial capital, technical expertise, and operational efficiency, while the public sector can provide guarantees, a consistent regulatory structure, and access to essential land resources. For example, a PPP could involve a private company building and operating a stormwater management system while the municipality provides land and tax incentives (O'Donnell and Thorne, 2020).

E. Tax Incentives and Subsidies:

Local governments can offer tax incentives or subsidies to private companies investing in blue-green infrastructure initiatives. This could potentially encourage private-sector engagement.

F. Cost-Sharing Mechanisms:

Collaborative contracts between municipalities, private companies, and financial institutions can be purposely decided to spread the upfront expenses and uncertainties associated with blue-green infrastructure (Austin, 2014).

G. Pricing of Water and Sanitation Services:

The inclusion of financing for blue-green infrastructure can be considered in the adjustment of tariffs for water and sanitation services. Tariff adjustments have the potential to generate revenue that can then be allocated to infrastructure projects (Leng et al., 2020; Livesley et al., 2016).

Tailoring each financing model and public–private partnership to each city's unique requirements and circumstances is imperative. Founding transparent, collaborative efforts between local governments, international organizations, financial institutions, and private companies is essential to fostering the development of effective and sustainable financing mechanisms (Peng et al., 2023; Nagenborg, 2019). The objective is to facilitate allocating of resources necessary to advance urban resilience, environmental sustainability, and improved quality of life in cities in developing countries.

8 Case Studies

A. Presentation of concrete cases of cities in developing countries that have overcome the challenges and taken advantage of the opportunities linked to Blue-Green infrastructure.

Curitiba, Brazil: Curitiba is widely cited as an example of sustainable urban planning (see Fig. 4a).The city has successfully built a state-of-the-art public transportation system centered on rapid buses while seamlessly integrating huge urban parks and protected natural areas (see Fig. 4b). These parks provide residents with green spaces and function as stormwater management areas to prevent flooding (Leng et al., 2020; Peng et al., 2023) (see Fig. 4d, 4e, 4f).

Mumbai, India: Mumbai has undertaken efforts to rejuvenate its mangrove ecosystems, which serve as a natural defence mechanism against the damaging effects of floods and storms (see Fig. 5a). Integrating wetland restoration measures into urban planning has been undertaken to address flood challenges and safeguard coastal cities effectively (O'Donnell and Thorne, 2020; Thorne, 2020) (see Fig. 5b).

Singapore: Singapore is internationally recognized as a leading global leader in implementing and developing blue-green infrastructure. The city has implemented advanced stormwater management strategies, including using rainwater harvesting techniques, creating retention ponds and buffer zones, and implementing green roof infrastructure (McPhillips et al., 2021; Shandas et al., 2020) (see Fig. 6). This intervention effectively addressed the challenges posed by the rapid urbanization process while simultaneously strengthening the city's capacity to resist and adapt to the impacts of climate change (Dhakal & Chevalier, 2017) (see Fig. 7).

Nairobi, Kenya: Nairobi has successfully implemented experimental blue-green infrastructure projects in slums, where uncontrolled stormwater problems pose significant challenges (see Fig. 8). To alleviate flooding and recover the well-being of residents, various community initiatives have been launched, including the creation of rain gardens and natural drainage systems (Fenner Andrew, 2017; Leng et al., 2020).

Bogota, Colombia: Bogota has invested in developing cycling infrastructure and greenways along river banks, facilitating sustainable transportation options and improving rainwater management strategies. Establishing these blue-green corridors has led to a notable transformation of the urban environment, making it more conducive to pedestrian and cycling activities(Liu et al., 2021a; Venkataramanan et al., 2019, Lehmann, 2021) (see Fig. 9).

The illustrations above demonstrate that cities in developing countries can effectively integrate blue-green infrastructure into urban planning despite financial limitations and demographic challenges (Cavalli et al., 2020; Livesley et al., 2016). Achieving this usually requires a forward-thinking perspective, strong political commitment, cooperation between different stakeholders, and the implementation of inventive financial frameworks. These approaches generate benefits beyond stormwater management, contributing to developing cities that are more resilient, sustainable, and conducive to a high quality of life for their inhabitants (Liao, 2019).

Fig. 4 **a** Curitiba, Brazil: the world's first sustainable city. *Source* https://urbanizehub.com/curitiba-brazil-worlds-first-sustainable-city/. **b** Dedicated Transit Lanes (garba.org/aboutme/pictures/curitiba). *Source* https://fr.scribd.com/document/279768693/Curitiba-Case-Study. **c** Parks in Curitiba. *Source*: https://pt.slideshare.net/ravenspawn77/curitiba-brazil?next_slideshow=true. **d** The Brazilian city of Curitiba has created many parks to reduce the risks of climate change. *Source* https://www.lesechos.fr/thema/articles/curitiba-saint-domingue-jinzhong-trois-villes-exemplaires-face-au-changement-climatique-1360145. **e** Green Spaces—Curitiba—Brazil. *Source* https://kenchucuritiba.weebly.com/green-spaces.html. **f** Parks to reduce climate change risks. *Source* https://blog.arboimoveis.com.br/cidades/bairros-mais-seguros-de-curitiba/

E. Analysis of Lessons Learned and Good Practices from These Case Studies

Cities in developing countries that have effectively integrated blue-green infrastructure into their urban development provide compelling illustrations of how the obstacles posed by increasing urbanization, climate change, and natural resource management can be overwhelmed (Baba et al., 2020; Jacobs, 2012). These urban areas have shown that using inventive strategies and a long-term perspective can improve their residents' well-being and ability to withstand growing environmental challenges (Venkataramanan et al., 2019). The success of blue-green infrastructure in

Fig. 4 (continued)

Fig. 5 **a** Living with water: integrating blue, green and grey infrastructure to manage urban floods. Source: (http://www.landprocess.co.th). **b** Creating a flood resilient Indian 'Sheher' through Water Sensitive Urban Design. *Source* https://wri-india.org/blog/creating-flood-resilient-indian-%E2%80%98sheher%E2%80%99-through-water-sensitive-urban-design

Fig. 5 (continued)

Fig. 6 Singapore faces storms: Published on July 27, 2017. *Source* https://www.toutsurmesservices.fr/Singapour-fait-face-aux-orages

Fig. 7 Aerial view of the central district of Singapore. *Source* Prom.: De Herde, André http://hdl.handle.net/2078.1/19679

Fig. 8 Aerial view of downtown Nairobi. *Source:* https://depositphotos.com/fr/photo/aerial-view-of-downtowm-nairobi-94130412.html

Fig. 9 Aerial view of the city of Bogot, Colombia. *Source* https://www.bnamericas.com/en/news/colombia-awards-us340mn-section-of-green-corridor

developing cities relies on several key pillars, including solid political commitment, comprehensive planning, active community engagement, technological advancements, collaborative efforts between the public and private sectors, monitoring and continuous assessment, flexibility and adaptability, educational initiatives, effective resource mobilization, and long-term strategic planning (Ahmed et al., 2019; Austin, 2014).

The transmission of these best practices has significant value for multiple reasons. First, it facilitates the dissemination of information and skills accumulated by prosperous cities, thereby avoiding the repetition of mistakes and unnecessary duplication of efforts (Sharifi & Yamagata, 2016). Furthermore, it has the potential to accelerate the transition towards sustainable and resilient urban areas, which is imperative given the current circumstances of rapid urban expansion and environmental threats that pose risks to the well-being of residents (Chelleri et al., 2015).

Disseminating best practices in blue-green infrastructure has the potential to generate global influence. Cities in poor countries have demonstrated the feasibility of implementing inventive solutions under challenging circumstances, thus serving as an example for developed countries (Billger et al., 2017; Bixler et al., 2020; Nguyen et al., 2019).

To facilitate the widespread adoption of these exemplary methodologies, it is imperative to establish strong information exchange networks, formulate comprehensive norms and standards, offer technical support, facilitate training initiatives and education, encourage financial investments, and promote research and development efforts (Bibri, 2021; Heinzlef & Serre, 2020; Ghofrani et al., 2017).

The examples of urban centers in developing countries that have effectively integrated blue-green infrastructure into their urban planning constitute compelling evidence that environmental sustainability, climate change resilience, and human well-being are not mutually exclusive goals. Exclusive. Instead, these goals are mutually reinforcing and can be achieved through deliberate strategic planning, innovative approaches, active community engagement, and unwavering political dedication (Brown et al., 2012; Marchese et al., 2018).

9 Conclusion

The ideas of urban resilience and sustainability will significantly impact how our cities develop in the years to come. To secure a prosperous and musical future, cities must embrace resilient and sustainable methods as urban populations grow and environmental and social concerns become more complicated.

Urban resilience means helping cities resist, adapt, and recover from natural and anthropogenic disasters. These include effective crisis management, intelligent urban planning, safeguarding water and energy resources, and establishing sustainable transport networks. The concept of urban resilience aims to protect both residents and infrastructure, ensuring the uninterrupted provision of essential services.

On the other hand, urban sustainability depends on creating sites that effectively meet the current population's needs while preserving the resources and well-being of future generations. Promoting energy efficiency, mitigating greenhouse gas emissions, preserving urban biodiversity, and encouraging sustainable lifestyles are all part of this initiative. Sustainable cities strive to achieve a harmonious balance between development's economic, social, and environmental dimensions.

The concepts of urban resilience and sustainability are closely linked and mutually reinforcing. Resilient urban areas are more prepared to face sustainability issues, while sustainable ones can manage disruptions and adapt flexibly to changes effectively. Designing resilient and sustainable cities requires government, urban planners, corporations, and civil society collaboration. These cities should improve everyone's quality of life while preserving the planet for future generations. Establishing a resilient and sustainable urban future is a primary goal to provide a more equitable and improved global condition.

Blue-green infrastructure is a powerful catalyst for fostering a more sustainable trajectory in urban environments. Integrating nature into urban planning can promote the development of resilient and environmentally friendly cities, providing many benefits to developing urban areas. This approach offers an innovative and sustainable water management and planning method. "Urban" refers to the characteristics, activities, or phenomena associated with cities. The current moment requires proactive measures to facilitate the adoption of sustainable practices in emerging countries, thereby facilitating the development of resilient and environmentally friendly cities with a focus on the well-being of future generations.

References

Acuti, D., Bellucci, M., & Manetti, G. (2020). Company disclosures concerning the resilience of cities from the Sustainable Development Goals (SDGs) perspective. *Cities, 99*, 102608.

Ahern, J., Cilliers, S., & Niemelä, J. (2014). The concept of ecosystem services in adaptive urban planning and design: A framework for supporting innovation. *Landscape and Urban Planning, 125*, 254–259.

Ahmed, S., Meenar, M., & Alam, A. (2019). Designing a Blue-Green Infrastructure (BGI) network: Toward water-sensitive urban growth planning in Dhaka, Bangladesh. *Land, 8*, 138.

Al-Azzawi, L. R., Abaas, Z. R. (2023) *Power between potentiality and actuality in urban sustainability and urban resilience*. AIP Conference Proceedings. AIP Publishing.

Anabaraonye, B., Okon, O. E., Dibia, S. I., Onwuzurike, U., Olisah, N. C., Ewa, B. O. (2018). Climate change education and multi-stakeholder partnership projects to scale up climate resilience in Africa. *environment (Barrios et al., 2006; Coulibaly et al., 2020; Rigaud et al., 2018)*, 11, 27.

Austin, G. (2014). *Green infrastructure for landscape planning: integrating human and natural systems*, Routledge.

Baba, K., Nagata, Y., Kawakubo, S. & Tanaka, M. (2020). A framework and indicators of resilience. *Resilient Policies in Asian Cities: Adaptation to Climate Change and Natural Disasters*, 3–45.

Bellezoni, R. A., Meng, F., He, P., & Seto, K. C. (2021). Understanding and conceptualizing how urban green and blue infrastructure affects the food, water, and energy nexus: A synthesis of the literature. *Journal of Cleaner Production, 289*, 125825.

Bhadouria, R., Tripathi, S., Singh, P., Joshi, P.K. and Singh, R. (2023). Urban Metabolism and Global Climate Change: An Overview. *Urban Metabolism and Climate Change: Perspective for Sustainable Cities*, pp. 3–22.

Bibri, S. E. (2021). Data-driven smart sustainable cities of the future: An evidence synthesis approach to a comprehensive state-of-the-art literature review. *Sustainable Futures, 3*, 100047.

Billger, M., Thuvander, L., & Wästberg, B. S. (2017). In search of visualization challenges: The development and implementation of visualization tools for supporting dialogue in urban planning processes. *Environment and Planning b: Urban Analytics and City Science, 44*, 1012–1035.

Birchall, S. J., & Bonnett, N. (2021). Climate change adaptation policy and practice: The role of agents, institutions and systems. *Cities, 108*, 103001.

Bixler, R. P., Lieberknecht, K., Atshan, S., Zutz, C. P., Richter, S. M., & Belaire, J. A. (2020). Reframing urban governance for resilience implementation: The role of network closure and other insights from a network approach. *Cities, 103*, 102726.

Boyd, E., & Juhola, S. (2015). Adaptive climate change governance for urban resilience. *Urban Studies, 52*, 1234–1264.

Boyd, D., Pathak, M., van Diemen, R., & Skea, J. (2022). Mitigation co-benefits of climate change adaptation: A case-study analysis of eight cities. *Sustainable Cities and Society, 77*, 103563.

Brown, A., Dayal, A., & Rumbaitis Del Rio, C. (2012). From practice to theory: Emerging lessons from Asia for building urban climate change resilience. *Environment and Urbanization, 24*, 531–556.

Bulkeley, H., Carmin, J., Broto, V. C., Edwards, G. A., & Fuller, S. (2013). Climate justice and global cities: Mapping the emerging discourses. *Global Environmental Change, 23*, 914–925.

Cariolet, J.-M., Vuillet, M., & Diab, Y. (2019). Mapping urban resilience to disasters–A review. *Sustainable Cities and Society, 51*, 101746.

Cavalli, L., Farnia, L., Lizzi, G., Romani, I. G., Alibegovic, M., Vergalli, S. (2020). The SDSN Italia SDGs city index two years later: update report.

Chan, F. K. S., Griffiths, J. A., Higgitt, D., Xu, S., Zhu, F., Tang, Y.-T., Xu, Y., & Thorne, C. R. (2018). "Sponge City" in China—a breakthrough of planning and flood risk management in the urban context. *Land Use Policy, 76*, 772–778.

Chelleri, L., Waters, J. J., Olazabal, M., & Minucci, G. (2015). Resilience trade-offs: Addressing multiple scales and temporal aspects of urban resilience. *Environment and Urbanization, 27*, 181–198.

Correa Ayram, C. A., Mendoza, M. E., Etter, A., Salicrup, D. R. P. (2016). Habitat connectivity in biodiversity conservation: A review of recent studies and applications. *Progress in Physical Geography,* 40**,** 7-37.

Dar, M. U. D., Shah, A. I., Bhat, S. A., Kumar, R., Huisingh, D., Kaur, R. (2021). *RETRACTED: Blue Green Infrastructure as a Tool for Sustainable Urban Development*. Elsevier.

de Macedo, L. S. V., Picavet, M. E. B., de Oliveira, J. A. P., & Shih, W.-Y. (2021). Urban green and blue infrastructure: A critical analysis of research on developing countries. *Journal of Cleaner Production, 313*, 127898.

Deely, J., Hynes, S., Barquín, J., Burgess, D., Finney, G., Silió, A., Álvarez-Martínez, J. M., Bailly, D., & Ballé-Béganton, J. (2020). Barrier identification framework for the implementation of blue and green infrastructures. *Land Use Policy, 99*, 105108.

Dhakal, K. P., & Chevalier, L. R. (2017). Managing urban stormwater for urban sustainability: Barriers and policy solutions for green infrastructure application. *Journal of Environmental Management, 203*, 171–181.

Drosou, N., Soetanto, R., Hermawan, F., Chmutina, K., Bosher, L., & Hatmoko, J. U. D. (2019). Key factors influencing wider adoption of blue–green infrastructure in developing cities. *Water, 11*, 1234.

Du, K., & Li, J. (2019). Towards a green world: How do green technology innovations affect total-factor carbon productivity. *Energy Policy, 131*, 240–250.

Fenner Andrew, R. (2017). Spatial evaluation of multiple benefits to encourage multi-functional design of sustainable drainage in blue-green cities. *Water, 9*, 953.

Gao, Z., Zhang, Q., Xie, Y., Wang, Q., Dzakpasu, M., Xiong, J., & Wang, X. (2022). A novel multi-objective optimization framework for urban green-gray infrastructure implementation under impacts of climate change. *Science of the Total Environment, 825*, 153954.

Ghofrani, Z., Sposito, V., Faggian, R. (2017). A comprehensive review of blue-green infrastructure concepts. *International Journal of Environment and Sustainability,* 6.

Hamann, F., Blecken, G.-T., Ashley, R. M., & Viklander, M. (2020). Valuing the multiple benefits of blue-green infrastructure for a Swedish case study: Contrasting the economic assessment tools B£ ST and TEEB. *Journal of Sustainable Water in the Built Environment, 6*, 05020003.

Heinzlef, C., & Serre, D. (2020). Urban resilience: From a limited urban engineering vision to a more global comprehensive and long-term implementation. *Water Security, 11*, 100075.

Hughes, S. (2013). Justice in urban climate change adaptation: criteria and application to Delhi. *Ecology and Society,* 18.

Ihlebæk, C., Næss, P., & Stefansdottir, H. (2021). Are compact cities a threat to public health? *European Planning Studies, 29*, 1021–1049.

Jacobs, M. (2012). *Green growth: Economic theory and political discourse*, Grantham Research Institute on Climate Change and the Environment London.

Jalkanen, J., Fabritius, H., Vierikko, K., Moilanen, A., & Toivonen, T. (2020). Analyzing fair access to urban green areas using multimodal accessibility measures and spatial prioritization. *Applied Geography, 124*, 102320.

Jayasooriya, V., Ng, A., Muthukumaran, S., & Perera, B. (2017). Green infrastructure practices for improvement of urban air quality. *Urban Forestry & Urban Greening, 21*, 34–47.

Kapetas, L., & Fenner, R. (2020). Integrating blue-green and grey infrastructure through an adaptation pathways approach to surface water flooding. *Philosophical Transactions of the Royal Society A, 378*, 20190204.

Keita, K., & Kourouma, S. (2023). Assessment of policy and legal frameworks of urban green infrastructure development: Republic of Guinea. *Buildings, 13*, 1945.

Kondo, M. C., Fluehr, J. M., McKeon, T., & Branas, C. C. (2018). Urban green space and its impact on human health. *International Journal of Environmental Research and Public Health, 15*, 445.

Kumar, A., Kumari, R., Saikia, P. (2022). Avenue plantation as a viable carbon confiscation tool: A blueprint for eco-smart cities. *Climate Change Alleviation for Sustainable Progression: Floristic Prospective and Arboreal Avenues as a Viable Confiscation Tool*, 333–346.

Lehmann, S. (2021). Growing biodiverse urban futures: Renaturalization and rewilding as strategies to strengthen urban resilience. *Sustainability, 13*, 2932.

Leitner, H., Sheppard, E., Webber, S., & Colven, E. (2018). Globalizing urban resilience. *Urban Geography, 39*, 1276–1284.

Leng, L., Mao, X., Jia, H., Xu, T., Chen, A. S., Yin, D., & Fu, G. (2020). Performance assessment of coupled green-grey-blue systems for Sponge City construction. *Science of the Total Environment, 728*, 138608.

Liao, K.-H. (2019). The socio-ecological practice of building blue-green infrastructure in high-density cities: What does the ABC Waters Program in Singapore tell us? *Socio-Ecological Practice Research, 1*, 67–81.

Liu, L., & Jensen, M. B. (2018). Green infrastructure for sustainable urban water management: Practices of five forerunner cities. *Cities, 74*, 126–133.

Liu, H.-Y., Jay, M., & Chen, X. (2021a). The role of nature-based solutions for improving environmental quality, health and well-being. *Sustainability, 13*, 10950.

Liu, S.-C., Peng, F.-L., Qiao, Y.-K., & Zhang, J.-B. (2021b). Evaluating disaster prevention benefits of underground space from the perspective of urban resilience. *International Journal of Disaster Risk Reduction, 58*, 102206.

Livesley, S., McPherson, E. G., & Calfapietra, C. (2016). The urban forest and ecosystem services: Impacts on urban water, heat, and pollution cycles at the tree, street, and city scale. *Journal of Environmental Quality, 45*, 119–124.

Marchese, D., Reynolds, E., Bates, M. E., Morgan, H., Clark, S. S., & Linkov, I. (2018). Resilience and sustainability: Similarities and differences in environmental management applications. *Science of the Total Environment, 613*, 1275–1283.

Matovnikov, S., & Matovnikova, N. (2016). Innovative urban planning methods for the urban landscape design in the Volgograd agglomeration. *Procedia Engineering, 150*, 1966–1971.

Maurya, S. P., Ohri, A., Singh, P. K., & Singh, R. (2023). Scenario-based evaluation of integrated urban water management framework: A decision making approach. *International Journal of Environment and Sustainable Development, 22*, 393–408.

McPhearson, T., Pickett, S. T., Grimm, N. B., Niemelä, J., Alberti, M., Elmqvist, T., Weber, C., Haase, D., Breuste, J., & Qureshi, S. (2016). Advancing urban ecology toward a science of cities. *BioScience, 66*, 198–212.

McPhillips, L. E., Matsler, M., Rosenzweig, B. R., & Kim, Y. (2021). What is the role of green stormwater infrastructure in managing extreme precipitation events? *Sustainable and Resilient Infrastructure, 6*, 133–142.

Mell, I. C. (2017). *Green Infrastructure: Reflections on Past, Present and Future Praxis*. Taylor & Francis.

Miles, A., & Ebrey, J. (2017). The village in the city: Participation and cultural value on the urban periphery. *Cultural Trends, 26*, 58–69.

Molavi, M. (2018). Measuring urban resilience to natural hazards. *TeMA-Journal of Land Use, Mobility and Environment, 11*, 195–212.
Mouratidis, K. (2019). Compact city, urban sprawl, and subjective well-being. *Cities, 92*, 261–272.
Nagenborg, M. (2019). Urban resilience and distributive justice. *Sustainable and Resilient Infrastructure, 4*, 103–111.
Nguyen, T. T., Ngo, H. H., Guo, W., Wang, X. C., Ren, N., Li, G., Ding, J., & Liang, H. (2019). Implementation of a specific urban water management-Sponge City. *Science of the Total Environment, 652*, 147–162.
O'donnell, E., Thorne, C. (2020). Urban flood risk management: the blue–green advantage. *Blue–green Cities: Integrating Urban Flood Risk Management with Green Infrastructure.* Ice Publishing.
Peng, C., Xiang, Y., Chen, L., Zhang, Y., & Zhou, Z. (2023). The Impact of the Type and Abundance of Urban Blue Space on House Prices: A Case Study of Eight Megacities in China. *Land, 12*, 865.
Pereira, L. M., Davies, K. K., den Belder, E., Ferrier, S., Karlsson-Vinkhuyzen, S., Kim, H., Kuiper, J. J., Okayasu, S., Palomo, M. G., & Pereira, H. M. (2020). Developing multiscale and integrative nature–people scenarios using the Nature Futures Framework. *People and Nature, 2*, 1172–1195.
Reischl, C., Rauter, R., & Posch, A. (2018). Urban vulnerability and adaptation to heatwaves: A case study of Graz (Austria). *Climate Policy, 18*, 63–75.
Ribeiro, P. J. G., & Gonçalves, L. A. P. J. (2019). Urban resilience: A conceptual framework. *Sustainable Cities and Society, 50*, 101625.
Romero-Lankao, P., Gnatz, D. M., & Sperling, J. B. (2016a). Examining urban inequality and vulnerability to enhance resilience: Insights from Mumbai, India. *Climatic Change, 139*, 351–365.
Romero-Lankao, P., Gnatz, D. M., Wilhelmi, O., & Hayden, M. (2016b). Urban sustainability and resilience: From theory to practice. *Sustainability, 8*, 1224.
Ruggerio, C. A. (2021). Sustainability and sustainable development: A review of principles and definitions. *Science of the Total Environment, 786*, 147481.
Sachs, J. D., Kroll, C., Lafortune, G., Fuller, G., & Woelm, F. (2022). *Sustainable development report 2022.* Cambridge University Press.
Shandas, V., Matsler, A. M., Caughman, L., & Harris, A. (2020). Towards the implementation of green stormwater infrastructure: Perspectives from municipal managers in the Pacific Northwest. *Journal of Environmental Planning and Management, 63*, 959–980.
Sharifi, A. (2019). Resilient urban forms: A macro-scale analysis. *Cities, 85*, 1–14.
Sharifi, A., Yamagata, Y. (2016). Urban resilience assessment: Multiple dimensions, criteria, and indicators. *Urban Resilience: A Transformative Approach,* 259–276.
Thorne, C. R., Lawson, E. C., Ozawa, C., Hamlin, S. L., & Smith, L. A. (2018). Overcoming uncertainty and barriers to adoption of Blue-Green Infrastructure for urban flood risk management. *Journal of Flood Risk Management, 11*, S960–S972.
Thorne, C. R. (2020). *Blue–Green Cities: Integrating urban flood risk management with green infrastructure*, ICE Publishing.
Tuofu, H., Qingyun, H., Dongxiao, Y., & Xiao, O. (2021). Evaluating the impact of urban blue space accessibility on housing price: A spatial quantile regression approach applied in Changsha, China. *Frontiers in Environmental Science, 9*, 164.
van Oijstaeijen, W., van Passel, S., & Cools, J. (2020). Urban green infrastructure: A review on valuation toolkits from an urban planning perspective. *Journal of Environmental Management, 267*, 110603.
Venkataramanan, V., Packman, A. I., Peters, D. R., Lopez, D., McCuskey, D. J., McDonald, R. I., Miller, W. M., & Young, S. L. (2019). A systematic review of the human health and social well-being outcomes of green infrastructure for stormwater and flood management. *Journal of Environmental Management, 246*, 868–880.
Verma, P., & Raghubanshi, A. S. (2018). Urban sustainability indicators: Challenges and opportunities. *Ecological Indicators, 93*, 282–291.

Weichselgartner, J., & Kelman, I. (2015). Geographies of resilience: Challenges and opportunities of a descriptive concept. *Progress in Human Geography, 39*, 249–267.
Yamagata, Y., Sharifi, A. (2018). Resilience-oriented urban planning. *Lecture Notes in Energy*, 65.
Zeng, X., Yu, Y., Yang, S., Lv, Y., & Sarker, M. N. I. (2022). Urban resilience for urban sustainability: Concepts, dimensions, and perspectives. *Sustainability, 14*, 2481.

From "Human-Centered" to "System-Oriented": Eco-Cultural Legacy of Feng Shui and Scientific Principles for Establishing Modern Resilient Cities

Wenjian Pan

Abstract Not only are cities confronting risks from global warming, but there are also multiple ecological challenges to achieving sustainable development goals (SDGs), particularly increasing urban heat island effects, more frequent heat waves, and the threat of flood hazards. However, blue-green infrastructure (BGI) plays a crucial role in mitigating these risks while maintaining the normal operation of cities. Feng Shui, a theorization of Chinese ancients' habitation experiences, offers an approach to interacting with nature and complying with the laws of natural ecosystems. It has provided many philosophical thoughts and scientific guidance on making habitats resilient in the face of environmental changes and multi-hazards. Using Hong Village as a standard application model for Feng Shui, this chapter reinterprets the naturally adaptable principles in Feng Shui and analyzes how its passive strategies can be adopted in dwelling systems. It finds that Feng Shui advocates co-developing with nature and treats human habitats as an organic living system that is interlinked with surrounding environments. Specifically, it emphasizes flowing air (*Feng*) and flowing water (*Shui*) as two fundamental natural elements to be integrated into construction activities, which can contribute to the accumulation and circulation of energies, resources, and dynamics. As such, a balance between humans' activities and the natural processes is critical in Feng Shui, making it a useful philosophy for this context. Based on these reinterpretations, the present chapter establishes a multi-scalar framework for scientific modeling and modern applications. Ultimately, the Cheonggyecheon urban restoration project in South Korea and the Yanweizhou ecological rehabilitation project in China are taken as two successful modern cases to showcase integrations of Feng Shui principles into BGIs in contemporary cities, and they aim to adapt to the emerging challenges of increasing urban heat and flood hazards, respectively. To conclude, a few nature-based, and soft intervention-oriented approaches are outlined for the construction and rehabilitation of BGIs to strengthen urban resilience and achieve the long-term goals of sustainable development.

W. Pan (✉)
School of Architecture and Urban Planning, Huazhong University of Science and Technology, Luoyu Road 1037, Hongshan District, Wuhan 430074, China
e-mail: wenjian-pan@connect.hku.hk; blueice3@163.com

P. K. Joshi et al. (eds.), *Blue-Green Infrastructure for Sustainable Urban Settlements*,
https://doi.org/10.1007/978-3-031-62293-9_7

Keywords Feng Shui · Blue-green infrastructure · Urban heat and flooding · Nature-based solutions · Soft intervention · Urban resilience

Glossary

Feng 风	Wind/Flowing air
Huo 火	Fire
I Ching 易经	The Book of Change
Jin 金	Metal
Kan Yu 堪輿	Geomancy/Deep investigation of land
Li Qi Pai 理氣派	Compass School
Luo Pan 罗盘	Chinese compass
Mu 木	Wood
Qi 气	Air
Ren Ding Sheng Tian 人定胜天	Man can conquer nature
Shui 水	Water
Shui Kou 水口	Water inlet/Water gap
Si Shui Gui Tang 四水歸堂	Rainwater flows from the pitched roofs to the courtyard
Tai Ji 太极	Tai Chi/The Great ultimate
Tian Jing 天井	Patio/Light well
Tian Ren He Yi 天人合一	Integration of humans and nature
Tu 土	Earth
Wu Xing 五行	Five elements
Xing Shi Pai 形勢派	Form School
Xue 穴	Lair, cave, and hole/construction site
Yang 阳	Light/Positive dimension
Yi Shan Bang Shui 依山傍水	Connecting to mountainous terrain and water areas
Yin 阴	Darkness/Negative dimension
Yin Di Zhi Yi 因地制宜	Adopting local methods and approaches
Zuo Bei Chao Nan 坐北朝南	Facing south and having mountains at the back

1 Introduction

From ancient times to the contemporary era, humans have faced a series of socio-economic crises, environmental changes, and multi-hazards (Guo et al., 2023a, b; McPhearson et al., 2021). The aspiration of establishing a good environment for human sustainability, has resulted in the expending of any efforts in the development of new technologies, dwelling patterns, as well as infrastructure forms that can produce high economic growth and support efficient city operations (Margolis & Robinson, 2007; Hauck et al., 2012; Kapoor et al., 2020; UNEP, 2021; Bhadouria et al., 2023). Not only are cities confronting risks from global climate change, but there are ecological challenges to achieving sustainable development goals (SDGs), specifically, the increasing urban heat island effect, intensifying air pollution, and frequent heat waves and flood hazards (Fig. 1). These have been accelerated by human activities and the radical changes of land use and land cover in surface of the earth, which have threatened the health of the ecosystems in cities (Chung et al., 2021; Romanello et al., 2023). Recent research reports reveal that the built environment already accounted for nearly 40% of global carbon emissions, within which the building industry and construction of infrastructures contributed significantly (Architecture, 2030, 2022). Moreover, the ever-highest global average temperature was detected during the near past 12 months (from Nov 2022 to Oct 2023) when comparing to the baseline of pre-industrial period of 1850–1900 (Wong, 2023). Many countries have taken measures to respond to the increasing urban heat and dramatic carbon emissions. Such an alarming situation has also been specifically discussed and debated in the COP28 (Burki, 2023). Under these circumstances, to make humans' habitats more adaptable and resilient has become a global consensus. There is an urgent need to reflect on current development modes and seek regeneration-oriented approaches. However, the blue-green infrastructure (BGI) plays a fundamental role in shaping the morphologies of cities and sustaining urban operations (Hauck et al., 2012; Chung et al., 2021; UNEP, 2023).

The "human-centered" development model has been upheld as a standard for sustainability for years. However, the world is complex, and it operates in a non-liner way with multiple layers of elements and connections working together (Bettencourt, 2021). This chapter argues that humans are part of the natural ecosystem and the construction and development of habitats (cities, towns, counties, villages,

Fig. 1 Contemporary ecological challenges in global to achieving sustainable development goals (reinterpreted and drawn by the author based on United Nations, 2022)

neighborhoods, buildings, etc.) should principally follow the rules and processes of nature. It is time to extend the connotation of "human-centered" development model to a more inclusive and complex "system-oriented" intervention paradigm. Feng Shui was originated from China's ancient agricultural civilization and a theorization of the ancients' dwelling experiences that how to interact with nature and comply with operating laws of eco-systems, so as to make the dwelling systems adaptable and resilient when confronting environmental changes and natural risks. Feng Shui theory treats the land (and also its surrounding environmental contexts) as a holistic living system and advocates co-developing with nature. It primarily aims to seek a balance state between humans' activities at the meso-scale to micro-scale and the natural processes at the macro-scale. Such a way of thinking is even more crucial for modern-day practices as cities are growing more complex and dynamic. Existing research has started to acknowledge the scientific values of Feng Shui while the modeling and analyses are limited to a single spatial scale and the simplification of the environmental contexts (Almodovar-Melendo & Cabeza-Lainez, 2018; Gao et al., 2007; Tang et al., 2012). Therefore, a multi-scalar reinterpretation of the ecological principles and strategies of Feng Shui provides new insights into the establishment of socio-ecologically resilient communities for contemporary society.

This chapter uses Hong Village as a standard application model for Feng Shui analysis. Through archival research, site survey and mapping, as well as informal interviews with indigenous residents, the chapter reinterprets the naturally adaptable principles in Feng Shui and analyses how its passive strategies can be effectively adopted in dwelling systems. Based on these reinterpretations, a multi-scalar framework is established for scientific modeling and modern applications. Meanwhile, the Cheonggyecheon urban restoration project in South Korea and the Yanweizhou ecological rehabilitation project in China are taken as two successful modern cases to demonstrate integrations of Feng Shui's eco-thoughts and scientific principles into the planning and construction of BGIs in cities. The Cheonggyecheon and Yanweizhou projects aim to respond to the emerging challenges of increasing urban heat and air pollution, as well as flood hazards, respectively. In the end, several nature-based approaches and soft intervention-oriented insights are outlined for both new construction and rehabilitation of BGIs to strengthen urban resilience.

2 Theoretical Framework

2.1 Environment *Matters—From "Human-Centered" to "System-Oriented" Paradigm*

Sustainable development refers to the "development that meets the needs of the present without compromising the ability of future generations to meet their own needs" (United Nations, 1987). Three key pillars are supporting the establishment of sustainable development goals (SDGs), namely, the "social," "economic," and "environmental" pillars (United Nations, 1987 and 2022). Although each of these pillars individually cater for specific demands of development, these three dimensions are interlinked and cross-functioning in a systematic way, in which the *environment* plays a fundamental and supporting role (Pan & Du, 2021).

From the last century until now, scholars from various disciplines have begun to view human communities as an organic part of ecosystem and advocated for human behaviors that should respect and comply with the material circulations in nature (Alberti, 2009; Browning et al., 2014; Groffman et al., 2017; Pan & Du, 2021; McHarg, 1995). As a response to this reflection, recent discussions and debates have focused on exploring and establishing an integrated framework for urban planning and architectural design that could effectively function with specific contexts rather than a homogenous paradigm. The emerging thoughts, concepts, and practical models for urban sustainability range from *green urbanism, landscape urbanism, clean urbanism, beautiful urbanism, infrastructure urbanism*, and *social urbanism*, to *ecological urbanism*. Representatively, the advocacy of *ecological urbanism* considers both the socio-economic and environmental impacts as well as their interrelations from the perspective of Felix Guattari's *Three Ecologies* (Guattari, 2000; Mostafavi, 2010; Pan, 2019). The philosophy of *Three Ecologies* (social ecology, mental ecology, and environmental ecology) advocates for a system-oriented perspective to evaluate the operation state of each habitat and make trade-off decisions for (re)development (Fig. 2).

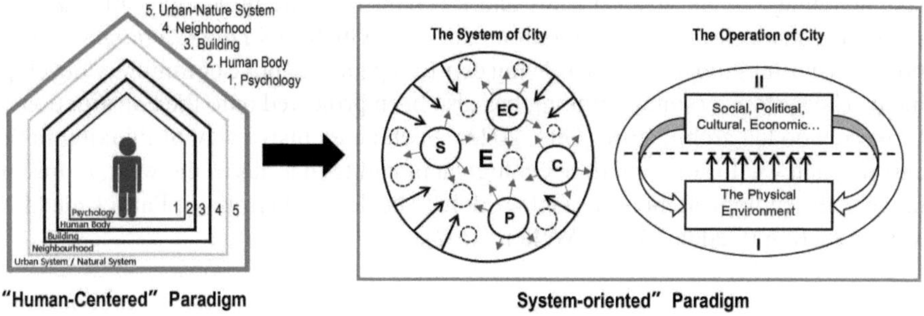

Fig. 2 Left: The "human-centered" paradigm (reinterpreted and redrawn by the author based on Marco & Burgess, 2015); Right: The "system-oriented" intervention as an extended paradigm of the "human-centered" paradigm—Conceptual graphics show the role of "environment" within city ecosystem and its relationship with other urban components (adapted from Pan, 2019)

Interpretations:
• ***The System of City***: Urban environment functions as a place/platform to contain various components of the city while it also reflects and exhibits the "status" of the city, which results from the interactions between multiple components within the city (**S**—social dimension, **EC**—economic dimension, **C**—cultural dimension, **P**—political dimension, etc.);
• ***The Operation of City***: The relationship between (I) and (II) is the *ecology* of human settlements and the city – (II) is based on (I); (I) supports (II), while (II) reshapes (I); (I) is the material carrier of (II), and; (II) develops from (I). Therefore, the "environment" plays a very critical role in (trans)forming the city.

Pickett et al. (2016) have theorized three intervention paradigms of "ecology" for urban investigations, namely, ecology *in* the city, ecology *of* the city, and ecology *for* the city, among which the complexity increases sequentially. The philosophy of *Three Ecologies* along with the emerging theoretical paradigms on *urban ecology* have inspired a resonance between urbanists, planners, architects, landscape designers, social reformers, and policy makers that we should "think ecologically" to understand the constitutional characteristics (ecology *in* the city) and operational mechanism (ecology *of* the city) of our dwelling system and be concerned about the fluidity of natural resources, elements, and populations, so as to seek suitable and adaptable models for its long-term sustainability (ecology *for* the city) (Pan & Du, 2021; Pickett et al., 2024). Specifically, the conceptualizations of "one health" (Guo et al., 2023a, b) and "nature-based solutions" (NbS) (Cohen-Shacham et al., 2016) are put forward and are being hotly discussed in recent years.

Although there was a discourse of *Ren Ding Sheng Tian* (man can conquer nature) in Chinese traditional culture, the ancient Chinese highly respected mother nature and its functioning logics. They focused on the integration of built elements and natural environments with the aim to construct feasible, impressive, appreciable, and livable while also poetic dwellings. The Chinese philosophy of *Tian Ren He Yi* (integration of humans and nature) is a master guide for them to conduct activities and normalize daily life. The *Tian Ren He Yi* is a "system-oriented" thinking which treats humans as part of the natural ecosystem on earth. Humans play an active role in the natural systems and should comply with the natural processes to the most extent and take advantage of the natural rules and mechanisms. The philosophy of *Tian Ren He Yi* also shares a similar connotation with Felix Guattari's *Three Ecologies*, which provides master guidance for achieving eco-living and adaptive community development. The system-oriented thinking has also been projected onto the construction of Chinese wooden architecture (Guo, 2023). Such an inclusive way of thinking in the east resonates with some modern ecological planning theories in the west, representatively, the ecological thoughts and wisdom to be delivered from Sir Patrick Geddes' Valley Section (Geddes, 1923; Welter, 2002).

2.2 *Feng Shui—A Grounded Culture about Dealing with* Environment

Feng Shui is a cultural phenomenon in China and is primarily grounded in Chinese agricultural civilization and sciences of human settlements (Wu, 2017). It summarized and theorized the Chinese ancestors' holistic observations, practical recordings, and perceptual understandings of nature and habitations (Bruun, 2008). Initially, Feng Shui was called *Kan Yu* (Geomancy) in Chinese traditional culture, meaning conducting a deep investigation on the environment and ecologies of a site in order to meet an ideal and sustainable model of residence and activity (Ding, 2011). The implementation of Feng Shui demonstrates Chinese ancestors' beautiful vision to establish a poetic habitation and maintain a harmonious relationship with nature (Paton, 2021; Yoon, 1989). There are two major schools of Feng Shui, namely the Compass School (*Li Qi Pai*) and the Form School (*Xing Shi Pai*), which generally offer philosophical basis and practice guides of Feng Shui, respectively (Luo & He, 2008; Mak & Ng, 2005; Wang, 2005).

- ***The Compass School—Integration of humans and nature***

The Compass School Feng Shui provides a deep interpretation of the philosophy of *Tian Ren He Yi* (integration of humans and nature), which is mainly based on the theories in *The Book of Change* (*I Ching*, written by King Wen of Zhou Ji Chang in Spring and Autumn period, 770–476 BC/403 BC). It argues that all elements in our world are interacted, inter-influenced, and inter-transformed, and should be treated as an organic entity to form the operation of the world (Wilhelm & Wilhelm, 1995). Specifically, the totem of Tai Chi (*Tai Ji*) in Compass School shows that all the elements in the universe contain two distinguished characteristics, namely, the *Yin* (darkness/negative dimension) and *Yang* (light/positive dimension) characteristics. The *Yin* and *Yang* co-exist and interact with each other within an entity ("all in one") in the world. The Tai Chi symbolizes the mutual compensation of the negative and positive aspects of things, and they finally should achieve a dynamic but balanced state, maintaining normal operations of all things in the world (Fig. 3). Therefore, the *Yin* and *Yang* characteristics within the entity can be transformed in specific conditions, meaning the elements whether with negative or positive characteristics can be utilized for creating good environments for dwelling under different contexts when taking an effort to maintain the balance between these two dimensions. The discourse of *Yin* and *Yang* indicates a connotation of "shared ecologies" when humans confront changes and shocks in the modern society (Pan, 2023). On the other hand, the Compass School also further sub-categorized the world's elements into five distinct natural attributes, which are the well-known "Five Elements" (*Wu Xing*), including Metal (*Jin*), Wood (*Mu*), Water (*Shui*), Fire (*Huo*), and Earth (*Tu*). The "Five Elements" represents five major types of materials and resources in the world. They are interlinked and inter-functioning, together sustain their circulations and maintain the world's operation (Wang, 2005, Fig. 3). What is more, the Compass School usually use *Luo Pan* (Chinese compass) to compose and examine time and space elements (Mak & Ng, 2005). In a nutshell, the core argument of the Compass School Feng Shui is to understand multiple aspects and diverse characteristics of each thing as well as its connection with and impacts on surrounding environments within the ecosystems.

- ***The Form School—Nature-based solutions based on air (Feng) and water (Shui)***

As mentioned, the primary vision of implementing Feng Shui principles is to create livable and sustainable habitation. Correspondingly, Feng Shui practices are mainly about dealing with environments and rehabilitating habitats in a soft manner. Thus, the nature-based solutions (NbS) are primarily adopted to conserve the natural

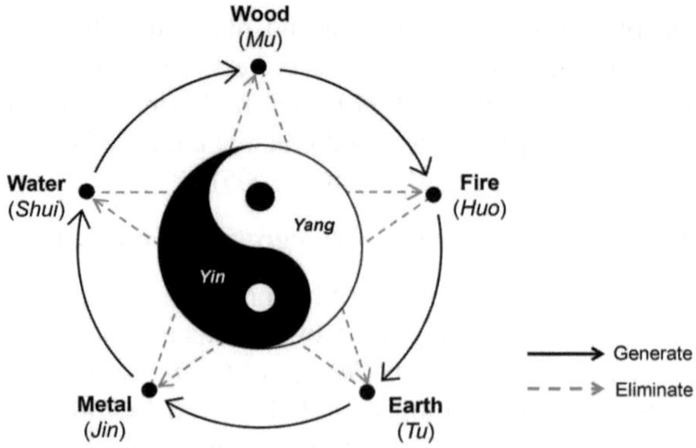

Fig. 3 The totem of Tai Chi and the inter-relationships and circulations among the "Five Elements" in Feng Shui (reinterpreted and redrawn by the author, based on Wang, 2005)

ecosystems. Even in modern China, the *environment* is still among the most important themes to be examined during urban planning and governance (Pan & Du, 2021; Xu & Chung, 2014). While the Compass School provides the philosophical basis of Feng Shui, the Form School is more related to the implementation and practical aspects. It provides specific principles to guide humans' activities. The Form School Feng Shui is primarily based on the *Book of Burial* (*Zang Shu*, written by Guo Pu in the East Jin Dynasty, AD 276–324) (Mak & Ng, 2005). We can find multiple regulations and principles in the *Book of Burial* that contains eco-thoughts for practices. There is a core statement in the book that "*Qi* will disperse under strong wind and be stopped by water" (*Qi Cheng Feng ze San, Jie Shui ze Zhi*) (Zhang, 2004). Here, the *Qi* represents the flowing air (*Feng*), which is a typical form of natural energy (Porkert, 1978). This statement provides important insights for ancient Chinese that the planning and construction of dwellings should enable the accumulation of *Qi* (air) to form and maintain energy flows in the land (construction site). In the meantime, water (*Shui*) is another state of natural energy, which works as a visible and touchable carrier to deliver materials and resources to the land. Water helps to circular vitality in between nature and human habitats. Thus, the element of water should be introduced into dwelling systems as it works together with *Qi* (air) to circular energies and resources: when *Qi* (air) meets water, it integrates into water and then being purified. This means that the introduction of water to residential areas contributes to generating fresh air and maintaining neighborhood dynamics (Luo & He, 2008). Such a discourse actually well-summarizes the general laws of the natural mechanisms and dynamic circulation processes. Therefore, the ancient Chinese treated *Qi* (air) and water as two fundamental forms of natural energies to be investigated and utilized during their daily activities, and according to the Form School Feng Shui, an ideal construction site should "sustain air and gain circulating water." Several key scientific principles towards the establishment of nature-friendly and resilient habitation in terms of site selection, settlement construction, and maintenance of community vitality are presented as follows (Wang, 2005; Fig. 4).

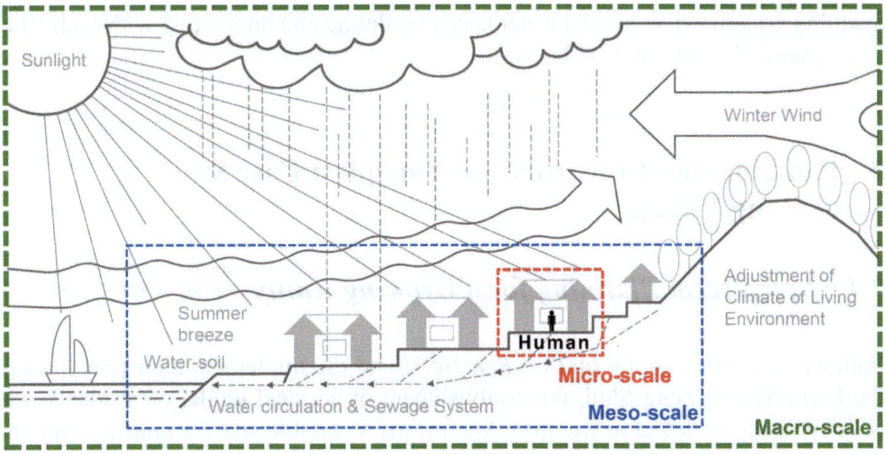

Fig. 4 Eco-wisdom and applications of Chinese Feng Shui at multiple scales (reinterpreted and redrawn by the author, based on Wang, 2005)

System-oriented Planning:

• **LOCALITY**—Local Approaches and Flexibility (*Yin Di Zhi Yi*)	Adopting architectural interventions that fit specific surrounding natural environments and processes
• **INTEGRITY**—Contextualized and System-oriented Strategies	Investigating both the micro-environment of the site and its macro-environments and background context
• **EQUILIBRIUM**—Trade-off and Seeking Dynamic Balanced State	Seeking equilibrium among multiple elements of the dwelling system through soft interventions

Nature-based Solutions:

• **GROUNDEDNESS**—Connecting to Natural Infrastructural Systems (*Yi Shan Bang Shui*)	Attaching to mountainous terrain and water areas
• **RECYCLABILITY**—Deep Assessment of Geological and Hydrological Conditions	Assessing the qualities of earth and circulation of water to guarantee sustainability of the site in a long run
• **PHOTOTROPISM**—Making Public Spaces Opened and Buildings Face South and Have Mountains at the back (*Zuo Bei Chao Nan*)	Gaining sufficient natural daylight and creating conditions to promote passive air ventilation

In short, most of the environmental issues can be essentially attributed to issues related to flowing air (*Feng*) and flowing water (*Shui*) in Form School Feng Shui. The theoretical model provides practical guidance to establish a natural adaptable dwelling system where air and water keep circulating and interacting with each other in a dynamic balance state (Fig. 4).

3 Planning and Construction Principles: Case Study of Hong Village

3.1 Long-Term Planning for a Growing Entity

Following the philosophy of *Tian Ren He Yi* and the implementation principles of the Form School Feng Shui, the establishment of an ideal model of residence and activity needs to detect a *Xue* first. *Xue* means a suitable site for construction and development, it could mean a lair, a cave, a niche, or a hole in ancient Chinese context (Yu et al., 2023). According to the Form School Feng Shui, a *Xue* is an area that possesses good conditions to introduce and accumulate *Qi* (air) and enables it easily to flow and circulate. Thus, a qualified *Xue* should follow the corridors or paths of *Qi* (flowing air, wind) and allow it to flow across spaces, from the sky and run to the ground (Ru, 1993). Usually, a *Xue* is located at the converging focus of a clearly demarcated serpentine chain of hills or mountains (Ding, 2011; Kang, 1999; Fig. 5-up). The dwelling system of Hong Village in China's Anhui Province is a good example that complies with the theoretical model of Feng Shui, where the natural elements of flowing air (wind) and flowing water were systematically utilized during the construction and operation of the village's infrastructures from ancient times to the present (Fig. 5-bottom).

The Hong Village is located in the northeast area of Yi County in China's Anhui Province (117° 38' E, 30° 11' N). It occupies a total land area of 188.95 km^2 (including the core dwelling zones of 420 acres and the surrounding natural areas) and has been included in the World Cultural Heritage List by UNESCO since 2000 (Xia et al., 2022). Currently, there are 1,850 residents live in Hong Village, including 1,368 indigenous villagers. According to the *Wang Genealogy*, the history of Hong Village can be dated back to the Southern Song Dynasty (AD 1131) when it was initially established by the Wang Clan 800 years ago (Chinese Traditional Village Museum, 2020; Jin, 2005). After centuries of development, the macro-environment of Hong Village was then carefully investigated and planned by the Feng Shui master He Keda and the Wang Clan in the 15th century (Pan, 2004). From the initial planning to the completion of construction, Hong Village has been gradually transformed into an organic dwelling system where living *Qi* (flowing air) is being accumulated and fresh-water is being introduced to keep circulating. By fully utilizing air (*Feng*) and water (*Shui*), all the components and spatial patterns within the village were intentionally designed to match the theoretical model of Feng Shui and well organized into the geographic terrains and ecosystems of the wider context of the range of Mount Huang.

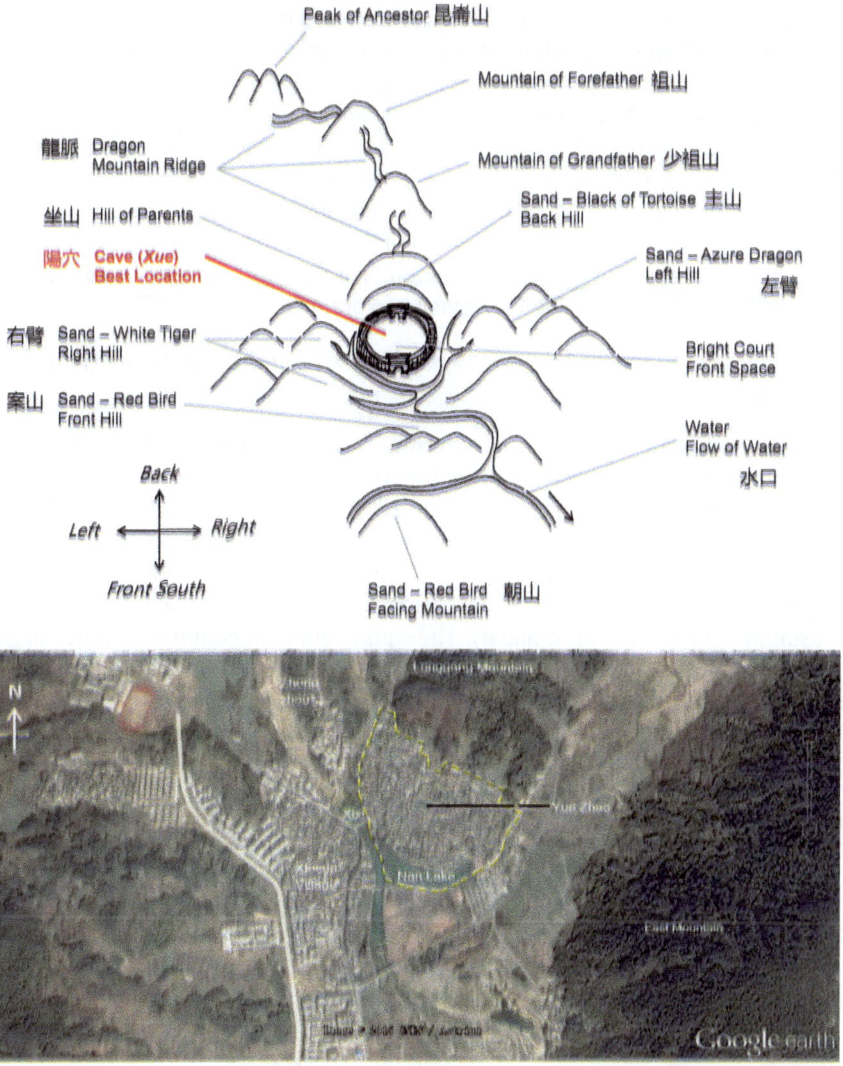

Fig. 5 Up: The theoretical model of Feng Shui (redrawn by the author, based on Wang, 2005 and Mak & So, 2011); Bottom: Overview of the master plan of Hong Village (adapted from Google Earth map)

3.2 Feng—*The Principles of Air*

- ***System morphology: Facing south and having mountains at the back***

Nearly all the buildings in Hong Village have a southwest orientation rather a due south orientation in order to obtain enough natural daylight (Fig. 6-a). According to the "Five Elements" (*Wu Xing*) in Compass School Feng Shui, a south orientation is related to the element of Fire (*Huo*), while Fire could melt Metal (*Jin*). One belief that many of the villagers have done business across regions of China since the ancient times. Such an economic phenomenon is related to the element of Metal. In order to resolve this "conflict", buildings was not designed to face due south (Jie, 2006). However, the principle of "Facing South and Having Mountains at the back" (*Zuo Bei Chao Nan*) should also be followed. Thus, villagers finally chose the southwest orientation for their buildings as it not only well complied with the local topographical conditions but also matched the direction of the local monsoon wind, so as to create favorable conditions for natural ventilation. Through this way, the trade-off among the cultural norms and ecological demands were met.

- ***Neighborhood space: Deep alleys enable breathability, connectivity, and walkability***

The construction of residential areas in Hong Village adopted a compact morphological pattern. Most of the buildings are low-rise (one to three floors) and most of the alleys only one to two meters wide (Shan, 2009). Consequently, the external walls of these buildings on both sides of the alleys are six to eight meters high with a spatial aspect ratio (H/W) of 3.0~8.0, forming a deep alley system across the whole village (Chen et al., 2013; Coseo & Larsen, 2015). These alleys are usually shaded during hot summer days (Fig. 6-b, c and d). The significant temperature difference between the shaded and unshaded areas in these alleys contributes to promoting the flow of air and then generating effective outdoor cooling effect (Ng, 2010; Salmond et al., 2012; Turner et al., 2023). Meanwhile, these deep alleys with different geometric

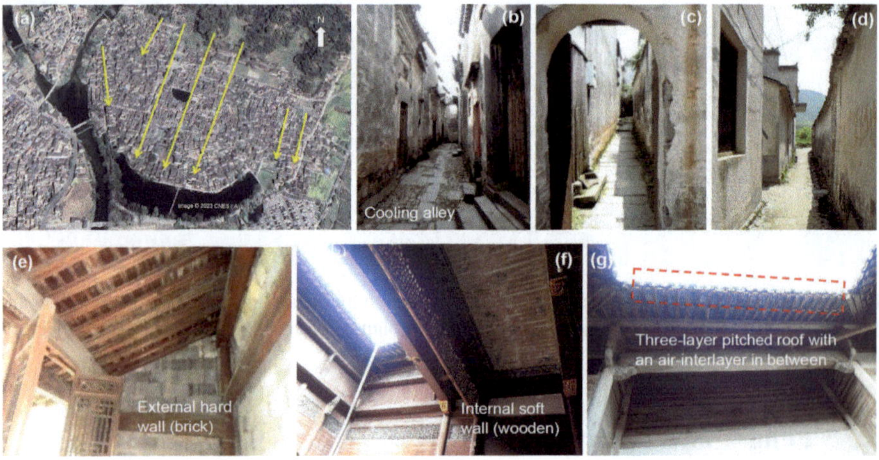

Fig. 6 The adopted principles of *air* at multiple scales in Hong Village (images **b~g**: photography by the author). **a** Spatial structure of Hong Village and its building layouts (adapted from Google Earth map); **b~d** High aspect ratio of the deep alleys in high-density Hong Village are shaded and generating cooling effect in strong sunny daytime; **e~g** The sandwich wall and pitched roof in Hong Village's building, showing three layers, including the air-interlayer

characteristics not only make the whole village a high connectivity and walkability to benefit residents' daily life, but also function as corridors for flowing air, well connecting with each building's external doors and windows. Through such spatial planning and design, cross ventilation is promoted, making the whole village function as a breathable dwelling system.

- ***Building construction: Air-interlayers—a passive micro-air circulation system***

In terms of building scale, the ancestors of Hong Village adopted a timber structural system (constituted and framed by wooden beams and pillars) for their houses. Both the external walls and pitched roofs in these houses were designed as sandwich forms which contain a unique air-interlayer (Shan, 2009). The external walls were constituted by three layers, including the brick layer (outside), the wooden layer (inside), and the air-interlayer (in between). The brick layer walls were constructed to have a more solid protection function to prevent fire risk, heat, and pollutants (Fig. 6-e), while the wooden layer walls were designed as soft elements to fit a warm home atmosphere (Fig. 6-f). As a result, the air-interlayer works as a buffer layer or a cavity to strengthen heat preservation and insulation while naturally making air flow circulate around the house. This air circulation mechanism also functions in the pitched roofs (Fig. 6-g). Such design strategies make the house a breathable architectural system at a micro-scale as well.

3.3 Shui—*The Principles of Water*

- ***System morphology: Operating as a living organism***

The master plan of Hong Village took references from ancient Chinese geomancy and bionics (Yan et al., 2023). The village is widely known as "buffalo-shaped village" as its morphology generally looks like the shape of a buffalo, in particular, the morphology of its water system (Shan, 2009; Fig. 5). The introduction of fresh-water is very important for human settlements as it contributes to preventing fire risk while meeting the need of agricultural plant irrigation. Through complying with the geographical characteristics and topography, the villagers constructed a circulating water system according to the biological structures of the animal buffalo (Duan & Jie, 2009). To analyze the bio-morphological constitution of Hong Village from an aerial view, Chinese cultural geographers vividly describe that the Leigang Mountain is the head of the "buffalo," the trees grow in the mountain are the horns of the "buffalo," the Xixi Stream is the tone of the "buffalo," and the water canals go through the whole village represent for the intestinal tract of the "buffalo." Moreover, the Moon Pool (*Yue Zhao*) located in the center of the village symbolizes the stomach of the "buffalo," while the South Lake (*Nan Hu*) implies the tripe of the "buffalo" (Duan & Jie, 2009; Jin, 2005; Pan, 2004).

- ***Neighborhood space: Diverse water forms and well-connected water channels***

In terms of artificial water systems, they were planned to integrate with the surrounding natural landscapes. Hong Village is surrounded by many water areas in its west, south, and east directions, while the Leigang Mountain is in the village's north. The water inlet (*Shui Kou*) is located between the Jiyang and Dongshan mountains (Fig. 7-a). The overall morphological structure of Hong Village is primarily based on its water systems as they were planned at the initial stage and established first (Liu, 2000). Then it provided a general morphological framework for later detailed construction of buildings and community public spaces. This has resulted in a compact built form as well as a high-efficiency land use pattern. Existing historical-morphological research reveals that the establishment of the water systems was step by step and has experienced three main rounds of transformation, including changing the watercourse of Xixi Stream (western water area), creating the Moon Pool (central water area) to sustain living *Qi* (Fig. 7-b), and digging the southern area to form the South Lake (southern water area) in order to circulate the water and improve the village's micro-climate (Duan & Jie, 2009; Pan, 2004; Wang & Jia, 2011). In addition, the villagers introduced fresh-water into their residential buildings from the Xixi Stream, Moon Pool, and South Lake to extend the circulation of fresh-water (Fig. 7-c). This has also promoted the interior cooling effect and enhanced thermal comfort during hot summer season. All in all, the overall water systems of Hong Village were designed for long-term sustainable use from ancient times to the present.

Fig. 7 The adopted principles of *water* at multiple scales in Hong Village (photography by the author). **a** Macro-scale: Water inlet (*Shui Kou*) of Hong Village, well connecting to surrounding natural environments and accumulating living *Qi* and enabling dynamic water circulation; **b** Neighborhood scale—Moon Pool (*Yue Zhao*), located in the central area of Hong Village, which is the village's most important public space embedded in the high-density building clusters, which helps to accumulate living *Qi*; **c** Neighborhood scale—circulating water systems surround the village; **d**~**f** Building scale—the patio/courtyard enables circulation of water

- ***Building construction: Courtyard and patio***

In the Chinese context, courtyard is another crucial architectural element to symbolize the philosophy of *Tian Ren He Yi* (integration of humans and nature). A courtyard is a transitional space between the interior living areas and the exterior activity areas in Chinese vernacular architecture. It functions as a spatial medium for humans to interact with nature and helps to directly receive sunshine, fresh air, energy, and rainwater (Ou, 2001; Shou et al., 2019). There is a cultural discourse on courtyards in China, that is, *Si Shui Gui Tang*, meaning the rainwater flows from the sky and pitched roofs to the courtyard. This symbolizes that the energies, fortune, and luck flow and accumulate into the building (Kang, 1999; Shan, 2009). Nowadays, many modern buildings in China still inherit this beautiful cultural vision. From this perspective, the courtyard is a critical architectural element that helps to fulfill both cultural beliefs and ecological effects. From the northern region to the southern part of China, the shape and size of courtyards in buildings are becoming smaller and narrower due to variations of climate and geographical conditions (Wang et al., 2019; Wu, 2014). Considering local climatic characteristics, people adopt local approaches to construct different forms and sizes of courtyards in their buildings in order to create suitable conditions to enable sufficient daily lighting, ventilation, and drainage, while preventing pollutants diffuse to indoor areas and keeping the air clean (Fig. 7-d, e and f). Patio or light well (*Tian Jing*) is a typical form of courtyard in eastern and southern China as these regions share a high solar elevation and receive more monsoon wind in the summertime (Zhong et al., 2023). As Hong Village is located in southeast China, the courtyards in buildings have been formed as narrow horizontal rectangles with a relatively small size. Therefore, these "tiny courtyards" are usually called patio or light well (Ou, 2001). On one hand, the patio enables the interior spaces of the building to gain enough daylighting for daily living while reducing the receiving of strong solar radiation. On the other hand, it allows the building to generate thermal pressure ventilation, which promotes the air flow and brings more fresh air into the interior living spaces (Zhong et al., 2023). Moreover, the patio contributes to collecting rainwater and circulating it to the water systems of the whole village (Shou & Xue, 2022; Fig. 7-b, c and e).

3.4 Synthesis: A Multi-Scalar Framework for Modern Applications

The above analyses reveal that, from macro-scale (master plan and overall morphological arrangement) to meso- and micro scales (architecture and building envelopes), the Hong Village functions as a complex dwelling system with the circulation of flowing air (*Feng*) and flowing water (*Shui*) to maintain the village's social-ecological dynamics. To synthesize, the flowing air (*Feng*) and flowing water (*Shui*) can be utilized as two fundamental and transforming natural elements to be integrated into the constructions of blue-green infrastructures and daily life activities of humans to

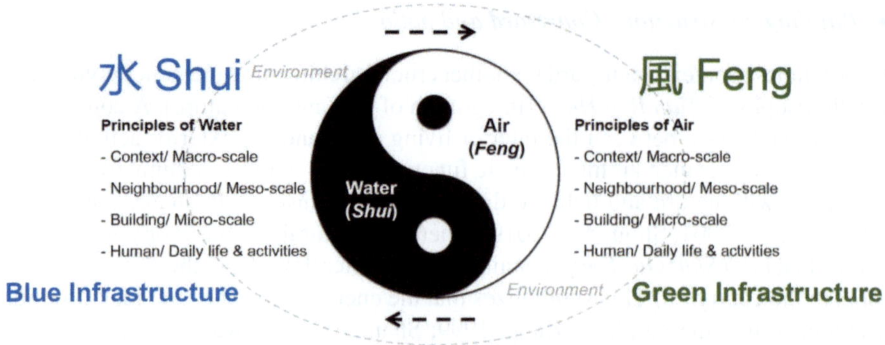

Fig. 8 A multi-scalar framework for modern applications—the inter-relationships among the two fundamental natural elements in Feng Shui and the framework of implementation principles (drawn by the author)

promote the accumulation and circulation of energies, resources, and dynamics at multiple inter-connected spatial scales (Fig. 8).

4 Implementation in Modern Cities

In recent years, many modern cities in the world have sought solutions from Feng Shui to respond to the crises of increasing environmental changes, in particular, the intensifying urban heat island effect, air pollution, and the threat of frequent flood hazards. The Cheonggyecheon urban restoration project in South Korea and the Yanweizhou ecological rehabilitation project in China are the two representative cases to showcase the successful integrations of Feng Shui's eco-thoughts into urban blue-green infrastructure constructions, specifically, the implementations of principles of air and water at macro- and meso-scales.

4.1 Cheonggyecheon: Reintroducing Natural Channel and Making City Breathable

Cheonggyecheon was originally a natural stream flowing through the urban center of Seoul in South Korea. The stream covered a drainage area of 59.83 km^2 and now is 6 km long (Chung et al., 2012; Zhang, 2019; Fig. 9). The transformation of Cheonggyecheon offers a critical lens to examine the histories of Seoul. With the demands for urban development and economic growth in the 1950s~1960s, the Cheonggyecheon Stream was covered to be an underground canal. Since the 1970s, the local government had built elevated expressways above the stream to enlarge the volume of urban transportation. In mid 1970s, the stream was totally covered and surrounded by modern hard infrastructures, including city roads, hard and impermeable pavements, and large-scale modern constructions (Deng et al., 2006; Chung et al., 2012). Consequently, the Cheonggyecheon Stream degenerated its natural properties and eco-functions while the city lost one of its key water systems and ecological corridors. Subsequently, the water quality of the stream became poor due to wastewater discharge and lacking effective circulation. Even worse, the polluted water along with the increasing exhaust emissions from motor vehicles, as well as the intensified traffic congestion accelerated the metropolitan area's air pollution issue, threatening citizens' health and urban environmental sustainability (Seoul Development Institute, 2003; Chung et al., 2012).

With the realization of the negative impacts of air pollution and urban heat island effect that resulted from rapid urbanization, the local government proposed a "Seoul Green Network Plan" in 1997 and then officially implemented in 2003 (Robinson & Ji, 2022). The main intervention strategies of this plan are to introduce natural elements into the urban areas and make the fragmented green patches and wildlife species' habitats to be reconnected. Finally, this plan aims to establish an urban landscape that integrates the artificial built environments and the natural elements (Robinson & Ji, 2022; Sun-Kee et al., 2007). The restoration of Cheonggyecheon Stream is one of the key projects of this plan that a series of ecological interventions had been taken, including:

- To demolish the large-scale motor lanes and elevated expressways in order to open the the linear space and restore the original water channels of the stream;
- To recover the nature-based and soft blue-green infrastructures along the stream in order to minimise artificial interventions while maximise the natural characteristics and eco-functions of the stream. The adopted soft blue-green infrastructures include natural embankments, small wetlands, pocket parks, soft river banks, and small-scale walkable bridges, etc.

The restoration was completed in 2005 (Deng et al., 2006). As a result, the restored Cheonggyecheon Stream has become an eco-medium that connects various parts of the city and promotes the circulation of fresh air into the high-density urban areas in Seoul (Park et al., 2019). Nowadays, the stream is functioning as a dynamic artery of the city ecosystems, which contains diverse urban functions and contributes to air pollution reduction, urban heat mitigation, micro-climate amelioration, as well as social well-being maintenance (Fig. 9).

The Cheonggyecheon urban restoration project illustrates the importance of constructing soft and adaptable infrastructures rather than hard and fixed facilities in cities. The project systematically evaluated the ecological problems of the whole city and aimed to address the air pollution and urban heat issues by adopting nature-based solutions that re-introducing circulating water and opening corridors for the flow of air. It has captured the interacting and transforming characteristics of flowing air (*Feng*) and flowing water (*Shui*) that emphasised in Feng Shui and seeks to create spaces to enable interactions between these two key natural elements and maintain a balanced state of them in the city.

Fig. 9 Up: Location of the Cheonggyecheon Stream in Seoul Metropolitan Area (adapted from Google Earth map); Bottom: Landscape of the Cheonggyecheon Stream with the remained piers of previous elevated expressways in the center of the photo (permission obtained from VCG.COM)

4.2 Yanweizhou: Adapting to Floods and Circulating Water

Yanweizhou Island is located in the west of Dongshi Street in the Duohu District of Jinhua City in China's Zhejiang Province (Fig. 10-up). The island has an area of 26 hm^2 and is surrounded by modern building blocks (Turenscape, 2017). Located in the subtropical region of eastern China, the city of Jinhua is influenced by a strong oceanic monsoon climate with distinct dry and rainy seasons, which confronts frequent floods during May to September of the year. In order to obtain more land for urban development, the local government developed a large number of floodplains in urban areas. A few cement-high embankments were built along the banks of the Yiwu River and Wuyi River to prevent floods, while separating citizens, natural vegetation, and urban built-up areas from the rivers (Turenscape, 2017). During the past few decades, the local government built two flood control embankments (as hard infrastructure) in some areas of the Yanweizhou Island, which had damaged the hydrophilicity of the sandbar park and narrowed the water channels of surrounding rivers (Li & Wang, 2019). Consequently, the destructive power of floods were strengthened, making urban waterlogging disasters frequently occurred. To mitigate this crisis and strengthen the city's resilience (La Loggia et al., 2020), the Jinhua Natural Resources and Planning Bureau cooperated with TURENSPACE during 2010~2013 to re-organize the ecological landscape pattern of the Yanweizhou Island. They systematically examined the overall water systems of the city as well as identified the role of Yanweizhou within the systems. In the end, the project was completed in May 2014 (Turenscape, 2017), and a series of nature-friendly rehabilitation strategies were designed and implemented (Fig. 10-bottom), including:

- To demolish the hard flood control embankments and replace them with adaptive flood embankments;
- To recover the original natural riverbanks while introduce adaptive vegetation and local aquatic carbon sequestration plants;
- To use permeable pavements while conservate the wetland eco-system on the island to sustain freshwater circulation.

Regarding education and entertainment, a series of elevated walkways constructed from local recycling materials (e.g., soft and permeable wooden plates and strips as well as crushed stone particles) were installed above the island with a group of small supported piers to minimise the interference with the natural landscape patterns and ecological processes on the island. This enables citizens to experience nature closely (fresh air, vegetations, and wildlife animals, etc.) on the soft elevated walkways on the one hand, while leaving the ground level of the island as a buffer zone for wildlife's habitats and the city's water systems during both dry and flood seasons. The strategy of creating the buffer zone on ground has also widely adopted in countries with low altitudes, such as the "water plazas", "integrated carparks", and "water reservoirs" in the Netherlands (Al, 2022).

The Yanweizhou ecological rehabilitation illustrates the effectiveness of the approaches of "negative planning" to some extents (Yu et al., 2008; You, 2023). By complying the "principles of water" in Feng Shui and through taking soft and nature-based interventions, the frequency of flood hazard in the city has been significantly reduced as the Yanweizhou Island has been integrated into the water systems of the whole city rather than an enclave that divides water bodies (Fig. 10). The success of Yanweizhou eco-rehabilitation has been promoted as an iconic project in the website of Jinhua Municipal Government (www.jinhua.gov.cn) and made it an effective mode for cities with similar situations across China to adapt to urban floods. Since its completion, the project has generated multiple positive impacts on the city, including:

Fig. 10 Multiple soft interventions to be adopted in Yanweizhou to respond to frequent flood hazards. Up: Area of Yanweizhou Island (adapted from Baidu satellite map); Bottom: Soft intervention approaches to minimize the impacts of artificial built elements and set apart buffer spaces for wildlife habitats and allowing water circulation (photograph by the author)

- Developing a complete natural eco-system
- Recovering soft-natural friendly blue-green infrastructures
- Enabling sustainable water circulation during both dry and flood seasons
- Cooling air and mitigating urban heat island effect
- Fostering local biodiversity
- Contributing de-carbonization through wetland and local plants

- Providing natural education and entertainment functions for citizens
- Strengthening urban adaptability and maintaining city ecological security
- Achieving city health and socio-cultural sustainability.

5 Conclusion and Outlook

Taking Hong Village as an example, this chapter has reinterpreted the ecological principles in Feng Shui and analyzed a few typical passive strategies to be applied in Chinese vernacular settlements. A multi-scalar framework is established by synthesizing the circulation and transforming mechanisms of air (*Feng*) and water (*Shui*) in nature. Feng Shui advocates co-development with nature. It treats human settlements as an organic living system that is interlinked with its surrounding environments. Specifically, it emphasizes flowing air (*Feng*) and flowing water (*Shui*) as two fundamental natural elements to be integrated into construction activities and daily living of humans, which can contribute to the accumulation and circulation of energies, resources, and dynamics. As such, adopting system-oriented thinking and seeking a balanced state between humans' activities and the natural processes is the primary objective in Feng Shui practices. To identify real problems and figure out efficient solutions for cities, we need to observe the inter-relationships among multiple spatial scales rather than concentrating on a single scale. The general principles in Feng Shui can be adapted for modern eco-planning and urban governance, including:

- Integration of humans, habitats, and nature
- Nature-based solutions
- Utilizing the circulation and transformation mechanisms of air and water
- Soft intervention and soft infrastructure
- Behaviors comply with natural laws and processes
- Diversity of locality and dynamic balance.

The deterioration of natural ecosystems and our living habitats, ranging from the increase of heat waves, air pollution, tropical storms and flooding, as well as the pandemic, has been significantly affected and accelerated by humans' activities and continues to threaten public health and social well-being. While the proposed development models and planning paradigms, such as *low-carbon city*, *eco-city*, *green city*, *landscape city*, *sponge city*, *smart city*, *healthy city*, *adaptable city*, as well as *resilient city*, were utilized to mitigate these ecological crises by adopting technologies, the local ecologies, specific needs, and habitus of residents have yet to be sufficiently evaluated and involved. The philosophy of *Tian Ren He Yi* and ecological principles in Feng Shui could provide alternative solutions and adaptive interventions to these emerging crises. Reinterpreting the eco-cultural legacy of Feng Shui also inspires urban designers and city decision-makers to reflect on the impacts of the past and present ongoing activities and then identify the essence of the problems, so as to figure out more sustainable solutions.

The culture and theory of Feng Shui is extensive and profound. While there are many deep insights of Feng Shui into the establishment of adaptive and resilient habitats in modern society, the present chapter only demonstrates the author's perspective of reinterpretation on the eco-cultural legacy and scientific values in Feng Shui. Further case studies at meso- and micro-scales are needed to fully capture the scientific values of Feng Shui for modern applications. In the contemporary era of digital revolution and data-driven development, re-interpreting and modeling the scientific principles in Feng Shui allow more efficient planning and achieve smart governance to benefit socio-cultural sustainability.

Acknowledgements This chapter was partially developed from the author's individual project on urban ecologies and environmental sustainability during his doctoral research in the Department of Architecture and working in the Urban Ecologies Design Lab at the The University of Hong Kong. The author thanks for Prof. Juan Du and Prof. Cole Roskam for the discussions at the initial stage of the study. Parts of the contents in this chapter have been presented in the 18th Multi-Hazards Symposium 2023 at Nanyang Technological University, Singapore.

References

Al, S. (2022). Multi-functional urban design approaches to mange floods: Examples from Dutch cities. *Journal of Urban Design, 27*(2), 270–278.

Alberti, M. (2009). *Advances in Urban Ecology: Integrating Humans and Ecological Processes in Urban Ecosystems*. Springer.

Almodovar-Melendo, J., & Cabeza-Lainez, J. (2018). Environmental features of Chinese architectural heritage: The standardization of form in the pursuit of equilibrium with nature. *Sustainability, 10*, 2443.

Architecture 2030. (2022). Why the built environment? Retrieved August 15, 2023, from https://architecture2030.org/why-the-building-sector/.

Bettencourt, L. M. A. (2021). *Introduction to Urban Science: Evidence and Theory of Cities as Complex Systems*. MIT Press.

Bhadouria, R., Tripathi, S., Singh, P., Joshi, P. K., & Singh, R. (2023). Urban metabolism and global climate change: An overview. *Urban Metabolism and Climate Change: Perspective for Sustainable Cities*, 3–22.

Browning, W., Ryan, C., & Clancy, J. (2014). *14 Patterns of Biophilic Design: Improving Health & Well-being in the Built Environment*. Terrapin Bright Green LLC.

Bruun, O. (2008). *An introduction to Feng Shui*. Cambridge University Press.

Burki, T. (2023). Health at COP28: Three things to watch for. *The Lancet, 402*(10416), 1954.

Chen, X. Y., Zheng, B., & Fu, X. Z. (2013). Actual measurement analysis of lowering temperature in cooling alley of folk house. *Architectural Journal, 02*, 82–85.

Chinese Traditional Village Museum. (2020). *Hong Cun (Hong village)*. (Accessed: http://main.dmctv.com.cn/villages/34102310101/Index.html).

Chung, J. H., Hwang, K. Y., & Bae, Y. K. (2012). The loss of road capacity and self-compliance: Lessons from the Cheonggyecheon stream restoration. *Transport Policy, 21*, 165–178.

Chung, M. G., Frank, K. A., Pokhrel, Y., Dietz, T., & Liu, J. G. (2021). Natural infrastructure in sustaining global urban freshwater ecosystem services. *Nature Sustainability, 4*, 1068–1075.

Cohen-Shacham, E., Walters, G., Janzen, C., & Maginnis, S. (Eds.). (2016). *Nature-based Solutions to address global societal challenges* (xiii + 97pp.). Gland, Switzerland: IUCN.

Coseo, P., & Larsen, L. (2015). Cooling the heat island in compact urban environments: The effectiveness of Chicago's green alley program. *Procedia Engineering, 118*, 691–710.
Deng, Y., Yoshihiko, Y., & Kim, G. (2006). Urban space transformation informed by the naming changes of the Korean capital city. *Beijing Planning Review*, (04), 42–46.
Ding, Y. (2011). *Kan Yu: A Forgotten Intangible Heritage in China.* Southeast University Press.
Duan, J., & Jie, M. H. (2009). *Spatial Analysis of the Ancient Village of Hongcun World Cultural Heritage.* Southeast University Press.
Gao, Y. F., Cheng, J. J., & Wang, Z. W. (2007). Ecology physical environment computational simulation of ideal Feng Shui position model. *Building Science, 23*(6), 19–23.
Geddes, P. (1923). *The Valley Section from Hills to Sea.* New York City (United States).
Groffman, P. M., et al. (2017). Moving towards a new urban systems science. *Ecosystems, 20*, 38–43.
Guattari, F. (2000). *The three ecologies.* London; New Brunswick, N.J.: Athlone Press.
Guo, X. N., Cheng, J., Yin, C. L., Li, Q., Chen, R. S., & Fang, J. Y. (2023a). The extraordinary Zhengzhou flood of 7/20, 2021: How extreme weather and human response compounding to the disaster. *Cities, 134*, 104168.
Guo, Z. Y., Zheng, J. X., Li, S. Z., & Zhou, X. N. (2023b). Orientation of one health development: Think globally and act locally. *Science in One Health, 2*, 100042.
Guo, H. Y. (2023). Analysis and reflection on the difficulties of modern inheritance of chinese traditional wooden architecture. *New Architecture*, December 7.
Hauck, T., Keller, R., & Kleinekort, V. (2012). *Infrastructure Urbanism: Addressing the in-between.*
Jie, M. H. (2006). The ancient village in Huizhou: The research on influential factors of Hongcun village spatial structure. *Journal of Shanghai City Management Vocational and Technical College*, A42–A45.
Jin, Y. H. (2005). *Hong Cun (Hong village).* Jiangsu Education Publishing House.
Kang, L. (1999). *Feng Shui and architecture.* Baihua Literature and Art Publishing House.
Kapoor, V., Tripathi, S., Devi, R. S., Srivastava, P., & Bhadouria, R. (2020). Chapter 6–ecological economics of an urban settlement: an overview. *Urban Ecology*, 91–110.
La Loggia, G., Puleo, V., & Freni, G. (2020). Floodability: A new paradigm for designing urban drainage and achieving sustainable urban growth. *Water Resources Management, 34*, 3411–3424.
Li, Y. Q., & Wang, C. L. (2019). Yanweizhou ecological embankment strategy and its effects on flood control in the Jinhua River Basin. *Acta Ecologica Sinica, 39*(16), 5955–5966.
Liu, Y. S. (2000). Shui Kou: Huizhou Minju de Zezhi Guannian (Water gap: a concept of site selection for residential buildings). *Wenshi Zhishi*, 76–80.
Luo, Y., & He, X. X. (2008). *Zhongguo Feng Shui Shi (The history of Chinese Feng Shui).* Jiuzhou Publishing House.
Mak, M. Y., & Ng, S. T. (2005). The art and science of Feng Shui–a study on architects' perception. *Building and Environment, 40*(3), 427–434.
Mak, M. Y., & So, A. T. (2011). *Scientific Feng Shui for the Built Environment: Fundamentals and Case Studies.* City University of Hong Kong Press.
Marco, E., & Burgess, S. (2015). Chapter 28-healthy housing. In *The routledge Handbook of Planning for Health and Well-being: Shaping a Sustainable and Healthy Future* (pp. 403–416). London; New York: Routledge, Taylor & Francis Group.
Margolis, L., & Robinson, A. (2007). *Living Systems: Innovative Materials and Technologies for Landscape Architecture.* Birkhäuser Architecture.
McHarg, I. L. (1995). *Design with nature.* Wiley.
McPhearson, T., Raymond, C. M., et al. (2021). Radical changes are needed for transformations to a good Anthropocene. *Npj Urban Sustainability, 5*, 1–13.
Mostafavi, M. (Ed.). (2010). *Ecological urbanism.* Lars Müller Publishers.
Ng, E. (Ed.). (2010). *Designing High-density Cities for Social and Environmental Sustainability.* London Sterling.

Ou, Y. W. (2001). Simple analysis on features of courtyard space structure in Huizhou traditional residences. *Journal of Beijing Institute of Civil Engineering and Architecture,* 15–17.

Pan, W. J. (2019). *Diverse Environmental Performances of Urban Villages and Insights for Enhancing Quality of Urban Renewal in Shenzhen.* The University of Hong Kong.

Pan, W. J. (2023). Self-adpative hybrid urban morphologies community (HUMC): Its shared environment and soft intervention for sustainable urban governance. *Building and Environment, 236*, 110251.

Pan, W. J., & Du, J. (2021). Towards sustainable urban transition: A critical review of strategies and policies of urban village renewal in Shenzhen, China. *Land Use Policy, 111*, 105744.

Pan, M. W. (2004). Buffalo-shaped village in Southern Anhui Yi County—Totem and Feng Shui in the planning of ancient Hong village. *Journal of Kunming University of Science and Technology (Science and Technology), 29*(2), 103–106.

Park, C. Y., Lee, D. K., Asawa, T., Murakami, A., Kim, H. G., Lee, M. K., & Lee, H. S. (2019). Influence of urban form on the cooling effect of a small urban river. *Landscape and Urban Planning, 183*, 26–35.

Paton, M. J. (2021). Science and *Fengshui*: The concept Shi, rationality and emotion, and the ritualisation of knowledge. *Science & Education, 30*, 1371–1386.

Pickett, S. T. A., Cadenasso, M. L., Childers, D. L., McDonnell, M. J., & Zhou, W. Q. (2016). Evolution and future of urban ecological science: Ecology *in*, *of*, and *for* the city. *Ecosystem Health and Sustainability, 2*(7), e01229.

Pickett, S. T. A., Frantzeskaki, N., Andersson, E., Barau, A. S., Childers, D. L., Hoover, F., Lugo, A. E., McPhearson, T., Nagendra, H., Schepers, S., & Sharifi, A. (2024). Shifting forward: Urban ecology in perspective. *Ambio, 53*, 890–897.

Porkert, M. (1978). *Theoretical Foundations of Chinese Medicine–Systems of Correspondence.* The MIT Press.

Robinson, T., & Ji, M. (2022). Freeway removal: From Cheonggyecheon to Seoullo 7017. In *Sustainable, smart and solidary Seoul: Transforming an asian megacity* (pp. 39–71). Springer.

Romanello, M., et al. (2023). The 2023 report of the Lancet Countdown on health and climate change: The imperative for a health-centred response in a world facing irreversible harms. *The Lancet, 402*(10419), 2346–2394.

Ru, J. F. (1993). Theoretical analysis of "Cave" in Feng Shui from the perspective of architectural ecology. *Time+Architecture, 1*, 20–22.

Salmond, J. A., Roth, M., Oke, T. R., Christen, A., & Voogt, J. A. (2012). Can surface-cover tiles be summed to give neighborhood fluxes in cities? *Journal of Applied Meteorology and Climatology, 51*(1), 133–149.

Seoul Development Institure. (2003). *Feasilibility Study and Master Plan of Cheonggyecheon (Non-technical part summary report) final report.* Cheonggyecheon Restoration Research Center, Seoul, Korea (in Korean).

Shan, D. Q. (2009). *Anhui Min Ju (Vernacular architecture of Anhui Province).* China Architecture & Building Press.

Shou, T., & Xue, D. (2022). Resilient containers: An exploration into the tectonic logics of the water environment regulation in Huizhou's traditional dwellings. *Architectural Journal, 12*, 84–91.

Shou, T., Zhang, T., & Liu, Q. (2019). The research of architectural thermodynamic model of Huizhou vernacular architecture. *Jian Zhu Shi (Architects), 6*, 35–44.

Sun-Kee, H. C., Song, I. J., & Wu, J. G. (2007). Fengshui theory in urban landscape planning. *Urban Ecosystems, 10*(3), 221–237.

Tang, L., Nikolopoulou, M., Zhao, F. Y., & Zhang, N. (2012). CFD modeling of the built environment in Chinese historic settlements. *Energy and Buildings, 55*, 601–606.

Turenscape. (2017). Resilient landscape–Yanweizhou Park, Jinhua. *Urban Environment Design*, (03), 327–329.

Turner, V. K., Middel, A., & Vanos, J. K. (2023). Shade is an essential solution for hotter cities. *Nature, 619*, 694–697.

UNEP (United Nations Environment Programme). (2021). *International Principles of Good Practice for Sustainable Infrastructure: Integrated, System-level Approaches for Policymakers.* https://wedocs.unep.org/20.500.11822/34853.
UNEP (United Nations Environment Programme). (2023). *Nature-based Infrastructure: How Natural Infrastructure Solutions Can Address Sustainable Development Challenges and the Triple Planetary Crisis.* https://wedocs.unep.org/20.500.11822/44022.
United Nations. (1987). *Report of the World Commission on Environment and Development: Our Common Future.*
United Nations. (2022). *The Sustainable Development Goals Report 2022.* Casemate Publishers and Book Distributors, LLC.
Wang, D. G., Lyu, Q. Y., Wu, Y. F., & Fan, Z. Q. (2019). The characteristic of regional differentiation and impact mechanism of architecture style of traditional residence. *Journal of Natural Resources, 34*(9), 1864–1885.
Wang, Q. H. (2005). *Feng Shui Lilun Yanjiu (Study of Feng Shui theory).* Tianjin University Press.
Wang, X. B., & Jia, D. (2011). The analysis of contemporary on constructing strategy in traditional settlements and architecture space: The thoughts of water system of landscape and architecture space changes in Huizhou Hongcun. *Huazhong Architecture*, 83–85.
Welter, V. M. (2002). *Biopolis: Patrick Geddes and the city of life.* MIT Press.
Wilhelm, H., & Wilhelm, R. (1995). *Understanding the I Ching: The Wilhelm Lectures on the Book of Changes.* Princeton University Press.
Wong, C. (2023). Earth just had its hottest year on record–climate change is to blame. *Nature*, 10 November.
Wu, L. Y. (2014). *Zhongguo Minju Shi (A history of Chinese vernacular dwellings).* China Architecture & Building Press.
Wu, L. Y. (2017). Development tendency of human settlements sciences. *Journal of Urban and Regional Planning, 9*(02), 1–14.
Xia, Y., Guo, J., & Zou, Y. (2022). Preservation strategy and soundscape survey of tourism-oriented traditional villages: Taking Hongcun as an example. *Huazhong Architecture, 40*(10), 173–177.
Xu, J., & Chung, C. K. L. (2014). "Environment" as an evolving concept in China's urban planning system. *International Development Planning Review, 36*(4), 391–412.
Yan, M. K., Gu, Y., & Lyu, L. S. (2023). A review of bionics applications in urban planning. *Huazhong Architecture, 41*(05), 11–16.
Yoon, H. K. (1989). The theory on the origin and development of ancient Chinese geomancy (Feng-Shui). *Studies in the History of Natural Science, 8*(1), 84–89.
You, X. Y. (2023). Flood-prone areas are hotspots for urban development. *Nature*, October 04.
Yu, G., Gan, H. G., & Guo, Y. Q. (2023). Location and fortune: An exploration of the Buddhism and Daoism roles of geomancy in the song dynasty. *Religions, 14*, 859.
Yu, K. J., Li, H. L., & Li, D. H. (2008). The negative approach and ecological infrastructure: The smart preservation of natural systems in the process of urbanization. *Journal of Natural Resources, 23*(6), 937–958.
Zhang, J. W. (2004). *A Translation of the Ancient Chinese: The Book of Burial (Zang Shu) by Guo Pu (276–324).* Edwin Mellen Press.
Zhang, Z. Y. (2019). The revival practice and spatial planning inspiration of Cheonggyecheon River in South Korea. *Urban Planning Forum, 06*, 120.
Zhong, W. Z., Xiao, W., & Zhang, T. (2023). Numerical investigations on natural ventilation in atria of China's Southern Yangtze vernacular dwellings. *Sustainable Cities and Society, 89*, 104341.

The Role of Community Engagement in Building Sustainable Blue-Green Infrastructure–Best Practices and Case Studies

Sharmila Jagadisan and Joy Sen

Abstract The concept of blue-green infrastructure (BGI), an essential integral component for a liveable city, is relatively new and finds increasing attention in many cities and countries across the globe. Infrastructure planning should be designed more sensitively to ecological considerations by deploying more nature-driven solutions to address multiple challenges induced by rapid urbanization and climate emergencies in urban areas. BGI promotes an integrated and coordinated approach that integrates blue interventions (emphasizing the physicality of water itself) with green, which represents the network of landscape components including vegetation systems and water bodies helping to promote equity and resilience for sustainable urban futures. BGI are multifunctional and adaptable systems that integrate the principles of hydrology and ecology to design urban features that create win–win solutions for the environment, community and economy. Community participation which embraces the bottom-up approach engaging multiple stakeholders is crucial for effective and successful BGI adoptions. Incorporating the principles of BGI offers projected benefits, such as substantial improvement on water and air quality, mitigating the urban heat island effect, habitat protection, biodiversity conservation, positive impacts on health and well-being etc. The purpose of this chapter is to synthesize and review all successful BGI case studies at different scales across the globe to demonstrate that meaningful community participation is critical to improve BGI sustainability. The intent of this chapter to develop a framework based on the breadth of evidence to demonstrate how local community engagement, local knowledge, opinions and decision-making in BGI projects are recognised as being important for community well-being. This study may help policymakers, environmentalist, urban planners to better understand how the concept of nature-based landscape resilience can potentially help their decision making to prioritise and embrace a more holistic and social inclusive water sensitive urban planning.

S. Jagadisan (✉)
School of Architecture, Vellore Institute of Technology (VIT), Vellore, Tamil Nadu 632014, India
e-mail: sharmila.jagadisan@vit.ac.in

J. Sen
Department of Architecture and Regional Planning, Indian Institute of Technology (IIT), Kharagpur, West Bengal 721302, India

P. K. Joshi et al. (eds.), *Blue-Green Infrastructure for Sustainable Urban Settlements*,
https://doi.org/10.1007/978-3-031-62293-9_8

Keywords Blue-green infrastructure · Community participation · Case studies · Inclusive · Local community · Nature-based Solutions · Human centered

Abbreviations

BGI	Blue-Green Infrastructure
GVNML	Gram Vikas Navyuvak Mandal Laporiya
UNDP	United Nations Development Programme

1 Introduction

Climate change, a global phenomenon is widespread, rapid, and intensifying in the recent past. An increase in average temperature associated with unprecedented change in weather patterns, frequent droughts, and biodiversity loss poses an existential threat to all life on Earth.

At the global level urban areas in 2020 are exposing to major human-induced climate risks and serious threats resulting in environmental disaster (Udas-Mankikar & Driver, 2021).

Relevant scientific studies show multiple lines of evidence that anthropogenic activities such as massive land cover changes, burning of fossil fuels, and deforestation alter the climatic system. According to UN-Habitat, cities and metros are the driving force of economies and prosperity as they contribute more than 80% of the global GDP (ADB (Asia Development Bank), 2019). Unfortunately, cities account for the larger share of global energy consumption of more than 78% and emit more than 60% of greenhouse gas emissions (UN, 2021). In recent times there has been a perceived agreement among experts that there is a complex relationship between geography, economics, demography and the environment which are uniquely sensitive to a multitude of impacts of climate change. All these factors should therefore duly be taken into account while looking forward to an adaptation strategy. Currently, many countries are moving fast setting out increasingly ambitious action plans to combat climate crises. We are in need of target-specific plans with more tangible benefits to attain the goal of sustainable development to make cities more liveable and healthier.

The Sustainable Development Goals (SDGs) 11.6 and 11.7 emphases in providing universal access to safe, accessible, and inclusive, green and public spaces by reducing the adverse per capita environmental impact of cities (UN Environment Programme, 2021). Despite the expected population increase, urban infrastructure resilience and retrofitting are very important as it examine city's overall capacity and performance comprehensively to withstand disaster risks and manage future shocks (Bahadur et al., 2016).

The intent of this chapter is to demonstrate how community engagement acts as a catalyst to develop a framework on the continuum of evidence identifying local agents to mobilize resources, local knowledge. It is a strategic process in recognizing marginalized groups opinions and impacts decision-making in BGI projects (in terms of policies and practices) which are pre-requisite before the implementation of projects leading to a more sustainable outcome.

1.1 Why BGI is Important?

In India, over many decades, the term infrastructure has been primarily associated with the traditional 'grey infrastructure systems, which have storm water drains, pumps, collection systems, gutters, etc. (Goswami & Anand, 2020). Though these systems proved to be effective in collecting and managing stormwater runoff and draining it from the city, reliance on them has led to numerous challenges and unintended negative repercussions due to extreme weather conditions in light of climate change. Sometimes heavier downpours overwhelm such systems and exceed the capacity of combined sewer systems for short periods of time. And also, urbanisation has increased impervious land cover and created higher surface run-off with higher peak discharge rates, causing flooding (Feng et al., 2021). On the other hand, during extended periods of dry weather, soils, water mains, and sewers are affected by drying, cracking, and shrinking, making them vulnerable to the infiltration and exfiltration of water and wastewater (IWA, 2010). A combination of high temperatures, a higher pollution load, longer retention times, and sediments due to higher run-off may lead to corrosion and structural failure of sewers, and higher operation costs.

By 2050 it is predicted that India's current urbanization rate is expected to double 0.25% by 2050 and this rate has continued steadily with the annual average economic growth of approximately 8% over the past 15 years (Ramaiah & Avtar, 2019). Though urbanization offers better opportunities and improved standard of living, ever-growing population with an increase in air pollution, congestion of spaces due to urban sprawl and increased scarcity of land and resources due to its overexploitation, poor quality of planning are some of the other issues which need to be addressed. Overcrowding in metropolitan cities especially in developing countries has caused a tremendous burden on the management of energy, water, and transportation (Uttara et al., 2012).

The equitable distribution and quality of the urban green cover remain a key issue with most cities increasing in population has adversely affected the green cover especially in urban India (Hwang et al., 2020). For instance, as per the UN recommended standards—Chennai and Mumbai we have a meagre 0.46 m^2 and 0.12 m^2 of green space per capita, respectively, as compared to the UN recommended standard of 9 m^2 of green space per capita (Ramanan & Osman, 2021). Bengaluru, as per 2020 data has 2.2 m^2 *of parks and playgrounds per person* and has seen 925% increase in a built-up area between 1973 and 2013. Bengaluru green features have decreased from

68 to 14%, and blue features from 3% to less than 1% (Ramachandra et al., 2017). Similarly, from 1977 to 2017, Mumbai witnessed a 60% loss in vegetation and 65% decrease in waterbodies (Shantaram et al., 2021). In urban areas, residential gardens are in great decline which adds to environmental degradation. In Kozhikode between 2000–2010 there was a significant decrement in the area of residential areas and the quantum of decrease was 47.35 m^2 for higher income group and 14.16 m^2 for the middle-income group, and 2.43 m^2 for the low-income group respectively (Imam & Banerjee, 2016).

A paradigm shift is needed where the focus of the conventional approach of grey infrastructure has to be shifted towards a holistic and integrated approach of blue-green infrastructure. Currently, many cities in India are turning towards BGI approach to enhance climate resilience as well as restore the health of ecosystems at the national, regional and municipal levels (Udas-Mankikar & Driver, 2021). Adopting a more nature-friendly alternative of sustainable water management by adopting the promising innovative alternative of blue green integration can strengthen climate resilience in the future.

2 What is Blue-Green Infrastructure?

BGI is a strategically planned network of "*…natural and semi-natural areas, ranging in size from rain gardens right up to green streets, that are designed and managed to deliver a wide range of environmental, economic, and social benefits including improved water quality*" (Brears, 2023). The term BGI has become so prevalent in contemporary development discourse and the below table explains how the term BGI has been summarized and categorized by other authors.

S. no	Blue-green infrastructure definitions	Authors
1	A concept that is principally structured by a hybrid hydrological/drainage network; complementing and linking relic green areas with built infrastructure that provides ecological functions (Ahern, 2007)	Ahern (2007)
2	Blue features have been considered in most definitions specifically pertaining to green infrastructure, and thus cannot simply be categorized as 'green' but must be looked at a broader perspective as 'blue and green' definitions. The European Commission defines "green infrastructure as strategically planned network of natural and semi-natural areas with other environmental features designed and managed to deliver a wide range of ecosystem services such as water purification, air quality, space for recreation and climate mitigation and adaptation" (Arcidiacono & Ronchi, 2021)	European Commission (Arcidiacono & Ronchi, 2021)

(continued)

(continued)

S. no	Blue-green infrastructure definitions	Authors
3	Blue and green infrastructure in a distinct manner, and confines with respect to the field of urban planning: "Blue' infrastructure refers to water bodies like rivers, canals, ponds, wetlands, floodplains, and water treatment facilities; while 'Green' stands for trees, lawns, hedgerows, parks, fields, and forests. The concept refers to urban planning where water bodies and land are inter-dependent and grow with the help of each other while offering environmental and social benefits	Delhi Development Authority (Delhi Development Authority (DDA), 2021)
4	Blue-Green Infrastructure (BGI) is a collective term used when blue and green applications are integrated to mitigate urbanization and adapt to climate change by providing multiple benefits for urban areas affected by unmanaged stormwater and high heat	(Versini et al., 2018a, b)
5	Blue refers to rivers and water bodies whereas green refers to greening landscapes and build a drainage layout in urban areas that interweaves the natural environment with community characteristics and contemporary functions	Sustainability Report 2019–2020 (Drainage Services Department (DSD), 2021)
6	Nature—based solutions rely on blue and green infrastructure (BGI), or natural infrastructure which is the interconnected network of natural and semi-natural elements capable of providing multiple functions and ecosystem services, ranging from green spaces to riparian and coastal vegetation, street trees and engineered systems such as bio-retention or green roofs	(Hamel & Tan, Blue-Green Infrastructure for Flood and Water Quality Management in Southeast Asia: Evidence and Knowledge Gaps, 2022)
7	Green and blue infrastructure (GBI) as all the individual parcels of natural space and features within both our urban and rural spaces that when connected, deliver quality of life and environmental benefits for communities and the nature that thrives within them as a result	(Moran, 2023)
8	The concept of green and blue infrastructure (GBI), also referred to as blue-green infrastructure or BGI) includes water bodies and their functions in this system, and is mostly used by practitioners and researchers in hydrology and engineering studies, as well as within the climate adaptation and resiliency research frameworks	(Valente de Macedo et al., 2021)
	Blue–green infrastructure (BGI) typically refers to an interconnected network of multifunctional green spaces and blue spaces that are strategically planned and managed to increase water retention and connectivity between natural areas to provide a range of ecological, social and economic benefits (Wagner, 2022b)	(Wagner, 2022a)

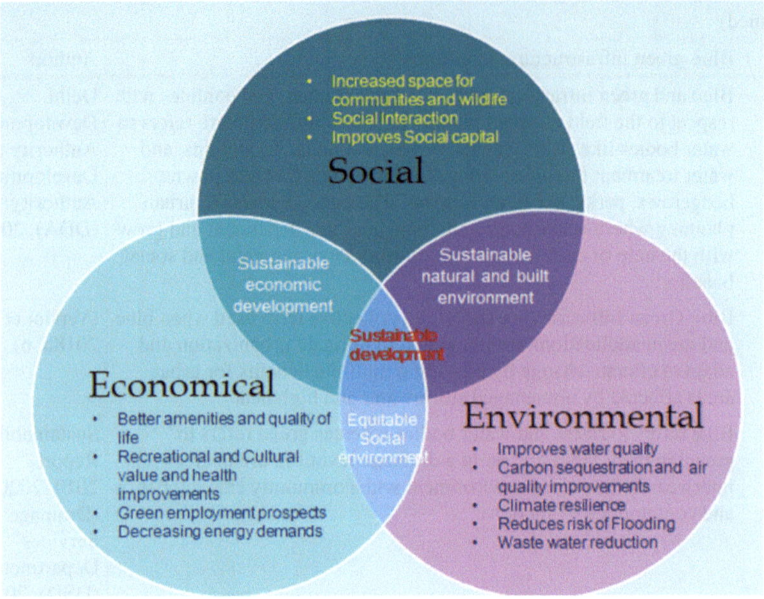

Fig. 1 Multiple benefits of BGI (*Source* author)

2.1 Benefits of Blue-Green Infrastructure

BGI refers to the interconnected network of natural and man-made green spaces and water bodies managing good water quality and ecosystem to create a healthy environment (Lamond & Everett, 2019).

In order to achieve a balance between all environmental elements, developing countries must take the initiative to incorporate BGI strategies into neighborhood retrofits and mainstream nature-based interventions in order to counteract and mitigate the negative impacts of climate hazards and manage urban water challenges (such as flood risk and urban storm water management) as shown in Fig. 1.

2.2 Growing Recognition of Blue-Green Infrastructure in Indian Cities

Although blue-green infrastructure is a new concept to many Indian cities, a few cities like Delhi, Bengaluru, Madurai, and Bhopal have begun the transition driven by exacerbated climate change and its impacts (refer Table 1).

To achieve a sustainable transformation, cities have to deploy the integration of blue-green components in their master or action plans in various sectors such as housing, transportation to shift towards low carbon economy. As part of climate

Table 1 Growing recognition of blue-green infrastructure in Indian context

Name of the Indian cities	Policy	Responsible authority	Objectives	Action plans/proposals
Delhi	Green–Blue Infrastructure—Master Plan for Delhi (2041)	DDA (Delhi Development Authority)	To develop a multi-pronged strategyto view both natural and planned green and blue assets asblue green infrastructure that protects its residents' health and is resistant to natural disasters and climate change (Rajput, 2020) To build a future-ready city that fosters sustainable, liveable and vibrant Delhi with efficient services and environmentally friendly transportation systems, and that would become a hub for the dynamic place in terms of creative, cultural, and economic aspects (MOUD, 2017)	Dust Management Plan at each construction site (Pillai, 2021) Delineation of Yamuna River to include 1 in 25 years flood plain Tree census to be maintained by local authorities to protect trees Allowing controlled development in greenbelt villages (Pillai, 2021) Development of green buffer along drains as green corridors A network of walking and biking paths through gardens. There will be also low-impact infrastructure installed such as community kitchen gardens, open-air theatres, like outdoor exercise areas, yoga gardens etc
Bengaluru	Blue-Green Action Plan that culminated Water and Sanitation Infrastructure		To mitigate the effects of climate change, encourage the shift to a low-carbon economy, alleviate poverty, and spur economic growth (Bhatt et al., 2019)	To do stormwater harvesting in lakes To reduce water leaks, increase water revenue, and enhance sanitation procedures and cleanliness requirements (Wankhade, 2014) To set up septage management To ensure universal access to household sanitation

(continued)

Table 1 (continued)

Name of the Indian cities	Policy	Responsible authority	Objectives	Action plans/proposals
Madurai	Blue-Green Future Proofing Action Plan	Madurai Municipal Corporation	To manage the interconnected set of problems associated with water resources management (surface and groundwater), water supply, water quality, sanitation, solid waste management, flooding and conservation of natural ecosystems in the context of fastest growth of trajectory of the city and its changing climate (Kumar & Vasimalai, 2018)	Sanitation, which includes repairing the sewer system Upgraded solid waste management; Rehabilitation of waterways and tanks and enhance green infrastructure development; Flood and surface water management; Water resources supply–demand balance; Governance and future-proofing planning
Bhopal	Blue-Green Master Plan (2021)	Bhopal Municipal Corporation and Bhopal Smart city development Corporation	To preserve and expand the green cover To further impact the people' lifestyle indicators to encourage the development of a people-centered city that is environmentally sustainable and serves as a hub for the arts, crafts, architecture, and natural heritage	To enhance the quality of life of cities and its surrounding natural resources; Provide evaluation and assessment To adopt different strategies to interconnect the public realm such as parks and water bodies across the cities—through the multifunctional integration of green- blue channels with a network of bike and pedestrian green routes Determine strategies to increase green cover, carbon fixations, and adopting ecologically sensitive energy saving options across different scales. These strategies may include urban agriculture, green roofs, and the preservation of water bodies in close proximity within the neighbourhoods

action plans the local authority should initiate the annual blue green audit to reduce the adverse impacts on water courses to future proof the cities of tomorrow (Udas-Mankikar & Driver, 2021).

2.3 Growing Recognition of Blue-Green Infrastructure in Indian Context

Unfortunately, the above-mentioned cities already have high density and suffer from areas of informal growth, overlapping jurisdiction among different agencies, spatial inequalities, indiscriminate land use conversion, **shrinkage of open spaces, and deteriorating environmental conditions** (Lamson-Hall et al., 2020). Prolonged and extreme heat waves along with heat stress are grappling major parts of India, creating cascading impacts on ecosystems, endangering nonhuman life and undermining the systems that support human health and survival (Raghunandan, 2023). The data indicating a rise in the deadliest heatwaves in recent decades is increasingly alarming. This poses a huge challenge and demand in terms of coping capacity and emergency needs. During 2021 summer season 11 people in Chennai died from heat exhaustion in a single day while queuing at the overwhelmed hospital (UN, 2022).

Heatwaves is one of the most dangerous environmental threats that will lead to the water crisis, climate-induced disasters, and burden and health and emergency services in the near future. Governments and communities should work together towards SDG 13.1 (To build resilience and to mitigate climate-related hazards and natural disasters across different nations) and come up with radical solutions in a systematic but proactive manner. One way to mitigate and combat climate change and environmental degradation requires **highly efficient and eco-friendly solutions from BGI for resilient and sustainable settlements** (Almaaitah et al., 2021). Green and blue infrastructure refers to nature-based interventions to protect and benefit nature through natural and modified ecosystems, specifically green and blue areas and habitat corridors that improve their spatial and functional connectivity and enhance overall environmental quality (Pochodyła-Ducka et al., 2021). Green landscape elements include orchards, forest margins, mosaics of grasslands, and ecological parks. They can refer to landscape characteristics at numerous spatial scale levels, from tree lines to entire bushes. Water is associated with blue landscape features such pools, pond systems, waterfalls, cascades, and man-made buffer basins, among others. As a result, their integration results in a network and nested system known as the "green=blue infrastructure" (Sowińska-Świerkosz et al., 2021).

3 Community Participation and Blue-Green Infrastructure

In the current context, it is widely accepted that community engagement and local resident's involvement in neighbourhoods with a greater sense of ownership of BGI are critical for an attractive and liveable environment (Hosseini et al., 2021). The potential value of deepening the community's engagement will heighten their awareness and give them a better understanding of the functions and benefits of more sustainable solutions in blue-green landscapes (DeVille et al., 2021). Above all, fostering community involvement is accepted as a good thing to develop a creative, open, inclusive, and collaborative design process. Community involvement is crucial when installing retrofit BGI components since these components have significant impact on how space is utilized in terms of functional and aesthetic aspects. In certain cases the communities beliefs and aspirations do not align with BGI interventions (Surrey County Council, 2023). According to Everett et al. (2015) study, "BGI may be negatively perceived sometimes for example, if people were financially or physically liable for upkeep and maintenance, or if they have lack of clarity to understand its purpose and functions or if they did not like choices of fauna and flora, or if components suffered from mismanagement or maltreatment, reduced functionality and negative aesthetics" (Everett et al., 2015). So, it is required to bridge the gap by adopting engagement strategies and tools to motivate citizens and local officials to work harmoniously and interact with each other to share their knowledge and build mutual trust (Khwaja, 2004). There is widespread evidence that if societies could cooperate with natural systems and processes, they would be more adaptive. Therefore, in developing countries there has been a growing concern to address the overarching challenges such as rising food insecurity, public health, and climate change through integrated blue-green infrastructure incorporating nature-based solutions local government (Macedo et al., 2021). Community participation will lead to a better process and help shape our project vision, thus giving a clearer picture of community characteristics and its potential pitfalls and assets. The involvement of the community is of paramount importance even after project completion for continued operation and successful maintenance of project facilities (Brennan et al., 2005).

3.1 Implementing BGI at Different Spatial Scales

To be successful BGI strategies must become more sensitive to ecological considerations where planning and interventions need to be carried out at different scales for their cumulative impact on urban areas (Barker et al., 2019).

Figure 2 demonstrates how to develop and adapt nature-based solutions at the macro-level (city wider/urban context), meso-level (Neighbourhood or local community), and micro-level (site and building). The increasing popularity of BGI is largely due to its multifunctionality and ability to provide multiple co-benefits beyond flood

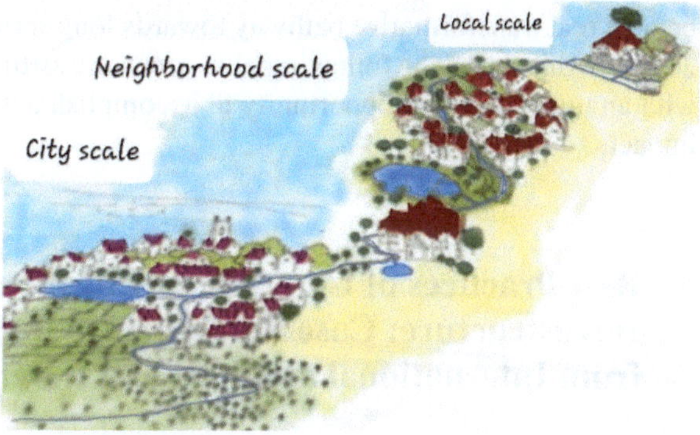

Fig. 2 Multi-scale blue-green infrastructure impacts (*Source* Author)

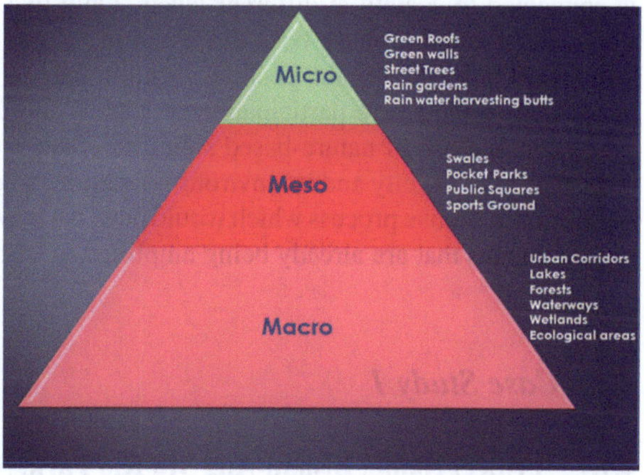

Fig. 3 BGI proposal implementation at different scales

and water risk management, since it may improve both the environment and society (O'Donnell et al., 2020).

Several cities started to realize that many of the unintended, detrimental side effects of grey infrastructure can be mitigated by leveraging the ingredients of nature and ecosystem-based solutions (Demuzere et al., 2014). BGI can reduce the impacts of climate related risks such as heatwaves, droughts, extreme events suc as storms and enhances air and water quality, water security, wildlife and biodiversity, and urban greenspace and landscape connectivity etc. (Sturiale & Scuderi, 2019). In addition, it also has positive association and contributes to overall physical health and mental well-being; creates attractive urban greenspace and ensures positive place-making; supports water resource management to be sustainable, and lengthens the service provision of ongoing service of ageing grey infrastructure to operate beyond its design life (Venkataramanan et al., 2019). Engaging with stakeholders is *central to improve accountability and a strong base for the BGI project. This can be achieved by fostering greater* civic *engagement* which emphasises a citizen-based approach at the grassroots level. Only when people include nature as part of their identity, will cities be on a real transformative pathway towards long-term, sustained change (Ingaruca, 2022). In this sense, the following blue-green infrastructure case examples provide us with an unprecedented opportunity to accomplish a major social and environmental impact.

4 Best Practices of Community Engagement in Blue-Green Infrastructure: Case Studies Have Been Selected from International and National Levels

The below listed case studies (refer to Table 2) explore different ways in which BGI is connected to society at different levels. Only two case studies (1 from International and 1 national) have been discussed in greater detail. In the context of urban sustainability transition, the positive social impact in these specific contexts should be viewed as the active participation of society through effective planning, implementation and use of nature-based solutions (Santos et al., 2022). To avoid further degradation of the city and its environs we require a people-centred approach with a highly collaborative process which would pave the way forward to effectively address the challenges that are already being amplified by climate risk.

4.1 Case Study 1

4.1.1 Project Name: Bishan-Ang Mo Kio Park, Singapore

Location: The 62-hectare Bishan-Ang Mo Kio Park was initially created in 1988 as a leisure destination and green space between the Ang Mo Kio and Bishan housing estates. North: Central Business District.

East: Central Water Catchment, in close proximity to Lower Pierce Reservoir.

South: Sin Ming Avenue.

West: Upper Thomson Road.

4.1.2 Project Description

Rapid population growth and modernized urban development of Singapore in the 1970s, incurs major environmental changes such as increased biodiversity losses along with increased flood risk. More concrete drains and canals were constructed in a number of strategic locations to allow excess water from heavy monsoons to drain quickly and to reduce widespread flooding (Koh et al., 2022). Kallang River is the nation's longest river and serves as a drinking water source to the city. Originally the river flowed through the centre of the island for ten kilometres stretch from Lower

Table 2 Successful BGI case studies from international and national context

International case studies at different scales and community participation

Name of the cities	Name of the project	Objectives	Action plans/proposals	Extent of the project/ scale
Singapore	Bishan-Ang Mo Kio Park from Concrete Canal To Natural Wonderland (Schaefer & Spirn, 2014)	To break down the utilitarian drain to create a natural meandering watercourse	To provide an integrated ecological and social infrastructure for effective stormwater management, bio-diversity, and recreation, and to create a vibrant, high-quality public space for the community	62 hectares
Medellin, Columbia	Units of Articulated Life Medellín, Colombia (AAUATF, 2022)	To turn unsafe dark urban voids in the city to vibrant green–blue public spaces that engage neighboring communities	To enhance social cohesion and bonding between communities that were separated from each other A pedagogic urbanism strategy was implemented to unite all community people by a common public space which was planned for approximately 162,000 m^2 with proposed new buildings of 40,000 m^2 To develop a sense of civic pride by providing high-end architecture and connectivity solutions to the city's less developed areas and to create a liveable environment	16.2 hectares
Sudan, North Africa	Wadi El Ku Catchment Project Darfur, Sudan (AAUATF, 2022)	To establish climate-resilient livelihoods and reduce natural resource conflicts and displacement due to the loss of livelihoods	Improvised agro-ecosystem resilience by incorporating sustainable intensification of agriculture and natural resource management supported by robust cooperation mechanisms To encourage and improve the livelihoods of women by encouraging them through community engagement activities which will improve their value chain productivity and provide incomes and support more girls to go to school	27,000 km^2

(continued)

Table 2 (continued)

International case studies at different scales and community participation				
Name of the cities	Name of the project	Objectives	Action plans/proposals	Extent of the project/ scale
National case studies at different scales and community participation				
Lapodia, Dudu Block of Jaipur Rajasthan (Singh, 2022)	Ana Sagar Lake—Introducing "Chauka System"- An innovative method for water harvesting	To reduce runoff velocity of the water across a slope, minimizing erosion To manage and direct excess water into the ponds within the catchment area for storage To maintain and and increase natural regeneration of vegetation cover To prioritize more for water security and manage during droughts To promote sustainable utilization of marginal areas	The innovative Chauka system has served 400 bhigas (one bigha is equivalent to 1 acre) and has been instrumental in restoring pastureland, boosting agriculture, grazing and recharging the groundwater table, serving 320 families	1.87 km^2
Madurai	Future-proofing of Indian cities as part of the Madurai smart city proposal (Sundaresan, 2014)	Through the process of innovative inter-regional collaboration and engagement of multiple stakeholders in short and long term programs will contribute in strengthening blue-green infrastructure, based on the selected themes	To rehabilitate and modernize irrigation tanks and drainage channels to improvegreen infrastructure development Flood and surface water infrastructure improvements 'Water walks' (also termed as '*Nathivalam*' in the vernacular) that established a link between community members with area based community representatives from the government and local organisations to focus attention in managing the degraded river ecological corridor and associated issues (World Resource Institute (WRI), 2011)	3,100 acres (complete smart city project proposal area)

Peirce Reservoir to the recently constructed Marina Reservoir, and it ran through the park for 2.7 kms along the Southern edge of the park in a straight fenced concrete channel.

By the early 2000s prior to redevelopment, the canal was in dire need of an upgrade and repair with regard to drainage problems as many of its concrete parts were crumbling and there were safety hazards during the period of heavy rain (Rinaldi, 2013). Another issue was the detachment of the park was apparent with the drainage canal demarcating a harsh line. The design team of Public Utilities Authorities (PUB, Singapore) with Singapore's National Parks Service (NParks) teamed up with the design firm Atelier Dreiseitl and the engineering firm CH2M HILL to adopt transformative approaches of infrastructure to maximise land, income opportunities and human resources. Thus, as part of the ABC (Active, Beautiful and Clean) Water Programme initiated by PUB, retaining its functionality this park was slated for redesign.

4.1.3 Vision

"To provide a new model for tropical urban hydrology of Singapore using the instruments of landscape infrastructure, addressing nation's dual need for water supply independence and flash flood management while creating access to a thriving riverine ecology within the dense city".

4.1.4 Objectives of the Project

- To initiate long-term strategies that will transform Singapore's water bodies beyond their functions of drainage and water supply, into vibrant, creative community spaces for activities and recreation.

Fig. 4 Transformation of canal from concrete to naturalized one (*Source* Author)

- To break down the utilitarian drain into a natural meandering waterway (Fig. 4).
- To use of natural processes and materials to mitigate flood and curb soil erosion, to reduce pollutant discharge and purify water to create spaces fostering community interaction and with a sense of responsibility.
- To re-engineer the river cross-section by maximizing its width from 17–24 m to almost 100 m and increases the river's conveyance capacity by approximately 40%.

4.1.5 Community Engagement and Public Participation

Community engagement is a relatively new concept in Singapore during the last decade where most of the public-housing residents have not experienced or been exposed to the participatory planning process. But for this **Bishan-Ang Mo Kio Park** project communities were engaged in maintaining the park and also throughout its redesigning stage. While the test bed was being constructed, children were invited to discover and learn various flora and fauna of the streams and they also had a sensorial experience by touching and playing with water (Dreiseitl, 2012). The children also created sculptures that were later incorporated in the playground design, with an effort to foster a sense of belonging and bond to the park (Hauber, 2014). Currently, there are citizen *groups* such as Friends of Kallang River @ Bishan-Ang Mo Kio Park (FOBP) and *NGOs* who *have* taken the initiative to study his sense of ownership has spread to include all ages. Since the park opened, a number of organization have been working to keep the environment clean and educate the public about the impacts anthropos have on the river ecosystem (PUB Singapore, Our Water, Our Future, 2013). The park project is also now used for as a powerful educational tool in the more formal sense, builds classroom learning for surrounding schools and students. The ABC Waters Learning Trails Program is an example of how this has become institutionalized. A program of the PUB, the ABC Waters Learning Trails "is a series of place-based trails which encourage students to explore, discover and provide educational opportunities on subjects ranging from biodiversity and conservation to water quality and human stewardship of ABC Waters sites" (PUB Singapore, ABC Waters Learning Trails: About Us, 2014). The interactive, hands-on path in Bishan-Ang Mo Kio Park teaches kids about Singapore's water system through real-world examples. Figure 5 shows the overall benefits of the project in terms of Sustainability aspects.

4.1.6 Biodiversity

The naturalization of the river stream and redesign of the Bishan-Ang Mo Kio Park led to the revitalization of the community unleashing the potential for green public space. Physical restoration works include re-configuring the river streams (i.e., bringing back the curves of a natural river), with green-natural river banks, instead of banks encased in concrete, had helped aquatic biodiversity to thrive (refer Fig. 6). The area is now again home to rare 60 species of wildflowers, more than 50 species of bird,

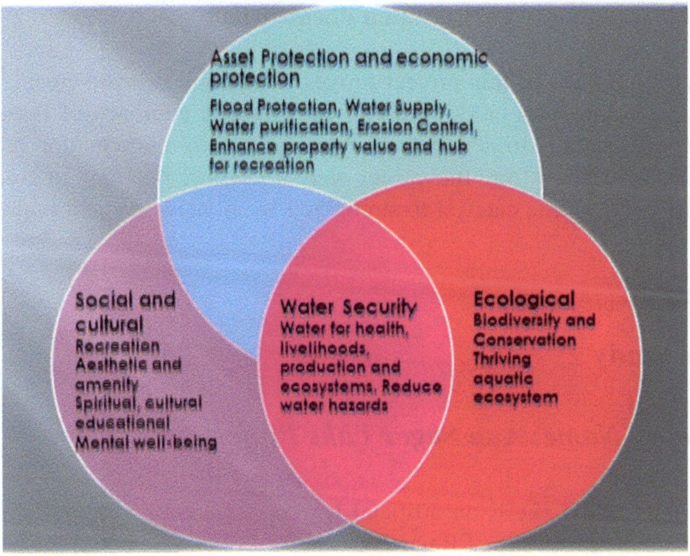

Fig. 5 Overall benefits of the Bishan Ang Mo Kio Park in terms of sustainability aspects (*Source* author)

and over 20 species of dragonfly, having larger fauna such as otters which are rare to see outside of a nature reserve (Schaefer & Spirn, 2014). Overall, 30% increase in biodiversity since the redesign of the park has been noticed (Hauber, 2014). This project sets a new model to demonstrate how hydrological and ecological restoration has created a wide variety of micro-habitats and increased the potential to be one of the biodiversity hotspots to achieve long-term survival ability.

The restoration of the river has created a huge variety of micro-habitats, which not only increases the biological diverseness but also the resilience of species within the park, making vast improvements to their long-term survival ability.

Fig. 6 Kallang River from a linear utilitarian concrete drainage channel to a meandering, natural river through the park (*Source* Author)

4.1.7 The Community Participation Phenomenon

The graceful outcome of this project proves that people's involvement forms the backbone for community empowerment. This holistic approach of team building is an inspiring example to demonstrate how community participation (convincing local residents, experts) from the master plan of land use planning to every level is meaningful and acts as a catalyst to strengthen social networks and improve social support.

5 Case Study 2

5.1 Project Name: Ana Sagar Lake Project

Location: Anasagar Lake is an artificial lake constructed by damming over Luni River situated in Ajmer city of Rajasthan State.

The climate of Ajmer city—is semi-arid with dry and hot summer and cool winter.

Lake water catchment area 53 km^2 (gross).

5.1.1 Project Description

Pressures of urbanization and anthropogenic activities have resulted in depletion and deterioration of Ana Sagar Lake. The increase in the population living around the lake poses threats such as pollution, dumping of solid waste etc.

5.1.2 Issues

(i) Massive depletion of aquatic species was attributed to over-exploitation of the lakes' valuable resources. Drastic decline in fish population in recent decades.
(ii) *Problem of Uncontrolled growth of water hyacinth resulting in degrading water quality.*
(iii) The lake can be seen with the huge amount of floating dry waste.
(iv) A vast settlement of residential development and industries are encroached in lake area results in further degradation
(v) Bathing and washing activities and the immersion of hundreds of idols after puja during Urs and Pushkar fair create a major impact on the lake ecosystem
(vi) The variety and number of migratory birds to Anasagar Lake have decreased with time.

5.1.3 Objectives of the Project

- To retain runoff water and reduce its velocity across a slope, minimizing erosion
- To make it easy to direct excess water to ponds for storage
- To maintain and regenerate natural vegetative cover
- To improve water security during droughts
- To have effective use of marginal areas.

5.1.4 Community Engagement and Public Participation

Under the leadership of Lakshman Singh who is the ingenious creator of the "*Chauka system*". "It's the bunds, channels and the pits that form the *chauka* system. A checkerboard-like pattern is made keeping in mind the existing natural slope of the land. Then a *chauka* is built with its embankments on all its three sides. Trees are planted on the periphery of the dykes to give additional support during rain. So, during heavy downpour water gets collected in one *chauka,* and flows through the down-gradient to its neighbouring *chauka* and on slowly and the excess water joins the monsoon drain" (Pani et al., 2021).

Ann Sagar when full to the brim can irrigate 250 hectares of land of Rabi and Kharif crops.

People of Lapodia gathered and wanted to name the pond "Ann Sagar," for it irrigated the land and helped them to grow food, the elixir of life.

6 Conclusion

All the above case studies show that *community participation is a determinant factor in successful sustainable blue-green projects*. It deepens the democracy by including diverse voices usually marginalized sectors or overlooked voices and empowers them to participate in decision making. A constructive relationship between different stakeholders leads to more inclusive, equitable, and better public decisions that precisely targets to satisfy community needs and aspirations.

References

AAUATF. (2022). *The social functions of green and blue infrastructure-international case studies and insights*. UK: Addis Ababa Urban Age Task Force.

ADB (Asia Development Bank). (2019). Creating liveable cities: Regional perspective. Mandaluyong City, Philippines. https://www.worldbank.org/en/topic/urbandevelopment/overview. https://doi.org/10.22617/TCS190465-2. (Asian Development Bank. Retrieved from The World Bank).

Ahern, J. (2007). *Green infrastructure for cities the spatial dimension*. IWA Publishing.

Almaaitah, T., Appleby, M., Rosenblat, H., Drake, J., & Joksimovic, D. (2021). The potential of blue-green infrastructure as a climate change adaptation strategy: A systematic literature review. *Blue-Green Systems, 3*(1), 224–248. https://doi.org/10.2166/bgs.2021.016

Arcidiacono, A., & Ronchi, S. (2021). *Ecosystem services and green infrastructure-perspectives from spatial planning in Italy* (1st ed.). Switzerland AG: Springer Nature. https://doi.org/10.1007/978-3-030-54345-7.

Bahadur, A., Tanner, T., & Pichon, F. (2016). *Enhancing urban climate change resilience-seven entry points for action.* ADB Sustainable Development Working Paper Series. Phillipines. https://doi.org/10.13140/RG.2.2.15158.91204.

Barker, A., Clay, G. D., Morrison, R., Payne, S., Gilchrist, A., Rothwell, J., & Tantanasi, I. (2019). *Understanding green infrastructure at different scales: A signposting guide.* University of Manchester.

Bhatt, G., Singh, V., Honnungar, V., & Ravindra, P. (2019). Delivering a sustainable and climate resilient water future for Bengaluru. *Water (E-Journal) Australian Water Association, 4*(2).

Brears, R. C. (2023). Blue and green cities: The role of blue-green infrastructure in managing urban water resources. Retrieved from https://www.ourfuturewater.com/2018/11/17/blue-and-green-cities-the-role-of-blue-green-infrastructure-in-managing-urban-water-resources/. (Our Future Water).

Brennan, M., Kumaran, M. C., & Spranger, M. (2005). *IFAS community development: The importance of local community action in shaping development. EDIS.* Sociology, Agricultural and Food Sciences. https://doi.org/10.32473/edis-fy729-2005.

Delhi Development Authority (DDA). (2021). Master plan for Delhi-2041 (MPD-2041). New Delhi. https://dda.gov.in/master-plan-2041-draft. (Delhi Development Authority-Public Notice).

Demuzere, M., K. Orru, O., Heidrich, E., Olazabal, D., Geneletti, H., Orru, A., et al. (2014). Mitigatingand adapting to climate change: Multi-functional and multi-scale assessment of green urban infrastructure. *Journal of Environmental Management, 146*, 107–115. https://doi.org/10.1016/j.jenvman.2014.07.025.

DeVille, N., Tomasso, L., Stoddard, O., Wilt, G., Horton, T., Wolf, K., et al. (2021). Time spent in nature is associated with increased pro-environmental attitudes and behaviors. *International Journal of Environmental Research and Public Health, MDPI, 18*(14). https://doi.org/10.3390/ijerph18147498.

Drainage Services Department (DSD). (2021). *DSD sustainability report 2019–20: Innovation for sustainability of water purification and flood protection.* The Government of Hong Kong Special Administrative Region.

Dreiseitl, H. (2012). *My experience of working on the Bishan-Ang Mo Kio park project* (pp. 68–69). A + U: Architecture & Urbanism.

Everett, G., Lamond, J., Morzillo, A., Matsler, A., & Chan, F. (2015). Delivering green streets: An exploration of changing perceptions and behaviours over time around bioswales in Portland, Oregon. *Journal of Flood Risk Management, 11*, S973–S985. https://doi.org/10.1111/jfr3.12225

Feng, B., Zhang, Y., & Bourke, R. (2021). Urbanization impacts on flood risks based on urban growth data and coupled flood models. *Natural Hazards, 106*, 613–627. https://doi.org/10.1007/s11069-020-04480-0.

Goswami, S., & Anand, A. (2020, September). Living with water: integrating blue, green and grey infrastructure to manage urban floods. Retrieved August 2023, from https://www.wricitiesindia.org/content/living-water-integrating-blue-green-and-grey-infrastructure-manage-urban-floods. (WRI India-Ross Center)

Hamel, P., & Tan, L. (2022). Blue-green infrastructure for flood and water quality management in Southeast Asia: evidence and knowledge gaps. *Environmental Management, 69*(4), 699–718. https://doi.org/10.1007/s00267-021-01467-w

Hauber, G. (2014). It takes a city to raise a river park. *My Liveable City*, 54–61.

Hosseini, F., Sajadzadeh, H., Aram, F., & Mosavi, A. (2021). The impact of local green spaces of historically and culturally valuable residential areas on place attachment. *Land, 10*(4). https://doi.org/10.3390/land10040351.

Hwang, Y. H., Nasution, I. K., Amonkar, D., & Hahs, A. (2020). Urban green space distribution related to land values in fast-growing megacities, Mumbai and Jakarta–unexploited opportunities to increase access to greenery for the poor. *Sustainability, MDPI, 12*(12: 4982), 1–17.

Imam, A. U., & Banerjee, U. K. (2016). Urbanisation and greening of Indian cities: Problems, practices, and policies. *AMBIO A Journal of the Human Environment, 45*(4), 442–457. https://doi.org/10.1007/s13280-015-0763-4

Ingaruca, M. (2022). *Elevating meaningful youth engagement for climate action.* UN Development Programme.

IWA. (2010). *Perspectives on water and climate change adaptation-climate change and the water industry–practical responses and actions.* International Water Association.

Khwaja, A. I. (2004). Is increasing community participation always a good thing? *Journal of the European Economic Association, 2*(2–3), 427–436. https://doi.org/10.1162/154247604323068113

Koh, Y. F., Loc, H. H., & Park, E. (2022). Towards a "City in Nature": Evaluating the cultural ecosystem services approach using online public participatory GIS to support urban green space management. *Sustainability, MDPI, 14*(3), 1–19. https://doi.org/10.3390/su14031499

Kumar, M. A., & Vasimalai, M. (2018, November 29). Future proofing of an Indian city: Lessons from Madurai. In *Climate and development knowledge network* (p. 3). Retrieved from United Nations Office for Disaster Risk Reduction Prevention Web. https://www.orfonline.org/research/blue-green-infrastructure-an-opportunity-for-indian-cities/.

Lamond, J., & Everett, G. (2019). Sustainable blue-green infrastructure: A social practice approach to understanding community preferences and stewardship. *Landscape and Urban Planning*, 191. https://doi.org/10.1016/j.landurbplan.2019.103639.

Lamson-Hall, P., Agrawal, H., Pachisia, H. V., & Shah, K. (2020). *Urban growth in India: Horizontal, chaotic, and informal.* Mumbai: IDFC Institute.

Macedo, L. S., Picavet, M. E., Oliveira, J. A., & Shih, W. -Y. (2021). Urban green and blue infrastructure: A critical analysis of research on developing countries. *Journal of Cleaner Production.*

Moran, R. (2023). *SURREY'S* 2050 *place ambition; a collective vision for place leadership, infrastructure and good growth.* UK: Surrey Future Steering Board.

MOUD, M. O. (2017). *India smart cities: Success stories from mission cities.* New Delhi: Government of India.

O'Donnell, E., Thorne, C., Yeakley, J., & Chan, F. (2020). Sustainable flood risk and stormwater management in blue-green cities; an interdisciplinary case study in Portland, Oregon. *Journal of the American Water Resources Association, 56*(26), 757–775. https://doi.org/10.1111/1752-1688.12854

Pani, A., Ghatak, I., & Mishra. (2021). Understanding the water conservation and management in India: An integrated study. *Sustainable Water Resources Management, 7*(77). https://doi.org/10.1007/s40899-021-00556-2.

Pillai, S. (2021, June 10). Draft master plan lists measures to save 'green and blue' assets. Retrieved September 19, 2023, from //www.hindustantimes.com/cities/others/draft-master-plan-lists-measures-to-save-green-and-blue-assets-combat-pollution-101623263709223.html. (Hindustan Times).

Pochodyła-Ducka, E., Jaszczak, A., & Glińska-Lewczuk, K. (2021). Blue-green infrastructure as a new trend and an effective tool for water management in urban areas. *Landscape Online, 92*, 1–20. https://doi.org/10.3097/LO.202192

Raghunandan. (2023, April 15). India braces for prolonged and severe heat waves amid climate change and La Niña. Retrieved September 19, 2023, from https://www.newsclick.in/india-braces-prolonged-and-severe-heat-waves-amid-climate-change-and-la-nina. (NewsClick).

Rajput, A. (2020, September 8). Explained: What is the 'green-blue' policy proposed by Delhi master plan 2041? Retrieved August 23, 2023, from https://indianexpress.com/article/explained/delhi-master-plan-green-blue-policy-dda-6587961/. (The Indian Express).

Ramachandra, T., Aithal, B. H., Kulkarni, G., & Shivamurthy, V. (2017). Green spaces in Bengaluru; quantification through geospatial techniques. *Indian Forester, 143*(4), 307–320.

Ramaiah, M., & Avtar, R. (2019). Urban green spaces and their need in cities of rapidly urbanizing India: A review. *Urban Science, MDPI, 3*(3), 94. https://doi.org/10.3390/urbansci3030094

Ramanan, S., & Osman, M. (2021). *Urban green spaces-creating and proiding equitable access.* Science Reporter.

Rinaldi, B. M. (2013). The soft sell: in Singapore, Atelier Dreiseitl has given back a river people can love [BishanAng Mo Kio Park]. *Landscape Architecture Magazine, 103*(1), 76–89. ISSN 0023-8031.

Santos, F. D., Ferreira, P. L., & Pedersen, J. S. (2022). The climate change challenge: A review of the barriers and solutions to deliver a paris solution. *Climate, 10*(5). https://doi.org/10.3390/cli10050075.

Schaefer, C., & Spirn, P. (2014). *Bishan-ang Mo Kio Park-from concrete canal to natural wonderland.* Singapore: Ecological Urbanism.

Shantaram, P., Pushparaj, A., & Rajasekaran, A. (2021). *International perspectives: Spatial analysis of green cover change in Mumbai City, India* (pp. 1–8). Arboricultural Research and Education Academy. Retrieved from India Global Buisness.

Singapore, P. U. B. (2013). *Our water, our future.* Public Utilities Board, Singapore.

Singapore, P. U. B. (2014). *ABC waters learning trails: About us.* Public Utilities Board, Singapore.

Singh, I. S. (2022, April 11). Meet the water warrior, Laxman Singh, who invented a new method of water conservation. Retrieved September 22, 2023, from https://vikalpsangam.org/article/meet-the-water-warrior-laxman-singh-who-invented-a-new-method-of-water-conservation/. (Economics and Technologies).

Sowińska-Świerkosz, B., Wójcik-Madej, J., & Michalik-Śnieżek, M. (2021). An assessment of the ecological landscape quality (ELQ) of nature-based solutions (NBS) based on existing elements of green and blue infrastructure (GBI). *Sustainability, MDPI, 13*(21), 1–17. https://doi.org/10.3390/su132111674

Sturiale, L., & Scuderi, A. (2019). The role of green infrastructures in urban planning for climate change adaptation. *Climate, 7*(10). https://doi.org/10.3390/cli7100119.

Sundaresan, J. (2014). *Future proofing Indian cities: Spatial and social analysis for sustainable development of Madurai: Action research report.* University College London, Development Planning Unit.

Surrey County Council. (2023, January 21). Green and blue infrastructure: best practice and case studies. Retrieved September 19, 2023, from https://www.surreycc.gov.uk/community/climate-change/what-are-we-ng/green-and-blue-infrastructure#section-1. (Surreys Greener Future).

Udas-Mankikar, S., & Driver, B. (2021). Blue-green infrastructure: An opportunity for Indian cities. New Delhi. https://www.researchgate.net/publication/353616196_Blue-Green_Infrastructure_An_Opportunity_for_Indian_Cities. (Observer Research Foundation).

UN Environment Programme. (2021). Sustainable development goal 11-make cities and human settlements inclusive, safe, resilient and sustainable. Nairobi, Kenya. https://www.unep.org/explore-topics/sustainable-development-goals/why-do-sustainable-development-goals-matter/goal-11. (Head Research and Capacity Development Branch).

UN. (2021). Climate action. Retrieved August 21, 2023, from https://www.un.org/en/climatechange/climate-solutions/cities-pollution. (United Nations).

UN. (2022). Climate and environment; global perspective human stories. Retrieved September 19, 2023, from https://news.un.org/en/story/2022/04/1117272.

Uttara, S., Bhuvandas, N., & Aggarwal, V. (2012). Impacts of urbanisation on environment. *International Journal of Research in Engineering and Applied Sciences, 2*(2), 1645–1673. ISSN: 2249-3905.

Valente de Macedo, L. S., Picavet, M. B., Puppim de Oliveira, J. A., & Shih, W. -Y. (2021). Urban green and blue infrastructure: A critical analysis of research on developing countries. *Journal of Cleaner Production*, 313. https://doi.org/10.1016/j.jclepro.2021.127898.

Venkataramanan, V., Packman, A., Peters, D., Lopez, D., McCuskey, D., McDonald, R., et al. (2019). A systematic review of the human health and social well-being outcomes of green infrastructure for stormwater and flood management. *Journal of Environmental Management, 246*, 868–880. https://doi.org/10.1016/j.jenvman.2019.05.028.

Versini, P.-A., Kotelnikova, N., Poulhes, A., Tchiguirinskaia, I., Schertzer, D., & Leurent, F. (2018a). A distributed modelling approach to assess the use of blue and green infrastructures to fulfil stormwater management requirements. *Landscape and Urban Planning, 173*, 60–63. https://doi.org/10.1016/j.landurbplan.2018.02.001

Versini P. A., Kotelnikova N.,, Poulhes A., Tchiguirinskaia I., Schertzer D, & Leurent F. (2018). *A distributed modelling approach to assess the use of blue and green infrastructures to fulfil stormwater management requirements* (pp. 60–63). Landscape and Urban Planning.

Wagner, I. (2022a). Formally designate blue-green infrastructure for climate adaptation. *Nature, 607*, 657.

Wagner, I. K. (2022b). Formally designate blue-green infrastructure for climate adaptation. *Nature, 607*(7920), 657. https://doi.org/10.1038/d41586-022-02000-7

Wankhade, K. (2014). *Future proofing Indian cities: Bangalore action plan for water and sanitation infrastructure*. UCL, UK: WS Atkins International Ltd.

World Resource Institute (WRI). (2011). INSIDE STORIES on climate compatible development-climate. Retrieved May 21, 2023, from https://www.wri.org/inside-stories-climate-compatible-development.

Contributions of Vertical Greening to Micro-Climate Control: Exploratory Case Study in a Residential Environment

Olumuyiwa Bayode Adegun

Abstract Growing plants on buildings, known as vertical greening or green walls, supports the (re)integration of greening into the urban fabric and promoting adaptation to heat stress. This exploratory study reports the rapid evaluation of a vertical greening prototype installed on a residential building in Akure, Nigeria. Field measurements were made on a building with the green wall installed and on a typical building (control) without the green wall. This aimed to establish micro-climate contributions (temperature moderation) of the vertical garden. Up to 2.88 °C maximum temperature and 0.7 °C minimum temperature reduction were recorded during a 45-day field measurement campaign between September and October 2021. It can be concluded that the vertical offers temperature moderation. Future studies are needed to fully establish the scale of thermal control and other ecosystem services through vertical greening. Furthermore, policy initiatives should seek to promote vertical greening as part of green infrastructure in the densifying urban fabric within African cities.

Keywords Green housing · Green infrastructure · Heat adaptation · Thermal comfort · Vertical gardens

1 Introduction

Climate change projections for sub-Sahara Africa indicates a future with increasing warming trends, frequent extreme weather events and public health challenges that are linked to extremely high temperature and changing micro-climatic situations (Adelekan et al., 2022; Serdeczny et al., 2017). Green infrastructure is gaining prominence within the climate adaptation discourse globally. Growing plants along the

O. B. Adegun (✉)
Department of Architecture, Federal University of Technology, Akure, Nigeria
e-mail: muyiwaadegun@yahoo.co.uk; obadegun@futa.edu.ng

School of Architecture and Planning, University of the Witwatersrand, Johannesburg, South Africa

P. K. Joshi et al. (eds.), *Blue-Green Infrastructure for Sustainable Urban Settlements*,
https://doi.org/10.1007/978-3-031-62293-9_9

building façade or in the vertical interior elements, known as vertical greening or green walls, is therefore currently re-emerging as a technique for the (re)integration of greening into the urban fabric. Vertical greening is part of urban green infrastructure. It is important to urban ecology given the intertwined and increasing pace of climate change and urbanization, a duo which also threatens achieving the Sustainable development goals (Bhadouria et al., 2022). This is more important for Africa where urban ecosystem services are understudied, thus precluding input of relevant evidence-based information in decision-making for urban sustainability (Awoyemi & Ibanez-Alamo, 2023).

Among the many benefits (ecosystem services) derived from green infrastructure, especially vertical greening, cutting down the heat is notable and more appreciated. Vegetation mitigates urban heat island effects in cities by regulating excessive heat thus moderating the micro-climate (Wang & Er, 2016). Generally, green infrastructure, vertical or otherwise modulates indoor or outdoor temperature through four means identified by Hunter et al. (2014). These are interception of solar radiation, evaporative cooling, thermal insulation, and modification of wind pattern. Many studies have shown that green walls generally reduce the indoor air temperature by between 0.5 and 5 °C across different climatic regions (See for example Prihatmanti & Taib, 2017; Besir & Cure, 2018; Li et al., 2021). A simulation study in Cairo shows that the external surface of walls with vertical greening can be up to 10 °C cooler than an exposed wall (Momtaz, 2018). These micro-climate (thermal) corollaries can impact metabolic flows in cities. Most available real-life experiments or field studies of vertical greening with empirical evidence of temperature reduction emanate from within urban settings in global north and developed countries generally. Notably, there is still a significant gap in the situation within Africa.

This chapter reports a small study on vertical greening systems in the residential context in tropical sub-Sahara Africa. The study is part of a larger project focusing on climate adaptation strategies in the built environment. The objective here is to undertake a rapid evaluation of micro-climate contributions (thermal control) of vertical greening in a low-density residential setting. The study location is Akure, the capital city of Ondo State, Nigeria. Eliciting empirical evidence of temperature moderation in the residential environment can inform the development of sustainable urban housing.

2 Green Infrastructure and Vertical Greening in Urban Contexts Within Nigeria

While the growing body of literature on green infrastructure is advancing knowledge frontiers on possibilities, benefits, pathways to urban sustainability, it appears that little attention is paid to the residential environment especially low-income contexts within cities in developing countries (Adegun et al., 2021). Available and recent studies that have dealt with green infrastructure in residential settings in

Nigeria (See for example Dipeolu & Ibem, 2020, 2022; Alabi, 2020) generally highlight the quantitative and qualitative inadequacy of green spaces; and reiterate low patronage and accessibility problems. For example, in Dipeolu and Ibem's (2020) study, 62.5% of the residents across four local governments in the Lagos metropolis ranked the quality of green infrastructure in their neighbourhoods as 'poor'. The quality of green infrastructure is rated low when existing parks lack appropriate features to attract users, available streets trees are not well maintained, vegetation (e.g. shrubs) are indiscriminately damaged thus making the landscape unsightly. The contribution of green infrastructure to aesthetics, micro-climate control, and air quality is also being recognised by urban residents. Among the 322 Abuja residents surveyed by Emechebe and Eze (2019), good climate (91.6%), comfortability (87.8%), good air quality (86.0%), and good health (85.7%) were the top four factors motivating the need for green spaces in their residential areas.

Urbanisation and the accompanying growth in the housing (either formal or informal) stock generally involves the loss of green spaces. The residential density that ensues from urban development makes less land available for urban greening, hence the need to explore opportunities for vertical greening strategies. There is very little empirical information on the design, implementation and evaluation (real-life or simulated) of vertical greening for indoor or outdoor thermal control in Nigeria. Within available literature and knowledge corpus from practice, it appears more attention is paid to the urban parks and other on-ground green open spaces (Adegun et al., 2021).

Only a project is so far known to report an experiment on vertical greening (Akinwolemiwa et al., 2018). In this situation, vertical greening prototypes made from bamboo and High-density *polyethylene* (HDPE) were installed on the blank walls of a residential building in a Lagos neighbourhood. The experiment showed that thermal conditions inside the experimental buildings improved. Temperature was measured from May to September 2014 (covering both wet and dry seasons) and it showed an average 2.3 °C reduction in indoor air temperature. The study called for additional studies in order to fully understand the performance of vertical greening for the low-income settings within African cities.

Some recent studies sought to understand the level of awareness of vertical greening among stakeholders within the built environment in Nigeria. The studies also explored factors that hinder the adoption of vertical greening and ideas for speedy and effective uptake of vertical greening. In a survey of 324 Abuja residents, Jegede et al. (2021) found that lack of support from professionals, cost of maintaining green walls, fear that insects and small organisms can infiltrate their houses; weak socio-economic capacity hinder the uptake of vertical greening.

The study by Olusoga and Adegun (2022) and Adegoke et al. (2022) focussed on professionals. Adegoke et al.'s (2022) study only engaged property managers. An online survey was completed by 282 Lagos-based property managers and estate valuers. The property managers were only slightly aware of the range of benefits that vertical greening offers. The low awareness is more visible in terms of economic benefits, followed by social and environmental benefits. With this level of awareness, vertical greening will not be a top greening feature recommended by these professionals.

Complementing the study on property managers, Olusoga and Adegun (2022) explored the perceptions of Lagos-based built environment professionals. Eighty architects, urban planners, quantity surveyors, engineers, and builders were surveyed about benefits and what impedes the adoption of vertical greening in Nigeria. The respondents generally agree that vertical greening can reduce energy costs (through indoor and outdoor thermal control) and improve air quality. They indicate willingness to enlighten and encourage clients to adopt vertical greening. The professionals' responses on impediments align with those expressed by Abuja residents (in Jegede et al., 2021). The professionals agree that a shortage of technical personnel, high installation and maintenance costs are notable challenges.

3 Study Setting and Implementation

This study is situated within Akure, the capital and largest city in Ondo State, Nigeria. Projections from the latest National Population Census in Nigeria show that Akure contains around 600,000 persons. The city is located on latitude 7′25°N and longitude 5′20°E. Average daily temperature can be between 21.4 and 34.2 °C, although it could be higher in built-up areas due to urban heat islands. Heat stress-causing temperature of 41–55 °C have been reported in the city (Alaigba et al., 2018). According to Adegoke and Dombo (2019), over 50% of the population in Akure must have experienced one form of thermal discomfort, based on long-term air temperature analysis.

A vertical greening system was implemented on a bungalow residential building within a medium-density neighbourhood in Akure. It was implemented on a part of the façades directly facing east and west. Other parts of the facades were left without greening so that the wall could serve as a control in this study. The wall facades provide a boundary for rooms of the same size, similar interior finishing, having the same orientation in the same wing. This study design allows a comparative assessment of the contributions of vertical vegetation.

The prototype implemented (See illustration in Fig. 1) had been earlier designed through a crowd-sourcing approach described in Adegun et al. (2022) and takes a cue from Akinwolemiwa et al. (2018). The prototype is made from 100 m diameter HDPE pipes hung horizontally on sunken timber planks (See Fig. 2). The materials were sourced locally and the construction involved artisans within the vicinity. As at the time of implementation around August/September 2021, the typical 4-layer HDPE prototype costs approximately 33,460 naira, inclusive of materials and labour. Three indigenous leafy vegetables were planted in the layers of pipes. They are jute leaf (also known as 'ewedu' and botanically known as *Corchorus olitorius*), Lagos spinach (also known as 'shoko', botanically as *Celosia argentia*) and African spinach (also known as 'tete' and botanically as *Amaranthus viridis*).

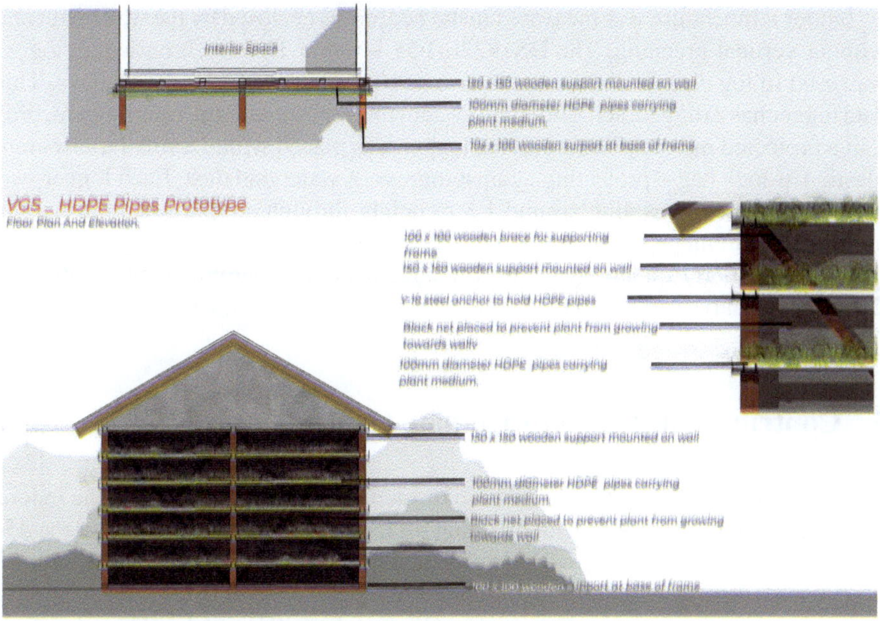

Fig. 1 Illustration of vertical greening prototype designed

Fig. 2 The prototype installed on the east-facing façade of a residential building in Akure, Nigeria. Façade with (left) and without (right) vertical greening. Pictures taken by Emmanuel Adeoti (October 2021), used with permission

4 Field Measurement for Micro-Climate Monitoring

After setting up the prototypes on both sides of the building, field measurement was undertaken between 1st September and 27th October 2021, a 45-day period towards the end of the annual rainy season in Akure. During the period, conditions are hot and humid with an average maximum *temperature at* 29 °C and average minimum *temperature of* 22 °C.

Indoor temperature was measured in the bedrooms enclosed by the walls with and without vertical greening. The DS1922L-F5# i-button Thermochron® data logger was used to log data (indoor temperature) at 10-min intervals for the 45 days. This data logger has a rugged, self-sufficient system that measures temperature and records it in a protected memory section. Each logger was placed within a small transparent plastic zip-lock bag—protecting against ingress of water and dust. Each logger was hung on the interior wall at around 1.8 m height through a repurposed plastic key holder. Measurements taken through the data loggers were harvested by plugging it into the USB port of a laptop and downloaded using the OneWireViewer software. Descriptive analysis of the logged data was done in Microsoft Excel software.

5 Contributions to Micro-Climate Control

The results from data logging in the case residential building showed that the indoor temperature in the bedroom with greening on its wall was lower throughout the 45-day field measurement period. This difference is illustrated in Fig. 3. Up to 2.88 °C maximum temperature and 0.7 °C minimum temperature difference were recorded.

Analysis of data for the hottest day (12 October 2021) and coldest day (3rd September 2021) identified during the measurement period show that the pattern of a lower temperature in the room with vertical greening was obtainable (See Figs. 4 and 5). The temperature difference was more obvious on the hottest day compared with the coldest day. The highest temperature difference of 4.23 °C at 17:00 h and 2.0 °C at 23:00 h were recorded on the hottest and coldest days, respectively.

Notwithstanding, the temperature difference between the hottest and coldest days, there is a similarity in both days in terms of maximum range of temperature reduction. This occurred between 00:00 h and 05:00 h, with over 3 °C and over 1.0 °C consistently recorded for the first five hours of the hottest and coldest days respectively. The general outlook is that the ambient temperature within the room with external

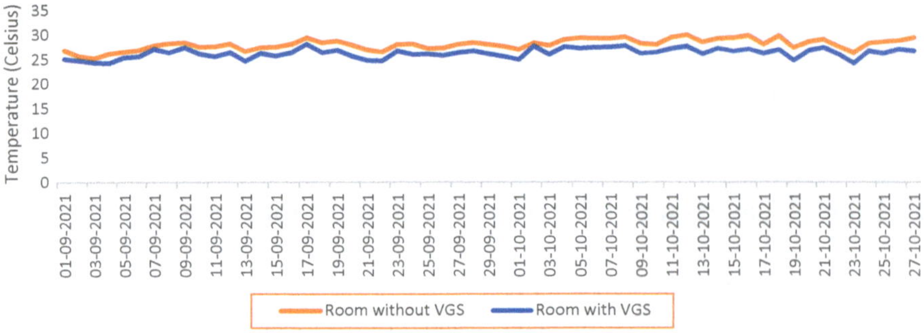

Fig. 3 Daily average Indoor temperature in the rooms with and without vertical greening on the wall

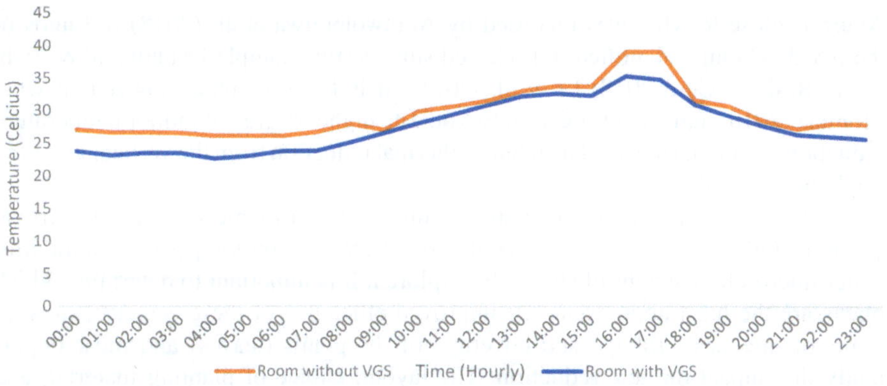

Fig. 4 Hourly average indoor temperature for hottest day (12th October 2021) in the rooms with and without vertical greening

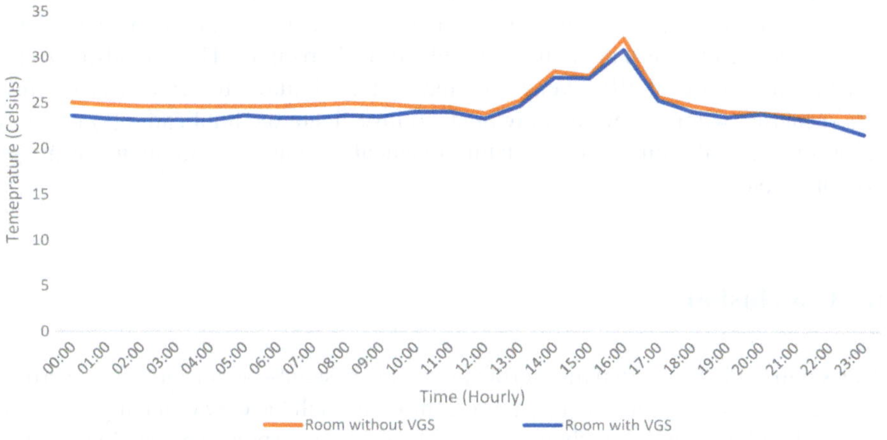

Fig. 5 Hourly average indoor temperature for the coldest day (3rd September 2021) in the rooms with and without vertical greening

wall greening is much lower, though at different levels on various days within the study period.

The results from this Akure study shows that vertical greening contributes to the reduction in indoor ambient temperature. The HDPE prototype implemented has demonstrated that vertical greening systems contribute to micro-climate control through the reduction in indoor air temperature in the study buildings. The principle that vertical greening as a component of green infrastructure enhances indoor thermal comfort and general regulation of the micro-climate has thus been advanced in this geographical setting.

These results align with findings from earlier international studies and the only one from within Nigeria. The 2.88 °C indoor temperature reduction recorded in

Akure is close to what was recorded by Akinwolemiwa et al. (2018) in Lagos or the 0.5–5 °C range identified in reviewed studies (for example Prihatmanti & Taib, 2017; Besir & Cure 2018; Li et al., 2021). It is however much lower that what Momtaz (2018) indicated for Cairo, Egypt. The higher degree of indoor temperature reduction highlights the need to enhance thermal reduction from the vertical greening implemented in this study.

Further studies are important in order to unravel how to achieve more with vertical greening. Other low-cost do-it-yourself vertical greening prototypes that might offer better micro-climate control should be explored. It is important to determine which plants are the most appropriate for improved efficiency of the green structures, in terms of thickness, foliage and coverage of the plants (leaves) and most importantly the impact on heat reduction. The layout, choice of planting material, and the composition of vegetation have important impacts in creating a desired thermal environment. Improved thermal performance of the green walls will go a long way at aiding passive cooling and energy efficiency in the low-income residential settings.

Long-term, cross-seasonal measurement across a range of indoor and outdoor spaces would be needed. This can help better establish the suite of variables and factors involved in thermal control through vertical greening. This can also provide full insights into cost-effectiveness of the system. Future studies can also simulate scenarios within relevant study areas to investigate the implication(s) not only regarding thermal control but also on improvement in air quality and saving on energy consumption.

6 Conclusion

This small study has shown that vertical greening systems—plants grown on vertical surfaces—can offer indoor temperature control through the case of a bungalow residential building. The case shows the need to further explore and exploit vertical greening systems to harness the range of ecosystem services possible. Given the potential to support climate adaptation, this environmental benefit motivates the need for government to encourage vertical greening in the course of optimising urban ecology. It provides an alternative and complement to parks and green spaces that are usually developed directly on the ground. Relevant incentives and policy initiatives can aid innovation on vertical greening and promote its adoption.

Acknowledgements The author acknowledges the research assistance and support of Mr Emmanuel Adeoti on the prototype implementation and data collection in this study.

References

Adegoke, O. O., & Dombo, T. P. (2019). Geospatial modeling of human thermal comfort in Akure Metropolis using thom's discomfort index. *International Journal of Environment and Bioenergy, 14*(1), 40–55.

Adegoke, A. S., Gbadegesin, J. T., Oluwafemi, T., Ayodele, S. E. A., Oyedele, J. B., Oladokun, T. T., & Ebede, E. O. (2022). Property managers' awareness of the potential benefits of vertical greenery systems on buildings. *International Journal of Construction Management*, 1–10.

Adegun, O. B., Ikudayisi, A. E., Morakinyo, T. E., & Olusoga, O. O. (2021). Urban green infrastructure in Nigeria: a review. *Scientific African*, e01044.

Adegun, O. B., Olusoga O. O., & Mbuya, E. (2022). Prospects and problems of vertical greening within low-income urban settings in sub-sahara Africa. *Journal of Urban Ecology.*

Adelekan, I., Simpson, N., Totin, E., & Trisos, C. (2022). IPCC sixth assessment report (AR6): Climate change 2022-impacts, adaptation and vulnerability: Regional factsheet Africa.

Akinwolemiwa, O. H., de Souza, C. B., De Luca, L. M., & Gwilliam, J. (2018). Building community-driven vertical greening systems for people living on less than£ 1 a day: A case study in Nigeria. *Building and Environment, 131*, 277–287.

Alabi, M. O. (2020). Sustainable urban form and challenges of open space utilization, Akure, Nigeria as a case study. *International Journal of Urban Sustainable Development, 12*(3), 328–339.

Alaigba, D. B., Fabiyi, O. O., & Akinnawo, O. S. (2018). GIS in the assessment of the impact of microclimatic variables on human comfort in Akure. *Nigeria. South African Journal of Geomatics, 7*(1), 64–74.

Awoyemi, A. G., & Ibáñez-Álamo, J. D. (2023). Status of urban ecology in Africa: A systematic review. *Landscape and Urban Planning, 233*, 104707.

Besir, A. B., & Cuce, E. (2018). Green roofs and facades: A comprehensive review. *Renewable and Sustainable Energy Reviews, 82*, 915–939.

Bhadouria, R., Upadhyay, S., Tripathi, S., & Singh, P. (Eds.). (2022). *Urban ecology and global climate change.* John Wiley & Sons.

Dipeolu, A. A., & Ibem, E. O. (2020). Green infrastructure quality and environmental sustainability in residential neighbourhoods in Lagos, Nigeria. *International Journal of Urban Sustainable Development, 12*(3), 267–282.

Dipeolu, A. A., & Ibem, E. O. (2022). The influence of green infrastructure on residents' connectedness with nature in Lagos Nigeria. *Urbani Izziv, 33*(1), 82–92.

Emechebe, L. C., & Eze, C. J. (2019). Integration of sustainable urban green space in reducing thermal heat in residential area in Abuja. *Environmental Technology & Science Journal, 10*(2), 24–32.

Hunter, A. M., Williams, N. S. G., Rayner, J. P., Aye, L., Hes, D., & Livesley, S. J. (2014). Quantifying the thermal performance of green facades: A critical review. *Ecological Engineering, 63*, 102–113.

Jegede, F. O., Adewale, B. A., Ajonye, G. O., & Adebajo, T. (2021, April). Evaluation of the use of green façade in residential buildings in Kubwa, Abuja, Nigeria. In *IOP Conference Series: Materials Science and Engineering* (vol. 1107, no. 1, p. 012185). IOP Publishing.

Li, J., Zheng, B., Chen, X., Qi, Z., Bedra, K. B., Zheng, J., Li, Z., & Liu, L. (2021). Study on a full-year improvement of indoor thermal comfort by different vertical greening patterns. *Journal of Building Engineering, 35*, 101969.

Momtaz, R. I. (2018). Vertical garden as a sustainable urban perspective in Cairo. *Journal of Engineering Sciences., 46*(2), 246–262.

Olusoga, O., & Adegun, O. (2022). Professionals' perception studies of vertical greening systems in Lagos, Nigeria. *International Journal of Building Pathology and Adaptation.* (ahead-of-print).

Prihatmanti, R., & Taib, N. (2017). Improving thermal comfort through vertical greeneries in transitional spaces for the tropical climate: A review. *GSTF Journal of Engineering Technology, 4*(3).

Serdeczny, O., Adams, S., Baarsch, F., Coumou, D., Robinson, A., Hare, W., Schaeffer, M., Perrette, M., & Reinhardt, J. (2017). Climate change impacts in Sub-Saharan Africa: From physical changes to their social repercussions. *Regional Environmental Change, 17*(6), 1585–1600.

Wang, C., Er, S. S., & Abdul Rahman, H. (2016). Indoor vertical greenery system in urban tropics. *Indoor & Built Environment., 25*(2), 340–356.

Landscape Planning Strategies Within the Scope of Development Plans and Implementation Problems: The Case of Türkiye

Nilüfer Kart Aktaş

Abstract Development plans are the plans that have been prepared by the government since 1963, aiming at development in issues such as economy, health, education, transportation, social security, justice, and determining the policy to be implemented in the public sector. With the 1961 Constitution, it was decreed that the Development Plans be prepared and the economic, social and cultural development be carried out according to this plan. The main purpose of this study is to investigate the planning and implementation problems of the prepared plans, which is one of the biggest problems of developing countries. In this context, the realization status of the landscape planning issues discussed in the plan process covering the years 1963–2023 were examined. Current urbanization problems, climate change, natural disasters and man-made disasters experienced especially within the scope of blue-green infrastructure in recent years have been researched and the decisions taken for disasters have been examined. Also, they were compared with plan decisions and the implementation of these plans. The research has been tried to be determined by a qualitative analysis made within the scope of development plans and current urban problems. As a result of the evaluations, it has been determined that the strategies for landscape planning are reflected in the development plans in parallel with the changing agendas in the world and Türkiye. It has been determined that there are problems in the detailed handling of the landscape planning strategies determined in the development plans in the sub-plans and especially in the implementation of these decisions. It will be possible to create resilient cities to climate change and disasters, with the correct implementation of planning decisions at the right time.

Keywords Blue and green infrastructure · Five-year development plans · Nature-based planning · Planning hierarchy

N. K. Aktaş (✉)
Department of Landscape Architecture, Faculty of Forestry, Istanbul University-Cerrahpaşa, İstanbul, Türkiye
e-mail: niluferk@iuc.edu.tr

P. K. Joshi et al. (eds.), *Blue-Green Infrastructure for Sustainable Urban Settlements*,
https://doi.org/10.1007/978-3-031-62293-9_10

1 Introduction

Today, especially metropolitan cities face many problems. The increasing population has led to continuous physical changes in cities. These changes are generally manifested in negative ways such as unplanned urbanization, increasing construction due to the need for housing and basic urban services (transportation, education, health, trade, etc.), increasing impervious surfaces and thus insufficient water resources, destruction and depletion of natural resources, increasing need for food, yet decreasing agricultural areas. In addition, today's global climate crisis is also affecting cities in a very serious way. With global warming caused by the intensive use of fossil fuels, deforestation, etc., disasters in different forms such as drought, floods, fires, etc. have started to be experienced very frequently in cities. The only condition to overcome these problems is proper planning (Aktaş, 2021; Rosenzweig et al., 2011; Tuğaç, 2021; Singh et al., 2021).

McHarg, a landscape architect and planner, stated that cities shaped and developed in the direction of economy-oriented decisions were kept away from natural sciences (McHarg, 1969; Çetinkaya & Uzun, 2014; Kaya & Uzun, 2019). As can be understood from the changes in cities, cities should be planned in harmony with nature, not by destroying nature. This will be possible with landscape planning.

Landscape planning aims to protect unique, rare and scarce resources, prevent hazards, conserve limited resources through controlled use, and identify areas suitable for development activities (Ahern, 2002). Landscape planning is a tool for incorporating ecological values into land use planning at urban and regional levels; creating an inventory of nature conservation, restoration and management decisions; developing habitat networks; identifying sensitive areas and conducting environmental impact assessments; creating spatial information; and providing participation in sectoral decisions (non-governmental organizations, society, public employees, academics, etc.) (Gruehn & Kenneweg, 1998; Luz, 2000; Reinke, 2002; Wende et al., 2005; Kaya & Uzun, 2019; Von Haaren & Albert, 2011). Therefore, cities planned with a holistic landscape planning approach will be sustainable, ecological, livable cities with a high quality of life, in harmony with nature, protecting natural resources and biodiversity, reducing the ecological footprints, having no problems in adapting to the effects of climate change, with appropriate land settlements (Aktaş, 2021; Singh et al., 2022; Bhadouria et al., 2022). For this reason, it is very important to make landscape planning as a basis for spatial planning in urban planning. In this way, appropriate land use distributions can be realized.

Since the 18th century, the relationship between the natural environment and human beings, which has changed with the Industrial Revolution and turned into uncontrolled destruction with the increasing consumption, production, population and volume of economic activities, has entered a new era in today's information society. When the first plans are examined, it is seen that the decisions were generally based on development, industrialization and economic welfare goals and that there were no decisions related to the environment, whereas today, sustainable use of cultural and natural resources, in other words, physical planning decisions are placed in an ecologically acceptable framework, rapidly gaining importance in today's spatial planning strategies, planned and environmentally sensitive policies are adopted (Ayhan & Hepcan, 2009; Dağdemir, 2015). These policies have evolved over time and the limits of production and consumption destroying a livable environment have been tried to be drawn (Erdem & Yenilmez, 2017).

Türkiye has entered a period of rapid change in line with the goals of reaching contemporary civilizations and modernization, which started with the establishment of the Republic in 1923 (FFYDP, 1963). Development moves, industrialization and accompanying internal migration have seriously affected urban and rural areas. Unfortunately, this process has negatively affected the development process of cities. Cities that received intense immigration began to urbanize, build and spread in an unplanned manner. Under these conditions, with the development plans that started to be prepared for five years, the issues on the agenda were different for each period.

Development plans are the plans that have been prepared by the government since 1963, aiming at development in issues such as economy, health, education, transportation, social security, justice, and determining the policy to be implemented in the public sector. With the 1961 Constitution, it was decreed that the Development Plans be prepared and the economic, social and cultural development be carried out according to this plan (1961 Constitution). In this context, development plans are a plan that plays a leading role in the development of our country.

The first five-year development plan, was almost mentioned nothing about urban planning and therefore landscape planning in the development goals explained under the main headings such as agriculture, manufacturing, mining, construction, energy and transportation (FFYDP, 1963). In the eleventh five-year development plan covering the years 2019–2023, landscape planning issues were included in a very comprehensive way, especially under the title of livable cities and sustainable environment (EDP, 2019).

Considering the fact that Türkiye's environmental priorities lag behind its development goals or have to be postponed, this study attempts to analyze the stages of this transformation trend within the framework of all development plans. This transformation shows how the perception of the environment has changed as well as the perspective on sustainability, ecology and disaster relations.

The main purpose of this study is to investigate the planning and implementation problems of development plans, which is one of the biggest problems of developing countries. In this context, the realization status of the landscape planning issues discussed in the plan process covering the years 1963–2023 were examined. In addition, the fact that development plans are at the top of the planning hierarchy and that all plans must comply with the objectives and decisions set out in development plans was a factor in the selection of development plans as the research material. Current urbanization problems, sustainability, nature conservation, water management, biodiversity, climate change, natural/man-made disasters have been researched especially within the scope of blue-green infrastructure. The research has been tried to be determined by a qualitative analysis made within the scope of development plans and current urban problems.

2 The Concept of Development

Development is conceptually defined as the material and spiritual progress of human life by changing and developing the economic, social and political structures of a country and, within this framework, increasing social welfare in general (Bardakçı et al., 2019; Çukurçayır & Tezcan, 2011; Öner, 2020).

In the classical capitalist model, economic development is constructed within the framework of an understanding of production based on unlimited production and consumption, with the logic of expanding consumption possibilities through continuous income growth. In fact, when the limited resources of society and nature are taken into account, actions that lead to excessive use of resources and deterioration of the environment and natural balance occur as a result of this fiction (Erdem & Bozan, 2021). Therefore, the development plan is a very important tool for protecting the natural environment and ensuring sustainability as well as providing economic development.

Development plans are macro plans that examine the economy with an integrated perspective that examines the relations between environmental, economic and social factors with a long-term future-oriented integrated perspective, provides guidance for the public and private sector in the long term, and take into account legal and technical rules (EDP, 2019). The development plan aims to improve and develop issues such as economy, health, education, transportation, environment, urbanization, housing, social security, justice, etc. and determines the policy to be implemented.

Today, with globalization being decisive in every field, periods are gone through in which risks and uncertainties for the future have emerged, societies of almost every country are experiencing painful changes/transformations, international balances have been reshaped, and existing balances between countries have been revised (Bardakçı et al., 2019). Global developments have not only diversified development goals but also revealed many problem areas such as climate change, global warming, solid waste, digital and biological pollution, etc. caused by development. It is precisely at this point that a world perspective discussing the phenomenon of development has been encountered, and under the influence and leadership of supranational organizations, the search for specific conceptual backgrounds and solutions such as combating climate change, global warming and regional/local roles, artificial intelligence and its authority over public/social life, digital pollution, sustainable development, recycling, etc. have begun (Öner, 2020).

Development plans, which determine the basic policies in our country in terms of solving existing social problems and building the future, started in the first quarter of the 1960s and primarily aimed to eliminate the development gap between regions (Çelikkaya et al., 2018). Until today, eleven development plans have been prepared and implemented in Türkiye. Although development plans come to the fore in socio-economic terms, they actually set goals and policies for all institutional structures. The development plans prepared in Türkiye have also focused on change and transformation in the field of local government (Öner, 2020).

3 Development Plans in Türkiye

3.1 First Five-Year Development Plan (1963–1967)

With the establishment of the Republic, social and economic movements took place at a great pace. The society, whose population grew rapidly, had especially great educational problems. While the illiterate population (60%) was very high, the shortage of educational institutions was very high. In this sense, society sought solutions to the issues of education and unemployment that emerged with rapid population growth (FFYDP, 1963).

In addition, the majority of the population was engaged in agriculture, and the desired yield in agriculture could not be achieved (FFYDP, 1963).

This plan, which aims to bring the people to a high standard of living in as wide a perspective as possible, was made for a five-year period but was prepared according to the targets to be reached in fifteen years. Accordingly, the main objectives were increase in national income and per capita income, fair income distribution, ensuring the education of the society, ensuring increased productivity in agriculture (modernization in agriculture), solving the employment issue, increasing immigration to cities and thus increasing housing need, prevention of slums, accelerating industrialization, training of sufficient and superior scientists and technical experts in every field necessary for the development of Türkiye and the realization of these goals in accordance with the principles of social justice (FFYDP, 1963).

When the issues related to natural resources are examined, some of the objectives are protecting and maintaining forests, reconstruction and rehabilitation of degraded forests, thus increasing the quality of today's forests, increasing their productivity and operating forests rationally, there are main objectives such as gaining new forest areas through afforestation and realizing the positive effects of forests on the natural and ecological balance of watersheds.

While this plan basically includes targets related to economic development and industrialization, it does not include environmental, disaster and sustainable planning issues.

3.2 Second Five-Year Development Plan (1968–1972)

In this plan, as in the first Plan, the main target is economic development. It is aimed to transition the country's economy from agriculture to a system dominated by the industrial sector. Efforts to modernize agriculture and increase the education level of the society continue. In this plan, unlike the first plan, urbanization, city planning and housing issues constitute an important agenda. The plan has objectives such as to benefit from the labor of the home builder, to solve the land ownership problems, to increase the production of housing, producing affordable housing in accordance with the conditions of the country, to make mass housing investments, to increase the housing resources and to determine the state's place in the housing market as "regulatory and helping those who build their own house" (SFYDP, 1968; Özlüer, 2021).

3.3 Third Five-Year Development Plan (1973–1977)

In the plan, the main problems of Turkish society were emphasized and these problems were classified as rapid population growth, lack of employment and productivity, low income, imbalance in income distribution, insufficient education and health services in terms of quality and quantity, irregular urbanization, scattered settlement, organizational and technological backwardness, foreign trade deficits and the structure of foreign trade, development a costly and slow-running business of public administration that cannot adequately contribute to its efforts. The purpose and principles are determined towards solutions to these problems; enhancing the quality of life, industrialization, reducing dependence on external resources, solving the employment problem and improving income distribution. A separate environmental chapter emphasizing the importance of the environment was first included. In this plan, it is stated that environmental policies that may impair development will not be accepted (TFYDP, 1973).

Intensive migration from rural areas to cities has led to the formation of scattered and unplanned large cities rather than balanced and gradual urbanization. The process of urbanization has been characterized by excessive concentration of industry in certain regions and population concentration in certain regions, which has led to problems for the development of cities. Urbanization, which is incompatible with economic development and industrialization and occurs in the form of overpopulation agglomeration, creates huge infrastructure needs and has led to the allocation of a large share of limited investment resources to these non-productive areas. On the other hand, land prices could not be controlled. Moreover, industrial units that cannot be established in a certain order in the close vicinity of cities lead to irrational land use and speculative movements, thus causing a loss of resources (TFYDP, 1973).

In this period, where disasters are mentioned for the first time, it is seen that disasters are only considered in relation to destroyed buildings and housing needs, and disaster management studies are not carried out (TFYDP, 1973).

In a part of the plan related to environmental problems, it is stated that it is not possible to see this issue apart from the development problems of developing countries. In addition, it was stated that slowing down the development efforts by asserting these problems would not be valid. In some industrial and dense urbanization areas, it is accepted that pollution of air, water and coasts is seen in a narrow sense and that this does not apply to the country as a whole. Also, it is accepted that issues such as erosion and environmental health have not been properly and sufficiently evaluated (TFYDP, 1973).

Within the framework of this understanding, principles and measures to be taken have been determined. The way to reach a rational balance between environment and human relations was expressed as social and economic development, and it was stated that no practice that would deviate Türkiye from its development goal by industrialization would be accepted in the work and legal regulations to be carried out at the international level regarding the solution of problems The way to achieve a rational balance between the environment and human relations was expressed as social and economic development, and it was stated that in the studies and legal arrangements to be carried out at the international level regarding the solution of the problems, no practice that would deviate Türkiye from the goal of development through industrialization would be accepted (Erdem & Yenilmez, 2017; TFYDP, 1973).

3.4 Fourth Five Year Development Plan (1979–1983)

In the plan, first of all, many problems experienced in the previous plan period are mentioned. In the plan, there are remarkable determinations about the directions for the zoning plans. Accordingly, it was stated in the plan that "zoning plans have become documents that are far from solving the physical problems of cities and regulating healthy spatial development, and that do not function beyond legalizing the unhealthy and irregular development that has occurred to a large extent" (FFYDP, 1979; Özlüer, 2021).

Environmental problems are also included under the title of "Urbanization and Municipalities" of the plan. In this context, firstly, the basic principle of environmental problems was expressed as solving them together with the process of social change. A kind of regional environmental policy was mentioned, stating that the problem would be solved in the prevention phase before it was created and that the most appropriate solutions would be selected for the socio-economic and ecological structure by considering the alternative solutions in the approaches to eliminate the existing environmental problems. It has been stated that the pollution seen in Türkiye is the environmental pollution seen in developed countries, besides, water and sewerage problems in the traditional sections, poor housing conditions in the slums, soil and forest erosion are also seen. Detections were made regarding the pollution of water, sea, air and soil erosion, and it was stated that although legal regulations were made to prevent water pollution, operability could not be achieved (Erdem & Yenilmez, 2017; FFYDP, 1979).

3.5 Fifth Five-Year Development Plan (1985–1989)

The 5th Development Plan is remarkable in that it establishes a cause-and-effect relationship between environmental pollution and disasters, as well as between environmental problems and changes in industrialization and agricultural production models, by stating that we are "facing environmental pollution as a result of urbanization, erosion and natural disasters, and environmental problems brought about by rapid industrialization and modernization in agriculture". On the other hand, the Zoning Law entered into force in 1985, but during this period, a zoning amnesty was also enacted that was contrary to the objectives set out in the zoning plan (FFYDP, 1984; Özlüer, 2021).

In the Plan, environmental problems in Türkiye are expressed as "environmental pollution caused by urbanization, erosion and natural disasters, and environmental problems brought on by rapid industrialization and modernization in agriculture". The basic approach of the plan is to utilize, protect and develop resources in the best possible way for the benefit of future generations (Erdem & Yenilmez, 2017; FFYDP, 1984).

The plan also includes a section on regional planning. In this section, topics such as accelerating the development in the regions, the economic situation of the region, its potentials, investment programs, and the development of small and large-scale projects are the main topics. The planning of natural resources at the regional scale was not considered (FFYDP, 1984).

3.6 Sixth Five-Year Development Plan (1990–1994)

In the plan, environmental problems, settlement-urbanization, housing and drinking water-sewage sub-titles were included in the title of environment and settlement. In the plan, re-planning of forests, legal regulations to eliminate environmental problems caused by mining activities, compliance of industrial organizations with environmental policies, production of vehicles that minimize environmental problems in automotive, taking legal measures regarding toxic exhaust gases, taking environmental principles as a basis in tourism investments and environmental issues, and health promotion issues (Erdem & Yenilmez, 2017; SFYDP, 1989).

This plan explicitly uses the concept of sustainable development. Some of the important headings from the principles, policies and targets determined in the plan are as follows;

- In order to reduce the negative effects of disasters such as deforestation, barrenness, soil erosion and flooding resulting from the irregular relations between people, nature and technology, protection, afforestation, rejuvenation, erosion control and rangeland rehabilitation works will be given priority.
- Forests will be planned, operated and developed on the basis of the principle of continuous, balanced and versatile use of resources and the protection of human health, plant-water-wildlife, forest landscape and natural values.
- In order to bring the love of forest and green to society as a culture and to promote the forested areas with their ecological, social, historical and scientific characteristics, it will be aimed that the information centers located in the forests provide enlightening and informative services within the practice of "nature school".

- The protection of natural beauty and cultural values will be given priority, certain areas will be protected and opened to tourism will be encouraged.
- Especially in the preparation of city development plans, it will be ensured that the size of the area per capita for green areas, sports facilities and children's playgrounds is in accordance with the standards of developed countries.
- In order to regulate the beneficial and economical use of the coastline and to consider the public interest as a priority in making use of the coastlines, a master plan will be prepared.
- Coordination between institutions will be ensured in the management of in-continental water resources and regular monitoring of the receiving water environment, and the concept of administration at the basin scale will be developed.
- Measures to prevent the destruction of wild plant genetic resources will be developed.
- It is essential to provide the population living in cities with a structure suitable for urban life and to create a quality and healthy living environment in settlements. It is essential to create a quality and healthy living environment in settlement areas.
- Physical plans will be followed in all kinds of settlement grading. These plans will be based on preserving historical, cultural and natural values and not increasing the current city density.
- Maintaining natural balance and human health, ensuring the management of natural resources in a way that will allow for continuous economic development, and leaving a natural, physical and social environment befitting human beings to future generations are the basic principles (SFYDP, 1989).

Despite all these goals, it is seen that the priority goal in the plan is development and the importance of the environment is behind the development priority.

3.7 *Seventh Five-Year Development Plan (1996–2000)*

In this plan, many problems were identified and policies were introduced to address these problems. First of all, the plan emphasized the intensive migration to cities and the pressures created by this migration. Attention was drawn to the problems of housing, settlement, transportation, infrastructure, environment, unemployment, education, health and public order. In addition, it is also stated that planning activities in cities have not been given the importance and priority they need, infrastructure and transportation problems in big cities have increased and slum areas have expanded (SFYDP, 1995).

In the plan, it is stated that there is a very low rate of 11 percent of good quality forest areas in our country; desertification, deforestation, flooding, landslides and soil erosion continue to increase; biological diversity is decreasing; cadastral works are not completed; management plans are inadequate in terms of operational purposes and multi-purpose use; annual afforestation and erosion control activities; the scarcity of National Parks, Protection and Reserve Areas; the inability to prevent forest fires; the increase in forest deaths due to acid rain; the lack of environmental protection, landscaping and ergonomic practices; and the low socio-economic status of the villages within the forests continue to be the main problems of the sector (SFYDP, 1995).

It is also stated in this plan that there are inadequacies in the protection, planning and management of groundwater and surface water resources in Türkiye and that unconstructed and uncontrolled construction, especially in the neighborhoods close to drinking water basins in large cities, adversely affects drinking water resources and quality (SFYDP, 1995).

The Plan frequently mentions legal and administrative gaps and inadequacy, uncertainty and overlaps in the division of authority and responsibility, and it is determined that this situation causes problems in the rational use of resources. There are serious deficiencies in planning (basin planning, management plans, planning and management of soil and water resources, etc.).

In accordance with these problems, some important headings from the objectives, principles and policies of the plan are as follows;

- A new structuring will be introduced for the management of water resources, which are of increasing importance today,
- In order to increase production, fragmentation of agricultural lands will be prevented, land consolidation and regularization of ownership will be ensured, land and water use planning studies will be carried out throughout the country and basin-based practices will be initiated,
- In line with the agricultural policy implemented, it will be essential to establish the infrastructure to ensure balanced, sustainable and environmentally friendly agricultural development and to prioritize projects that will contribute to the national economy in a short time,
- The fact that each region in the country has different opportunities, characteristics and problems necessitates a new planning approach in which sectoral preferences and spatial analysis are handled together,
- Urban land use patterns in practice, except for housing, are inadequate compared to the standards of developed countries. Cities are almost exclusively composed of residential areas with limited transportation, trade, recreation and public open spaces,
- In order to minimize the destruction of nature, cultural and natural assets, the existing natural conditions and ecological balance will be taken into consideration in determining the location of settlements and development areas,
- Considering that the province is the basic administrative unit, it will be strengthened in accordance with its importance in terms of delivering all public services to the public and provincial Development Plans will be prepared in line with regional planning decisions. It will be aimed to preserve the existing green areas in and around the cities, especially in the cities that are regional centers, and by proposing green areas, these areas will be afforested,
- The living environment and quality of life in cities will be developed and improved,

- Legislation on disasters will be reorganized according to the current conditions. Efforts will be made to prevent natural disasters and reduce their damages, and necessary measures will be taken to minimize the risk of disasters,
- In line with the sustainable development approach, the main strategy will be to ensure the management of natural resources by protecting natural balance and human health and to leave a decent physical, natural and social environment for future generations. Protection of the environment will be handled with a harmonized and integrated approach from economic, commercial, social and political perspectives in the modern sense.
- A National Environmental Strategy will be prepared for an effective environmental management,
- Information and awareness-raising activities for the protection of the environment will be linked to the needs of the target group and the principle of lifelong education will be observed, taking into account the strategic priority of environmental protection.
- Environmental Law, Forest Law, Tourism Encouragement Law, Zoning Law, Coastal Law, Cultural and Natural Heritage Protection Law will be revised (SFYDP, 1995).

Accordingly, in this period, in addition to legislative efforts, improvements were targeted in the protection, rational use and management of natural resources, ecological balance, protection of cultural assets, lifelong education on environmental protection, prevention of disasters and minimization of risks. During this plan (1996–2000), a National Environmental Strategy and Action Plan was prepared and published.

3.8 *Long-Term Strategy and Eighth Five-Year Development Plan (2001–2005)*

The National Biodiversity Strategy and Action Plan was prepared during this plan. The National Desertification Action Plan is among the targets. Some important headings from the objectives, principles and policies of the plan are as follows;

- Forests will be managed, protected and developed in line with the principles of continuity, multi-purpose utilization, participation, specialization, protection of biological diversity, water and wildlife and improvement of social stability within the ecosystem approach, taking into account the conditions of the growing environment, inter-sectoral dependency, productivity and carrying capacity, forest health and landscape, ecotourism, productivity, pollution, fire-insect-landslide-snow-avalanche-flood-freeze and drought realities and ergonomic factors,
- National parks, nature reserves and similar protected areas will be developed and expanded in order to protect biological diversity, water and wildlife, cultural and aesthetic values, to enable research on yet unknown benefits of forests, to prevent soil erosion, landslides and avalanches and to develop ecotourism. In

these studies, it will be taken as a basis that they are of sufficient size to protect the values contained in ecosystems,

- In order to reduce social pressure on natural forests for social, cultural and environmental reasons, the establishment of urban forests and memorial forests will be encouraged and expanded,
- It is deemed necessary to review legislation, strategy, organization, activity programs and product standards, to take measures to prevent the destruction of herbal resources, medicinal plants and natural fungi subject to domestic consumption and foreign trade, to establish the principles and criteria of the Green Certificate or Green Label and to put it into practice in our country,
- Priority will be given to the improvement of streams in preventing floods. Construction on natural stream beds will not be allowed in the zoning plans,
- National Biodiversity Strategy and Action Plan and National Desertification Action Plan will be prepared and studies will continue to participate in the framework Convention on Climate Change,
- For the purpose of increasing the quality of life, urban and rural infrastructure will be developed with a new and holistic approach, and the characteristic cultural patterns of the cities will be preserved,
- It is essential to establish a social, legal, institutional and technical structure for minimizing disaster damages. Continuous and systematic training will be provided in order to correctly perceive disasters and raise awareness of all segments of society. In addition, due importance will be given to vocational training on the subject. A disaster management system will be developed that will ensure a balanced distribution of tasks and effective coordination between central and local government, private sector and non-governmental organizations. Disaster-sensitive zoning plans will be prepared by preparing integrated disaster maps throughout the country. An effective building control system will be established (LTSEFYDP, 2001).

As part of the preparatory work of the Development Plan, the Natural Disasters Specialization Commission was established for the first time. In addition, for the first time, a separate section on disasters was included in the Development Plans, and legal, institutional and administrative dimensions related to the subject in different sectors were evaluated in detail (LTSEFYDP, 2001).

3.9 Ninth Development Plan (2007–2013)

The Ninth Development Plan was prepared with the vision of "Türkiye, a country of the information society, growing instability, sharing more equitably, globally competitive and fully completed her coherence with the European Union" (2001–2023).

In this plan, the principles of sustainable development and common but differentiated responsibility were included while Türkiye fulfills its international obligations.

Some important headings from the objectives, principles and policies of the plan are as follows;

- Conditions will be established for the protection and utilization of natural resources, taking into account the needs of future generations. Environmental management systems will be established to ensure that natural resources can be used by all,
- Researching, protecting and utilizing the biological diversity, natural and genetic resources activities and for transforming them into economic value will be accelerated in Türkiye,
- Works will be completed to make regulations on the use, improvement and protection of water resources and to establish an administrative structure,
- Ground and surface water resources will be protected and the use of treated wastewater in agriculture and industry will be encouraged,
- Multi-purpose and efficient management of forests will be targeted (NDP, 2006).

3.9.1 The Tenth Development Plan (2014–2018)

The Plan includes stable and inclusive economic growth, rule of law, knowledge society, human development, environmental protection and sustainable use of resources (TDP, 2013).

The issue of climate change, which has become the most important agenda of recent years and which is the determinant of all vital areas such as food, water, housing and agriculture, has been specifically examined under the title of "climate change and environment" in the Tenth Development Plan and it has been evaluated that "the frequency of disasters has increased as a result of climate change and their effects have reached serious dimensions"set principles and targets on climate change, water management and biodiversity (Keleş et al., 2015; TDP, 2013).

It was stated that the concept of "green growth" with clean production and eco-efficiency in production sectors, has gained importance in order to achieve sustainable development goals. In this context, priority was given to the prevention of pollution, protection and sustainable use of natural resources and biodiversity (TDP, 2013).

In the Plan, it was stated that Türkiye will reduce the pressure on the environment by adopting policies that prioritize pollution prevention efforts, protection and sustainable use of natural resources and biodiversity. In this context, it has been observed that almost all sub-headings in the plan, such as tourism, infrastructure, energy, logistics and transportation, and industry, include approaches that include the protection of the environment or prioritize the environment. It was stated that one of the most important dimensions of livable spaces is the protection of environmental quality and the adoption of a development and spatial development approach that will not reduce the welfare and happiness of future generations. This approach also shows the final form of the policies that Türkiye has developed over the years to solve environmental problems (TDP, 2013; Özlüer, 2021).

3.9.2 Eleventh Development Plan (2019–2023)

This plan focuses on especially competitiveness and efficiency in all fields. It has five main themes; these are; "livable cities and sustainable environment", "stable and strong economy", "qualified human and strong society", "competitive production and productivity", and "rule of law, democratization and good governance" (EDP, 2019).

It was stated that environment-friendly urbanization approach maintains its importance. This approach is very important for landscape-based urbanism. In the plan, it was stated that there would be reforms regarding local administrations. In addition, the people-oriented city; green city; smart city; quality service delivery; participation; transparency; accountability; optimum geographic area etc. concepts are included (EDP, 2019).

It is important that the issue of Disaster Management, which is included under the title of "Liveable Cities, Sustainable Environment", is addressed within the theme of "quality of life". In this section, space, environment, and rural–urban planning, which is a life that is resistant to disasters, sensitive to the environment, based on cultural values and based on human dignity, is explained among the indicators of development and welfare. In addition, disaster hazard maps according to disaster types, local disaster prevention projects, disaster preparedness, infrastructure facilities against disasters and communication infrastructure projects for disasters are also included in the plan (EDP, 2019).

The plan includes evaluations on the protection of cultural heritage in a way that takes into account disaster risks, especially during the climate change process, and the negative impact of development movements on urban identities (EDP, 2019).

In addition, the issues of urban transformation and urban climate have been an important topic in the plan. It is also very important to note that Türkiye is geographically affected directly and more by climate change in the form of increased sudden rains, floods, and drought disasters (EDP, 2019).

The Plan was stated that in order to ensure sustainable development of cities, there is a need for a disaster and climate-resilient infrastructure, implementation of effective disaster management, an accessible and connected urban transportation system, long-term integrated urban planning and design, prepared with all stakeholders. In this framework, reducing this pressure, preventing environmental pollution, protecting biodiversity and natural resources and sustainable use are prioritized (EDP, 2019).

4 Planning Hierarchy and Implementation Problems

Development plans, which are at the top of the planning hierarchy together with country and regional plans, form the basis for spatial plans. In our country, plans are listed from the top level to the bottom level in terms of the area they cover and their objectives, as shown in Fig. 1.

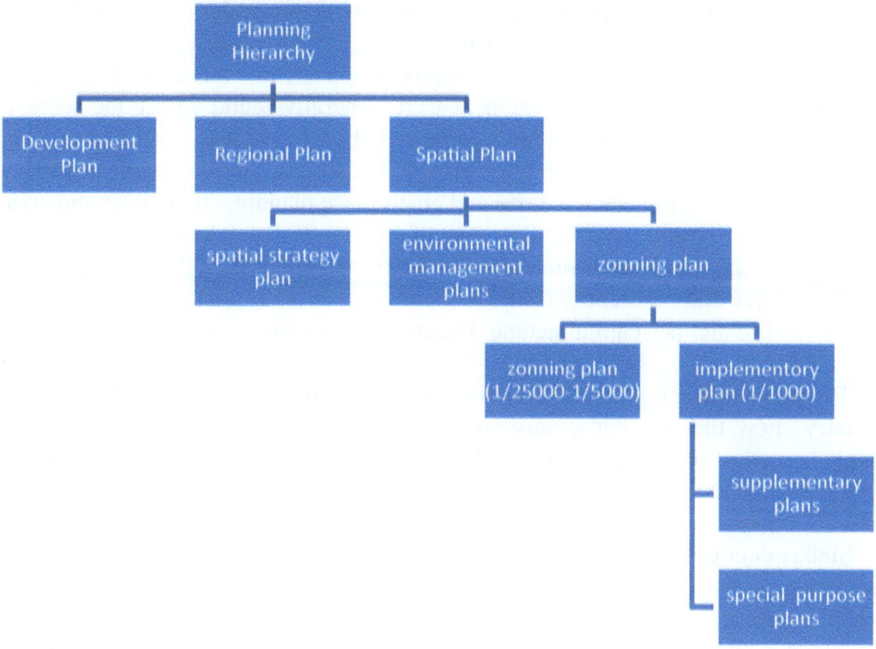

Fig. 1 Planning hierarchy in Türkiye

Landscape planning, which is not included in the planning hierarchy in our country, is at the top of the planning hierarchy in many developed countries. With the analyses made in the landscape planning process, area characteristics such as natural resources that need to be protected, areas that are not suitable for settlement, and areas suitable for agriculture are determined. Cities planned with this understanding are settled in appropriate places, the sustainability of natural resources is ensured and the city becomes resilient to today's problems. Green infrastructure, an important part of landscape planning, is vital for today's cities.

According to Benedict and McMahon (2002), green infrastructure is the combination of various natural and restored ecosystems and landscapes into a link or center. The content of green infrastructure is a synthesis of environmental management and ecological conservation thought within a holistic landscape planning approach (Benedict & McMahon, 2002, 2006). Although geographical diversity leads to the different development of green infrastructure in landscape planning approaches, the concepts of holistic planning, multifunctionality, accessibility, integration, social communication, long-term strategies, and connectivity are at its center. Green infrastructure includes practices that support the integrated ecological network system of regions and cities (Mell, 2017; Pauleit et al., 2011; Sarıarmağan and Var, 2019).

Many concepts were mentioned in the development plans that do not include concepts such as landscape planning and blue green infrastructure. Some concepts are emphasized in detail. When the plans are evaluated in the context of landscape and landscape planning terms, especially in the first three plans, the word sustainability was not used very often, but only in terms of protecting and ensuring the continuity of forest areas. Forests were mostly used for production and raw materials. Expressions related to water and water management were generally found within the scope of water utilization, irrigation and management. Urbanization was described in terms of industrialization, slums, population growth, housing, administrative structure and legislation, while topics such as open and green space planning, recreation and urban aesthetics were hardly covered. The issue of climate change was only raised for the first time in the 8th Plan in the early 2000s. In plans that include the concepts of resilient cities and resilience, resilient society is generally mentioned in terms of quality of life and per capita income. Green infrastructure was not included in any plan (Table 1).

The plans, laws, regulations, reports, guides and many projects prepared in our country show that we know very well what needs to be done. However, at the implementation stage, unfortunately, lack of funds, the understanding of saving the day, hurrying, the inability to make long-term policies and fragmented projects and practices reveal our difference from developed countries.

Since our country is an earthquake country, urban transformation is being carried out in many provinces of the country. Urban transformation is an opportunity for the reorganization of cities, but we do not learn lessons from disasters such as earthquakes, floods, fires and droughts that we have frequently experienced in our country in recent years. Again, problems such as planning mistakes (piecemeal plans instead of holistic, the establishment of settlements in places where there should be no settlements such as stream beds and earthquake fault lines, urban sprawl and population growth, inadequate quality and quantity of open and green spaces, etc.) and failure to take measures against the problems caused by climate change (rainwater management, promoting public transportation and bicycle use, increasing roof gardens, urban afforestation, increasing permeable surfaces, etc.) are seen in projects that are promoted as the most important projects.

5 Conclusion

When development plans, the highest plan in the planning hierarchy of the state, are analyzed, it is possible to understand that the approach of environment and urbanisation has undergone a transformation over the years. In line with the development targets, the economy, which was predominantly agricultural, was directed towards industrialization and the urban population of the country started to increase rapidly.

Although the plans, which were prepared in a comprehensive framework, made decisions to solve problems related to many issues such as population growth, industrialization, employment, agriculture and environment, especially development, in the next plan it was stated that these problems continued, that they could not be eliminated for various reasons, that there was a need for revision in sub-plans, and that confusion of authority and responsibility in legal-administrative processes was

Table 1 The presence of some terms of landscape and landscape planning

Development plans	Landscape planning									
	Sustainability/ sustainable environment/ sustainable cities	Landscape	Nature conservation/ biodiversity	Blue-green infrastructure	Water management	Climate change	Life quality/ public health/ social justice	Resillent cities	Disaster/ disaster management	Agriculture and forestry
FFYDP	+	−	+	−	−	−	+	−	−	+
SFYDP	+	−	+	−	−	−	+	−	−	+
TFYDP	+	−	+	−	−	−	+	−	−	+
FFYDP (1979)	+	−	+	−	−	−	+	−	−	+
FFYDP (1984)	+	−	+	−	−	−	−	−	−	+
SFYDP	+	+	+	−	+	−	−	−	−	+
SFYDP	+	+	+	−	+	−	+	−	+	+
LTSEFYDP	+	+	+	−	+	+	+	−	+	+
NDP	+	−	+	−	+	+	+	−	+	+
TDP	+	+	+	−	+	+	+	+	+	+
EDP	+	+	+	−	+	+	+	+	+	+

experienced or could not be eliminated. In addition, it has been determined that although solution proposals are developed with determination in the plans, in practice, it is essential to improve coordination among public institutions and to increase institutional, financial and technical capacity.

In addition, in all plans, the principles, objectives and policies set for important issues do not mention landscape planning as a solution to today's environmental problems. While concepts such as regional planning, basin planning, rural planning and urban planning are mentioned, the need to evaluate them from a holistic perspective is not considered. Landscape planning is an important tool that enables planning with a holistic and ecological analysis on top of these concepts. However, although this concept has an important place in academia, its importance has not yet been understood by plan makers and implementers (ministries, municipalities and various public institutions).

In this context, in our country, landscape planning should be included in the planning hierarchy and set the agenda for development plans. It is obvious that adaptation and mitigation of the global climate crisis can only be achieved through nature-based solutions in urban planning.

References

1961 Constitution, https://www.anayasa.gov.tr/tr/mevzuat/onceki-anayasalar/1961-anayasasi/.

Ahern, J. (2002). *Greenways as strategic landscape planning: Theory and application* (183 p.). Wageningen University. ISBN 90-5808-605-4

Aktaş N. K. (2021). Enerji etkin peyzaj uygulamaları ve mavi-yeşil altyapının önemi. Sanat, Tasarım ve Mühendislik Alanında Akademik Değerlendirmeler (Academic Evaluations in Art, Design and Engineering). In D. Erdirençelebi (Ed.), Publisher SRA (Strategic Researches Academy). ISBN: 978-625-7148-26-9.

Ayhan, Ç. K., & Hepcan, Ş. (2009). Özgün Peyzaj Karakteristiklerine Sahip Mekanlara Yönelik Bir Peyzaj Planlama Yönteminin Ortaya Konulması; Bozcaada Örneği. *Journal of Tekirdag Agricultural Faculty, 6*(1), 93–105. Retrieved from https://dergipark.org.tr/tr/pub/jotaf/issue/19049/201474.

Bardakçi, S., Karagöz, Y., & Dündar, S. (2019). On Birinci Kalkınma Planında Yer Alacak Öncelikli Kalkınma Politikalarına Yönelik Toplumsal Bakış Açısının İncelenmesi: Tr 72 Bölgesinde Bir Uygulama. *Elektronik Sosyal Bilimler Dergisi/Electronic Journal of Social Sciences, 18*(70), 798–817. https://doi.org/10.17755/esosder.458691.

Benedict, M. A., & McMahon, E. T. (2002). Green infrastructure: Smart conservation for the 21st century. *Renewable Resources Journal, 20*, 12–17. https://rnrf.org/wp-content/uploads/2020/01/RRJV20N4.pdf.

Benedict, M. A., & McMahon, E. T. (2006). *Green infrastructure: Linking landscapes and communities*. Island Press.

Bhadouria, R., Upadhyay, S., Tripathi, S., & Singh, P. (Eds.). (2022). *Urban ecology and global climate change*. John Wiley & Sons.

Çelikkaya, S., Yaman, H., Dağli, İ., & Çapik, E. (2018). Bölgesel Politikaların Kalkınma Planlarındaki Yeri Ve Zamana Bağlı Dönüşümü. *Süleyman Demirel Üniversitesi Vizyoner Dergisi, 9*(22), 21–31. https://doi.org/10.21076/vizyoner.440667.

Çetinkaya, G., & Uzun, O. (2014). Peyzaj Planlama (1st ed., p. 30). İstanbul, Türkiye: Birsen Publishing.

Çukurçayir, S., & Tezcan, K. (2011). Demokratikleşme ve Ekonomik Kalkınma: Etkileşim Analizi. *Bilgi Ekonomisi ve Yönetimi Dergisi, VI*(II), 48–76. Retrieved from https://dergipark.org.tr/tr/pub/beyder/issue/3475/47275.

Dağdemir, Ö. (2015). *Çevre Sorunlarına Ekonomik Yaklaşımlar ve Optimal Politika Arayışları* (3rd ed.). Gazi Publishing.

EDP. (2019). Eleventh development plan, https://www.sbb.gov.tr/wp-content/uploads/2022/07/On_Birinci_Kalkinma_Plani-2019-2023.pdf.

Erdem, M. S., & Yenilmez, F. (2017). Türkiye'nin Avrupa Birliği Çevre Politikalarına Uyum Sürecinin Değerlendirilmesi. *Optimum Journal of Economics and Management Sciences, 4*(2), 91–119. https://doi.org/10.17541/optimum.292768.

Erdem, M. S., & Bozan, M. (2021). Türkiyede Sürdürülebilir Kalkınma Çerçevesinde Ekolojik Turizmin Kalkınma Planları İçerisinde Değerlendirilmesi. *Sinop University Journal of Social Sciences 5*(1): 67–90. Retrieved from https://doi.org/10.30561/sinopusd.917549.

FFYDP. (1963). First five year development plan, https://www.sbb.gov.tr/wp-content/uploads/2022/07/Kalkinma_Plani_Birinci_Bes_Yillik_1963-1967.pdf.

FFYDP. (1979). Fourth five year development plan, https://www.sbb.gov.tr/wp-content/uploads/2022/08/Dorduncu-Bes-Yillik-Kalkinma-Plani_1979_1983.pdf.

FFYDP. (1984). Fifth five year development plan https://www.sbb.gov.tr/wp-content/uploads/2022/08/Besinci-Bes-Yillik-Kalkinma-Plani-1985-1989.pdf.

Gruehn, D., & Kenneweg, H. (1998). *Berücksichtigung der Belange von Naturschutz und Landschaftspflege in der Flächennutzungsplanung.* Landwirtschaftsverlag.

Kaya, M. Y., & Uzun, O. (2019). Ekosistem Hizmetleri ve Mekânsal Planlama İlişkisinin Peyzaj Planlama Çerçevesinde Değerlendirilmesi. *Düzce University Journal of Science &Technology, 7*(3), 2166–2193. https://doi.org/10.29130/dubited.546496.

LTSEFYDP. (2001). Long term strategy and eighth five year development plan, https://www.sbb.gov.tr/wp-content/uploads/2022/07/Long-Term_Strategy_and_Eight_Five_Year_Development_Plan_2001-2005.pdf.

Luz, F. (2000). Participatory landscape ecology–a basis for acceptance and implementation. *Landscape and Urban Planning, 50*(13), 157–166. https://doi.org/10.1016/S0169-2046(00)00087-6

McHarg, I. (1969). *Design with nature* (1st ed., pp.10–50). New York, USA: The Natural History Press, Garden City.

Mell, I. (2017). Green infrastructure: Reflections on past, present and future praxis. *Landscape Research. Routledge, 42*(2), 135–145. https://doi.org/10.1080/01426397.2016.1250875

NDP. (2006). Ninth development plan, https://www.sbb.gov.tr/wp-content/uploads/2022/07/Dokuzuncu_Kalkinma_Plani-2007-2013.pdf.

Öner, Ş. (2020). 11th Development plan in terms of local governance approaches (2019–2023). *Balıkesir University The Journal of Social Sciences Institute, 23*(43), 383–410. https://doi.org/10.31795/baunsobed.671928.

Özlüer, İ. Ö. (2021). İklim Krizi ve Afet Planlamasi (Türkiye'nin Kalkınma Planlarina Yansiyan Afet Stratejisi). *Memleket Siyaset Yönetim (MSY), 16*(36), 397–420. Retrieved from https://dergipark.org.tr/tr/pub/msydergi/issue/68237/1063492.

Pauleit, S., Liu, L., Ahern, J., & Kazmierczak, A. (2011). Urban ecology: Patterns, processes and applications. In J. Niemela, J. H. Breuste, G. Guntenspergen, N. E. McIntyre, T. Elmqvist, & P. James (Eds.), *Multifunctional green ınfrastructure planning to promote ecological services in the city* (s. 272–285). New York: Oxford University Press.

Reinke, M. (2002). *(2002) Qualität der kommunalen Landschaftsplanung und ihre Berücksichtigung in der Flächennutzungsplanung im Freistaat Sachsen.* Logos Verlag.

Rosenzweig, C., Solecki, W. D., Hammer, S. A., & Mehrotra, S. (2011). Urban climate change in context. In C. Rosenzweig, W. D. Solecki, S. A. Hammer, & S. Mehrotra (Eds.), *Climate change and cities: First assessment report of the urban climate change research network* (pp. 3–11). Cambridge: CambridgeUniversity Press.

Sariarmağan, Ş., & Var, M. (2020). Green infrastructures as protection method of urban biodiversity. In *International Conference on Rediscovering Cities* 2*K*20, 7–9 July, India.

SFYDP. (1968). Second five year development plan, https://www.sbb.gov.tr/wp-content/uploads/2022/07/Ikinci_Bes_Kalkinma_Plani-1968-1972.pdf.

SFYDP. (1989). Sixth five year development plan, https://www.sbb.gov.tr/wp-content/uploads/2022/07/Altinci_Bes_Yillik_Kalkinma_Plani-1990-1994.pdf.

SFYDP. (1995). Seventh five year development plan, https://www.sbb.gov.tr/wp-content/uploads/2022/07/Yedinci_Bes_Yillik_Kalkinma_Plani-1996-2000.pdf.

Singh, R., Verma, P., Kumar Singh, V., Srivastava, P., & Kumar, A. (2022). Urban ecology and climate change:Challenges and mitigation strategies. In R. Bhadouria, S. Upadhyay, S. Tripathi, & P. Singh (Eds.), *Urban ecology and global climate change*. John Wiley & Sons.

TDP. (2013). The tenth development plan (2014–2018), https://www.sbb.gov.tr/wp-content/uploads/2022/08/Onuncu_Kalkinma_Plani-2014-2018.pdf.

TFYDP. (1973). Third five year development plan (New strategy and development plan third five year), https://www.sbb.gov.tr/wp-content/uploads/2022/08/Yeni-Strateji-ve-Kalkinma-Plani_Ucuncu-Bes-Yil_1973_1977.pdf.

Tuğaç, Ç. (2021). An evaluation on the adaptation dimension of combatingclimate change and the relationship to health. In *The 1st Internatıonal Health Andclımate Change Congress. Full Text Book*. ISBN: 978-975-92020-4-0.

Von Haaren, C., & Albert, C. (2011). Integrating ecosystem services and environmental planning: Limitations and synergies. *International Journal of Biodiversity Science, Ecosystem Services & Management, 7*(3), 150–167. https://doi.org/10.1080/21513732.2011.616534

Wende, W., Herberg, A., & Herzberg, A. (2005). Mitigation banking and compensation pools: Improving the effectiveness of impact mitigation regulation in project planning procedures. *Impact Assess Proj Apprais, 23*(2), 101–111. https://doi.org/10.3152/147154605781765652

Blue-Green Infrastructure for Flood Resilience: Case Study of Indonesia

Ariyaningsih, Riyan Benny Sukmara, Rahmat Aris Pratomo, Nurrohman Wijaya, and Rajib Shaw

Abstract Due to climate change, there is an increasing risk to communities from urban floods especially in developing countries like Indonesia. Massive impacts of flooding can influence the damage cost of infrastructure as well. Furthermore, numerous literature reviews mentioned that the importance of blue-green infrastructure in disaster risk reduction has been recognized for managing urban floods by integrating with urban environmental aspects. Blue-green infrastructure provides numerous opportunities and advantages for tackling the diverse concerns related to the environment, societal well-being, and climate change. This chapter discusses the issues, gaps, opportunities, and implementation of Blue-Green Infrastructure for flood resilience especially city in Indonesia (Bandung, Balikpapan, Samarinda, and Semarang). Case studies highlighting blue-green infrastructure gaps, opportunities, successes, and threats from different cities in Indonesia. In detail, this chapter identifies the background and history of the blue-green policy in a specific area/city followed by the characteristics of the specific case. Moreover, this chapter explains how the selected cities implement blue-green infrastructure and explores its effectiveness and implications to the city.

Ariyaningsih (✉) · R. Shaw
Graduate School of Media and Governance, Keio University Shonan Fujisawa Campus, Fujisawa, Japan
e-mail: ayya19@keio.jp

R. Shaw
e-mail: shaw@sfc.keio.ac.jp

R. B. Sukmara
Department of Civil Engineering, National Central University, Taiwan and Institut Teknologi Kalimantan, Balikpapan, Indonesia
e-mail: bennysukmara@lecturer.itk.ac.id

R. A. Pratomo
Urban and Regional Planning, Institut Teknologi Kalimantan, Balikpapan, Indonesia
e-mail: r.a.pratomo@lecturer.itk.ac.id

N. Wijaya
Urban and Regional Planning, Institut Teknologi Bandung, Bandung, Indonesia
e-mail: nurrohman.wijaya@itb.ac.id

P. K. Joshi et al. (eds.), *Blue-Green Infrastructure for Sustainable Urban Settlements*,
https://doi.org/10.1007/978-3-031-62293-9_11

Keywords Blue-green infrastructure · Disaster risk reduction · Flood management policies · Indonesia · Urban flood risk

1 Introduction

The rapid process of urbanization and changes in land use that have occurred over the past century have had a significant impact on the world, resulting in many challenges in terms of development and the need to adapt to climate change (Wen et al., 2019). Furthermore, climate change which has many impacts on urban areas has been widely recognised for causing the high intensity and frequency of urban disasters (Reckien et al., 2018; Sahani et al., 2019; Swain et al., 2020). As a result, because of high-rate urbanisation and global climate change, the intensity of flooding (especially urban flooding) in the urban area is also estimated to be increasing (IPCC, 2013; Jago-on et al., 2009). Furthermore, the flood has many negative impacts, especially on socio-economic systems, which causes infrastructure and facilities damages and losses (Nofal and van de Lindt, 2022). Based on previous research (Ariyaningsih et al., 2021; Marfai & King, 2008; Hapsari & Zenurianto, 2016a, b), flooding, which happens in developing countries like Indonesia are triggered by five factors, including natural characteristics like heavy rain and coastal flooding, man-made factor including river channel construction error, land conversion, and also river silting. Moreover, the high population growth that induces climate change in this country contributes to flooding (Wells et al., 2016; Wijayanti et al., 2017; Ghozali et al., 2016). To support those statements, the World Bank also projected that the community in Indonesia will be exposed to urban flooding, and it is estimated to grow by over one million by 2035–2044 (Bank World, 2002; Stanton-Geddes & Vun, 2019).

Considering flooding impacts, urban capacities to face climate-related hazards like floods should be increased. The precipitation will be unpredictable due to climate change. In order to address these issues, numerous urban resilience development concepts have been suggested to mitigate flood hazards. These include the water supply city, sewered city, drained city, waterways city, water cycle city, and water-sensitive city (Kabisch et al., 2016; Adedeji et al., 2019). However, traditional flood management in Indonesia only focuses on grey infrastructures such as drainage pipes, storm sewers, water pumps, and detention basins (Dewi, 2007; Hapsari & Zenurianto, 2016a, b; Dwirahmadi et al., 2019). Furthermore, a potential approach for flood management in urban areas involves the implementation of grey, green, and blue-green infrastructure (Hamel & Tan, 2022; O'Donnell et al., 2021). However, nowadays, it is understood that those approaches (grey and green) only offer low sustainability while the BGI (Blue-Green Infrastructure) can provide numerous complementary benefits (O'Donnell, 2021).

In developed countries, BGI (Blue-Green Infrastructure) protects urban and other landscapes' ecological and hydrological values, providing resilient measures to address environmental threats, including disasters like floods (Kapetas & Fenner,

2020). For instance, the Netherlands is implementing BGI (Blue-Green Infrastructure) to reduce flood risks (Ghofrani et al., 2016). To mitigate flood damage in this scenario, structural interventions might be implemented in the landscape to create additional pathways for the circulation of flood water. Moreover, Green Infrastructure (GI), Sustainable Urban Drainage Systems (SUDS), and Water Sensitive Urban Design are all synonyms for the BGI in the United Kingdom (O'Donnell et al., 2017). In contrast, in the United States, BGI is also referred to as Low Impact Development (Pour et al., 2020) and stormwater Best Management Practices (BMPs) (Battemarco et al., 2022). International agencies like UN-Habitat are supporting Blue-Green Infrastructure actions. In addition, they always enhance collaboration with local municipalities, local authorities, and stakeholders working for urban water or water infrastructure to implement BGI (O'Donnell et al., 2020a, b; Drosou et al., 2019). As mentioned by some experts (Kaur & Gupta, 2022; Ariyarathna, et al., 2023; Ghofrani et al., 2016), The Blue-Green concept combines water management and green infrastructure in order to create a more natural water cycle in urban settings, making cities more resilient to floods and other water-related hazards. In addition, examples of Blue-Green Infrastructure include rain gardens, bioretention systems, planted drainage assets such as green roofs, as well as green walls, and controlled water storage areas (e.g., car parks, recreational areas). In addition, the primary roles of Blue-Green elements are to facilitate water reuse, detention, infiltration, transportation, water treatment, evapotranspiration, the delivery of local amenities, and the creation of feasible environments for local ecosystems (Kapetas & Fenner, 2020; Drosou et al., 2019).

BGI's multifunctionality is a key factor in its increasing popularity, as it is progressively acknowledged as enhancing people by providing multiple co-benefits in addition to water and flood risk management. BGI enables the possibility of supporting climate change adaptation to extreme weather events such as storms, heatwaves, and droughts. It also allows for the extension of the lifespan of aging infrastructure, improvement of wildlife and biodiversity, enhancement of air and water quality, improvement of landscape connectivity and access to green spaces, promotion of health and well-being, and creation of appealing landscapes that enhance the overall quality of a place (Kabisch et al., 2016; Adedeji et al., 2019; Demuzere et al., 2014).

Against this background and the benefits of implementing Blue-Green Infrastructure, this chapter tries to explore Blue-Green Infrastructure implementation in several cities in Indonesia. The next sections are structured as follows: Sect. 2 identifies flood characteristics in Indonesia and its impacts. Furthermore, Sect. 3 shows the integration of urban flood risk with Blue-Green Infrastructure. Section 4 explores the results, including implementing Blue-Green Infrastructure in selected cities in Indonesia. Moreover, Sect. 5 discusses the issues and barriers to delivering Blue-Green infrastructure in Indonesia, followed by the conclusions in Sect. 6.

2 Flood Characteristics and the Impacts in Indonesia

Flooding is a hydrological disaster that can arise from inadequate or inappropriate management of the hydrological cycle (Sahani et al., 2019). An imbalanced hydrological process can be a trigger to be a flooding hazard. High-intensity rainfall, topographical characteristics, and inadequate drainage capacity will cause a massive flood in certain areas. This is also exacerbated by climate change and population growth on a vast scale (Swain et al., 2020). Flooding become the most frequent natural hazard around the globe and threatening humans and the environment intensively (Marfai et al., 2008). In Asia, some countries are most vulnerable to flood hazards, and Indonesia is ranked third after China and India (Dewi, 2007). In terms of the total number of disasters in Indonesia, floods account for 32% of them (Yulianto et al., 2015). Based on recorded data from Indonesia's National Disaster Management Agency (Badan Penanggulangan Bencana Nasional, BNPB) there are 8116 flood events between the years 2014 to 2023 (BNPB, 2023).

With over 17000 islands and more than 1.9 million km^2 of land areas, Indonesia has diverse landscape conditions. Concerning the discussion of flood hazards, the various landscape conditions in each region will result in distinct flood characteristics. Low-land areas in Indonesia are most frequently to be flooded and most of the major cities in Indonesia are in this area. In general, Indonesia has several main types of flooding: pluvial flood, fluvial flood, and coastal flood. In Indonesia, fluvial flooding has become the most prevalent type, particularly in low-lying areas. This deluge in Indonesia results from insufficient river capacity to convey water without overflowing. Typically, sedimentation issues are the leading cause of diminished river conveyance capacity. Inadequate river maintenance and land use change or land cover change (LULCC) can both contribute to sedimentation. Land Use Change triggers soil erosion, which eventually leads to the accumulation of erosional sediments in rivers.

As an integrated system, the river is interconnected between upstream and downstream areas. Changes in the upstream area, such as a land use/land cover change (LULCC), will increase flood discharge in the downstream area. If the river's capacity in the downstream areas is insufficient to handle the growing discharge, fluvial flooding problems will occur. Several studies in Indonesia such as Java, Papua, Sumatra, and Borneo (Kalimantan) also reveal that the LULCC also contributes to the flooding issues in Indonesia (Dasanto et al., 2014; Emam et al., 2016; Mahmud et al., 2018; Novitasari & Kurdi, 2022; Sugianto et al., 2022; Tarigan, 2016; Wells et al., 2016). Extreme precipitation in the upstream area can also cause flooding downstream, even if there is a small amount or no rainfall in the downstream area.

An additional factor contributing to the decreasing river capacity is the existence of informal settlements along the riverbank, particularly in urban areas (Rizaldi et al., 2019). Numerous factors, including economics, topography, and psychology, contribute to the inclination of riverbank dwellers (Pramantha et al., 2021). Mostly, people who live along the riverbank are low-income families whose economic ability cannot afford the house price with proper location and mostly, they don't need to

(a) (b)

Fig. 1 **a** River sedimentation in Ciliwung River, Jakarta (POSKOTA, 2020); **b** Informal settlement in Ciliwung River, Matraman Segment (adapted from Detik.com, 2022)

pay property tax (*Pajak Bumi dan Bangunan*) to live there. Because of the limited space in the riverbank, they tend to build a house that jutted into the river, which will narrow the river's width (see Fig. 1).

Insufficient drainage systems, high rainfall intensity and decreasing river/channel capacity are major flooding causes. Moreover, concave areas suffer more due to runoff accumulation from surrounding higher elevations. In the northern region of Java Island, apart from pluvial and fluvial flooding, coastal flooding become the most experienced in Indonesia. Most of the cities in this area has experienced coastal flood (Budiyono et al., 2016; Isa et al., 2019; Marfai, 2014; Nashrrullah et al., 2013; Nirwansyah & Braun, 2019). Coastal floods in northern Java Island are caused by various sources, including storm surges, tides, sea level rise, and land subsidence. The occurrence of tidal floods in northern Java is periodic, happening from July to August, namely during the East Monsoon season (Khasanah et al., 2016), and occurs not only during high tide but also during regular tide (Solihuddin et al., 2021). Concerning to the occurrence of land subsidence, Jakarta and Semarang are an example city in northern Java that have severe coastal flooding and have significant issues with land subsidence. By launching an environmental monitoring satellite developed by the European Space Agency (ESA) called ERS-1, Researchers can utilize the interferometric synthetic aperture radar technique (InSAR) to examine the spatial–temporal patterns of ground subsidence in Jakarta and Semarang. The study reveals that the land subsidence in Semarang is up to 100 mm/year and Jakarta is around 60 mm/year. In addition, most large urban areas in Indonesia are in low-land areas, close to the coast, and social factors have contributed to the complexity of flood issues in Indonesia. Meanwhile, in the higher area with stepped-slope, flash flood.

Researchers have conducted several studies related to flood issues in Indonesia, revealing many factors that could be the cause of flooding. For instance, Jakarta as the current capital of Indonesia has a high frequency of flooding every year (Anugrahadi et al., 2020), especially in rainy season during October to March. According to historical data, in 2007 Jakarta was hit by major flood and regarded as a national disaster with total losses reach up to US$ 565 million (Wijayanti et al., 2017). A total 56 people died, over 3,40,000 people had to be evacuated, and over 74 thousand houses were submerged with flood depths reaching up to four meters (Garschagen et al., 2018). Additionally, Jakarta was hit by a major flood event in January 2020. Jakarta has experienced an extreme rainfall of up to 377 mm/day, the highest observational record since 1866, which forced more than 1,70,000 people to be evacuated, 66 died, more than 60% of residential areas were inundated and suffered economic loss over US$ 700 million (Lubis et al., 2022).

3 Integrating Urban Flood Risk Management with Blue-Green Infrastructure

BGI or Blue-Green Infrastructure refers to the benefits of green spaces serving various urban water management functions. The term is like "Green-Infrastructure" (GI), but with a "Blue" component added, such as water (Kabisch et al., 2016; Demuzere et al., 2014). Consequently, by combining current grey assets with newly produced or upgraded blue, green, and blue-green assets, an integrated network can effectively offer a diverse range of ecosystem services. Furthermore, urban green spaces and naturalized watercourses like green roofs, retention and detention ponds, bioswales, and rain gardens make up Blue-Green Infrastructure (BGI), which is widely acknowledged in international literature as an effective strategy for managing stormwater and flood risk (Alexander et al., 2019; Ghofrani et al., 2016; O'Donnell et al., 2017). The Blue-Green Infrastructure, also known as BGI, is a comprehensive approach to addressing various interconnected challenges associated with the management of water resources, including surface and groundwater, water supply, water quality, sanitation, solid waste management, flooding, and the preservation of natural ecosystems. BGI focuses on integrating these issues into urban development plans while considering the impacts of climate change (Agonafir et al., 2022; Drosou et al., 2019).

Blue-green infrastructure is well recognized as a favored solution used by numerous communities to address the impacts of climate change and degradation of the environment. The approach has numerous social benefits, including the development of effective water management systems, the decrease of average temperatures, and the enhancement of air quality. However, blue-green infrastructure is being developed in other areas, emphasizing flood resilience and ecosystem health restoration (Thorne et al., 2018). This approach includes water gardens, green roofs, and swales. Flood risk management is being reevaluated in cities all around the world. Urban areas are increasingly investing in conventional grey infrastructure, including flood walls, barriers, lined drainage channels, underground pipes, and detention tanks. However, there is a growing emphasis on transitioning from flood defense to water resilience. This involves adopting approaches like water sensitive urban design (Fawzy et al., 2020; Walsh et al., 2016), Sustainable drainage system (SUDS) (Caleiro & Ramos, 2016; O'Donnell et al., 2020a, b), green infrastructure (Zeng et al., 2022), and 'Sponge Cities'. These strategies vary slightly from one another, but they all represent the concept of a BGI, in which integrated water management and green infrastructure collaborate to restore a water cycle more closely aligned with nature, thereby reducing flood risk and providing additional environmental, social, and economic benefits (Ghofrani et al., 2016; Santos et al., 2020).

An example of the blue-green infrastructure concept is the Sustainable Drainage System (SuDS), which is an environmentally friendly drainage system. Extreme weather, such as floods and droughts, require drainage systems that can adapt to these conditions and handle them effectively while contributing to reducing carbon emissions. To implement BGI solutions that are suitable to the local context, flood risk management must be both inclusive and proactive in its approach, welcoming "bottom-up" community participation (Drosou et al., 2019; Pal et al., 2021). Another example of FRM (Flood Risk Management) and BGI (Blue-Green Infrastructure) is RWH (Rain Water Harvesting) (Kaur & Gupta, 2022; Lawton, 1961; Mamangkey et al., 2021). An alternate strategy concerning stormwater as a resource is highlighted by the prospective advantages of RWH on urban flood risk management and water supply augmentation. However, RWH (Rain Water Harvesting) systems are implemented in a variety of ways around the world, and many of these systems fail to fully capitalize on their benefits (Mamangkey et al., 2021). As a result, BGI addresses a wide range of concerns concurrently, including flood mitigation issues. Examples of these measures include mitigating urban runoff and promoting infiltration through permeable surfaces, improving water quality, and restoring aquatic biodiversity.

4 Blue Green Infrastructure Implementation for Flood Resilience in Indonesia

4.1 Case of Samarinda City, East Borneo (Kalimantan Island)

In 2019, the Government of Indonesia (GoI) decided to start their mega project to move the Capital City from Jakarta in Java Island (Praditya et al., 2023) to Borneo Island called Ibu Kota Nusantara (IKN) (Sensuse et al., 2022). The location is in Penajam Paser Utara Regency, adjacent to the two major cities in East Borneo Province, Balikpapan and Samarinda. Establishing a new capital city will have crucial implications for various aspects of urban development for its neighboring cities, especially Samarinda City (Fig. 2).

In the future, Samarinda, also the provincial capital of East Kalimantan, will become a very important city and play an important role as the primary buffer city of Nusantara because of the relocation project of the new capital city. However, despite its vital function, Samarinda faces numerous obstacles that impede its city development. Frequent flooding comes out as a significant issue among these obstacles. Over five floods occur annually in Samarinda, primarily during the rainy season. According to historical records, Samarinda experienced devastating floods in 1998, 2019, 2020 (Kaltim Today, 2020), and 2022, affecting many people (BNPB, 2022).

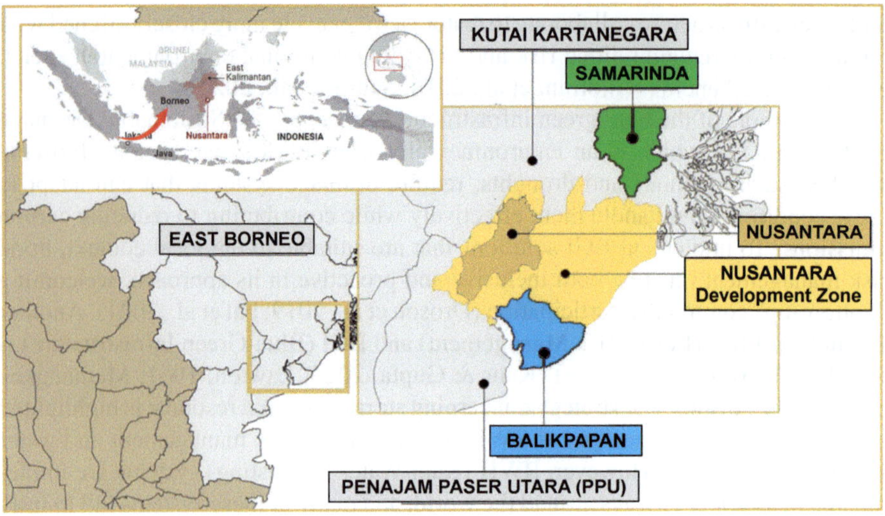

Fig. 2 Location of Samarinda (Neighboring city of Nusantara)

Floods in Samarinda occurred due to several causes, such as rapid urbanization which resulted in massive land use/land cover conversion becoming impermeable surface. Karang Mumus Watershed (KMW) is the biggest watershed in Samarinda and has significant experience in land use change between the years 2014–2020. Approximately, settlement area has increased by around 5.77 km^2, and bare land by around 2.71 km^2 (Fatimatuzahra Indra et al., 2021). This is in-line with the findings of the study done by Warsilan (2019), which revealed a 23.81% increase in stormwater runoff between the year 2000–2016 (Warsilan, 2019). An inadequate drainage system exacerbates the problem, impeding the efficient drain of excess water during high-intensity rainfall events. The social issue also contributes to the complexity of flooding issues. Residents who live near riverbanks contribute to reducing river/channel capacity and pose challenges for maintenance efforts. Moreover, there is a lack of awareness among people, particularly those living near rivers and drainage channels. Unfortunately, we still witness individuals irresponsibly discarding their garbage into these waterways, exacerbating the situation (Hartono et al., 2023) (Fig. 3).

Furthermore, another concern connected to the flooding issue in Samarinda is the presence of coal mining operations in and around Samarinda (see Fig. 4). These coal mines have altered the land use and land cover, resulting in reduced infiltration capacity. Consequently, there will be an increase in stormwater discharge, which will eventually cause overflow in urban channels.

Following the trend of flood measurement method, one of the new innovative approaches is Blue-Green infrastructure (BGI) concept. BGI is the opposite of the Grey infrastructure (GI) in the case of flood measurement. Grey infrastructure describes to the conventional human-made or engineered systems and structures

Fig. 3 Samarinda's flooding in 2019

used for urban development and managing water resources. It primarily addresses water-related challenges such as stormwater runoff, flood control, and wastewater treatment. Grey infrastructure typically consists of pipelines, drainage channels and canals, dams, water treatment facilities, and other man-made structures.

The implementation of BGI is a widely acknowledged approach to enhance urban resilience in the face of increasing climate change and calamity risks (Thorne et al., 2018). BGI heavily depends on the ecosystem services offered by urban green spaces and natural water flows to reduce urban risk and hazard (Dhyani et al., 2020). In addition, BGI is an effort to integrate natural elements and engineered systems to mitigate flood risk and promote sustainable urban development. Implementing BGI may entail the creation of green spaces, such as parks, gardens, and wetlands, as well as installing permeable surfaces and sustainable systems for stormwater management.

Referring to Blue-Green infrastructure (BGI), it can be said that implementing the BGI concept for Samarinda's flood management is still far from this concept. Flood management in Samarinda continuously relies on conventional approaches such as building a culvert, river dredging and reshaping (normalisasi sungai) and constructing flood control structures including temporary storage ponds (retention ponds), flood gates and pumping houses.

In addition, with the various structural approaches, the Government of Samarinda City (GoS) continuously increased the green space area (ruang terbuka hijau), both for recreational purposes or as infiltration zones to reduce stormwater runoff. Based on the regulation, the minimum allocation for green space area is 30% from the total urban area (Law No. 26 the Year 2007 On Spatial Management, 2007). However,

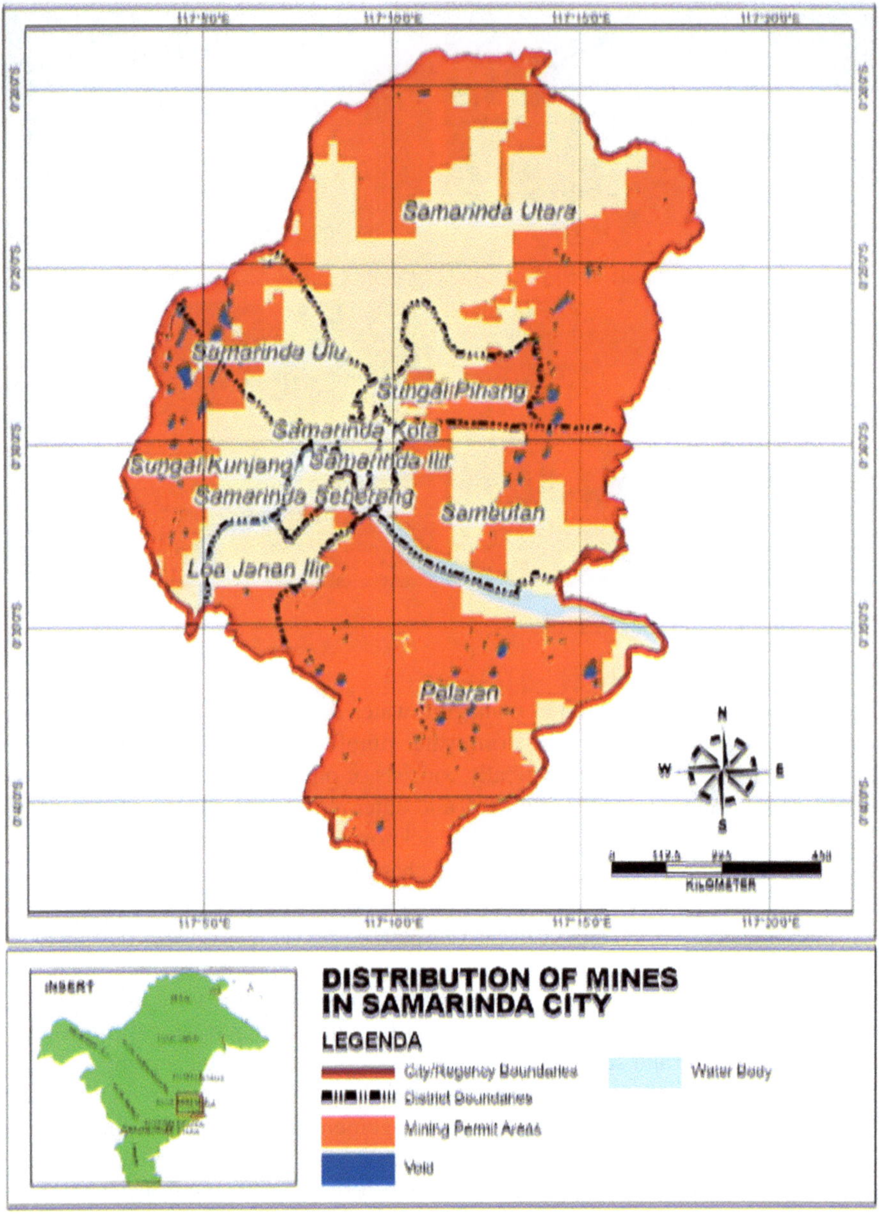

Fig. 4 Distribution of mine areas in Samarinda city. *Adopted from* (Damayanti et al., 2021)

Fig. 5 One of green spaces in Samarinda (Bebaya Park, Samarinda). *Source* (ProKaltim, 2022)

Samarinda City is presently unable to meet the requirement. Only 7% of Samarinda's total area, or approximately 4,308 hectares, comprises green space (Santoso et al., 2022) (Fig. 5).

According to the flood measurement regulation, Samarinda lacks specific flood management regulations. Samarinda has only local regulations (Peraturan Daerah) for governing spatial planning, while regulations for flood management, such as river buffer zones, flood warning system, and the obligation to provide retention ponds which refer to the National policy called "Zero Increase peak discharge" are not available yet.

4.2 *Case of Semarang City, Central Java (Java Island)*

Semarang, situated in the northern region of Java Island, serves as the capital city of Central Java province. With a population over 1.65 million individuals, it represents as an important urban center. Geographically, Semarang City has two characteristics: lowlands in the north of the city, which includes the coast, and highlands in the south, which is a hilly area (Abidin et al., 2013; Wijayanti & Pratomo, 2016; Yuwono et al., 2013). As a coastal city, Semarang City also suffers from typical problems experienced by other coastal cities (Pratomo et al., 2022, 2023), namely massive and irregular urban growth and flooding (Hadi, 2017; Harwitasari & van Ast, 2011). In terms of flooding, the City experiences two types of regular flooding. These are called river and tidal floods (Marfai & King, 2008). The river flood is mainly influenced by the high intensity of rainfall and non-optimal urban drainage system (Suliyati, 2014; Wijaya & Suhardi, 2022). However, in some cases, river floods in Semarang are also influenced by geographical conditions. Upstream flooding causes runoff to other downstream areas (Wismarini & Ningsih, 2010).

Unlike river floods, tidal floods generally occur in seaside or coastal areas (Suliyati, 2014). This kind of flood is exacerbated by land subsidence and sea level rise (Abidin et al., 2013; Harwitasari & van Ast, 2011; Marfai & King, 2008; Marfai et al., 2008), thus expanding the inundated area (Harwitasari & van Ast, 2011).

Evidence shows that land subsidence in Semarang City reached between 1.795–7.796 cm/year in 2018–2019 (Istiqomah et al., 2020). The trend in the frequency of flooding events in Semarang City is also influenced by climate change, which can increase the frequency and intensity of rainfall (Husnayaen et al., 2018; Kuehn et al., 2010; Pratomo et al., 2023). These conditions prioritize flood-related issues for Semarang City's government agenda (Drosou et al., 2019).

Semarang's initial involvement in climate change action occurred in 2009, when the Rockefeller Foundation introduced the Asian Cities Climate Change Resilience Network (ACCCRN). This program aimed to enhance the resilience capabilities of cities in South and Southeast Asia (Reed et al., 2015). Semarang served as a pilot project for ACCRN from 2009 to 2016, where a participatory method was introduced to the planning and decision-making process for climate resilience adaptation (Orleans Reed et al., 2013). The ACCCRN also facilitated the adoption of rainwater collection programs and the establishment of community-based flood warning systems (Sari & Prayoga, 2018; Yuniartanti et al., 2016).

Semarang participated in the Rockefeller Foundation's 100 Resilient Cities program from 2016 to 2018. This membership led the Semarang City Government to publish the Resilient Semarang Strategy Book entitled "RESILIENT SEMARANG: Moving Together towards a Resilient Semarang". The book outlines the City's potential risks and pressures and how to improve its resilience, focusing on six themes: water and new energy, new economy, disaster and disease, integrated mobility, public information transparency, and competitive human resources. Moreover, 1 of the 18 Semarang city resilient strategies emphasized in the book is sustainable water and energy, reflected by the development and implementation of Blue-Green infrastructure (BGI). BGI can be defined as an interconnected network of landscape components, either natural or man-made, that can provide environmental, economic, and social benefits (Ghofrani et al., 2017; O'Donnell & Thorne, 2020). BGI can be linked to inventive Flood Risk Management (FRM) strategies that integrate water management with green infrastructure (Ghofrani et al., 2017).

The BGI is considered one of the strategies to accommodate or retain flood in Semarang, which can be infiltration wells, biopores, retention ponds or artificial lakes, rainwater harvesting systems, and permeable pavement. In addition, there are some other BGI Implementations through several programs for instance Integrated Water Resources and Flood Management Project for Semarang, Water as Leverage (WaL), and Blue-Green Infrastructure in Semarang (BuGIS), as well as other programs established by or involving the local government.

Looking back at the Resilient Semarang Strategy Book, actions that reflected the implementation of BGI were rainwater harvesting as well as the use of biopores and infiltration wells, all of which were implemented to provide alternative water supply and reduce runoff in Semarang City. The program was initiated by Semarang City Environmental Agency, Semarang City PSDA/ESDM Agency, Semarang City Urban Planning and Housing Agency, and supported by the local community. It has also been mandated in Regional Regulation No. 7/2014 on the Master Plan of Semarang City Drainage System 2011–2031. Based on this regulation, rainwater harvesting is divided into two groups: rainwater harvesting for the central and northern parts of Semarang, which includes the Kecamatan of Tugu, Central Semarang, Genuk, Pedurungan, East Semarang, Gayamsari, North Semarang, and also West Semarang and infiltration wells for the southern part of Semarang or in the highland areas, which includes Mijen, Gunungpati, Banyumanik, Gajah Mungkur, South Semarang, Candisari, Tembalang, and Ngaliyan.

The implementation of rainwater harvesting and biopores was also carried out through the BuGIS program, which focused on *Kelurahan* Gedawang and *Kelurahan* Padangsari, which are close to the mountains and prone to flooding. In *Kelurahan* Gedawang, biopore construction was initiated by the Environmental Agency with *Kelurahan* officials' involvement. Until now, there have been no additional biopores in *Kelurahan* Gedawang since 2015, where the existing biopores have not been maintained even though they are located near the Kelurahan office. Meanwhile, *Kelurahan* Padangsari has 40 biopore hole-making tools that are used annually.

By 2020, rainwater harvesting, and infiltration wells have been implemented in 8 *Kelurahan* and 18 schools in Semarang City. Meanwhile, a total of 1,360 biopore pits were successfully made by the collaboration between the Semarang Environmental Agency and the communities in 14 *Kelurahan.* Some of them are shown in Fig. 1a and b. Furthermore, in 2018, Semarang City was selected as one of the cities in Asia for the Water as Leverage (WaL) program. WaL is a collaborative program between the Dutch Government and the Semarang City government that aims to formulate solutions for water problems in urban areas (Fig. 6).

In 2019, the WaL Program in Semarang City established five main concepts, including (1) Resilient *Kampung*, (2) Rechanneling the City, (3) Spongy Mountain, (4) Water Neutral Industry, and (5) Integrated Coastal Management which are then derived into various policy recommendations, programs, and development projects. The Cascading Semarang Team is one of the responsible teams for implementing the

(a) Rainwater Harvesting SMPN 42, *Kecamatan* Tembalang

(b) Rainwater Harvesting *Kelurahan* Cangkiran

Fig. 6 Rainwater harvesting in Semarang City **a** Rainwater harvesting SMPN 42, Kecamatan Tembalang, **b** Rainwater harvesting Kelurahan Cangkiran. *Source* Semarang Environmental Agency 2020 project. Based on the Cascading Semarang Phase II Report, by 2019, there are several BGI implementation programs in Semarang City, one of which is the construction of a polder. A polder system is one of the ways to deal with flooding. This system has physical structures such as drainage channels, retention ponds, sluice gates, and pumps controlled under one management unit (Sunaris & Tallar, 2019). Some existing polders in Semarang City are:

(1) *Kali Semarang Polder*, located in North Semarang with a length of 7.08 km and a total watershed area of 788.44 ha The Kali Semarang Polder is equipped with a 6.8 Ha retention pond and 8 pumps, 6 of which have a capacity of 5 m^3/s and 2 with a capacity of 2.5 m^3/s.
(2) *Banger Polder*, situated in East Semarang, is 13.326 km long with a watershed area of 462 Ha. This polder has a pond of 9.8 Ha and 5 pumps with a capacity of 2 m^3/s.
(3) *Tawang Polder*, located in Kota Lama, East Semarang, is used as a water reservoir equipped with 6 pumps with a capacity of 400 l/s each and 1 pump with a capacity of 100 l/s. The watershed area of this polder reaches 70 hectares with a pond area of 1 hectare that can accommodate up to 15,000 m^3 of water.

A visualization of these three polders is shown in Fig. 7 According to the measurement of polder effectiveness conducted by Pratiwi et al. (2016), Kali Semarang Polder has the highest value of 59%. Based on the measurement of polder effectiveness conducted by Pratiwi et al. (2016) Kali Semarang Polder has the highest value of 59% compared to other Semarang Polders. Another study by Dwiputri et al. (2021) explained that Tawang Polder still needs arrangement and improvement to meet its purpose and function as a flood management facility. Meanwhile, Banger Polder should to be preserved in terms of management and maintenance to ensure its benefits can be sustained into the future.

(a) Kali Semarang Polder, *Kecamatan* Semarang Utara (b) Banger Polder, *Kecamatan* Semarang Timur (c) Tawang Polder, *Kelurahan* Tanjung Mas

Fig. 7 Semarang City Polder **a** Kali Semarang Polder, Kecamatan Semarang Utara, **b** Banger Polder, Kecamatan Semarang Timur, **c** Tawang Polder, Kelurahan Tanjung Mas

4.3 Case of Bandung City, West Java (Java Island)

Bandung City, with a population of 2.4 million people, is the fourth largest city in Indonesia. It had a growth rate of 0.35% in 2022.It is also known as one of the most densely populated cities, accommodating 14,713 people per square kilometer in an area of 167.31 km^2 (Statistics of Bandung Municipality, 2023). As the capital city of West Java Province, this tropical monsoon city has faced a significant challenge related to urban development and the environment (Singh et al., 2021). Rapid urbanization and economic growth have driven some pressures on urban environments, such as uncontrolled land use change in periphery areas (see Wijaya, 2015), an increase in urban emissions (mainly from motorized vehicles), and an indirect impact on the decline of public health. Those drivers also affect population and asset concentration (e.g., public infrastructures and facilities) in risk-prone areas. Bandung City is naturally flood-prone due to its topography in the Citarum basin area. In the meantime, climate change also induces several natural disasters in the city. Based on Indonesian Disaster Data and Information (DIBI) from BNPB (2023), flooding is the most frequent disaster in the city, especially during the high intensity and frequency of the rainy season.

In responding to urban development and climate-related disaster issues, the municipal government and local communities have adopted several policies and initiatives related to Blue-Green Infrastructures (BGI) for flood disaster protection. For example, the municipality has considered BGI measures in its land use allocation and urban structure plans policy, which is the Spatial Plan of Bandung City 2022–2042 (Bandung Municipality Regulation No. 5 of 2022), including drainage system improvement, river area restoration to protect water catchment, water retention pond in flood-prone areas, constructed wetlands, infiltration wells, and rehabilitation of urban green space area. The municipality has adopted a sustainable urban drainage system to control groundwater use and provide rainwater infiltration facilities in flood-prone areas. However, integration between macro and micro drainage systems and the limited-service area coverage need to be resolved. Also, some retention ponds have been built in the city (e.g., Sirnaraga, Rancabolang, and Derwati Retention Ponds) to manage stormwater runoff and prevent flooding, but the capacity is still small to prevent surface runoff from the enormous land conversion in the upstream area (Atharinafi & Wijaya, 2021). In addition, the city currently has 12.25% of urban green space areas (Statistics of Bandung Municipality, 2023) in urban forests, parks, green streets, and cemetery areas. Although, it is still less than the minimum requirement of green space proportion according to the Law of Spatial Planning No. 26 of 2007, which is about 30%. The allocation of urban green spaces in Bandung City is unevenly distributed, thereby making these places inadequate for strengthening resilience in urban flood protection (Afriyanie et al., 2020).

Besides, socio-ecological technological approaches have been applied as part of BGI, such as bio-pores, infiltration wells, urban farming, and green verticals in buildings. Bio-pores are the alternative to stormwater absorption capacity and water inundation reduction in the rainy season. In 2013, the municipality launched a program called “A million of bio-pores movement” (Kompas, 2013). This initiative has involved the local government and communities of Bandung City in encouraging the citizens to implement bio-pores by providing tools to make the pores at the community level. The program successfully engaged the citizen because there was much enthusiasm. However, this program was insufficient to mitigate floods because of the pores’ limited capacity to contain stormwater. Besides that, this program needed to be more precisely regulated. Another initiative is known as Buruan Sae (Sehat Alami dan Ekonomis—Healthy, Natural, and Economical) as a part of an integrated urban farming program conducted in Bandung City to deliver an substitute for food facility in the city by the use of household yards for intensive development and empowerment of urban agriculture, livestock husbandry, farming residue usage for composting, and processing farming products to a high-quality and economic-valued food product (Margareth et al., 2021). The program focuses on diverse urban farming goods, mostly to address their food requirements, with some being marketed and distributed to local markets and communities. The primary goals of Buruan Sae are enhancing the standard of urban green areas, advocating for environmental education, and minimizing inefficient utilization of urban land by means of environmental restoration. It requires local government and community commitment and cooperation to sustain the program, including the financial and human resource capacity.

4.4 *Case of Balikpapan City, East Borneo (Kalimantan)*

Balikpapan is a big urban center within the East Kalimantan Province, having a large population as well as a large area, covering an area of 843.48 km^2. The city has an overall hilly topography (85%), with elevations varying from 0 to more than 100 m above sea level. The predominant soil types in Balikpapan consist of red-yellow podzolic, alluvial, and quartz sand, all of which are highly susceptible to erosion. Water issues such as flooding and clean water are a serious problem in Balikpapan. Balikpapan City gets clean water from Municipal Waterworks and surface water for daily activities (Suseno, 2017). Manggar Reservoir located in Karang Joang Village, North Balikpapan Sub-district, is the raw water source for the Municipal Waterworks with a 900 L/s capacity. Municipal Waterworks with a total of 900 L/s. The reservoir capacity is 16 million m^3 to accommodate rainfall. During the dry season, the baseflow of the Manggar River is very small (Rezagama & Tamlikha, 2016). In addition, flooding has been a priority environmental issue for the Balikpapan City Government since 2015, with a 13.34% increase in the land cover of residential areas in Balikpapan City from 2009 to 2015, but only 70% of the drainage capacity to accommodate inundation volumes every year.

The Balikpapan City Government has made various efforts to deal with flooding from structural and non-structural (Source). Still, flooding remains a disaster that causes many large losses from year to year. Balikpapan has several water polders or dams, namely Bendali I, II located in the Sepinggan area; Bendali III, IV located in Gunung Bahagia village, then Bendali V, sustainable board and Telaga Sari, the construction of a water polder with another purpose to overcome flooding because in overcoming flooding is not enough with ordinary drainage. Research from Ariyaningsih (2023) says that flooding in Balikpapan is caused by key drivers such as population growth, climate change, land-use change, and urbanization. However, handling flooding in Balikpapan city does not go directly to the root of the problem, focusing only on short-term strategies.

The implementation of BGI has been initiated by the private sector. For example, PT Kilang Pertamina Internasional (KPI) Balikpapan Unit launched the Rawa Bening Program or Rain Water Harvesting and Urban Farming. In addition to tackling flooding, there are also retention ponds to tackle flooding. Another function of retention ponds is to replace the role of infiltration land that is used as closed land/residential/office, so retention ponds can replace the function of infiltration. The function of this pond is to collect direct rainwater and flow from the system to be absorbed into the ground. So, this retention pond needs to be placed at the lowest part of the land.

Balikpapan has the potential to implement BGI. One type of Blue-Green Infrastructure (BGI) that can be applied is through the construction of permeable pavements and retention ponds through the utilization of pedestrian paths due to limited land in Balikpapan City. The Blue-Green Infrastructure (BGI) concept seeks to mitigate flooding and enhance the quality of stormwater discharge through the implementation of blue-green elements. These components effectively control stormwater by utilizing infiltration, evapotranspiration, retention, detention, and slow transit mechanisms. Additionally, they offer numerous advantages to the city, including conservation of local water resources and enhanced quality of life, as well as social and cultural benefits.

5 Issues and Barriers to Delivering Blue-Green Infrastructure in Indonesia

BGI is acknowledged as a robust approach and mechanism to developing urban resilience to respond to increasing urban vulnerabilities due to climate-related disaster risks (Drosou et al., 2019), particularly for controlling urban floods. Yet, a shortage of integrated approaches, lack of public acceptance, insufficient understanding and knowledge of experts and related stakeholders, limited land availability, and insufficient management and control are critical challenges to mainstream BGI in rapidly developing urban areas (Makopondo et al., 2020; Mukherjee & Takara, 2018; Thorne et al., 2018). Urban planning should incorporate a greater amount of locally-focused practical knowledge in order to integrate BGI (Blue-Green Infrastructure) more effectively. Additionally, there is a need for more thorough examination and collection of primary data on the hydrological, sociological, and environmental effects of BGI (Hamel & Tan, 2021). Also, all related stakeholders, such as urban planners, local governments, policymakers, and academics, should be aware of urban resilience and sustainability in decision-making (Zhang & Li, 2018), especially in the local development and spatial plans.

Implementing Blue-Green Infrastructure in Indonesia still faces several problems, especially the lack of integration of the BGI concept in spatial planning. As a result, the role of BGI in disaster mitigation is still not optimal (Astuti et al., 2020; Caesarina & Rahmani, 2019). In addition, the Asian Development Bank (ADB), through its 2016 report entitled "River Basin Management Planning in Indonesia, Policy and Practice," explained that funding is one of the main problems in implementing BGI in Indonesia. Drosou et al. (2019) further explained that this had become a concern for policymakers because of the uncertainty regarding the costs that must be spent, and the benefits obtained in implementing BGI. In Semarang City, the government is mostly responsible for BGI. This limits public access to information and understanding of implementing BGI (Drosou et al., 2019; Hamel & Tan, 2022). Another issue related to implementing BGI in Indonesia is the maintenance and continuity of the BGI program. Reflecting on Semarang City, some BGI implementations did not continue for a long time. This can be seen from the condition of the bio pores installed in 2015, which are no longer maintained. This is also indicated because of the incorrect approach (Drosou et al., 2019), resulting in the unsuitability of infrastructure implementation to the community's capabilities and needs. This condition must be a concern for various parties to optimize facilitation and monitoring as well as post-infrastructure development programs to ensure the implementation of the BGI concept runs continuously.

Despite the current initiatives and policies on BGI, Bandung City is still facing the challenge of getting the co-benefits from the result of BGI due to some internal and external factors, such as the regulation, financial aspect, institutional matters including the cooperating among the related stakeholders, and the sustainability of initiatives at the community level (leadership role and behavioral aspect). BGI's purpose is not only to improve water management but also to benefit the socio-economic aspect. BGI offers additional advantages such as mitigating the risks associated with disasters and adapting to climate change. The objectives encompass mitigating flood hazards, enhancing water purity, lowering temperatures, and safeguarding urban ecosystems and biodiversity (Liao et al., 2017). Thus, it is essential to consider co-benefits in contributing to the role of BGI in urban resilience building and disaster risk mitigation, significantly reducing flood disaster risk. Also, it is necessary to get much attention and consideration in an integrated manner, be included in the local development planning and spatial planning process making and reproduce nature in the city through BGIs due to its beneficial implications for urban dwellers. Because of a lack of financial capacity, the municipality of Bandung City needs to involve all multifunctional spaces, communities, private sectors, and other participating actors interested in getting involved in maintaining these BGI, for instance, by looking at the potential of non-conventional public finance.

In the case of the Indonesian program, especially regarding the adaptation measures for urban areas, fragmented governance is to be also evident. By taking both the horizontal and vertical decision-making boundaries into consideration, multi-level and multi-actor governance can be encouraged (García Sánchez et al., 2018; Pauleit et al., 2021). To mitigate and adapt to climate change on urban sprawl, strong political and administrative action needs to be taken on a regional scale. It is crucial to acknowledge that local-level climate change adaptations may pose novel and unfamiliar challenges for local governments, potentially surpassing their current capabilities in terms of risks, understanding, financial needs, and legal obligations. This phenomenon is predominantly noticed in emerging and poor nations, such as Indonesia (Garschagen et al., 2011). This is possible if the program is not just viewed as an endeavor to achieve a particular goal, but as an integrated element of a broader urban agenda, such as climate change adaptation and mitigation, reaching SDGs, and improving quality of life and urban productivity. Meanwhile, collaboration among stakeholders, particularly between government sectors, is critical for mitigating flood events. This cooperation is still missing between the government (such as Bandung City and the other governmental districts outside the city) due to the difference of authority for different administrative areas.

6 Conclusions

Blue Green Infrastructure is recognized as a method for mitigating the effects of flooding in metropolitan settings. The adoption of BGI has been observed in Indonesian cities such as Samarinda, Balikpapan, Bandung, and Semarang City, however its optimization is yet to be achieved. The government's significant role in flood management, achieved via the application of Blue green Infrastructure (BGI), is crucial due to the community's limited resources and knowledge. This chapter highlights that:

- The implementation of Blue-Green Infrastructure (BGI) in Indonesia is hindered by various challenges, particularly the insufficient incorporation of the BGI concept into spatial planning.
- The selected case studies provide a significant example for comprehending the difficulties that arise from the rapid urbanization in implementing BGI. This highlights the major obstacles in effectively implementing BGI (Big Data, Genomics, and Informatics) and offers a chance for BGI supporters to create new ideas and modify current programs to fit different situations.
- Aside from establishing a knowledge within the community of relevant stakehokders, a crucial long-term strategy for promoting the adoption of BGI is to enhance public awareness regarding the advantages of BGI and provide education on suitable BGI technology and sustainable urban planning.

The government plays a major role in regulating conditions that are prone to disasters. Hence, it is imperative for the government to cultivate proficiencies in the areas of strategic planning, readiness, operational management, collaboration, and logistical support when confronted with calamities. Indonesia, a developing nation, is facing numerous challenges in integrating BGI within its community or governmental framework. The concept of Building Green Infrastructure (BGI) is highly effective in enhancing urban resilience, particularly in the context of flood management. By effectively integrating both blue infrastructure (such as dams and levees) and green infrastructure (such as wetlands and green roofs), the potential for flooding can be significantly reduced.

References

Abidin, H. Z., Andreas, H., Gumilar, I., Sidiq, T. P., & Fukuda, Y. (2013). Land subsidence in coastal city of Semarang (Indonesia): Characteristics, impacts and causes. *Geomatics, Natural Hazards and Risk, 4*(3), 226–240. https://doi.org/10.1080/19475705.2012.692336

Adedeji, T., Proverbs, D., Xiao, H., Cobbing, P., & Oladokun, V. (2019). Making Birmingham a flood resilient city: Challenges and opportunities. *Water (Switzerland), 11*(8). https://doi.org/10.3390/w11081699.

Agonafir, C., Pabon, A. R., Lakhankar, T., Khanbilvardi, R., & Devineni, N. (2022). Understanding New York City street flooding through 311 complaints. *Journal of Hydrology, 605*(November 2021), 127300. https://doi.org/10.1016/j.jhydrol.2021.127300.

Alexander, K., Hettiarachchi, S., Ou, Y., & Sharma, A. (2019). Can integrated green spaces and storage facilities absorb the increased risk of flooding due to climate change in developed urban environments? *Journal of Hydrology, 579*(October), 124201. https://doi.org/10.1016/j.jhydrol.2019.124201.

Anugrahadi, A., Khomsiyah, H. F. Y., & Triany, N. (2020). The Jakarta flood disaster mapping January 2020. *International Journal of Advanced Science and Technology, 29*(5), 602–610.

Ariyaningsih, R. B. Sukmara, & Pradita, L. (2021). The Sendai framework for disaster risk reduction: Insight from Covid-19 in Balikpapan City, Indonesia. *IOP Conference Series: Earth and Environmental Science, 896*(1). https://doi.org/10.1088/1755-1315/896/1/012068.

Ariyarathna, I. S., Abeyrathna, W. P., & Jamei, E. (2023). *Areview of the application of blue–green infrastructure (BGI) as an effective urban flood mitigation strategy for livable and healthy cities in Australia* (pp. 461–76).

Ariyaningsih, Sukhwani, V., & Shaw, R. (2023). Vulnerability assessment of Balikpapan (Indonesia) for climate change-induced urban flooding. *International Journal of Disaster Resilience in the Built Environment, 14*(3), 387–401. https://doi.org/10.1108/IJDRBE-08-2021-0111

Astuti, K. D., Sariffuddin, S., & Pangi, P. (2020). Integrasi Ruang Biru Pada Rencana Tata Ruang Wilayah Sebagai Instrumen Mitigasi Bencana Banjir di Kota Semarang. *TATALOKA, 22*(2), 236–248. https://doi.org/10.14710/tataloka.22.2.236-248.

Atharinafi, Z., & Wijaya, N. (2021). Land use change and its impacts on surface runoff in rural areas of the upper citarum watershed (case study: Cirasea subwatershed). *Journal of Regional and City Planning, 32*(1). https://doi.org/10.5614/jpwk.2021.32.1.3

Bank World. (2002). THE WORLD BANK ANNUAL REPORT 2011 ER R ON I CÔT CÔ E D' I V O I R E He A R A B R E P U B L I C O F E G Y P T, the Volume t S LO VA K R E P U B L I C.

Battemarco, B. P., Tardin-Coelho, R., Veról, A. P., de Sousa, M. M., da Fontoura, C. V. T., Figueiredo-Cunha, J., Barbedo, J. M. R., & Miguez, M. G. (2022, August). Water dynamics and blue-green infrastructure (BGI): Towards risk management and strategic spatial planning guidelines. *Journal of Cleaner Production*, 333. https://doi.org/10.1016/j.jclepro.2021.129993.

BNPB. (2022). *Banjir Masih Ganggu Aktivitas Warga Kota Samarinda*. Badan Penanggulangan Bencana Nasional (BNPB).

BNPB. (2023). *Data Bencana Indonesia*. Badan Penanggulangan Bencana Nasional (BNPB).

Budiyono, Y., Aerts, J. C. J. H., Tollenaar, D., & Ward, P. J. (2016). River flood risk in Jakarta under scenarios of future change. *Natural Hazards and Earth System Science, 16*, 757–774. https://doi.org/10.5194/nhess-16-757-2016

Caesarina, H. M., & Rahmani, D. R. (2019). Penyediaan Ruang Terbuka Hijau dengan Pendekatan Kota Hijau pada Perkotaan Martapura. *Jurnal Planoearth, 4*(1), 11. https://doi.org/10.31764/jpe.v4i1.712.

Caleiro, F., & Ramos, H. M. (2016). *Smart water management (SWM): Flood control and water uses* (no. July).

Damayanti, A., Arifianto, F., & Indra, T. L. (2021). Development area for floating solar panel and dam in The Former Mine Hole (Void) Samarinda City, East Kalimantan Province. *International Journal on Advanced Science, Engineering and Information Technology, 11*(5), 1713–1720. https://doi.org/10.18517/ijaseit.11.5.14097.

Dasanto, B. D., Pramudya, B., Boer, R., & Suharnoto, Y. (2014). Effects of forest cover change on flood characteristics in the upper Citarum watershed. *Jurnal Manajemen Hutan Tropika, 20*(3), 141–149. https://doi.org/10.7226/jtfm.20.3.141

Demuzere, M., Faehnle, M., Orru, K., Olazabal, E., Geneletti, D., Heidrich, O., Orru, H., & Feliu, E. (2014). *Evidence on the contribution of green urban infrastructure to climate change mitigation and adaptation.*

Detik.com. (2022). *Warga Matraman Tewas Kejeblos Dari Toilet, Ini Kondisi Rumah Bantaran Ciliwung*. Detik.com.

Dewi, A. (2007). *Community-based analysis of coping with urban flooding: a case study in Semarang, Indonesia.* M.Sc. thesis, International Institute for Geo-information Science and Earth Observation, Netherland.

Dhyani, S., Karki, M., & Gupta, A. K. (2020). Opportunities and advances to mainstream nature-based solutions in disaster risk management and climate strategy. In A. Dhyani, M. Karki, & A. K. Gupta (Eds.), *Nature-based solutions for resilient ecosystems and societies* (pp. 1–26). Singapore, Springer. https://doi.org/10.1007/978-981-15-4712-6_1.

Drosou, N., Soetanto, R., Hermawan, F., Chmutina, K., Bosher, L., & Hatmoko, J. (2019). Key factors influencing wider adoption of. *Water, 11*(6), 1–20. https://www.mdpi.com/2073-4441/11/6/1234.

Dwiputri, M., Aisha, S., & Mentari, R. (2021). EVALUASI SISTEM DRAINASE POLDER KOTA SEMARANG DILIHAT DARI SUDUT PANDANG KOTA RAWAN BENCANA BANJIR. *RUSTIC Jurnal Arstektur, 1*(2), 1–9. http://ojs.itb-ad.ac.id/index.php/RUSTIC.

Dwirahmadi, F., Rutherford, S., Phung, D., & Chu, C. (2019). Understanding the operational concept of a flood-resilient urban community in Jakarta, Indonesia, from the perspectives of disaster risk reduction, climate change adaptation and development agencies. *International Journal of Environmental Research and Public Health, 16*(20). https://doi.org/10.3390/ijerph16203993.

Emam, A. R., Mishra, B. K., Kumar, P., Masago, Y., Fukushi, K. (2016). *Impact assessment of climate and land-use changes on flooding behavior in the upper Ciliwung River, Jakarta, Indonesia. Water (Switzerland), 8*(12). https://doi.org/10.3390/w8120559.

Fatimatuzahra Indra, T. L., Damayanti, A., & Dimyati, M. (2021). Monitoring land cover change in Karang Mumus Watershed, Samarinda City in 2006, 2014, and 2020. *IOP Conference Series: Earth and Environmental Science, 716*(1). https://doi.org/10.1088/1755-1315/716/1/012020.

Fawzy, S., Osman, A. I., Doran, J., & Rooney, D. W. (2020). Strategies for mitigation of climate change: a review. *Environmental Chemistry Letters, 18*, 2069–2094. https://doi.org/10.1007/s10311-020-01059-w

García Sánchez, F., Solecki, W. D., & Ribalaygua Batalla, C. (2018). Climate change adaptation in Europe and the United States: A comparative approach to urban green spaces in Bilbao and New York City. *Land Use Policy, 79*, 164–173. https://doi.org/10.1016/J.LANDUSEPOL.2018.08.010

Garschagen, M., Kraas, F., Garschagen, M., & Kraas, F. (2011). Urban climate change adaptation in the context of transformation: Lessons from Vietnam. In *Resilient cities* (pp. 131–139). https://doi.org/10.1007/978-94-007-0785-6_13

Garschagen, M., Surtiari, G. A. K., & Harb, M. (2018). Is Jakarta's new flood risk reduction strategy transformational? *Sustainability, 10*(8), 4–7. https://doi.org/10.3390/su10082934

Ghofrani, Z., Sposito, V., & Faggian, R. (2016). Designing resilient regions by applying blue-green infrastructure concepts. *The Sustainable City XI, 1*(Sc), 493–505. https://doi.org/10.2495/sc160421.

Ghofrani, Z., Sposito, V., & Faggian, R. (2017). A comprehensive review of blue-green infrastructure concepts. *International Journal of Environment and Sustainability, 6*(1). https://doi.org/10.24102/ijes.v6i1.728

Ghozali, A., Sukmara, R. B., & Aulia, B. U. (2016). A comparative study of climate change mitigation and adaptation on flood management between Ayutthaya City (Thailand) and Samarinda City (Indonesia). *Procedia-Social and Behavioral Sciences, 227*(November 2015), 424–429. https://doi.org/10.1016/j.sbspro.2016.06.096.

Hadi, S. P. (2017). In search for sustainable coastal management: a case study of Semarang, Indonesia. *IOP Conference Series: Earth and Environmental Science, 55*(1), 12054. https://doi.org/10.1088/1755-1315/55/1/012054

Hamel, P., & Tan, L. (2021). Blue–green infrastructure for flood and water quality management in Southeast Asia: Evidence and knowledge gaps. *Environmental Management, 69*(4), 699–718. https://doi.org/10.1007/S00267-021-01467-W

Hamel, P., & Tan, L. (2022). Blue–green infrastructure for flood and water quality management in southeast Asia: Evidence and knowledge gaps. *Environmental Management, 69*(4), 699–718. https://doi.org/10.1007/s00267-021-01467-w

Hapsari, R. I., & Zenurianto, M. (2016a). View of flood disaster management in Indonesia and the key solutions. *American Journal of Engineering Research (AJER), 5*(3), 140–151.

Hapsari, R. I., & Zenurianto, M. (2016b). View of flood disaster management in Indonesia and the key solutions. *American Journal of Engineering Research (AJER), 5*(3), 140–151.

Hartono, Y. C., Rebecca Milka, N. B., & Setyawan, P. B. (2023). Making education support media about flood hazard in Samarinda in the form of digital comic. *Kata, 25*(00), 126–131. https://doi.org/10.9744/kata.25.00.126-131

Harwitasari, D., & van Ast, J. A. (2011). Climate change adaptation in practice: People's responses to tidal flooding in Semarang, Indonesia. *Journal of Flood Risk Management, 4*(3), 216–233. https://doi.org/10.1111/j.1753-318X.2011.01104.x.

Husnayaen, R. A. B., Osawa, T., Parwata, I. N. S., As-syakur, A. R., Kasim, F., & Astarini, I. A. (2018). Physical assessment of coastal vulnerability under enhanced land subsidence in Semarang, Indonesia, using multi-sensor satellite data. *Advances in Space Research, 61*(8), 2159–2179. https://doi.org/10.1016/j.asr.2018.01.026.

IPCC. (2013). Climate Change 2013 the physical science basis. In T. F. Stocker, D. Qin, G. K. Plattner, M. M. B. Tignor, K. Simon (Eds.), *Working group I contribution to the fifth assessment report of the intergovernmental panel on climate change*

Isa, M., Fauzi, A., & Susilowati, I. (2019). Flood risk reduction in the Northern Coast of Central Java Province, Indonesia: An application of stakeholder's analysis. *Jamba: Journal of Disaster Risk Studies, 11*(1): 1996–1421. https://doi.org/10.4102/jamba.v11i1.660.

Istiqomah, L. N., Sabri, L., & Sudarsono, B. (2020). ANALISIS PENURUNAN MUKA TANAH KOTA SEMARANG METODE SURVEI GNSS TAHUN 2019. *Jurnal Geodesi Undip, 9*(2), 208–216. http://inaCORS.big.go.id.

Jago-on, K. A., Bianet, S. K., Fujikura, R., Fujiwara, A., Imai, T., Matsumoto, T., Zhang, J., et al. (2009). Urbanization and subsurface environmental issues: An attempt at DPSIR model application in Asian cities. *The Science of the Total Environment, 407*(9), 3089–3104. https://doi.org/10.1016/j.scitotenv.2008.08.004

Kabisch, N., Frantzeskaki, N., Pauleit, S., Naumann, S., Davis, M., Artmann, M., Haase, D., et al. (2016). Nature-based solutions to climate change mitigation and adaptation in urban areas: Perspectives on indicators, knowledge gaps, barriers, and opportunities for action. *Ecology and Society, 21*(2). https://doi.org/10.5751/ES-08373-210239.

Kaltim Today. (2020). Menguak Catatan Sejarah Banjir Besar Yang Menerjang Samarinda. Kaltim Today

Kapetas, L., & Fenner, R. (2020). Integrating blue-green and grey infrastructure through an adaptation pathways approach to surface water flooding. *Philosophical Transactions of the Royal Society A: Mathematical, Physical and Engineering Sciences, 378*(2168). https://doi.org/10.1098/rsta.2019.0204.

Kaur, R., & Gupta, K. (2022). Blue-green infrastructure (BGI) network in urban areas for sustainable storm water management: A geospatial approach. *City and Environment Interactions, 16*(June), 100087. https://doi.org/10.1016/j.cacint.2022.100087

Khasanah, I. U., Heliani, L. S., & Basith, A. (2016). Trends and seasonal to annual sea level variations of North Java Sea derived from tide gauges data. In *Proceedings of the First International Conference on Technology Innovations Society, Padang, West Sumatra, Indonesia, 21 July 2016–22 July 2016* (pp. 308–316). Padang, Indonesia: ITP Press. https://doi.org/10.21063/ICTIS.2016.1049.

Kompas. (2013). *Ridwan Kamil Luncurkan Gerakan Sejuta Biopori Halaman all—Kompas.com*. Kompas. https://regional.kompas.com/read/2013/12/20/1214497/Ridwan.Kamil.Luncurkan.Gerakan.Sejuta.Biopori?page=all

Kuehn, F., Albiol, D., Cooksley, G., Duro, J., Granda, J., Haas, S., Hoffmann-Rothe, A., & Murdohardono, D. (2010). Detection of land subsidence in Semarang, Indonesia, using stable points network (SPN) technique. *Environmental Earth Sciences, 60*(5), 909–921. https://doi.org/10.1007/s12665-009-0227-x

Lawton, M. P. (1961). "Wolman!" *Contemporary Psychology: A Journal of Reviews*. https://doi.org/10.1037/006636.

Law No. 26 Year 2007 On Spatial Management. (2007). Republic of Indonesia.

Liao, K. H., Deng, S., & Tan, P. Y. (2017). Blue-green infrastructure: New Frontier for sustainable urban stormwater management. In *Advances in 21st century human settlements* (pp. 203–226). https://doi.org/10.1007/978-981-10-4113-6_10/TABLES/9

Lubis, S. W., Hagos, S., Hermawan, E., Respati, M. R., Ridho, A., Risyanto, Paski, J. A. I., et al. (2022). Record-breaking precipitation in Indonesia's Capital of Jakarta in Early January 2020 linked to the Northerly Surge, Equatorial Waves, and MJO. *Geophysical Research Letters, 49*(22). https://doi.org/10.1029/2022GL101513.

Mahmud, M., Kusumandari, A., Sudarmadji, S., Supriyatno, N. (2018). A study of flood causal priority in Arui Watershed, Manokwari Regency, Indonesia. *Jurnal Manajemen Hutan Tropika, 24*(2), 81–94. https://doi.org/10.7226/jtfm.24.2.81.

Makopondo, R. O. B., Rotich, L. K., & Kamau, C. G. (2020). Potential use and challenges of constructed wetlands for wastewater treatment and conservation in game lodges and resorts in Kenya. *Scientific World Journal*. https://doi.org/10.1155/2020/9184192

Mamangkey, E. T., Sukmara, R. B., & Ariyaningsih. (2021). Rainwater harvesting as an alternative of freshwater supply in Balikpapan City-a case study of Institut Teknologi Kalimantan. *IOP Conference Series: Earth and Environmental Science, 896*(1). https://doi.org/10.1088/1755-1315/896/1/012040.

Marfai, M. A. (2014). Impact of sea level rise to coastal ecology: A case study on the Northern Part of Java Island, Indonesia. *Quaestiones Geographicae, 33*(1), 107–114. https://doi.org/10.2478/quageo-2014-0008

Marfai, M. A., & King, L. (2008). Coastal flood management in Semarang, Indonesia. *Environmental Geology, 55*(7), 1507–1518. https://doi.org/10.1007/s00254-007-1101-3

Marfai, M. A., King, L., Sartohadi, J., Sudrajat, S., Budiani, S. R., & Yulianto, F. (2008). The impact of tidal flooding on a coastal community in Semarang, Indonesia. *The Environmentalist, 28*(3), 237–248. https://doi.org/10.1007/s10669-007-9134-4

Margareth, Abdullah, O. S., Cahyandito, M. F., & Safitri, K.I. (2021). Buruan Sae in the city of Bandung: Realizing ideas from urban society. *IOP Conference Series: Earth and Environmental Science, 918*(1). https://doi.org/10.1088/1755-1315/918/1/012051

Mukherjee, M., & Takara, K. (2018). Urban green space as a countermeasure to increasing urban risk and the UGS-3CC resilience framework. *International Journal of Disaster Risk Reduction, 28*, 854–861. https://doi.org/10.1016/J.IJDRR.2018.01.027

Nashrrullah, S., Pasaribu, J. M., Hazarika, M. K., & Samarakoon, L. (2013). Study of flood inundation in Pekalongan, Central Java. *International Journal of Remote Sensing and Earth Sciences, 10*(2). https://doi.org/10.30536/j.ijreses.2013.v10.a1845.

Nirwansyah, A. W., & Braun, B. (2019). Mapping impact of tidal flooding on solar salt farming in Northern Java using a hydrodynamic model. *ISPRS International Journal of Geo-Information, 8*(10). https://doi.org/10.3390/ijgi8100451.

Nofal, O. M., & van de Lindt, J. W. (2022). Understanding flood risk in the context of community resilience modeling for the built environment: Research needs and trends. *Sustainable and Resilient Infrastructure, 7*(3), 171–187. https://doi.org/10.1080/23789689.2020.1722546

Novitasari, N., & Kurdi, H. (2022). Flood mitigation in Banjar Regency, South Kalimantan, Indonesia in 2021: Between hydro-meteorological factor and anthropogenic factor. *IOP Conference Series: Earth and Environmental Science, 999*(1). https://doi.org/10.1088/1755-1315/999/1/012010.

O'Donnell, E. C., Lamond, J. E., & Thorne, C. R. (2017). Recognising barriers to implementation of blue-green infrastructure: A newcastle case study. *Urban Water Journal, 14*(9), 964–971. https://doi.org/10.1080/1573062X.2017.1279190

O'Donnell, E. C., Thorne, C. R., Alan Yeakley, J., & Chan, F. K. S. (2020a). Sustainable flood risk and stormwater management in blue-green cities; an interdisciplinary case study in Portland, Oregon. *Journal of the American Water Resources Association, 56*(5), 757–775. https://doi.org/10.1111/1752-1688.12854

O'Donnell, E., Thorne, C., Ahilan, S., Arthur, S., Birkinshaw, S., Butler, D., Dawson, D., et al. (2020b). The blue-green path to urban flood resilience. *Blue-Green Systems, 2*(1), 28–45. https://doi.org/10.2166/bgs.2019.199

O'Donnell, E. C., & Thorne, C. R. (2020). Drivers of future urban flood risk. *Philosophical Transactions of the Royal Society A: Mathematical, Physical and Engineering Sciences, 378*(2168), 20190216. https://doi.org/10.1098/rsta.2019.0216

O'Donnell, E. C., Netusil, N. R., Chan, F. K. S., Dolman, N. J., & Gosling, S. N. (2021). International perceptions of urban blue-green infrastructure: A comparison across four cities. *Water (Switzerland), 13*(4), 1–25. https://doi.org/10.3390/w13040544

O'Donnell, Emily. (2021). Delivering multiple co-benefits in blue-green cities. https://doi.org/10.55203/pclw1513

Orleans Reed, S., Friend, R., Toan, V. C., Thinphanga, P., Sutarto, R., & Singh, D. (2013). "Shared learning" for building urban climate resilience-experiences from Asian cities. *Environment and Urbanization, 25*(2), 393–412. https://doi.org/10.1177/0956247813501136

Pal, I., Doydee, P., Utarasakul, T., Jaikaew, P., Razak, K. A. B., Fernandez, G., Huang, T., & Chen, C. S. (2021). System approach for flood vulnerability and community resilience assessment at the local level–a case study of Sakon Nakhon Province, Thailand. *Kasetsart Journal of Social Sciences, 42*(1), 107–116. https://doi.org/10.34044/j.kjss.2021.42.1.17.

Pauleit, S., Vasquéz, A., Maruthaveeran, S., Liu, L., Cilliers, S. S., Vasquéz, A., Maruthaveeran, S., Liu, L., & Cilliers, S. S. (2021). *Urban green infrastructure in the global South* (pp. 107–143). https://doi.org/10.1007/978-3-030-67650-6_5

POSKOTA. (2020). *Duh, Kali Ciliwung Di Kwitang Alami Pendangkalan*. POSKOTA.

Pour, S. H., Abd Wahab, A. K., Shahid, S., Asaduzzaman, M., & Dewan, A. (2020). Low impact development techniques to mitigate the impacts of climate-change-induced urban floods: Current trends, issues and challenges. *Sustainable Cities and Society, 62*(June), 102373. https://doi.org/10.1016/j.scs.2020.102373.

Praditya, E., Suprapto, F. A., Ali, Y., Surjaatmadja, S., & Duarte, R. (2023). Nusantara capital city (IKN): Threats and defense strategies for Indonesia's new capital. *The Journal of Indonesia Sustainable Development Planning, 4*(1), 21–34. https://doi.org/10.46456/jisdep.v4i1.420.

Pramantha, R. Q., Agustian, E., Suminar, L., Refnitasari, L. (2021). The characteristics of riverbank slum settlement in Indonesia. Case study: Depok, Palembang, Surabaya, and Surakarta. *IOP Conference Series: Earth and Environmental Science, 916*(1). https://doi.org/10.1088/1755-1315/916/1/012012.

Pratiwi, M. A. N., Nuraini, N., Nugroho, H., & Kurniani, D. (2016). Kajian Kinerja Sistem Polder dengan Balance Scorecard. *Jurnal Karya Teknik Sipil, 5*(1), 79–89.

Pratomo, R. A., Samsura, D. A. A., & van der Krabben, E. (2022). Living on the edge: Comparing the quality of life transformation of local communities induced by new town development in different Peri-Urban areas. *Sustainability (Switzerland), 14*(20). https://doi.org/10.3390/su142013435.

Pratomo, R. A., Samsura, D. A. A., & van der Krabben, E. (2023). Towards sustainable life of local communities in coastal cities: A longitudinal study of new town development in Makassar. In S. Roitman, & D. Rukmana (Eds.), *Routledge hanbook of urban Indonesia* (1st ed., p. 14). Routledge.

ProKaltim. (2022). Samarinda Menuju Kota Pusat Peradaban Di Tengah Persoalan Banjir, Benahi Tata Ruang, Perbanyak Kawasan Hijau. ProKaltim.

Reckien, D., Salvia, M., Heidrich, O., Church, J. M., Pietrapertosa, F., De Gregorio-Hurtado, S., d'Alonzo, V., et al. (2018). How are cities planning to respond to climate change? Assessment of local climate plans from 885 cities in the EU-28. *Journal of Cleaner Production, 191*, 207–19. https://doi.org/10.1016/j.jclepro.2018.03.220.

Reed, S. O., Friend, R., Jarvie, J., Henceroth, J., Thinphanga, P., Singh, D., Tran, P., & Sutarto, R. (2015). Resilience projects as experiments: Implementing climate change resilience in Asian cities. *Climate and Development, 7*(5), 469–480. https://doi.org/10.1080/17565529.2014.989190

Rezagama, A., & Tamlikha, A. (2016). Identifikasi Pencemar Waduk Manggar Kota Balikpapan. *Jurnal Pengembangan Kota, 4*, 40. https://doi.org/10.14710/jpk.4.1.40-48

Rizaldi, A., Moe, I. R., Farid, M., Aribawa, T. M., Bayuadji, G., & Sugiharto, T. (2019). Study of flood characteristic in Cikalumpang River by using 2D flood model. *MATEC Web of Conferences, 270*, 04010. https://doi.org/10.1051/matecconf/201927004010

Sahani, J., Kumar, P., Debele, S., Spyrou, C., Loupis, M., Aragão, L., Porcù, F., Shah, M. A. R., & Di Sabatino, S. (2019). Hydro-meteorological risk assessment methods and management by nature-based solutions. *Science of the Total Environment, 696*, 133936. https://doi.org/10.1016/j.scitotenv.2019.133936

Santos, A., Sousa, N., Kremers, H., & Bucho, J. L. (2020). Building resilient urban communities: The case study of Setubal Municipality, Portugal. *Geosciences (Switzerland), 10*(6), 1–13. https://doi.org/10.3390/geosciences10060243

Santoso, E. B., Rahmadanita, A., & Ryandana, M. D. (2022). Ruang Terbuka Hijau Di Kota Samarinda: Pencapaian, Permasalahan Dan Upayanya. *Jurnal Ilmu Pemerintahan Widya Praja, 48*(1), 103–26. https://doi.org/10.33701/jipwp.v48i1.2828.

Sari, A. D., & Prayoga, N. (2018). Enhancing citizen engagement in the face of climate change risks: A case study of the flood early warning system and health information system in Semarang City, Indonesia. In S. Hughes, E. K. Chu, & S. G. Mason (Eds.), *Climate change in cities: Innovations in multi-level governance* (pp. 121–137). Springer International Publishing. https://doi.org/10.1007/978-3-319-65003-6_7.

Sensuse, D. I., Putro, P. A. W., Rachmawati, R., & Sunindyo, W. D. (2022). Initial cybersecurity framework in the New Capital City of Indonesia: Factors, objectives, and technology. *Information (Switzerland), 13*(12), 1–10. https://doi.org/10.3390/info13120580

Singh, K. K., Katewongsa, P., Wijaya, N., & Kwan, S. C. (2021). Green, open spaces and transport for healthy and sustainable cities in Asian developing countries. *Malaysian Journal of Medicine and Health Sciences, 17*(3), 2636–9346.

Solihuddin, T., Husrin, S., Mustikasari, E., Heriati, A., Kepel, T. L., Salim, H. L., Risandi, J., & Dwiyanti, D. (2021). Coastal inundation and land subsidence in North Coast of West Java: A new hazard? *IOP Conference Series: Earth and Environmental Science, 925*(1), 0–14. https://doi.org/10.1088/1755-1315/925/1/012015.

Stanton-Geddes, Z., & Vun, Y. J. (2019). Strengthening the disaster resilience of Indonesian cities. *Time to ACT: Realizing Indonesia's Urban Potential*, no. September, 161–171. https://doi.org/10.1596/978-1-4648-1389-4_spotlight1.

Statistics of Bandung Municipality. (2023). Bandung municipality in figures 2023. https://bandungkota.bps.go.id/publication/2023/02/28/13fdfc9d27b1f2c450de2ed4/kota-bandung-dalam-angka-2023.html

Sugianto, S., Deli, A., Miswar, E., Rusdi, M., & Irham, M. (2022). The effect of land use and land cover changes on flood occurrence in Teunom Watershed, Aceh Jaya. *Land, 11*(8). https://doi.org/10.3390/land11081271.

Suseno, N. V., Widyastuti, M. (2017). Analisis Kualitas Air PDAM Tirta Manggar Kota Balikpapan. *J Bumi Indones 6*(1):1–8.

Suliyati, T. (2014). Penataan Drainase Perkotaan Berbasis Budaya dalam Upaya Penanganan Banjir di Kota Semarang. *HUMANIKA, 19*(1), 59–69.

Sunaris, M. L., & Tallar, R. Y. (2019). Green polder system: Kajian Konsep Infrastruktur Berkelanjutan Pada Wilayah Pesisir. *Seminar Nasional Inovasi Dan Aplikasi Teknologi Di Industri, 2019*, 372–376.

Swain, D. L., Wing, O. E., Bates, P. D., Done, J. M., Johnson, K. A., & Cameron, D. R. (2020). Increased flood exposure due to climate change and population growth in the United States. *Earth's Future, 8*(11). https://doi.org/10.1029/2020EF001778.

Tarigan, S. D. (2016). Land cover change and its impact on flooding frequency of Batanghari Watershed, Jambi Province, Indonesia. *Procedia Environmental Sciences, 33*, 386–392. https://doi.org/10.1016/j.proenv.2016.03.089

Thorne, C. R., Lawson, E. C., Ozawa, C., Hamlin, S. L., & Smith, L. A. (2018). Overcoming uncertainty and barriers to adoption of blue-green infrastructure for urban flood risk management. *Journal of Flood Risk Management, 11*, S960–S972. https://doi.org/10.1111/jfr3.12218

Walsh, C. J., Booth, D. B., Burns, M. J., Fletcher, T. D., Hale, R. L., Hoang, L. N., Livingston, G., et al. (2016). Principles for urban stormwater management to protect stream ecosystems. *Freshwater Science, 35*(1), 398–411. https://doi.org/10.1086/685284

Warsilan, W. (2019). Dampak Perubahan Guna Lahan Terhadap Kemampuan Resapan Air (Kasus: Kota Samarinda). *Jurnal Pembangunan Wilayah & Kota, 15*(1), 70. https://doi.org/10.14710/pwk.v15i1.20713.

Wells, J. A., Wilson, K. A., Abram, N. K., Nunn, M., Gaveau, D. L., Runting, R. K., Tarniati, N., Mengersen, K. L., & Meijaard, E. (2016). Rising floodwaters: Mapping impacts and perceptions of flooding in Indonesian Borneo. *Environmental Research Letters, 11*(6). https://doi.org/10.1088/1748-9326/11/6/064016.

Wen, K., Ren, G., Li, J., Zhang, A., Ren, Y., Sun, X., & Zhou, Y. (2019). Recent surface air temperature change over Mainland China based on an urbanization-bias adjusted dataset. *Journal of Climate, 32*(10), 2691–2705. https://doi.org/10.1175/JCLI-D-18-0395.1

Wijaya, N. (2015). Deteksi Perubahan Penggunaan Lahan dengan Citra Landsat dan Sistem Informasi Geografis: Studi Kasus di Wilayah Metropolitan Bandung, Indonesia. *Geoplanning: Journal of Geomatics and Planning, 2*(2), 82–92. https://doi.org/10.14710/GEOPLANNING.2.2.82-92

Wijaya, U. A. A., & Suhardi, D. (2022). Penanganan Permasalahan Banjir di Kota Semarang. *Seminar Keinsinyuran, 2022*, 181–183.

Wijayanti, P., Zhu, X., Hellegers, P., Budiyono, Y., & van Ierland, E. C. (2017). Estimation of river flood damages in Jakarta, Indonesia. *Natural Hazards, 86*(3), 1059–1079. https://doi.org/10.1007/s11069-016-2730-1

Wijayanti, W. P., & Pratomo, R. A. (2016). Adaptation of social-economic livelihoods in coastal community: The case of Mangunharjo Sub-District, Semarang City. *Procedia—Social and Behavioral Sciences, 227*, 477–484. https://doi.org/10.1016/j.sbspro.2016.06.103

Wismarini, Th. D., & Ningsih, D. H. U. (2010). Analisis Sistem Drainase Kota Semarang Berbasis Sistem Informasi Geografi dalam Membantu Pengambilan Keputusan bagi Penanganan Banjir. *Jurnal Teknologi Informasi DINAMIK, XV*(1), 41–51.

Yulianto, F., Sofan, P., Zubaidah, A., Sukowati, K. A. D., Pasaribu, J. M., & Khomarudin, M. R. (2015). Detecting areas affected by flood using multi-temporal ALOS PALSAR remotely sensed data in Karawang, West Java, Indonesia. *Natural Hazards, 77*(2), 959–985. https://doi.org/10.1007/s11069-015-1633-x

Yuniartanti, R. K., Handayani, W., & Waskitaningsih, N. (2016). Monitoring and evaluation effectiveness in flood early warning system project in Semarang City. *International Journal of Society Systems Science, 8*(1), 49–77. https://doi.org/10.1504/IJSSS.2016.076004

Yuwono, B. D., Abidin, H. Z., & Hilmi, M. (2013). Analisa Geospasial Penyebab Penurunan Muka Tanah Di Kota Semarang. *Prosiding Seminar Nasional Sains Dan Teknologi, 4*, 2013.

Zeng, X., Yu, Y., Yang, S., Lv, Y., & Sarker, M. N. I. (2022). Urban resilience for urban sustainability: Concepts, dimensions, and perspectives. *Sustainability (Switzerland), 14*(5), 1–27. https://doi.org/10.3390/su14052481.

Zhang, X., & Li, H. (2018). Urban resilience and urban sustainability: What we know and what do not know? *Cities, 72*, 141–148. https://doi.org/10.1016/J.CITIES.2017.08.009

Wismarini, TH. D., & Ningsih, D. H. U. (2010). Analisis Sistem Drainase Kota Semarang Berbasis Sistem Informasi Geografi dalam Membantu Pengambilan Keputusan bagi Penanganan Banjir. Jurnal Teknologi Informasi DINAMIK, XV(1), 41–51.
[illegible]
[illegible]
[illegible]
[illegible]

Scaling and Assessment for Urban Planning

Modeling Climate Change-Resilient Urban Agriculture in Developing Nations: A Case Study

Alba Lombardía, Thomas Schroepfer, María Teresa Gómez-Villarino, and Carlos Bañón

Abstract The generalized expansion of urbanization and population requires the implementation of sustainable practices in cities, particularly in developing nations more vulnerable to climate change impacts. Blue-green infrastructures can be viable solutions to climate and sustainability emergencies in the built environment. In this context, urban areas possess the opportunity to create climate change-resilient agricultural systems in order to meet the "No Poverty" (SDG 1) and "Zero Hunger" (SDG 2) goals included in the United Nations (UN) Sustainable Development Goals (SDGs). Consequently, with 70% of the population residing in urban areas, food systems can close the gap between production and consumption to ensure local food security. The substantial growth of cities in Asia, South America, and Africa renders localized urban agriculture essential for feeding its inhabitants. As the Food and Agriculture Organization (FAO) indicates, 6,000 tons of food are imported daily to feed cities with more than 20 million residents, such as Sao Paolo or Mexico City. Food chains and imports could be drastically mitigated by utilizing existing urban surfaces (roofs, facades, and ground areas) to produce crops. For this reason, there is a rising demand for research to assist urban planners and governments in strategizing local agriculture production. This chapter addresses food security in developing cities

A. Lombardía (✉) · T. Schroepfer · C. Bañón
Department of Architecture and Sustainable Design, Singapore University of Technology and Design, 8 Somapah Road, Singapore 487372, Singapore
e-mail: alba_lombardia@mymail.sutd.edu.sg

T. Schroepfer
e-mail: thomas.schroepfer@sutd.edu.sg

C. Bañón
e-mail: carlos_banon@sutd.edu.sg

T. Schroepfer
Singapore-ETH Centre Future Cities Laboratory Global, 1 Create Way, Create Tower, #06-01, Singapore 28040, Singapore

M. T. Gómez-Villarino
Department of Agroforestry Engineering, School of Agricultural, Food and Biosystems Engineering, Polytechnic University of Madrid, 28040 Madrid, Spain
e-mail: teresa.gomez.villarino@upm.es

P. K. Joshi et al. (eds.), *Blue-Green Infrastructure for Sustainable Urban Settlements*,
https://doi.org/10.1007/978-3-031-62293-9_12

by introducing a sunlight-based computational approach for designating the agricultural potential of three-dimensional urban surfaces according to crop-centric environmental suitability. Building-integrated agriculture (BIA) can deliver crops throughout the entire year, minimizing the externalities that endanger food production, such as climate variations (droughts and floods) and climate-induced pests (from human and soil pathogens). Critics of high-tech indoor developments in hydroponics, aeroponics, and aquaponics argue that these systems are energy-intensive. In response, hybrid lighting systems offer alternative solutions that can efficiently reduce energy requirements by integrating sunlight data analysis. Blending sunlight to optimize lighting can have remarkable energy savings of up to 70% in developing nations with constricted energy infrastructures. Therefore, the availability of an appropriate amount of solar radiation on three-dimensional surfaces is a critical factor in selecting BIA locations.

Keywords Agricultural simulation · Agriculture urban planning · Food security · Green infrastructure · Photosynthetically active radiation · Sustainable development

Abbreviations

BIA	Building-Integrated Agriculture
CEA	Controlled Environment Agriculture
DLI	Daily Light Integral
EPW	EnergyPlus Weather
FAO	Food and Agriculture Organization
GNI	Gross National Income
HDI	Human Development Index
IMF	International Monetary Fund
NEE	Newly Emerging Economy
PAR	Photosynthetically Active Radiation
RGB	Red, Green, Blue
SDGs	Sustainable Development Goals
TMY	Typical Meteorological Year
UA	Urban Agriculture
UN	United Nations
WEA	Daysim Weather

1 Introduction

One of the significant challenges of the 21st century includes the evolution towards sustainable and efficient forms of agricultural production that can adapt to climate change. These transformations will benefit many developing countries that depend on traditional agriculture and are affected by global warming. The consequences of climate change in agriculture are diverse and have significantly affected food production across products, locations, and periods (Frona et al., 2021; Malhi et al., 2021).

Essential objectives for the sustainability of a growing population include food stability, low environmental impacts, and resilience to climate change. In developing countries, meeting the nutritional needs of their population is even more critical. The strong growth of the built environment in America, Asia, and Africa makes urban agriculture an essential element for feeding their inhabitants. As the Food and Agriculture Organization (FAO) of the United Nations (UN) indicates, to supply cities with more than 20 million inhabitants, such as Sao Paolo or Mexico City, it is necessary to import 6,000 tons of food daily.

Moreover, to address urbanization and food security issues, it is required that public and private sectors operate together. The participation of local governments is essential in this context. However, municipal managers and urban planners have often excluded agricultural systems when designing, planning, or managing cities. Generally, urban stakeholders have relied on food imported from rural or foreign areas. Nevertheless, recent pandemic and energy crises manifested shared risks in developing and more developed countries, particularly in food import-dependent nations such as islands and city-states. Consequently, there is a growing worldwide cognizance that urban resilience and food security are critical targets in determining the sustainable development of cities.

There is a demand for sustainable local food production in developing cities, and Urban Agriculture (UA) has the potential to achieve food security. UA can assume various forms in urban environments and may become an effective tool to complete one of the most significant global challenges identified in the United Nations' Sustainable Development Goals (SDGs). Enhancing agriculture in cities and urban settings can contribute to several interlinked and mutually dependent objectives (Lombardía & Gómez-Villarino, 2023). In addition, UA contributes to poverty reduction (SDG1), food security with improved nutrition (SDG2), social cohesion and cooperation (SDG10), sustainability through the efficient use of natural resources (SDG12), climate change mitigation (SDG13), biodiversity (SDG15), and resilient and sustainable cities (SDG 11).

For a long time, traditional farmers have developed agricultural systems and cultivation techniques adapted to hostile environments since their subsistence depended on guaranteed food, goods, and services (Koohafkan et al., 2012). Such adaptation was based on utilizing the available resources: native species, organic fertilizers, rainwater use, and local construction materials. Similarly, UA has adapted to often hostile urban environments by employing traditional techniques while developing new ones, such as Controlled Environment Agriculture (CEA). CEA presents another clear example of adaptation developed by ultramodern and intensive agricultural practices to produce food with the smallest possible footprint and water requirements. This innovative type of agriculture is reminiscent of industrial production through soilless cultivation, aquaponics, aeroponics, and controlled atmospheres that produce elevated yields. Additionally, CEA bears the social potential to be part of urban green spaces based on more conventional UA, such as community gardens (Briz & De Felipe, 2016; Gómez-Villarino et al., 2021).

CEA technologies offer advantages over traditional agriculture. These innovations allow the optimization of the environmental factors that determine the development of plants, such as light, temperature, carbon dioxide concentration, humidity, and nutrients. The mentioned controlled optimal conditions improve crop productivity and quality (SharathKumar et al., 2020; Dsouza et al., 2023). Additionally, these systems present other advantages since the food-travel kilometers are reduced due to their proximity to urban consumers, whereas the environmental externalities of transportation are minimized (Kalantari et al., 2018; Dsouza et al., 2023). Overall, CEA is a novel climate-resilient infrastructure that can facilitate food production within urban environments in a sustainable manner, which has generated substantial research interest (Al-Kodmany, 2018; van Delden et al., 2021; Dsouza et al., 2023).

Climate change and agricultural degradation impacts continue to be felt, particularly in developing countries. Methods for sustainable agricultural development must be explored to concurrently address climate change adaptation, rapid urbanization, and ecosystem degradation. Remarkably, people already in poverty are profoundly affected by ecological and anthropogenic climatic changes (Rockström et al., 2009; Zari et al., 2020). In this regard, the aim of this chapter is to (1) trace areas in developing nations under miscellaneous changing climates and high food risk that offer sufficient environmental data for future assessment; (2) present a climate-change resilient model driven by an environmentally-informed methodology for urban food security planning; (3) explore how the proposed methodology and model can be applied in the context of selected urban areas; and (4) briefly report on results from the case study in order to propose future refinement and development.

In traditional agricultural practice, the cultivated crop varieties depend on empirically measured Photosynthetically Active Radiation (PAR) and Daily Light Integral (DLI) quantities. However, there have been limited efforts to leverage radiation modeling to guide urban agriculture decisions. Consequently, this paper empirically assesses and validates DLI simulations to inform the sustainable selection of urban spaces for the BIA purpose. The agricultural model is applied and discussed in developing nations experiencing changing climates. Monthly simulation-based DLI maps are generated to provide guidelines and identify sites for urban-scale agriculture applications. Additionally, the generated simulations suggest suitable crop species and prospective crop yield values adapted to specific environmental sunlight constrictions.

The computational methodology proposed herein can considerably boost food security and equity in local communities. The outcome of this chapter establishes a validated model for future urban agriculture studies in the context of changing climates and thereby establishes a methodology to accelerate the process of urban food security planning. Developing countries that require climate change-resilient solutions will considerably benefit from climate immutable agricultural practices.

2 Mapping Agricultural Potential in Developing Nations

A developing country can be defined as a sovereign state with a lower developed industrial base and Human Development Index (HDI) relative to other nations. However, no general agreement exists on which countries fit this definition. Other economic-related terms are emerging and interchangeably used, such as "Newly Emerging Economy" (NEE). Nevertheless, as defined by previous studies, human development goes beyond HDI and should consider broader indicators (Ranis et al., 2006). In this chapter, four main classifications of data have been taken into account for the selection of assessment areas: (1) the economic designation by the International Monetary Fund (IMF) (International Monetary Fund, 2022); (2) the level of development indicated by the United Nations (UN) (UN, 2023); (3) the World Food Program Hunger Map (World Food Program, 2020); and (4) the Köppen-Geiger Climate Classification (Beck et al., 2018), which divides climates based on seasonal precipitation patterns, temperature, and vegetation types occurring in a given region.

The above data from developing countries is contrasted with the availability and reliability of environmental solar radiation information on an open-source climate data site (Climate.OneBuilding, 2023) to apply the research methodology presented below. The selected countries (See Fig. 1) fulfill two requirements: (1) Be considered developing nations according to the IMF and the UN, and (2) have a malnutrition rate between 5.0 and 14.9% in consonance with the World Food Program, which is a high proportion compared with developed countries with an undernutrition percentage of less than 2.5%. Consequently, with sufficient economic and social capacity, their capitals possess the potential to foster urban agriculture infrastructure that could substantially improve their food security. The selected cities and their climates, according to the Köppen-Geiger Climate Classification, are as follows: Mexico City with a subtropical dry-winter highland climate (Cwb); Lima with a hot arid desert climate (Bwh); Cape Town with a hot summer Mediterranean climate (Csa); Muscat with a hot arid desert climate (Bwh); New Delhi with a dry-winter and humid monsoon subtropical climate (Cwa).

The selection of assessment areas followed the above classifications and data to illustrate a mutual innovation ground in developing countries facing common challenges that can be addressed through well-planned urban agriculture and sustainable and resilient green infrastructure. According to the UN, these shared challenges are: (1) High poverty levels based on Gross National Income (GNI) per capita; (2) human resource weaknesses that encompass nutrition and health indicators; (3) economic vulnerability, including instability of agricultural production. Therefore, urban agriculture provides an opportunity to tackle these issues under a holistic approach that fosters an increase of GNI per capita, minimum coverage of nutrition levels, and

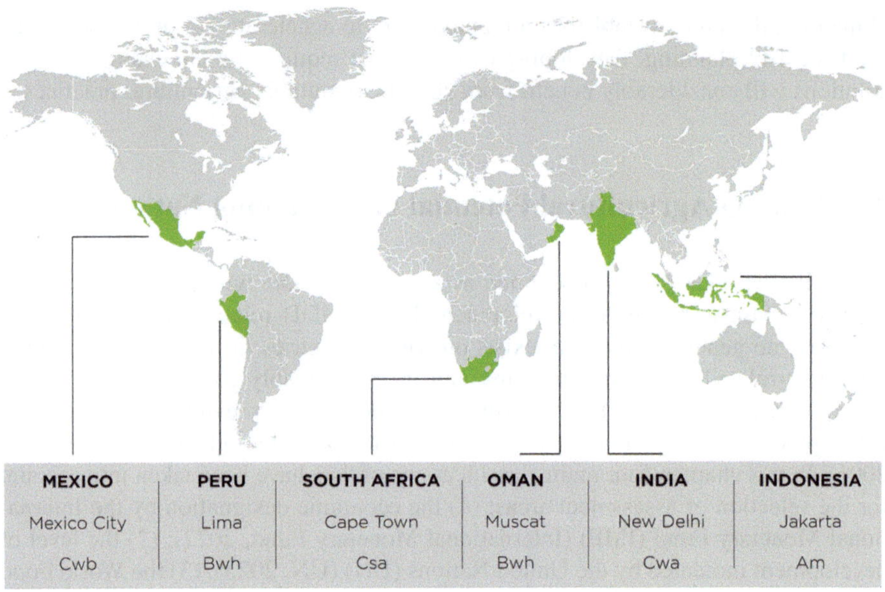

Fig. 1 Selection of cities under assessment and their Köppen-Geiger Climate Classification. Figure developed by author A. Lombardia for this publication

stable, resilient food production achieved by sustainable agriculture implemented in local urban communities and households.

Contemplating the miscellaneous changing climates in developing countries is essential to examine while creating and validating a sustainable agriculture model. Although developed countries are significant contributors to carbon emissions and developing countries have not been the primary cause of climate change, these countries are more vulnerable to unstable climate adaptation challenges due to the interconnection of several factors, namely, high climate vulnerability, low economic status, restrictions on the availability of technology, limited infrastructure, and inaccessible financial resources (Althor et al., 2016). Consequently, as expressed by the Climate Vulnerability Monitor, climate vulnerability is signaled by four impact areas: health, extreme weather, habitat loss, and economic stress (DARA and the Climate Vulnerable Forum, 2010). This climate vulnerability causes an elevated number of climate-related deaths due to hunger and disease in developing countries that require emergent solutions for food security and "climate justice."

In general, most studies indicate that climate change harms crop production and yields. Climate-related disasters, including droughts, floods, and tropical storms, significantly drive food insecurity (Porter et al., 2015; Hadley et al., 2023). Due to traditional agriculture's dependence on temperature and precipitation stability, food production is susceptible to climate. In this regard, CEA offers a climate-change resilient solution to produce year-round harvests. Another significant advantage of cultivation in controlled environments is the low usage of irrigation water (up to 95%

less than traditional agriculture), which can contribute significantly to countries with insufficiently managed water systems or drought-vulnerable nations.

Furthermore, the emplacement of food production close to consumption in city centers and the creation of local food systems (Pigford et al., 2018) establishes resilience and sustainable consumption cycles. Food security is when "all people, at all times, have physical, social and economic access to sufficient, safe and nutritious food that meets their dietary needs and food preferences for an active and healthy life" (Pérez-Escamilla, 2017). With 70% of the population living in urban areas, it is essential to realize that closing the gap between production and consumption will ensure food security in developing cities.

3 Climate Change-Resilient Agricultural Model

3.1 Sunlight-Based Computational Approach

The effects of climate change on traditional agricultural production will be diverse and location-specific. In the last decade, climate-related impacts have more frequently occurred, and long-term extrapolations have revealed significant future impacts on the developing world (Chinowsky et al., 2011). Temperature and rainfall are the key elements determining crop productivity (Altieri et al., 2015). Low-latitude developing countries possess sufficiently warm climates that are especially susceptible to changes affecting the preceding two factors (Mendelsohn, 2008). Consequently, crop yields fall as warming increases, producing vulnerabilities in food security for developing cities. Against this backdrop, CEA can deliver crops throughout the year, minimizing the externalities that endanger food production, such as climate variations (droughts and floods) and climate-induced pests (from human and soil pathogens).

Nonetheless, the main disadvantage of CEA's widespread implementation in cities originates from energy-intensive configurations (Proksch et al., 2019; Engler & Krarti, 2021). As a result, the scalability of these high-tech systems at the urban level can generate problems due to excessive consumption of resources as indicated by CEA life cycle assessments (Proksch et al., 2019; Engler & Krarti, 2021). Excessive consumption rates exert stress on constrained energy infrastructures in developing nations that require universal energy access and the implementation of renewable energy systems. The agriculture model introduced in this chapter tenders an alternative solution to efficiently reduce CEA energy requirements by locating agricultural systems on sunlight-impacted surfaces. Described as "the way forward for modern agriculture to achieve both productivity and sustainability" (Neo et al., 2022), hybrid lighting systems that include sunlight attain significant energy savings of up to 70% (Hamm, 2015) while ensuring optimal crop growth.

Hence, the availability of an adequate amount of solar radiation on three-dimensional surfaces is a critical condition in the sustainable selection of Building-Integrated Agriculture (BIA) locations. In traditional agricultural practice, the selection of crop varieties depends on empirically measured quantities of Photosynthetically Active Radiation (PAR) and Daily Light Integral (DLI). However, applying radiation modeling to guide urban agriculture decisions is uncommon, and crop-specific requirements are often unleveraged.

For these reasons, the chapter addresses food security in developing cities by introducing a sunlight-based computational approach to designate the agricultural feasibilities of three-dimensional urban surfaces according to crop-centric environmental suitability. In addition, the model, in combination with CEA, manifests the possibility of reducing electrical usage, even to zero. Consequently, the proposed agriculture model, energy efficient and resilient to climate changes, concurrently harmonizes with the UN 2030 Agenda for Sustainable Development, which targets economic, social, and environmental stability in developing cities (Lombardía & Gómez-Villarino, 2023). The model mainly focuses on the achievement of "No Poverty" (SDG 1) (Bhattacharya et al., 2022), "Zero Hunger" (SDG 2) (Arora & Mishra, 2022), and the "Sustainable Cities and Communities" (SDG 11) (Franco et al., 2020) goals included in the SDGs initiative. Climate-change resilient and energy-efficient agricultural systems demand investment to stabilize less developed countries by augmenting food security and establishing resource-effective practices toward food production.

3.2 *Photosynthetically Active Radiation Urban Modeling*

3.2.1 Conceptuality

Photosynthetically Active Radiation (PAR) is a fundamental factor to consider in the environmental suitability of urban agriculture. PAR is utilized in botanical applications, generally measured employing DLI values. DLI corresponds to the total amount of PAR that impacts a square meter over 24 h. The total solar radiation received by three-dimensional surfaces is addressed by current urban studies and validated through computer applications such as Ladybug (Roudsari & Pak, 2013). However, most urban simulation methods lack green infrastructure light specifications and neglect the differences between total solar radiation and PAR. Solar radiation comprises three spectral ranges: infrared, visible (PAR), and ultraviolet. Of the light that reaches the Earth's surface, infrared radiation corresponds to 49.4%, visible light to 42.3%, and ultraviolet radiation to 8% (Fondriest Environmental, 2014). Therefore, total solar radiation is inaccurate for urban agriculture calculations since the quantity of light exploited during plant photosynthesis is markedly less.

Contemporary literature studies solar accessibility in cities since it englobes relevant applications, such as energy harvesting (Zhu et al., 2022). Recent findings highlight the value of digital models for analyzing the solar capacity of urban surfaces (Zhu et al., 2022). Generally, urban studies exclusively consider solar radiation values. However, as previously mentioned, in the case of urban agriculture, it is known that crops cannot absorb the total amount of solar radiation but assimilate a part of it. Accordingly, this modeling study scrutinizes visible light, also known as PAR, the wavelength of light feasible for photosynthesis that ranges from 400 to 700 nm (nm).

The angle relative to the sun, which depends on location, influences the amount of light reaching a particular surface on Earth. Recent literature indicates that outdoor sunlight DLI varies from 5 to 60 mol $\cdot$ m^{-2} $\cdot$ d^{-1} (Torres & Lopez, 2012). Accordingly, previous studies have provided country-based DLI maps for the mainland US (Korczynski et al., 2002; Faust & Logan, 2018) and China (Sun et al., 2015).Nevertheless, related research for developing countries is inexistent. Ultimately, applying digital models to establish accurate surface DLI values in developing cities has extraordinary potential for prospective urban agriculture and food security planning.

Contemporary urban research that contemplates PAR or DLI is rare and usually employs empirical formulas obtained from experimental sensor data (Weiss & Norman, 1985) to calculate values extracted from the total incoming solar radiation (Subramaniam et al., 2020; Palliwal et al., 2021; Kyropoulou et al., 2021). However, this study exhibits more precise results by computationally extracting the visible light (also called PAR) values directly from the meteorological dataset utilized in the simulation, avoiding subsequent conversions.

3.2.2 Modeling

The sunlight modeling procedure is influenced by the Ladybug plugin radiation simulation methodology, which is built on top of several validated engines (Roudsari & Mackey, 2013): Radiance (Larson et al., 1988), EnergyPlus/OpenStudio, Therm/Window, and OpenFOAM. The analysis of incident radiation in modeled geometry first requires a matrix generation that restrains the radiation values of the sky-dome. Hence, the Radiance's Gendaymtx function (Ashdown & Ward, 2013) is utilized to calculate the radiation for each patch of the sky (See Eq. 1).

Eventually, the Gendaymtx function uses the Perez all-weather model (Perez et al., 1993) to compute a matrix of sky patch values, adding the weather tape as input and setting other parameters related to sky resolution (Reinhart, Tregenza), column matrix, sky color, sun-sky-ground interactions, horizon, and rotation among

```
gendaymtx [ -v ][ -h ][ -A ][ -d|-s|-n ][ -D sunfile [ -M sunmods ]][ -r deg ][ -m N ][ -g r g b ][ -c r g b ][ -
o{f|d} ][ -O{0|1} ] [ tape.wea ]
```

Eq. 1 Gendaymtx function synopsis (Ashdown & Ward, 2013)

others. Under option −O1, the function obtains as output the total solar radiance, archetypal in urban and architectural research. Instead, the opposite functionality, excluding the −O1 option in the function, derives a weather tape array with visible radiation, also known as PAR, prototypically suitable for crop-centric requirements.

The Gendaymtx function outputs radiation values in RGB format. Consecutively, these estimates are converted to broadband radiation by applying RGB conversion constants. In addition, each row of the Gendaymtx RGB data in $W/sr/m^2$ is parsed and converted to radiation values in KWh/m^2, where "sr" corresponds to "steradian", the unit that measures three-dimensional angles. The parsing includes the application of broadband weighting to the RGB bands, multiplication by steradians of each patch, and proliferation by the time duration (hours) represented in the sky matrix. The default values of the RGB conversion constants, if no color correction factor is specified, correspond to 0.265074126, 0.670114631, and 0.064811243 for red, green, and blue, respectively, approximating the standard photometric curve (Ehrlich et al., 2001). Additionally, it is required to generate two consecutive calls to Gendaymtx separately, one for the direct radiation and another for the diffuse radiation, which is subsequently aggregated to obtain the total visible radiance (PAR). Direct radiation appertains to the solar energy that can be absorbed directly at a surface, while indirect radiation corresponds to the light that has been scattered upon entering the atmosphere.

The environmental solar radiation data is obtained from an open-source climate data site (Climate.OneBuilding, 2023) comprising EnergyPlus Weather (EPW) files, where data can be extracted in WEA (Daysim Weather) format. The website contains climate data explicitly designed to support building simulations. The included files are Typical Meteorological Years (TMY) published by various organizations. The TMYx dataset created by the authors of this website contains data that includes the one derived from the most recent 15 years (2007–2021) in a global range encompassing all World Meteorological Organization regions. TMY data is a synthesized file that includes data from representative years over a multi-year period. Each set contains 8,760 values for each meteorological parameter, corresponding to the number of hours in a year. Data collection is sourced from weather stations worldwide, including gridded solar radiation satellite data. Real-time weather data spawns datasets with hourly values of different parameters, including solar radiation, temperature, wind speed, and wind direction. Subsequently, the weather datasets are analyzed to construct the representative year that utilizes the TMY format.

Classic Greek geometry stated that a hemisphere can be divided entirely into zones of equal size and shape in limited ways. Currently, luminance measurements ponder two general approximations for subdivision across the celestial hemisphere, one with a simplified number of patches and another with a higher density, called the Tregenza and the Reinhart models. The presented model employs the abbreviated Tregenza subdivision for the calculations (Tregenza, 1987) since, according to recent literature, different levels of sky discretization do not indicate any impact on accuracy (Subramaniam & Mistrick, 2017). This distribution of circular patches over a simple grid covers approximately 70% of the total surface. The Tregenza model divides the hemisphere into bands with 151 resulting zones. Consecutively, to provide overall symmetry in the subdivision, the patches are reduced to 145, 144 plus one patch located on the zenith. Consequently, the Tregenza patches per row used in the simulator are (30, 30, 24, 24, 18, 12, 6) plus (1) corresponding to the zenith location.

The volume of daylight that collides with a surface depends on the sky's luminance and the surrounding elements. The luminance across the celestial hemisphere oscillates independently from one viewing angle to another. For instance, a moving cloud can substantially alter brightness for a small sky area while the rest remain unaffected. Therefore, the application of daylight coefficients that consider the angular size (steradians) defined by the altitude, azimuth, and angular width of the patches of each row of the Tregenza sky dome is required (Tregenza & Waters, 1983). The following coefficients correspond to the solid angles that align with the dome vectors (0.0435449227, 0.0416418006, 0.0473984151, 0.0406730411, 0.0428934136, 0.0445221864, 0.0455168385, 0.0344199465) and are applied in the Tregenza calculations.

Overall, PAR simulation involves two steps: First, discretize the sky vault using the Tregenza scheme with 145 patches. Second, apply the Perez all-weather luminance distribution model to predict the radiance/luminance at the centroid of the patches and aggregate the results for the period of interest, including a given climate file and coordinates. The above-explained cumulative sky approach for simulation calculation demonstrates general benefits compared to other methods, including computational efficiency with relatively small simplification errors (Robinson & Stone, 2004).

4 Application in the Context of Developing Nations

Vital knowledge of urban food security coupled with urban planning will be optimized with shared frameworks and insights from developed and developing countries. This section illustrates how the presented climate change-resilient agricultural model, initiated and substantiated in Singapore, can be administered in distinct contexts to withstand the test of time and diverse climates. Although each urban and topographic scenario is unique in urban food planning, prospective resilient and sustainable CEA solutions share similarities.

4.1 Photosynthesis Potential in Developing Cities

There is finite academic research on the planning and design of building-integrated agriculture for the gradation from energy-intensive systems to energy-efficient infrastructure capable of scaling and being implanted in the context of developing nations. Additionally, urban studies for CEA scalability favor space availability (Goodman &

Minner, 2019) over space suitability (Nadal et al., 2017), and crop-specific requirements are routinely dismissed. The proposed climate change-resilient agricultural model aims to amplify CEA efficiency through a passive sunlight optimization framework by installing agricultural systems on solar-impacted surfaces and designating their uniquely interlinked suitability with specific DLI requirements of crop species. This chapter investigates possible solutions for urban agriculture that incorporate emergent technologies and passive systems by integrating parametric architectural design workflows (Bañón & Raspall, 2021) with the latest computational techniques promoting sustainable urban development (Schröpfer, 2012).

The climate change-resilient agricultural model computes the selected cities' three-dimensional photosynthesis potential (See Fig. 3). The environmental data WEA files provisioned for model calculations are formatted as TMYx, the latest and most reliable format encompassing data from 2007 to 2021. A hypothetical comparison of horizontal and vertical surfaces, including north, south, east, and west orientations, is effectuated. The theoretical surface corresponds to a ten-by-ten-meter area (100 m^2). A manageable scale is appropriate when launching CEA in communities and households at public and private levels. The computation is expressed in DLI quantities measured in $\text{mol} \cdot \text{m}^{-2} \cdot \text{d}^{-1}$. Likewise, future comparisons utilize the same unit for evaluating the suitability of crop species and surface photosynthesis potential.

The results indicate high levels of annual average DLI at horizontal (ground and roof) and vertical (façade) levels in the designated developing cities. Primarily, the calculations express differences related to the geolocation of cities and position in the northern or southern hemisphere. The preceding factors solely affect DLI quantities since the simulations transpire on a hypothetical surface. The morphology of specific locations in the urban environments has been dismissed as it does not govern the computation displayed. For this reason, DLI values are immutable on each simulation, being the lowest and the highest coincident for the yearly average. The eminent DLI quantity and photosynthesis capacity at the horizontal level is detected in Muscat with 52 ($\text{mol} \cdot \text{m}^{-2} \cdot \text{d}^{-1}$). While in a vertical arrangement, Cape Town has inherent potentiality, with 32, 11, 26, and 27 ($\text{mol} \cdot \text{m}^{-2} \cdot \text{d}^{-1}$) at the north, south, east, and west orientations.

4.2 Crop Suitability and Food Production Simulations

The effect of solar radiation on crop growth has been comprehensively studied and modeled for extensive traditional outdoor agriculture (Trnka et al., 2007). Conversely, research that clarifies proper crop sunlight exposure is nascent and still emerging. Additionally, the topic has been studied on the façade (Palliwal et al., 2021), rooftop, and ground levels (Saha & Eckelman, 2017) but not from the scope of three-dimensional surfaces. Likewise, urban studies frequently neglect crop species and their requirements. The presented climate change-resilient agricultural model aids this research disparity, contrasting the optimal sunlight requirements of 26 crop

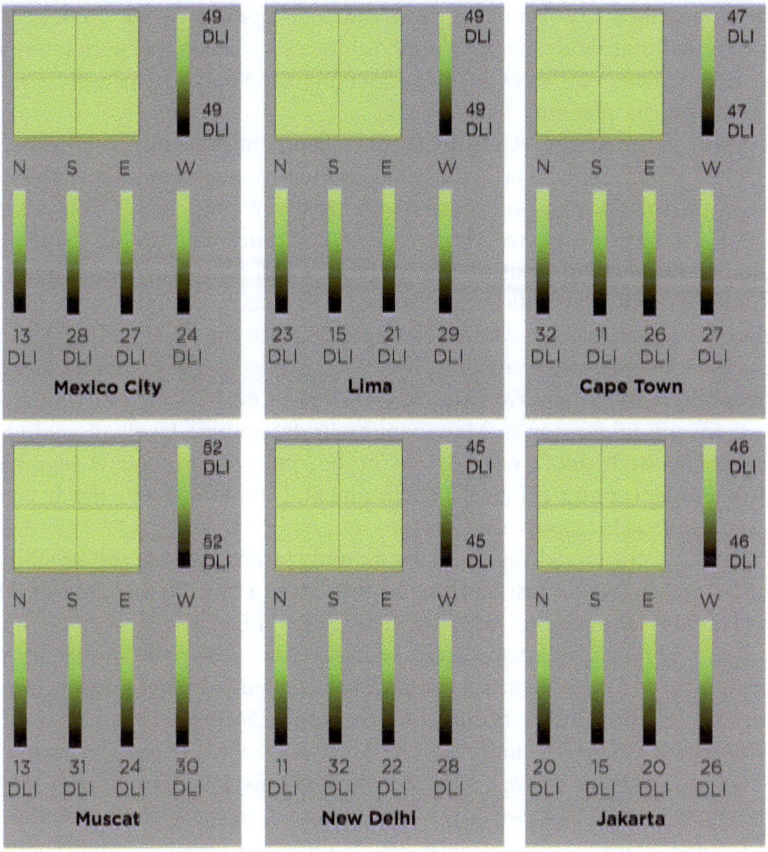

Fig. 3 Cities' Photosynthesis Potential calculated through the presented climate change-resilient agricultural model. DLI quantities are expressed in (mol $\cdot$ m^{-2} $\cdot$ d^{-1}). Figure developed by author A. Lombardia for this publication

species with the outdoor DLIs of urban surfaces. The unique correlation connecting the sunlight values of urban surfaces and crop insolation demands fosters lighting efficiency in agricultural urban planning. In addition to lighting assessments, the agricultural model reveals the potential yield attainment for the examined urban area. The related data for the computation and the scientific literature references are included in the "Dataset on Crop Species Daily Light Integrals (DLIs) and Yields" (Lombardia, 2023).

The megacity of Delhi, which includes India's capital, New Delhi, is one of the largest cities in the world. Delhi has an estimated population of 32,941,309 and still has a prospective growth of up to 43,345,059 inhabitants by 2035 under a 2.73% growth rate, according to the UN World Urbanization Prospects (UN, 2018). India is home to an estimated 200 million malnourished people, with an elevated prevalence of food insecurity, as expressed in a recent systematic review (McKay et al., 2023).

Considering Delhi's exponential growth and food insecurity, the climate change-resilient agricultural model surveys opportunities for sustainable agriculture development and unlocks pathways for implementing crop-specific measures to enable urban food security planning.

According to the Köppen-Geiger Climate Classification, winters are perceptively dry in Delhi, while summers bring turbulent stormy weather conditions. The tropical air masses provoke a monsoon season, and the dry winters result from a subtropical high-pressure atmosphere. The referenced climatic features are evident in the computed monthly simulations of crop suitability and food production (See Fig. 4). Consequently, dry winters reflect an absence of cloud cover, displaying higher levels of outdoor DLI, reaching a maximum average of 33 mol $\cdot$ m^{-2} $\cdot$ d^{-1} DLI in December. Whereas the monsoon season exhibits lower sun exposure due to remarkably cloudy skies, reaching a minimum average DLI of 10 mol $\cdot$ m^{-2} $\cdot$ d^{-1} in July. In addition, the pre-monsoon midseason in May manifests intermediate levels of sunlight with an average DLI of 13 mol $\cdot$ m^{-2} $\cdot$ d^{-1}. Consequently, the climate change-resilient agricultural model contrasts the DLI environmental conditions with the diverse light requirements of each crop species.

Subsequently, the model simulates the monthly projections of crop suitability and production yields for the test surface. In this case, a typological area of 31.6 by 31.6 m (1000 m^2) is analyzed, with the surface positioned as a façade bearing south orientation. The results indicate steady, suitable crops grown during the dry season from September to March, including cucumber (*Cucumis sativus*), pepper (*Capsicum annuum*), strawberry (*Fragaria x ananassa*), tomato (*Solanum lycopersicum*), black bean (*Phaseolus vulgaris*), sunflower (*Helianthus*), quinoa (*Chenopodium quinoa*) and Asian rice (*Oryza sativa*). Convenient midseason (pre-monsoon) crops for May encompass basil (*Ocimum basilicum*), lettuce (*Lactuca sativa*), arugula (*Eruca sativa*), raspberry (*Rubus idaeus*) and spinach (*Spinacia oleracea*). Finally, monsoon-appropriate crops, from June to August, include mint (*Mentha spicata*), oregano (*Origanum vulgare*), rosemary (*Salvia rosmarinus*), cilantro (*Coriandrum sativum*), dill (*Anethum graveolens*), parsley (*Petroselinum crispum*), Swiss chard (*Beta vulgaris*), kale (*Brassica oleracea* var. *sabellica*), pac choi (*Brassica rapa* var. *chinensis*), cabbage (*Brassica oleracea* var. *sabauda*). The model estimates the suitability of the crops based on sunlight optimization to coalesce BIA, light and energy efficiency without assessing other crop production components, such as irrigation or nutrition.

Additionally, in surface-integrated CEA, the light transmissivity of the cover envelope is one of the main components influencing the level of sunlight that reaches the crop's plant canopy. For this reason, estimates for the light reductions due to material transmittance are applied in the model. In this case, plastic coverage is selected for its economical price compared to glass, which is more appropriate in developing nations with constricted economies. With attention to the plastic typology, an eco-friendly biodegradable plastic (PLA) that can degrade in the environment is designated in the calculation, with evidence showing excellent optical properties of up to 70% transmittance (He et al., 2020).

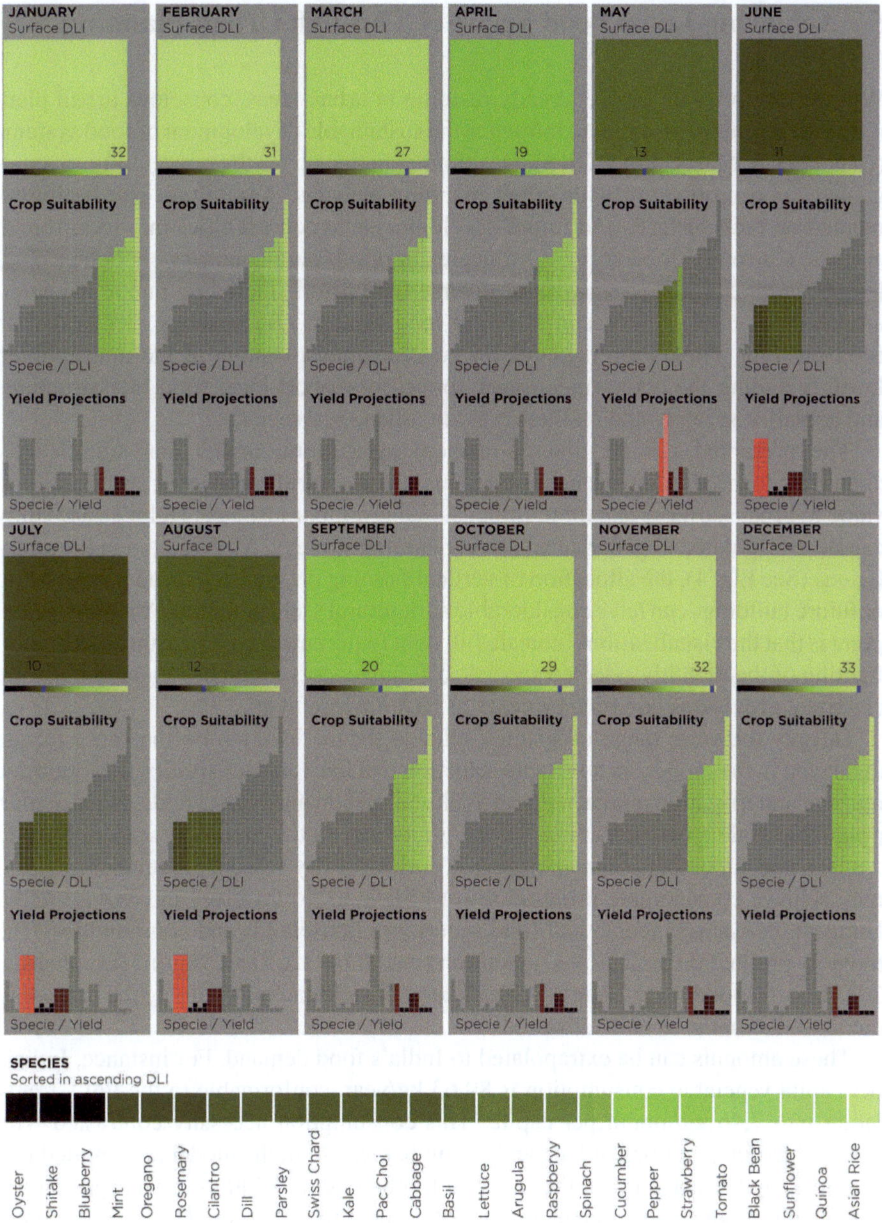

Fig. 4 Delhi's crop suitability and food production monthly simulations. From top to bottom, the figure indicates surface monthly DLI, crop suitability, and monthly yield projections. On the bottom part, detailed crop species included in the model are sorted in ascending DLI in correspondence to the "x" coordinate of the crop suitability and yield projection diagrams. In the crop suitability diagrams, DLI quantities in the "y" coordinate are expressed in (mol · m^{-2} · d^{-1}). In the yield projection diagrams, harvest values in the "y" coordinate are expressed in (kg · m^{-2}). Figure developed by author A. Lombardia for this publication

5 Achieving Local Food Security Through Urban Planning

With the majority of people already residing in urban areas, conscious urban planning will be vitally essential to influence the sustainable development of food systems (Cabannes & Marocchino, 2018). Developing nations have the opportunity to amalgamate food production with urban planning agendas. Agriculture can be implemented on built surfaces and future developments to convert cities into food-reliant entities. Currently, food security strategies in developing countries encounter challenges under a complex network of diverse food actors, including public agencies, agribusiness, civil society, and informal food vendors (Boossabong, 2018). Urban food planning integration depends on governmental planning at the metropolitan level, including the actors mentioned above, to support local food infrastructures and actualize security and resilience in the urban food sector.

The presented climate change-resilient agricultural model can significantly contribute to the aforementioned local food infrastructure by assigning an agricultural value to urban surfaces at a three-dimensional level. For instance, Delhi's case study can be extrapolated to generate an urban planning strategy. As exhibited in the simulations (See Fig. 4), the allocation of vertical farming on southern facades of existing or future buildings can have considerable agricultural yield potential. Another critical point is that the visualizations indicate full light requirement coverage through natural lighting of the monthly selected species, offering energy efficiency and reducing the electrical provisions and requirements of BIA environments.

Despite the wide range of species suitable for the location on the south façade displayed in the model, an advisable selection of a low range of species will optimize nutrient and irrigation intake in the agricultural system and boost yields. In particular, producing Kale (*Brassica oleracea* var. *sabellica*) in the monsoon season, May to August, and cucumber (*Cucumis sativus*) during the dry season, September to April, may achieve considerable progress toward SDG 1 (No Poverty) and SDG 2 (Zero Hunger) increasing income and food security in households and communities. The monthly production projections for kale and cucumber are 5 and 6 kg/m^2 (Lombardia, 2023), corresponding to 20 and 48 kg/m^2 annual yields acknowledging the preceding mentioned months of cultivation.

These amounts can be extrapolated to India's food demand. For instance, India's per capita vegetable consumption is 89.63 kg/year, conformable to the FAO, translating into 7.46 kg/month per capita. This consumption necessity contrasted with the production quantities of kale and cucumber shown in the model and applied to a south-facing vertical surface the size of a football field (7,140 m^2). The agricultural area would produce a sufficient amount of aliment to fully provide self-reliance per capita intake of vegetables for 214 people during four months in the monsoon season while covering the nutrition of 513 citizens for eight months in the dry season.

The extrapolation of the model explicitly demonstrates the framework's scalability toward specific urban planning measures that could address food security in developing cities. Indeed, the state-of-the-art literature suggests that harvests are more substantial than conventional agricultural yields (Led et al., 2023). Based on the estimated performance of CEA soilless systems, the potential for food production on urban surfaces appears profound and compelling.

6 Conclusion and Future Perspectives

In conclusion, the presented climate change-resilient urban agriculture model can assist developing nations in the urban planning of BIAs. The implementation of building-integrated CEA in existent or future buildings will have not only the advantage of achieving the energy efficiency of these systems but also climate change-resiliency, producing 24/7 for 365 days a year regardless of climate, with pesticide-free food production, 95% less use of fresh water compared to conventional agriculture, and at least 200 times the yield. All these factors translate into higher production and revenue. Under those circumstances, a green infrastructure that contributes to building sustainable and resilient urban settlements in developing countries in the face of climate change can be generated. The transcendence of the framework fosters not only the establishment of local food security in developing urban settlements but also the mitigation of the repercussions of the global food system that considerably contribute to climate change, including deforestation, greenhouse gas emissions, ocean dead zones, and species extinction.

The introduced sunlight-informed model is an estimation grounded on scientific data and computational simulations. Consequently, specific real-time conditions will generate results that may differ from the simulated values. The model should be adapted and improved with real-time data once the simulated results are contrasted through sensors and built agricultural developments in the future. Furthermore, yield values are considered essential data for planning future agricultural developments. However, it must be considered that the included yields are based on scientific literature, with harvests obtained under optimal conditions of light, nutrients, and humidity in vertical CEA systems. Therefore, the results obtained in specific urban developments may differ from the simulated data considering optimal conditions. The model has a high potential to evolve towards developing an open-source tool that allows knowledge sharing among developed and developing nations, fostering alliances between decision-makers and stakeholders to define sustainable pathways for urban food security.

References

Al-Kodmany, K. (2018). The vertical farm: A review of developments and implications for the vertical city. *Buildings, 8*, (2), 24. https://doi.org/10.3390/BUILDINGS8020024.

Althor, G., Watson, J. E. M., & Fuller, R. A. (2016). Global mismatch between greenhouse gas emissions and the burden of climate change. *Science and Reports, 6*, 1–6. https://doi.org/10.1038/srep20281

Altieri, M. A., Nicholls, C. I., Henao, A., & Lana, M. A. (2015). Agroecology and the design of climate change-resilient farming systems. *Agronomy for Sustainable Development, 35*, 869–890. https://doi.org/10.1007/s13593-015-0285-2

Arora, N. K., & Mishra, I. (2022). Current scenario and future directions for sustainable development goal 2: A roadmap to zero hunger. *Environmental Sustainability, 5*, 129–133. https://doi.org/10.1007/s42398-022-00235-8

Ashdown, I., & Ward, G. (2013). Gendaymtx. Generate an annual Perez sky matrix from a weather tape. *Radiance,* 1–2

Bañón, C., Raspall, F. (2021). *3D Printing architecture: Workflows, applications, and trends* (Springer Nature).

Beck, H. E., Zimmermann, N. E., McVicar, T. R., et al. (2018). Present and future Köppen-Geiger Climate Classification maps at 1-km resolution. *Scientific Data, 5*, 1–12. https://doi.org/10.1038/sdata.2018.214

Bhattacharya, S., Lal, H., & Sachdev, B. K. (2022). Vertical farming a hope for India to eradicate the crisis of food. *Galaxy International Interdisciplinary Researcharch, 9*, 529–535.

Boossabong, P. (2018). Articulating public agencies, experts, corporations, civil society and the informal sector in planning food systems in Bangkok. In *Integrating food into urban planning* (pp. 60–79). London: UCL Press; Rome: FAO.

Briz, J., & De Felipe, I. (2016). La naturación: instrumento base para ciudades inteligentes sostenibles. *Ambienta,* 84–97

Cabannes, Y., & Marocchino, C. (2018). *Integrating food into urban planning*. London, UCL Press.

Chinowsky, P., Hayles, C., Schweikert, A., et al. (2011). Climate change: Comparative impact on developing and developed countries. *Engineering Project Organization Journal, 1*, 67–80. https://doi.org/10.1080/21573727.2010.549608

Climate.OneBuilding. (2023). Climate.OneBuilding.Org. https://climate.onebuilding.org/.

DARA and the Climate Vulnerable Forum. (2010). Climate vulnerability monitor 2010. https://daraint.org/climate-vulnerability-monitor/climate-vulnerability-monitor-2010/.

Dsouza, A., Newman, L., Graham, T., Fraser, E. D. G. (2023). Exploring the landscape of controlled environment agriculture research: A systematic scoping review of trends and topics. *Agricultural Systems, 209*, 103673.

Ehrlich, C., Papamichael, K., Lai, J., et al. (2001). Simulating the operation of photosensor-based lighting controls. In: *7th International Building Performance Simulation Association Conference* (pp. 1–9). Rio de Janeiro.

Engler, N., & Krarti, M. (2021). Review of energy efficiency in controlled environment agriculture. *Renewable and Sustainable Energy Reviews, 141*, 1–12. https://doi.org/10.1016/j.rser.2021.110786

Faust, J. E., & Logan, J. (2018). Daily light integral: A research review and high-resolution maps of the United States. *HortScience, 53*, 1250–1257. https://doi.org/10.21273/HORTSCI13144-18.

Fondriest Environmental Inc. (2014). Solar radiation and photosynthetically active radiation. In *Fundamentals of environmental measurements.* Retrieved October 27, 2022, from https://www.fondriest.com/environmental-measurements/parameters/weather/solar-radiation/.

Franco, I., Ellennderbyshire, T., & Tracey, J. (Eds.). (2020). *Science for sustainable societies actioning the global goals for local impact*

Frona, D., Szenderak, J., & Harangi-Rakos, M. (2021). Economic effects of climate change on global agricultural production. *Nature Conservation, 44*, 117–139. https://doi.org/10.3897/NATURECONSERVATION.44.64296.

Gómez-Villarino, M. T., Urquijo, J., Gómez Villarino, M., & García, A. I. (2021). Key insights of urban agriculture for sustainable urban development. *Agroecology and Sustainable Food Systems, 45*, 1441–1469. https://doi.org/10.1080/21683565.2021.1917471.

Goodman, W., & Minner, J. (2019). Will the urban agricultural revolution be vertical and soilless? A case study of controlled environment agriculture in New York City. *Land Use Policy, 83*, 160–173. https://doi.org/10.1016/j.landusepol.2018.12.038

Hadley, K., Wheat, S., Rogers, H. H., et al. (2023). Mechanisms underlying food insecurity in the aftermath of climate-related shocks: A systematic review. *Lancet Planet Health, 7*, e242–e250. https://doi.org/10.1016/S2542-5196(23)00003-7

Hamm, M. (2015). Feeding cities-with indoor vertical farms? In *Center for regional food systems*. https://www.canr.msu.edu/news/feeding_cities_with_indoor_vertical_farms.

He, H., Gao, M., Illés, B., & Molnar, K. (2020). 3D Printed and electrospun, transparent, hierarchical polylactic acid mask nanoporous filter. *International Journal of Bioprinting, 6*, 1–9. https://doi.org/10.18063/ijb.v6i4.278.

International Monetary Fund. (2022). World economic and financial surveys world economic outlook database—WEO groups and aggregates information. Retrieved September 27, 2023, from https://www.imf.org/external/pubs/ft/weo/2022/01/weodata/groups.htm.

Kalantari, F., Tahir, O. M., Joni, R. A., & Fatemi, E. (2018). Opportunities and challenges in sustainability of vertical farming: A review. *Journal of Landscape Ecology (Czech Republic), 11*, 35–60. https://doi.org/10.1515/JLECOL-2017-0016

Koohafkan, P., Altieri, M. A., & Holt Gimenez, E. (2012). Green agriculture: Foundations for biodiverse, resilient and productive agricultural systems. *International Journal of Agricultural Sustainability, 10*, 61–75. https://doi.org/10.1080/14735903.2011.610206

Korczynski, P. C., Logan, J., & Faust, J. E. (2002). Mapping monthly distribution of daily light integrals across the contiguous United States. *Horttechnology, 12*, 12–16. https://doi.org/10.21273/horttech.12.1.12.

Kyropoulou, M., Subramaniam, S., & Hoffmann, S. (2021). An automated workflow for informing urban greening based on photosynthetic radiation modelling. *Journal of Physics: Conference Series*, 2069. https://doi.org/10.1088/1742-6596/2069/1/012060

Larson, G. W., Shakespeare, R. A., Apian-Bennewitz, P., et al. (1988). Rendering with radiance: The art and science of lighting visualization. Retrieved August 11, 2023, from https://floyd.lbl.gov/radiance/book/.

Led, B., Lights, G., Yield, H., et al. (2023). *Hydroponics yield versus soil yield: Lessons from Dr. Howard Resh*. Retrieved February 3, 2023, from https://uponics.com/hydroponics-yield/.

Lombardia, A. (2023). Dataset on crop species daily light integrals (DLIs) and yields. Mendeley Data V1. https://doi.org/10.17632/3shw2jx2wk.1.

Lombardía, A., & Gómez-Villarino, M. T. (2023). Green infrastructure in cities for the achievement of the UN sustainable development goals: a systematic review. *Urban Ecosystems*, 1–15. https://doi.org/10.1007/s11252-023-01401-4

Malhi, G. S., Kaur, M., & Kaushik, P. (2021). Impact of climate change on agriculture and its mitigation strategies: A review. *Sustainability, 13*(3), 1318. https://doi.org/10.3390/SU13031318.

McKay, F. H., Sims, A., & van der Pligt, P. (2023). Measuring food insecurity in India: A systematic review of the current evidence. *Current Nutrition Reports, 12*, 358–367. https://doi.org/10.1007/s13668-023-00470-3

Mendelsohn, R. (2008). The impact of climate change on agriculture in developing countries. *Journal of Natural Resources Policy Research, 1*, 5–19. https://doi.org/10.1080/19390450802495882

Nadal, A., Alamus, R., Pipia, L., & Ruiz, A. (2017). Urban planning and agriculture. Methodology for assessing rooftop greenhouse potential of non-residential areas using airborne sensors. *Science of the Total Environment Elsevier, 601–602*, 493–507. https://doi.org/10.1016/j.scitotenv.2017.03.2140048-9697

Neo, D. C. J., Ong, M. M. X., Lee, Y. Y., et al. (2022). Shaping and tuning lighting conditions in controlled environment agriculture: A review. *ACS Agricultural Science and Technology, 2*, 3–16. https://doi.org/10.1021/acsagscitech.1c00241

Palliwal, A., Song, S., Tan, H. T. W., & Biljecki, F. (2021). 3D city models for urban farming site identification in buildings. *Computers, Environment and Urban Systems, 86*, 1–16. https://doi.org/10.1016/j.compenvurbsys.2020.101584

Perez, R., Seals, R., Michalsky, J., & Sciences, A. (1993). All-weather model for sky luminance distribution: Preliminary configuration and validation. *Solar Energy, 50*, 235–245.
Pérez-Escamilla, R. (2017). Food security and the 2015–2030 sustainable development goals: From human to planetary health. *Current Developments in Nutrition, 1*, 1–8. https://doi.org/10.3945/cdn.117.000513
Pigford, A. A. E., Hickey, G. M., & Klerkx, L. (2018). Beyond agricultural innovation systems? Exploring an agricultural innovation ecosystems approach for niche design and development in sustainability transitions. *Agricultural Systems, 164*, 116–121. https://doi.org/10.1016/j.agsy.2018.04.007
Porter, J.R., Xie, L., Challinor, A.J., Cochrane, K., Howden, S.M., Iqbal, M.M., Lobell, D.B., & Travasso M.I. (2014). Food security and food production systems. In: Climate Change 2014: Impacts, Adaptation, and Vulnerability. Part A: Global and Sectoral Aspects. Contribution of Working Group II to the Fifth Assessment Report of the Intergovernmental Panel on Climate Change [Field, C.B., V.R. Barros, D.J. Dokken, K.J. Mach, M.D. Mastrandrea, T.E. Bilir, M. Chatterjee, K.L. Ebi, Y.O. Estrada, R.C. Genova, B. Girma, E.S. Kissel, A.N. Levy, S. MacCracken, P.R. Mastrandrea, and L.L. White (eds.)]. Cambridge University Press, Cambridge, United Kingdom and New York, NY, USA, pp. 485–533.
Proksch, G., Ianchenko, A., & Kotzen, B. (2019). Aquaponics in the built environment. In *Aquaponics food production systems* (pp. 523–558). Springer.
Ranis, G., Stewart, F., & Samman, E. (2006). Human development: beyond the human development index. *Journal of Human Development, 7*, 323–358. https://doi.org/10.1080/14649880600815917
Robinson, D., & Stone, A. (2004). Irradiation modelling made simple: The cumulative sky approach and its applications. In *PLEA-passive and low energy architecture* (pp. 19–22). Eindhoven, The Netherlands.
Rockström, J., Steffen, W., Noone, K., et al. (2009). A safe operation space for humanity. *Nature, 461*, 472–475.
Roudsari, M., & Mackey, C. (2013). Ladybug tools. Retrieved December 15, 2022, from https://www.ladybug.tools/ladybug.html.
Roudsari, M. S., & Pak, M. (2013). Ladybug: A parametric environmental plugin for grasshopper to help designers create an environmentally-conscious design. In *Proceedings of BS 2013: 13th Conference of the International Building Performance Simulation Association* (pp. 3128–3135). https://doi.org/10.26868/25222708.2013.2499
Saha, M., & Eckelman, M. J. (2017). Growing fresh fruits and vegetables in an urban landscape: A geospatial assessment of ground level and rooftop urban agriculture potential in Boston, USA. *Landscape and Urban Planning, 165*, 130–141. https://doi.org/10.1016/j.landurbplan.2017.04.015
Schröpfer, T. (2012). Ecological urban architecture: Qualitative approaches to sustainability. Birkhäuser: Basel/Berlin/Boston.
SharathKumar, M., Heuvelink, E., & Marcelis, L. F. M. (2020). Vertical farming: Moving from genetic to environmental modification. *Trends in Plant Science, 25*, 724–727. https://doi.org/10.1016/j.tplants.2020.05.012
Subramaniam, S., & Mistrick, R. (2017). A more accurate approach for calculating illuminance with daylight coefficients. In *Illuminating Engineering Society Annual Conference* 2017 (p. 18).
Subramaniam, S., Kyropoulou, M., & Hoffmann, S. (2020). Radiation modeling strategy for incorporating vegetation in urban microclimate simulations. SIMAUD 2020, Vienna, Austria.
Sun, Y., Sun, Y., & Sun, Z., et al. (2015). Mapping monthly distribution of daily light integrals across China. In: *Annual Conference of ASHS*. New Orleans, LA.
Torres, A. P., & Lopez, R. G. (2012). Measuring daily light integral in a greenhouse. *Purdue Extension Horticulture HO-238-W*, 1–7.
Tregenza, P. R. (1987). Subdivision of the sky hemisphere for luminance measurements. *Lighting Research & Technology, 19*, 13–14. https://doi.org/10.1177/096032718701900103

Tregenza, P. R., & Waters, I. M. (1983). Daylight coefficients. In *CIBS National Lighting Conference* (pp. 65–71). Warwick.

Trnka, M., Eitzinger, J., Kapler, P., et al. (2007). Effect of estimated daily global solar radiation data on the results of crop growth models. *Sensors, 7*, 2330–2362. https://doi.org/10.3390/s7102330

UN. (2018). World urbanization prospects. In Department of Economic and Social Affairs, United Nations. Retrieved September 29, 2023, from https://population.un.org/wup/.

UN. (2023). List of least developed countries. Chrome-extension://efaidnbmnnnibpcajpcglclefindmkaj/. Retrieved September 27, 2023, from https://www.un.org/development/desa/dpad/wp-content/uploads/sites/45/publication/ldc_list.pdf.

van Delden, S. H., SharathKumar, M., Butturini, M., et al. (2021). Current status and future challenges in implementing and upscaling vertical farming systems. *Nature Food, 2*(12), 944–956. https://doi.org/10.1038/s43016-021-00402-w.

Weiss, A., & Norman, J. M. (1985). Partitioning solar radiation into direct and diffuse, visible and near-infrared components. *Agricultural and Forest Meteorology,* 205–213.

World Food Program. (2020). Hunger map. Retrieved September 27, 2023, from https://www.wfp.org/publications.

Zari, M. P., Blaschke, P. M., Jackson, B., et al. (2020). Devising urban ecosystem-based adaptation (EbA) projects with developing nations: A case study of Port Vila, Vanuatu. *Ocean & Coastal Management, 184*, 105037. https://doi.org/10.1016/j.ocecoaman.2019.105037

Zhu, R., Wong, M. S., Kwan, M. P., et al. (2022). An economically feasible optimization of photovoltaic provision using real electricity demand: A case study in New York city. *Sustainable Cities and Society, 78*, 1–15. https://doi.org/10.1016/j.scs.2021.103614

Redeveloping Open and Urban Green Spaces for Climate Resilience; A Case Study of Chennai

Dhanapal Govindarajulu, Francisco García Sánchez, and Pradeep Chaudhry

Abstract Open and Urban green spaces (UGS) are important assets of urban sustainability, provide essential ecosystem services, climate resilience and contribute to human well-being. However, planning open and UGS for climate resilience requires appropriate land use planning aided by urban policies and planning guidelines. In this chapter, we present a GIS (Geographical Information System) based framework to identify open and UGS for developing climate resilience using Chennai as a case city. Using high-resolution satellite imageries, we have identified 63 sq. km of potential land parcels that can be developed as open and UGS in Chennai city that will provide climate adaptation benefits against floods and urban heat island effect. We debate the importance of identifying and protecting open areas and UGS in urban development master plans and further discuss some of the current opportunities enabled by recent urban policies and missions such as the smart city and AMRUT mission and barriers, particularly on master planning including capacity gaps and urban governance for redeveloping urban green spaces in Indian cities for climate resilience.

Highlights:

1. High-resolution open-source satellite imageries like Sentinel are useful to map potential areas for UGS.
2. Understanding the city climate risks allows for better identification and evaluation of open spaces for urban resilience.
3. Recent urban policies in India and schemes like Smart City Mission, AMRUT provide opportunities for redeveloping open and green spaces
4. Constraints include poor master planning and implementation, limited capacities of urban planners in planning green spaces in the context of climate change and weak urban governance.

D. Govindarajulu (✉)
Global Development Institute, University of Manchester, Manchester, UK
e-mail: dhanapal.cws@gmail.com

F. García Sánchez
University of Cantabria, Av. de los Castros S/N, 39005 Santander, Spain

P. Chaudhry
Indian Institute of Forest Management, Bhopal 462003, India

P. K. Joshi et al. (eds.), *Blue-Green Infrastructure for Sustainable Urban Settlements*,
https://doi.org/10.1007/978-3-031-62293-9_13

Keywords AMRUT · Chennai · Planning · Smart city mission · Urban Floods

1 Introduction

Urban Green Spaces (UGS) and open areas have multifunctional importance such as recreation, social interaction, aesthetics, cultural heritage, and ecological functions (Mell, 2009; Haaland & van den Bosch, 2015). In recent years, there has been an increasing recognition of the potential of UGS for environmental sustainability, climate adaptation, and disaster risk reduction (Cheng et al., 2021; Chondrogianni & Stephanedes, 2022; García et al., 2018; Govindarajulu, 2020). The cities of the Global South, dragging specific local contexts of a neoliberal character and marked by colonial administration systems (Baruah et al., 2017), have reached maturity with enormous inequalities and environmental problems (Onodugo & Ezeadichie, 2020). This has resulted in unplanned urban sprawl, land use change and loss of UGS (Haaland & van den Bosch, 2015; Caldeira, 2017). Urbanisation in the Global South has been much rapid and the cities there have been unable to plan and develop green spaces adequately (Rigolon et al., 2018; Myers et al., 2021). The development of UGS and recreational open areas in the Global South is lagging due to a lack of importance given to their sociocultural and economic services, furthermore, there is a lack of urban planning guidelines in Global South to improve UGS for climate resilience and sustainability (Pauleit et al., 2021).

Urban green space planning in the Indian context offers important lessons for redeveloping urban green spaces in fast urbanising cities of the Global South. India has witnessed rapid urbanisation in the last two decades with the urban population raising from 285 million in 2001 to 377 million in 2011 and is expected to cross 800 million by 2050 (IIHS, 2018; UN, 2014). India will be home of some of the world's largest cities in the twenty-first century (Kundu & Pandey, 2020). In the last published national census, India had 468 urban agglomerations with a population of over 100,000 of which 53 had a population of more than one million (Census, 2011). In 1996, the Government of India released its first Urban Development Plans Formulation and Implementation (UDPFI) Guidelines based on which the State governments are responsible for preparing urban development master plans for their cities. Master plans are statutory instruments to guide and regulate the development of cities and are critical for managing urbanization and sustainability. However, 65% of the 7,933 urban settlements in India do not have any master plan (GoI, 2021). Unplanned urbanisation in over 42 large cities has caused land use changes, where more open spaces including wetlands were converted to residential, commercial, and industrial areas (Avasiha & Garg, 2016; Ramaiah & Avtar, 2019).

This loss of green and open spaces has indeed increased climate risks of many Indian cities to urban floods and the urban heat island effect. To take the case of Bangalore City urban sprawl has affected its drainage network, local hydrology, and groundwater table levels (Kumar et al., 2016). In the Mumbai floods in 2005, inappropriate land use planning, and extreme rainfall event combined with lack of natural drainage was studied as important reasons for inundations (Ranger et al., 2011). During the Chennai floods in 2015, it was observed that several of flood plains had been converted to buildings thereby affecting the drainage flow (CAG, 2017). Cities like Ahmedabad have been witnessing extreme urban heat island effect in recent years (Mohammad et al., 2022).

In recent years, particularly after 2014, there has been increased recognition of UGS for climate adaptation and resilience in India mainly due to policy shifts and increasing climate risks in cities (García & Govindarajulu, 2023). however, there are methodological challenges on planning and developing urban green spaces and financing them. In this article, we present a framework for urban planners and administrators to identify potential urban green spaces using remote sensing techniques that can be developed for to improve the city's climate resilience against urban floods and urban heat island effect. We demonstrate this framework using Chennai city as a case study. We further discuss some of the opportunities and challenges in redeveloping urban green spaces in Chennai and other Indian cities.

Globally, there are different definitions of green space adopted. Lovell and Taylor (2013, p. 1448) define UGS as "An undeveloped piece of land located within the context of a city and open to the public". Moreover, UGS is defined as "all urban land covered by vegetation of any kind on private and public grounds, irrespective of size and function" (WHO, 2017, p. 2). Informal green spaces such as aquatic fronts and urban habitats such as abandoned industrial sites and overgrown gardens have also been considered as UGS (Tzoulas et al., 2007; Venn & Niemelä, 2004). UGS do not necessarily include only green and include blue areas like riparian zones and lakes that support the growth of greenery (Cvejić et al., 2015; WHO, 2017). For this study, we use urban green space as defined by the Government of India in its Urban Greening Guidelines and this includes parks with trees, streetscapes with avenue trees and forests (GoI, 2014) and open areas that may not have trees but used for recreational and other ecosystem services purposes like flood mitigation. The objective of this study is to map the current open and UGS and identify suitable land parcels that can be developed for UGS through using satellite imageries. We further discuss the opportunities available in urban planning guidelines and schemes like AMRUT and smart city missions to develop these UGS and for climate resilience purpose.

2 Study Area

Chennai city lies on the Southeast coast of India. Chennai's urban agglomeration population grew from 2.64 million in 1971 to 4.68 million in 2011 (Census, 2011). It is currently estimated to have exceeded 9 million (IIHS, 2018, p. 24) and will reach 14 million by 2030 (UN, 2014). Its administrative boundary was expanded from 176 sq. km to 426 sq. km in 2011, bringing in smaller municipalities and census towns into the larger ULB (Urban Local Bodies) of the Chennai Municipal Corporation (CMC, 2019). The urban green space in Chennai city when its administrative boundary was

176 sq. km was only 366 ha. This makes it one of the cities with very low green spaces in India (Govindarajulu, 2014).

Chennai city located in a coastal area is vulnerable to cyclones, urban floods (Arabindoo, 2016), sea level rise and urban heat island effects (Amirtham, 2016; Ilayaraja et al., 2017; Kotharkar et al., 2018). The disaster risk varies spatially; and is particularly high in the low-lying areas along coastal, flood plains and river banks (Joerin et al., 2014). Chennai city has lost of much of its urban green space to the rapid urban sprawl in the city, this includes the conversion of wetlands including rivers, creeks, lakes, and ponds (Gupta & Nair, 2011). As suggested by Gupta and Nair (2011), the lack of green space is likely to affect the capacity of the land to hold water, resulting in increased surface water run-off and reduced groundwater recharge rates. Built-up area increased by 88% during the period 1997 to 2017 (IIHS, 2018) and the repetitive floods in 1985, 2002, 2005, 2015 are due to the city's drainage system failure (Bremner, 2020). In 2015, Chennai witnessed one of urban India's worst flood situations with large parts of the city submerged in 1.0–1.5-m-high water levels, leading to high economic losses and distress for its citizens. Part of this growth has been in flood-prone areas or with high environmental values, substantially reducing the vegetation surfaces, open spaces, and water bodies (Fig. 1). Chennai also ranks high in risks of extreme heat stress and dangerous heat strokes in compared to other metropolitan cities in India (Kumar et al., 2022).

The open spaces and water bodies combined have reduced by around 20% during the period 1997 to 2017 (IIHS, 2018). This has led to the city having inadequate green space to meet the social and environmental needs. Rapid urbanization and population growth have placed a strain on the city's critical infrastructure systems, such as sewage, water supply, and electricity. For instance, the sewage system is below capacity to allow free storm water run-off (Govindarajulu, 2020). Another critical challenge facing Chennai is uneven urbanization, which has led to increased

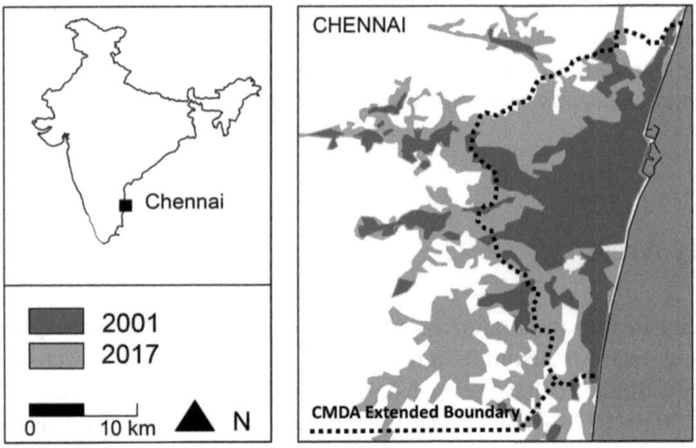

Fig. 1 Changes in urban growth, Chennai 2001–2017. Compiled by authors based on IIHS (2018)

vulnerability to droughts and floods (Srinivasan et al., 2013). The revised Master Plan for Chennai 2026 prepared by the Chennai Metropolitan Development Authority (CMDA, 2008), covers over 1,150 sq. km area of Greater Chennai, which is the expected city growth. However, the master plan does not specifically mention how much UGS will be targeted to be developed and for what benefits- recreational, environmental or resilience, etc. This also presents an opportunity to identify possible UGS that can be developed and integrated into the future master plan for climate adaptation given the city's climate risks.

3 Methods

One of the challenges urban planners face on planning UGS is to identify potential land parcels that can be developed to open and UGS and for the purpose of climate adaptation and resilience in cities. Here we present a framework that can help urban planners to use the open-source satellite imageries in Google Earth engine to map and plan UGS for environmental sustainability in the context of climate adaptation. The methods involve first mapping the current UGS in Chennai city using remote sensing techniques, second is to identify the spatial distribution of climate risks using a digital elevation model and from other secondary data sources like published risk assessments for Chennai and third to identify potential UGS from remotely sensed landcover classes for development as climate adaptation measure. The workflow is illustrated in Fig. 2. Finally, based on analysis of current urban planning guidelines and policies, we discuss the options of redeveloping and financing UGS.

The advancements in remote sensing techniques and high-resolution satellite imageries have made the mapping and planning of UGS much easier for urban planners (Shahtahmassebi et al., 2021). Tools like the Google Earth engine further allow the processing of large sets of satellite imageries across time to map land cover and its changes (Kumar & Mutanga, 2018). For this study, we use Sentinel 2A satellite

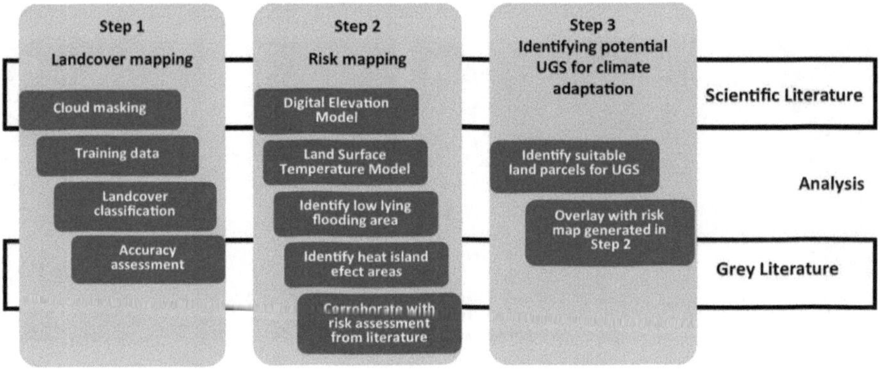

Fig. 2 Workflow of the research methodology. *Source* Compiled by the authors

imagery produced by the European Space Agency (Drusch et al., 2012). The Sentinel 2A comes with a high resolution of 10 m and with 13 spectral bands in visible and shortwave infra-red domains (Gascon et al., 2017) and recent years have been found to be the most accurate imagery to produce land cover classification (Hu et al., 2021).

Ten cloud free Sentinel 2A images across different seasons of the year were mosaiced in the Google Earth engine for the study area for the year 2020. The image was then classified using the Spectral Temporal Matrix approach and Random Forest Classification. This method has been widely used in urban studies (e.g., Ruiz Hernandez & Shi, 2018). We used Bands B2, B3, B4, B8, B11 and B12 of the Sentinel satellite imageries, which correspond to Blue, Green, Red, Near Infra-Red, Short Wave Infra-Red1 and 2 respectively. In addition to the bands, we also used the Normalized Difference Vegetation Index (NDVI) and Land Surface Water Index (LSWI) as additional indices. By using high-resolution Google Earth imageries, training points were generated for the year 2020. The training data was collected for four different land cover categories: Built-up, Green Spaces, Barren Land and Water Bodies. For the training data purpose, we considered barren land as a parcel of land of 0.5 ha size without trees or other vegetation, this includes unconstructed open areas, flood plains and agricultural land that are cultivated or left fallow. Figure 3 shows the selection of training data for green and barren land.

The land cover classification was validated using the accuracy assessment tool' in *ArcGIS pro* by validating the classification with high-resolution validation points for the four land cover classes collecting using Google Earth image and computing the confusion matrix, user, and producer accuracy. We followed the standard procedure for accuracy assessment as described in Olofsson et al. (2014).

For mapping climate risks on urban flooding for which Chennai city is very prone to, topography data has been used extracted from Google Earth vectorial imageries. A Digital Elevation Model (DEM) was derived for the Chennai city from the CartoDEM satellite data, which is available at 30 m resolution. The CartoDEM has been widely used in developing elevation, slope, and hydrological watershed models to assess urban flood risks (e.g. Jain et al., 2018). From the DEM, we delineated areas that fall below two meters in elevation to map flood risk areas, this was corroborated

Fig. 3 Training data examples; **a** a patch of forest used as green space in training data; **b** a patch of barren land used as training data

with secondary literature evidence of places of flooding and flood risks in Chennai (Andimuthu et al., 2019; Joerin et al., 2014). Evidence of past floods such as that of the 2015 Chennai floods (CAG, 2017) was also used to map out the flood risk areas. To identify heat island effect areas, we used Landsat 8 thermal band and composed the peak temperature for the Chennai city in the month of May 2020. We followed a standard procedure to convert the thermal band of Landsat 8 satellite imagery as described in Ermida et al. (2020).

From the Landcover mapping done in the first step, open land parcels that are suitable for development as UGS were identified, this was then spatially overlaid with the risk assessment map. The potential of developing these UGS for environmental sustainability and climate adaptation was analysed in the current urban development and policy context. We further discuss some of the opportunities and barriers for redeveloping UGS in Indian cities.

4 Results

4.1 Distribution of Urban Green Spaces and Climate Risks

The landcover analysis from the Sentinel 2A satellite imagery for the year 2020, showed that the total urban green spaces including marshlands, trees on streets and parks are 68.2 square kilometres (sq. km), which is around 15% of the city area. The two large green spaces in Chennai city are the Guindy National Park and Pallikaranai Marshland (Fig. 4). The supervised classification of the Sentinel image for four land cover classes gave better results and accuracy of (0.71 and Kappa- 0.86) as in Fig. 5.

Several studies in the past have assessed land use and land cover change in Chennai city using remote sensing techniques (Mathan & Krishnaveni, 2020), however, these were incomparable with the current study due to differences in the boundary used. Kaaviya and Devadas (2021) analysed Chennai Municipal Corporation Boundary as in this study focusing on water resilience, mapping around 78% as built-up area based on Landsat imageries, 12 points higher than in our study. Chanu et al. (2021) also used Landsat imageries, which is of 30 m resolution but the results show the high difference in land cover values. This could be due to Landsat imageries limitation to assess green cover. The study developed by Vigneswaran and Kumar (2018) use Sentinel imageries and an NDVI (Normalized Difference Vegetation Index) approach to identify forest cover and the land cover results are similar to this study (see Table 1).

Our analysis shows the largest green space in the city to be Guindy National Park which covers about 270 ha. and the Pallikaranai marshes are about 1,500 ha. Other large green spaces are government land under the Ministry of Defence (590 ha.) and land parcels under the Integral Coach Factory of the Indian Railways (136 ha.) as shown in Fig. 4, which are of no access to the public for recreational purposes. The other few large existing green areas are located in academic institutions like the Madras Christian College and the Besant Gardens managed by the Theosophical

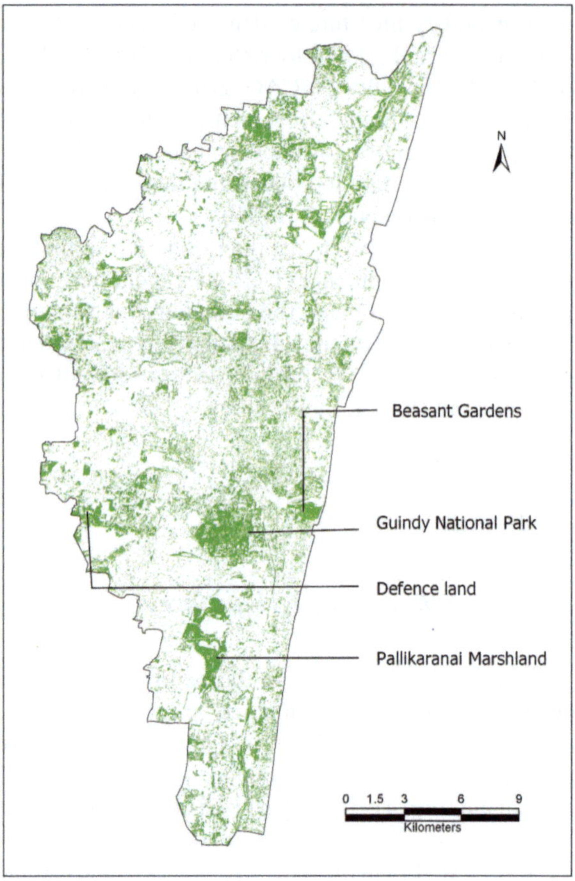

Fig. 4 Urban green spaces as per NDVI analysis (Sentinel imagery)

Society. As per the Chennai Municipal Corporation records in 2015, the number of designated parks was only 332 covering a surface area of 252 ha (CMC, 2019). Approximately half of the green area identified in the GIS analysis is located in small landscaped spaces connected to main streets and natural green corridors associated with the city's lakes and ponds. These large and small urban green spaces of Chennai city were never planned for climate adaptation purposes.

The increasing climate risks in Chennai city need to be analysed in the current context of available green spaces. Chennai city has been frequently affected by urban floods (Joerin et al., 2014) and most severe being in 2015 (Govindarajulu, 2020). The topography of the city is gently undulating with two major rivers Adyar and Cooum river draining through the city. Further, a large part of the city towards the coast lies in low elevation of less than two meters above mean sea level and are prone to urban floods (Fig. 6). The south Chennai region lying in lower elevation was severely flooded in 2015. Chennai City is also prone to the urban heat effect when the summer temperatures exceed 40 degrees Celsius. The low-elevation areas

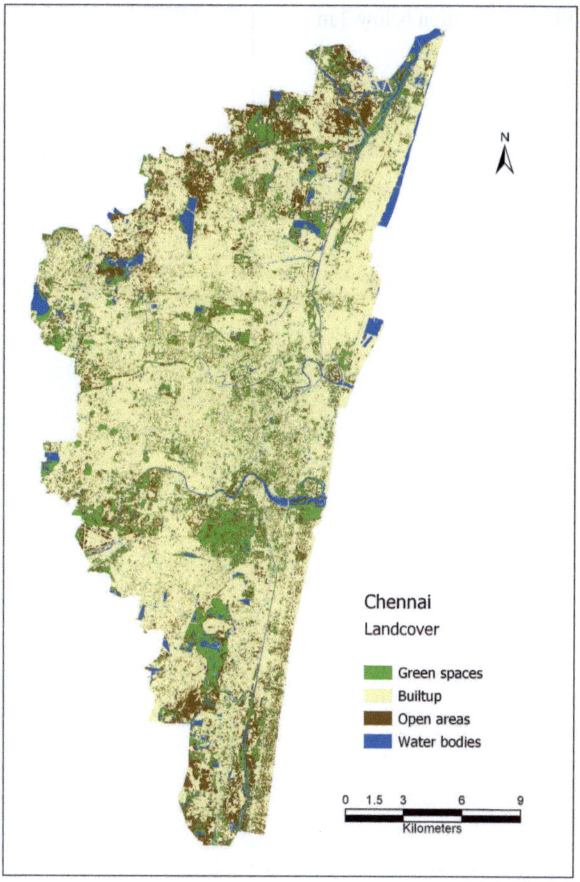

Fig. 5 Land cover map as on May 2020 showing green spaces and marshland classes

Table 1 Compared land cover classes obtained based on analysis of 2020 Sentinel image

Land cover classes	Results Area in sq. km	Results Area in %	Area based on Sentinel (Vigneswaran & Kumar, 2018)	Area in %	Area based on Landsat (Chanu et al., 2021)	Area in %
Waterbodies	***19.6***	***4.4***	22.5	5.2	25.0	5.6
Built-up	***298.7***	***66.4***	244.9	56.9	363.9	82.0
Open areas	***62.9***	***14.0***	109.3	25.4	37.2	8.4
Green spaces	***68.2***	***15.2***	54.1	12.5	17.9	4.0
Total surface area	***449.4***	***100.0***	430.8	100.0	444.0	100.0

Fig. 6 Elevation below 2 m

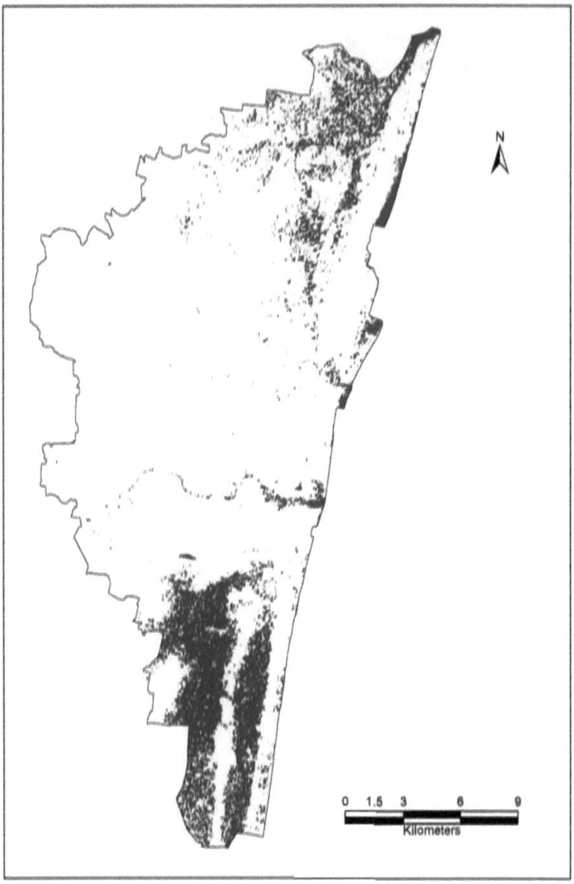

that have a high risk of flooding and areas that experience high urban heat island effect are presented in the below maps.

Observing Fig. 7, we can relate how large green infrastructures considerably reduce the urban heat island. Pallikaranai Marshland and Guindy National Park located south of Chennai favour a noticeable reduction in surface temperature. To the north, Red Hills Reservoir and other blue infrastructure wetlands temper the urban heat island. However, the central and western areas of the city, with sectors of greater population density, are highly disturbed by the heat island effect.

4.2 Potential Open and UGS in Chennai

Analysis of satellite imagery for the year 2020 (Table 1) shows that about 62.9 sq. km of open areas are available, which has the potential for developing UGS for climate

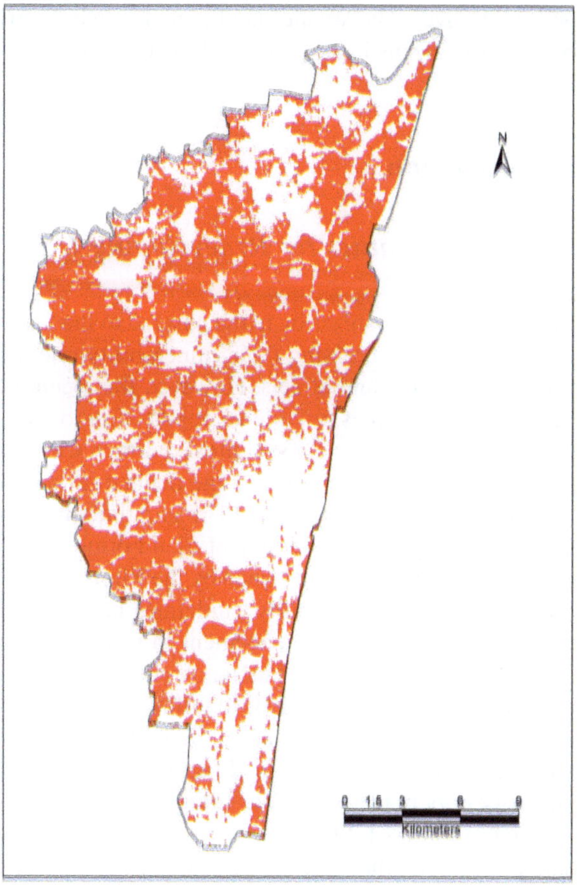

Fig. 7 Heat-island effect of areas that show higher than 40 °C in May 2020

adaptation. Much of the available open spaces are distributed in the north and south region of Chennai, which are fast urbanising areas. Bearing in mind that according to the Chennai Metropolitan Development Authority, the agricultural lots in the study area are supposed 1,229 ha. (CMDA, 2008), the actual availability of open areas is approximately 50 sq. km. These open areas are further laid in low elevation of less than two meters above mean sea level, there by presenting an opportunity to preserve these for flood mitigation purposes.

This although a very limited land available given the rapid urbanisation rate that Chennai city is facing, has the potential to be promoted as UGS for climate adaptation. The current availability of open areas is around 14% of the city area but still holds high potential to develop these as designated urban green spaces for climate adaptation like urban flood and urban heat island effect. While large open areas offer suitable spaces for creating flood mitigation efforts such as the Pallikaranai Marshlands or open waterbodies, the smaller open areas provide scope for development as urban parks with tree cover. The smaller land parcels once developed will provide recreational benefits to citizens in addition to climate adaptation.

5 Discussion

5.1 Opportunities for Redeveloping UGS

The unplanned urbanization in the past led to the loss of open and UGS in most cities in India thereby increasing their vulnerability to climate change impacts like urban floods and heat island effects. However, the recent policy developments and urban development schemes provide an opportunity to redevelop open and UGS in Indian cities under a climate change context. Further, Chennai shows some innovative ways in which it has managed to leverage finance for developing open and green spaces.

The Ministry of Housing and Urban Affairs in 2014 published new guidelines on Urban and Regional Development Plan Formulation and Implementation (URDFI) guidelines. This document issued by the Ministry of Urban Development has a separate chapter on sustainability guidelines covering climate change, environmental protection and disaster management in urban development and planning (MoUD, 2014, Chap. 6). The URDFI recommends land use and ecosystem-based approaches to resilience planning for ULBs to follow, emphasizing the increase of open spaces in urban areas to reduce heat island effects. Urban green space planning should consider the various social and ecological factors including land use, ecology, soil, elevation, etc. It is important to regulate future urban growth that is compatible with land-use and considers future climatic risks. In the same year, the Government of India also published Urban Greening Guidelines for city planners to undertake strategies for supporting green infrastructure (GoI, 2014). In these guidelines, the Ministry of Urban Development defines the hierarchy of open spaces and promotes action to mainstream green solutions into urban planning. This period represents a turning point and a shift towards more resilient planning models, however, planning tools cannot adapt to these guidelines fast enough that cities need.

Coincident in time, the Smart City Mission launched in 2015 also provides scope and funds for redevelopment of urban green and recreational spaces (AMRUT, 2015). AMRUT scheme also provides for the formulation of Geographical Information System (GIS) based master plans for 500 cities to aid the formulation of a master plan for effective land use management and utilisation, spatial growth management, project planning and urban management, all of which could contribute to integrating urban green spaces mapped through GIS and remote sensing techniques into urban plans. Chennai has benefitted with funds to improve basic infrastructure services (water supply, sewerage, stormwater drains) and redevelop green spaces and small parks inside the city. Among other strategies for the protection of urban green spaces, it stands out the Coastal Regulation Zone (CRZ), defines Ecologically Sensitive and Protected areas and takes into consideration climate change. This legal framework developed by the Master Plan also provides opportunities for city planners to mark coastal zones that are to be protected from urban development (GoI, 2019).

The Chennai Resilience Strategy (2019) document also echoes on increasing green cover through plantation drives across the city (CMC, 2019). This strategy represents a milestone in defining intervention actions on green infrastructure from the perspective of resilience to climate change. With support from 100 Resilient Cities—The Rockefeller Foundation, the document identifies over 52 water bodies and associated UGS including rivers, canals, marshlands and lakes for restoration and rejuvenation. Among the measures proposed by the Strategy, the development of legislative changes related to urban planning stands out, for example, compensating through Transferable Development Rights for private urban developments that are environmentally sensitive, or assigning 10% open space reserve areas for plots less than 10,000 square meters, today these plots do not have the obligation or can financially compensate this requirement.

Chennai Resilient Strategy promotes actions to mitigate climate disasters such as the recovery of Ennore Creek in the north of the city, a highly sensitive area that requires flood prevention and dredging work during the pre-monsoon period. Other measures such as the mapping of blue infrastructure for resilient planning, the massive planting of local tree species, urban horticulture, etc., are actions that mark a new model of city management where community participation is now essential (CMC, 2019, p. 113). The open areas available in the city and detected in the GIS analysis carried out, can be effectively incorporated into these strategies.

In September 2021, the Chennai Municipal Corporation launched the Singara Chennai Project 2.0. One of the project objectives is to redevelop urban green and recreational spaces and has identified several open spaces for redevelopment into parks and recreational areas. Between them highlight 42 sponge park projects, they have been proposed with artificial ponds and rain gardens (TUFIDCO, 2022, p. 3). These projects will help to combat inundation during the rainy season and contribute to groundwater by directing rainwater runoffs into the ground. There are similar success stories on the redevelopment of urban open spaces in Chennai, focusing on climate change adaptation actions, such as the redevelopment of Buckingham Canal and Pallikaranai Marshland (García & Govindarajulu, 2023) and the restoration of Adyar River. One of the first restoration actions promoted by the city in coordination with the German Society for International Cooperation has been the Buckingham Canal redevelopment project. In collaboration with the residents and different teams of urban designers, a proposal for the recovery of a sector of the canal has been made from the perspective of climate resilience (GIZ, 2019). The Adyar Creek restoration was taken up during 2010, however, improvement actions are currently being developed with river-mouth opening, shore plantation, resettlement and rehabilitation of affected families, and setting up of parks. Pallikaranai Marshland was improved recently to enhance the water holding capacity. The Chennai Rivers Restoration Trust was entrusted with Adyar Creek restoration and Cooum River, also, Tamil Nadu Urban Infrastructure Financial Services Ltd. (TNUIFSL), a public–private partnership company, participated in leveraging and financing urban infrastructure projects including for stormwater drains. This helped improve the flood capacity of the river as well and the creation of Adyar Ecopark and Estuary in 145 ha of land is a major recreational facility for residents of Adyar. Institutions such as the Chennai River Restoration Trust, and Pallikaranai Marshland Trust that incorporate government officials, civil society organisations and researchers have helped planning restoration activities in the city. This also brings in immense local community participation in city development (García & Govindarajulu, 2023).

Undoubtedly, these processes of redevelopment of open areas require specific financing formulas that guarantee, on the one hand, the transformation of these possible surfaces into green spaces, and on the other hand, the long-term maintenance of this new network of urban green spaces.

5.2 Barriers to Redeveloping Open and UGS

The key lessons in developing open and UGS in Chennai and other Indian cities is that first, the future urban development master plans must recognise the importance of open spaces and UGS for climate adaptation against urban floods and urban heat island effect. Second, current policies and schemes on urban development such as the AMRUT and Smart City Mission along with other financing options discussed here must best be used to develop UGS, and third, to keep UGS as an important environmental sustainability indicator in the urban development and monitor its development, these variables must be incorporated in planning processes. Focusing all these initiatives through adequate green infrastructure planning must be supported by the information that new open-access technologies can offer to cities in the Global South.

The detection of open spaces through cartographic analysis shows the opportunity it offers to the city of Chennai to structure new working methods with the aim of expanding the area dedicated to UGS for a resilient response to the challenges of climate change. However, there are several barriers in realising this potential in Chennai and other Indian cities. First is the limited skills of urban planners in using geospatial data in urban planning, a problem identified in the AMRUT mission too (AMRUT, 2015) and specific capacity-building programmes were made for urban planners in using geospatial data.

The Second is the urban governance. Many ULBs in India are almost non-functional and some of the urban development and administration works are carried by the State agencies than the elected local bodies (Ahluwalia, 2019) and even in cases where there is an elected ULB, their capacities to develop and implement a climate resilient master plan is limited due to lack of skilled climate experts and planners and with technical knowledge in the ULBs (Bhardwaj and Khosla, 2021). In the case of Chennai, this is very evident. The lack of focus on protecting urban green spaces and open areas from the first master plan led to unplanned urbanisation as identified by an audit report by the Government of India (CAG, 2017). The CMDA had allowed 439 land use conversions for urban development reducing agriculture zone, open spaces and recreation areas, which was one of the major reasons for heavy flood impacts in 2015.

The Second Master Plan originally prepared by Chennai Metropolitan Development Authority in 1995, was finally approved by the State government only in 2008. The second master plan was to guide the development of the greater Chennai region until 2026. The second master plan was to focus on developing a network of open spaces to provide green environment to be used as flood moderators during critical months of the year (CMDA, 2008). The CMDA recommends for development of greenery and tree planting in public and private spaces, protection of trees and enhancement of biodiversity through specific action plans including community-based environment improvement projects.

Third is the financial limitation of the cities in redeveloping open and UGS. A scientific analysis related to the financial health of Chennai City Corporation found the city had limited revenue to invest in urban infrastructure projects (Subramanian & Jana, 2019). In the absence of such finance, the municipal corporation should explore alternate financing options like Public Private Partnership (PPP), private sector engagement and Corporate Social Responsibility (CSR). For instance, in Fiscal Year 2020–2021, Chennai received 3.5 million US dollars in CSR funds for environmental projects (CSR Journal, 2022). The restoration of a mid-sized waterbody, the Annaikeni Pond in Sholinganallur represents an example of new opportunities for public–private collaboration in order to get to cater to the needs of around 8,000 to 10,000 individuals directly and indirectly within the vicinity. Corporate Social Responsibility (CSR) funding for the development of parks and gardens in cities can be a viable option of securing sustainable and liveable city culture.

6 Conclusion

Urban planning and development in Chennai prior to 2014, shows disregard for urban green spaces, which has resulted in the loss of several UGS thereby increasing the city's climate risks to flood and urban heat island effect. Despite its rapid urbanisation, the current land use of Chennai shows opportunities for redeveloping urban green spaces for environmental sustainability. By protecting and redeveloping these open spaces and blue-green infrastructure, in the low-lying areas flood mitigation and redevelopment of blue-green infrastructure can be made for climate adaptation against urban floods and urban heat island effects. Smaller available open spaces across the cities provide opportunities for developing socially accessible urban green spaces like parks and other recreational water sports activities as well. Our method on identifying suitable land parcels using remote sensing techniques for redeveloping open and UGS for climate adaptation and environmental sustainability can be replicated in other cities. As demonstrated in this study, open-source data like the Sentinel image in Google Earth Engine can be used to identify potential land parcels suitable for development of open and UGS for climate resilience using the classification approach described in this study. We find that the analysis gives a good classification accuracy on the available open spaces, although ground truthing will be essential. For this purpose, capacity building of urban planners on both using geospatial data and on understanding climate risks is essential along with strengthening urban governance institutions and urban finances. This study presents a useful case for urban planners in India and cities of the global south on the need and method for redeveloping open and UGS for climate change adaptation and environmental sustainability.

References

Ahluwalia, I. J. (2019). Urban governance in India. *Journal of Urban Affairs, 41*(1), 83–102.

AMRUT. (2015). *Atal mission for rejuvenation and urban transformation. Mission statement & guidelines* (76 pp.). Ministry of Urban Development, Government of India. Retrieved April 14, 2023, from http://amrut.gov.in/content/innerpage/guidelines.php.

Amirtham, L. R. (2016). Urbanization and its impact on urban heat Island intensity in Chennai Metropolitan Area, India. *Indian Journal of Science and Technology, 9*(5), 1–8. https://doi.org/10.17485/ijst/2016/v9i5/87201.

Andimuthu, R., Kandasamy, P., Mudgal, B. V., Jeganathan, A., Balu, A., & Sankar, G. (2019). Performance of urban storm drainage network under changing climate scenarios: Flood mitigation in Indian coastal city. *Scientific Reports, 9*(1), 7783. https://doi.org/10.1038/s41598-019-43859-3.

Arabindoo, P. (2016). Unprecedented natures? An anatomy of the Chennai floods. *City, 20*(6), 800–821. https://doi.org/10.1080/13604813.2016.1239410.

Avasiha, V.K., & Garg, A. (2016). Urban infrastructure and governance mission under JNNURM: Have the reforms delivered? *Economic and Political Weekly, 51*(2): 41–43, 49–57. https://www.jstor.org/stable/44003080.

Baruah, N.G., Henderson, J.V., & Peng, C. (2017). Colonial legacies: Shaping African cities. In *SERC Discussion Papers (SERCDP226).* London, UK: Spatial Economics Research Centre, London School of Economics and Political Science. http://eprints.lse.ac.uk/86574/.

Bhardwaj, A., & Khosla, R. (2021). Superimposition: How Indian city bureaucracies are responding to climate change. *Environment and Planning E: Nature and Space, 4*(3), 1139–1170.

Bremner, L. (2020). Planning the 2015 Chennai floods. *Environment and Planning e: Nature and Space, 3*(3), 732–760. https://doi.org/10.1177/2514848619880130.

CAG. (2017). *Report No.4 of 2017—Performance audit of flood management and response in Chennai and its suburban areas Government of Tamil Nadu.* Comptroller and Auditor General of India. Retrieved April 14, 2023, from https://cag.gov.in/ag1/tamil-nadu/en/audit-report/details/45722.

Caldeira, T. P. (2017). Peripheral urbanization: Autoconstruction, transversal logics, and politics in cities of the global south. *Environment and Planning d: Society and Space, 35*(1), 3–20. https://doi.org/10.1177/0263775816658479.

Census. (2011). *Census of India 2011. Provisional population totals. Urban agglomerations and cities.* Retrieved from: http://censusindia.gov.in/.

Chanu, C. S., Elango, L., & Shankar, G. R. (2021). A geospatial approach for assessing the relation between changing land use/land cover and environmental parameters including land surface temperature of Chennai metropolitan city, India. *Arabian Journal of Geosciences, 14*, 132. https://doi.org/10.1007/s12517-020-06409-0.

Cheng, Y. D., Farmer, J. R., Dickinson, S. L., Robeson, S. M., Fischer, B. C., & Reynolds, H. L. (2021). Climate change impacts and urban green space adaptation efforts: Evidence from US municipal parks and recreation departments. *Urban Climate, 39*, 100962. https://doi.org/10.1016/j.uclim.2021.100962.

Chondrogianni, D. V., & Stephanedes, Y. J. (2022). Evaluation of urban planning methods toward bioclimatic and resilient urban spaces. *Environment and Planning b: Urban Analytics and City Science, 49*(5), 1354–1370. https://doi.org/10.1177/23998083211063220.

CMC (2019). *Resilient Chennai strategy* (p. 134). Chennai Municipal Corporation—100 Resilience Cities Initiative. Retrieved April 14, 2023, from https://resilientchennai.com/wp-content/uploads/2019/07/Resilience-Strategy_20190703.pdf.

CMDA. (2008). *Chennai revised master plan 2026*. Chennai Metropolitan Development Authority. Retrieved April 14, 2023, from http://www.cmdachennai.gov.in/smp_main.html.

CSR Journal. (2022, December 14). Top CSR projects in Chennai. *CSR Journal*. Retrieved April 14, 2023, from https://thecsrjournal.in/top-csr-chennai-madras-tamil-nadu-corporate-social-responsibility/.

Cvejić, R., Železnikar, Š, Nastran, M., Rehberger, V., & Pintar, M. (2015). Urban agriculture as a tool for facilitated urban greening of sites in transition: A case study. *Urbani Izziv, 26*, S84–S97. https://doi.org/10.5379/urbani-izziv-en-2015-26-supplement-006.

Drusch, M., Del Bello, U., Carlier, S., Colin, O., Fernandez, V., Gascon, F., Hoersch, B., Isola, C., Laberinti, P., Martimort, P., Meygret, A., Spoto, F., Sy, O., Marchese, F., & Bargellini, P. (2012). Sentinel-2: ESA's optical high-resolution mission for GMES operational services. *Remote Sensing of Environment, 120*, 25–36. https://doi.org/10.1016/j.rse.2011.11.026.

Ermida, S. L., Vasco, P., Göttsche, F., & Trigo, I. (2020). Google earth engine open-source code for land surface temperature estimation from the landsat series. *Remote Sensing, 12*(9), 1471. https://doi.org/10.3390/rs12091471.

García, F., & Govindarajulu, D. (2023). Integrating blue-green infrastructure in urban planning for climate adaptation: Lessons from Chennai and Kochi, India. *Land Use Policy, 124*, 106455. https://doi.org/10.1016/j.landusepol.2022.106455.

García, F., Solecki, W. D., & Ribalaygua, C. (2018). Climate change adaptation in Europe and the United States: A comparative approach to urban green spaces in Bilbao and New York City. *Land Use Policy, 79*, 164–173. https://doi.org/10.1016/j.landusepol.2018.08.010.

Gascon, F., Bouzinac, C., Thépaut, O., Jung, M., Francesconi, B., Louis, J., Lonjou, V., Lafrance, B., Massera, S., Gaudel-Vacaresse, A., Languille, F., Alhammoud, B., Viallefont, F., Pflug, B., Bieniarz, J., Clerc, S., Pessiot, L., Trémas, T., Cadau, E., … Fernandez, V. (2017). Copernicus sentinel-2A calibration and products validation status. *Remote Sensing, 9*, 584. https://doi.org/10.3390/rs9060584.

GIZ. (2019). *Eyes on the canal—Reimagining Chennai's Buckingham canal generating collective ownership through participatory planning*. Deutsche Gesellschaft für Internationale Zusammenarbeit (GIZ) GmbH.

GoI. (2014). Urban greening guidelines, Ministry of Urban Development, Government of India. Retrieved April 14, 2023, from https://mohua.gov.in/upload/uploadfiles/files/G%20G%202014(2).pdf.

GoI. (2019). *Coastal regulation zone notification* (pp. 30–56). Published in the Gazetteer of the Government of India. Retrieved April 14, 2023, from https://www.environment.tn.gov.in/crznotification.

GoI. (2021). *Reforms in urban planning capacity in India*. NITI Aayog. Retrieved April 14, 2023, from https://www.niti.gov.in/sites/default/files/2021-09/UrbanPlanningCapacity-in-India-16092021.pdf.

Govindarajulu, D. (2014). Urban green space planning for climate adaptation in Indian cities. *Urban Climate, 10*, 35–41. https://doi.org/10.1016/j.uclim.2014.09.006.

Govindarajulu, D. (2020). Strengthening institutional and financial mechanisms for building urban resilience in India. *International Journal of Disaster Risk Reduction, 47*, 101549. https://doi.org/10.1016/j.ijdrr.2020.101549.

Gupta, A.K., & Nair, S.S. (2011). Urban floods in Bangalore and Chennai: Risk management challenges and lessons for sustainable urban ecology. *Current Science, 100*(11), 1638–1645. https://www.jstor.org/stable/24077767.

Haaland, C., & van den Bosch, C. K. (2015). Challenges and strategies for urban green-space planning in cities undergoing densification: A review. *Urban Forestry & Urban Greening, 14*(4), 760–771. https://doi.org/10.1016/j.ufug.2015.07.009.

Hu, B., Xu, Y., Huang, X., Cheng, Q., Ding, Q., Bai, L., & Li, Y. (2021). Improving urban land cover classification with combined use of sentinel-2 and sentinel-1 imagery. *ISPRS International Journal of Geo-Information, 10*, 533. https://doi.org/10.3390/ijgi10080533.

IIHS. (2018). *Study for the XV finance commission on urban infrastructure and resilience* (110 pp.). Draft Final Report. Indian Institute for Human Settlements. https://fincomindia.nic.in/writereaddata/html_en_files/fincom15/StudyReports/Urban%20infrastructure%20development%20and%20resilience%20building%20by%20ULBs.pdf.

Ilayaraja, K., Reza, W., Kumar, V., Paul, S., & Chowdhary, R. (2017). Estimation of land surface temperature of Chennai Metropolitan Area using landsat images. *International Journal of Civil Engineering and Technology, 8*(8), 1450–1456. http://iaeme.com/Home/issue/IJCIET?Volume=8&Issue=8.

Jain, A. O., Thaker, T., Chaurasia, A., Patel, P., & Singh, A. K. (2018). Vertical accuracy evaluation of SRTM-GL1, GDEM-V2, AW3D30 and CartoDEM-V3. 1 of 30-m resolution with dual frequency GNSS for lower Tapi Basin India. *Geocarto International, 33*(11), 1237–1256. https://doi.org/10.1080/10106049.2017.1343392.

Joerin, J., Shaw, R., Takeuchi, Y., & Krishnamurthy, R. (2014). The adoption of a climate disaster resilience index in Chennai, India. *Disasters, 38*(3), 540–561. https://doi.org/10.1111/disa.12058.

Kaaviya, R., & Devadas, V. (2021). Water resilience mapping of Chennai, India using analytical hierarchy process. *Ecological Processes, 10*, 71. https://doi.org/10.1186/s13717-021-00341-1.

Kotharkar, R., Ramesh, A., & Bagade, A. (2018). Urban heat island studies in South Asia: A critical review. *Urban Climate, 24*, 1011–1026. https://doi.org/10.1016/j.uclim.2017.12.006.

Kumar, L., & Mutanga, O. (2018). Google earth engine applications since inception: Usage, trends, and potential. *Remote Sensing, 10*(10), 1509. https://doi.org/10.3390/rs10101509.

Kumar, P., Geneletti, D., & Nagendra, H. (2016). Spatial assessment of climate change vulnerability at city scale: A study in Bangalore, India. *Land Use Policy, 58*, 514–532. https://doi.org/10.1016/j.landusepol.2016.08.018.

Kumar, P., Rai, A., Upadhyaya, A., & Chakraborty, A. (2022). Analysis of heat stress and heat wave in the four metropolitan cities of India in recent period. *Science of the Total Environment, 818*, 151788. https://doi.org/10.1016/j.scitotenv.2021.151788.

Kundu, D., & Pandey, A. (2020). Exploring rural-urban inequality in india in the post-economic reform period. *Environment and Urbanization ASIA, 11*(1), 102–122. https://doi.org/10.1177/0975425320906278.

Lovell, S. T., & Taylor, J. R. (2013). Supplying urban ecosystem services through multifunctional green infrastructure in the United States. *Landscape Ecology, 28*(8), 1447–1463. https://doi.org/10.1007/s10980-013-9912-y.

Mathan, M., & Krishnaveni, M. (2020). Monitoring spatio-temporal dynamics of urban and peri-urban land transitions using ensemble of remote sensing spectral indices-a case study of Chennai Metropolitan Area, India. *Environmental Monitoring and Assessment, 192*, 1–11. https://doi.org/10.1007/S10661-019-7986-Y.

Mell, I. C. (2009). Can green infrastructure promote urban sustainability? *Proceedings of the Institute of Civil Engineers: Engineering Sustainability, 162*(1), 23–34. https://doi.org/10.1680/ensu.2009.162.1.23.

Mohammad, P., Goswami, A., Chauhan, S., & Nayak, S. (2022). Machine learning algorithm based prediction of land use land cover and land surface temperature changes to characterize the surface urban heat island phenomena over Ahmedabad city, India. *Urban Climate, 42*, 101116. https://doi.org/10.1016/j.uclim.2022.101116.

MoUD (2015). Urban and Regional Development Plans Formulation and Implementation (URDPFI) guidelines, 2014. Government of India. http://tcpo.gov.in/sites/default/files/TCPO/schemes/URDPFI.pdf.

Myers, G. (2021). Urbanization in the global south. In C. Shackleton, S. Cilliers, E. Davoren, & M. du Toit (Eds.), *Urban Ecology in the Global South* (pp. 27–49). Springer.

Olofsson, P., Foody, G. M., Herold, M., Stehman, S. V., Woodcock, C. E., & Wulder, M. A. (2014). Good practices for estimating area and assessing accuracy of land change. *Remote Sensing of Environment, 148*, 42–57. https://doi.org/10.1016/j.rse.2014.02.015.

Onodugo, A. V., & Ezeadichie, H. N. (2020). Future planning of global south cities with inclusive informal economic growth in perspective. In A. Almusaed, A. Almssad, & L. Truong-Hong (Eds.), Sustainability in Urban Planning and Design. IntechOpen. https://doi.org/10.5772/intechopen.89145.

Pauleit, S., Vasquéz, A., Maruthaveeran, S., Liu, L., & Cilliers, S. (2021). Urban green infrastructure in the Global South. In: *Urban ecology in the Global South* (pp. 107–143). Springer.

Ramaiah, M., & Avtar, R. (2019). Urban green spaces and their need in cities of rapidly urbanizing India: A review. *Urban Science, 3*(3), 94. https://doi.org/10.3390/urbansci3030094.

Ranger, N., Hallegatte, S., Bhattacharya, S., Bachu, M., Priya, S., Dhore, K., Rafique, F., Mathur, P., Naville, N., Henriet, F., Herweijer, C., Pohit, S., & Corfee-Morlot, J. (2011). An assessment of the potential impact of climate change on flood risk in Mumbai. *Climatic Change, 104*(1), 139–167. https://doi.org/10.1007/s10584-010-9979-2.

Rigolon, A., Browning, M., Lee, K., & Shin, S. (2018). Access to urban green space in cities of the global south: A systematic literature review. *Urban Science, 2*(3), 67. https://doi.org/10.3390/urbansci2030067.

Ruiz Hernandez, I. E., & Shi, W. (2018). A random forests classification method for urban land-use mapping integrating spatial metrics and texture analysis. *International Journal of Remote Sensing, 39*, 1175–1198. https://doi.org/10.1080/01431161.2017.1395968.

Shahtahmassebi, A., Li, C., Fan, Y., Wu, Y., Lin, Y., Gan, M., Wang, K., Malik, A., & Blackburn, G. (2021). Remote sensing of urban green spaces: A review. *Urban Forestry & Urban Greening, 57*, 126946. https://doi.org/10.1016/j.ufug.2020.126946.

Srinivasan, V., Seto, K. C., Emerson, R., & Gorelick, S. M. (2013). The impact of urbanization on water vulnerability: A coupled human–environment system approach for Chennai, India. *Global Environmental Change, 23*(1), 229–239. https://doi.org/10.1016/j.gloenvcha.2012.10.002.

Subramanian, D., & Jana, A. (2019). Evaluating the cultural ecosystem services of India: Comparison of budget allocations to improve the use value of recreational open spaces. *Ecosystem Services, 38*, 100960. https://doi.org/10.1016/j.ecoser.2019.100960.

TUFIDCO. (2022). *Singara Chennai 2.0. Implementation of projects of Greater Chennai Corporation*. Tamil Nadu Urban Finance and Infrastructure Development Corporation. Retrieved April 14, 2023, from https://www.tufidcoschemes.co.in/media/pdf/G.O_D_No.422-Dt.17.11.22-Rs.131.22_cr-Singa_Che_2.0-C.No.17714-2022.pdf.

Tzoulas, K., Korpela, K., Venn, S., Yli-Pelkonen, V., Kazmierczak, A., Niemelä, J., & James, P. (2007). Promoting ecosystem and human health in urban areas using green infrastructure: A literature review. *Landscape and Urban Planning, 81*, 167–178. https://doi.org/10.1016/j.landurbplan.2007.02.001.

UN. (2014). *World urbanization prospects: The 2014 revision, highlights (ST/ESA/SER.A/352)* (27 pp.). United Nations, Department of Economic and Social Affairs, Population Division. https://esa.un.org/unpd/wup/Publications/Files/WUP2014-Highlights.pdf.

Venn, S., & Niemelä, J. (2004). Ecology in a multidisciplinary study of urban green space: The URGE project. *Boreal Environment Research, 9*, 479–489. http://www.borenv.net/BER/archive/pdfs/ber9/ber9-479.pdf.

Vigneshwaran, S., & Kumar, S. V. (2018). Extraction of built-up area using high resolution sentinel-2a and google satellite imagery. *The International Archives of Photogrammetry, Remote Sensing and Spatial Information Sciences, XLII-4/W9*, 165–169. https://doi.org/10.5194/isprs-archives-XLII-4-W9-165-2018.

WHO (2017). *Urban green spaces: A brief for action* (24 pp.). UN City, Denmark: World Health Organization Regional Office for Europe. Retrieved April 14, 2023, from https://apps.who.int/iris/handle/10665/344116.

Spatial Accessibility Disparities and Users' Opinion Assessment of Khartoum State Public Green Spaces

Ahmed Alhuseen, Huda Omer, Eva Pauditšová, Hussein Sulieman, Mária Kozová, and Pavel Cudlin

Abstract Khartoum State, the capital of Sudan is situated in the Sahelian and Sudanese arid zone and it is expected to face excessive dry climate conditions due to climate change. In such an eventuality, accessible Urban Green Spaces (UGSs) represent a traditional adaptation measure for its growing population. This study aims to describe and explain the spatial disparities and potential accessibilities to public green space of Khartoum State and to determine users' opinions about them using Geographic Information Systems (GIS) and questionnaires. It assesses the potential accessibility to 44 urban parks, community gardens, and city forests in Khartoum State for both drivers and pedestrians in 5, 10, and 15 min. The results showed inequitable access opportunities to UGSs, mostly in favour of first-class residential areas. Within a 5-min walking interval, the UGSs service zone covered <1% of the population. Furthermore, users' opinion assessment revealed that 80% of the parks require entrance fees; however, 75% of the respondents show the willingness to bear extra travel costs and time to access well-equipped parks. The study concludes that potential accessibility to UGSs does not correspond to the international quality of life indices.

A. Alhuseen (✉) · P. Cudlin
Department of Ecosystem Functional Analysis of the Landscape, Global Change Research Institute, Academy of Sciences of the Czech Republic, Lipová 1789/9, 370 05 Ceské Budejovice, Czech Republic
e-mail: alhussen.a@czechglobe.cz

H. Omer
Ministry of Infrastructure and Transportation, P.O. Box: 818, Khartoum, Sudan

E. Pauditšová
Department of Landscape Ecology, Faculty of Natural Sciences, Comenius University in Bratislava, Ilkovicova 6, Mlynska Dolina, 842 15 Bratislava, Slovakia

H. Sulieman
Faculty of Agricultural and Environmental Sciences, University of Gadarif, Centre for Remote Sensing and GIS, P.O. Box 449, Al Gadarif, Sudan

M. Kozová
Faculty of Education, Catholic University in Ruzomberok, Hrabovska Street 1, 03401 Ruzomberok, Slovakia

P. K. Joshi et al. (eds.), *Blue-Green Infrastructure for Sustainable Urban Settlements*,
https://doi.org/10.1007/978-3-031-62293-9_14

Keywords International quality of life indices · Potential accessibility · Residential area · Urban green spaces · Urban parks

1 Introduction

Research on the adaptation of urban centres to the challenges of climate change is topical all around the globe. Urban Green Spaces (UGSs) are seen as a traditional adaptation measure for urban areas. Therefore, it is necessary to investigate the status quo of UGSs adequacy within the urban fabric and their potential accessibility to build and reinforce inhabitants' resilience to the foreseen climate change hazards (Reháčková & Pauditšová, 2004).

The different forms of private and public green spaces in cities offer positive environmental, health, and economic returns to the cities' inhabitants (Lee et al., 2015). Green spaces sequestrate carbon dioxide (CO_2), generate oxygen, dissipate urban heat islands, and stabilize the soil. In addition, they conserve biodiversity, supporting wildlife (sanctuary, food, living space). Furthermore, public green spaces provide opportunities for physical activities, and recovery from stress; they facilitate social contacts and provide recreational and aesthetic functions (Reháčková & Pauditšová, 2004; Lee et al., 2015). Markevych et al. (2017), have named three domains of potential pathways that link greenspaces to health, which emphasize three general functions of greenspace: reducing harm (e.g., reducing exposure to air pollution, noise, and heat), restoring capacities (e.g., attention restoration and physiological stress recovery) and building capacities (e.g., encouraging physical activity and facilitating social cohesion). Moreover, UGSs help to reduce energy consumption and bring natural solutions to urban heatwaves and flood risks; hence, adequate UGSs and equitably distributed within the urban fabric are considered a precondition for promoting the well-being of a city's inhabitants and for enhancing the city's adaptive capacity to the growing threats of climate change. Consequently, equal access to UGSs and gross per capita of green space areas are indices used to determine the performance of city services and quality of life. Public green spaces are one of the 14 categories of basic community needs that measure the performance of smart community infrastructures (The International Organization for Standardization, 37,151: 2015).

Various literature sources have shown the diverse nature of the term "urban green spaces". According to Cvejić et al. (2015), UGSs encompass all vegetation found in the urban environment, including blue spaces such as lakes or rivers and their adjacent greenery. Nevertheless, Burgess et al. (1988) pointed out that the public can perceive green spaces differently from trained planners. According to Baycan-Leven et al. (2002), urban green spaces are defined as "*open spaces in urban areas that are primarily covered by vegetation which can be public or private*" (in So, 2016).

Since the types of green spaces are diverse, it has been necessary to design an inventory to comprehend their functions, services, and connectivity to residential neighbourhoods. For this purpose, several classifications have been suggested. Bell et al. (2007) distinguished 15 UGSs types, while Swanwick et al. (2003) identified 25 UGSs types falling into 10 subgroups and four main groups (amenity green space, functional green space, semi-natural habitats, and linear green space).

Commonly Geographic Information Systems (GIS) are equipped with a service area analysis function under the network analysis tool. It can calculate the service area coverage of UGSs, and it can estimate the actual and potential accessibility patterns from different parts of the city and for different groups of population. Previous studies have used GIS and network analysis for this purpose. Van Herzele and Wiedemann (2003) applied a spatial GIS-based indicator system to cities in Belgium to plan accessible and attractive green spaces. Germann-Chiari and Seeland (2004) used GIS and regression analysis to assess the distribution and access of UGSs by social target groups in three cities in Switzerland. Balram and Dragićević (2005) used combined methods involving semi-structured interviews and a collaborative GIS workshop for accessibility assessments.

Accessibility is an important factor to consider when studying UGSs. Van Herzele and Wiedemann (2003) have argued that accessibility is a determinant of attraction to public green spaces. Accessibility can be defined as "*the ability of the environment to provide all people an equal opportunity to use in a direct, immediate, permanent and as autonomous way as possible*" (Figueiredo et al., 2016). Additionally, the term "accessibility" includes actual accessibility and potential accessibility; whereas the former emphasizes the actual use of green space, the latter highlights the availability of green space in an area (Yuan & Xu 2015). In this study, accessibility is defined as the walking and/or driving distance between the centroid of the public green space and the residential areas of Khartoum State. The term "drivers" refers to the persons who use either private cars or public transportation to reach green spaces in the study area. According to Bannaga (2018), the percentage of passengers using public transport in Khartoum State was 70% and about 15.5% use motorcycles, private vehicles, and public transport provided by private and government institutions and there were 171,700 vehicles registered as private cars in 2011. Another study by El Tayeb et al. (2015) showed the disparities in the use of modes of transport used to reach formal healthcare facilities, cross-tabulated by socioeconomic status and mode of living urban/rural. The study showed that in the low-income group, 16% walk or use public transport to reach a healthcare facility, whereas, in the high-income group, 53% uses the private car. According to El Tayeb et al. (2015), nearly a quarter of the injured from both urban (26%) and rural (27%) residents used private cars to reach the health facility. While taxi or public transport was used by 59% of people in the middle socio-economic group. The study presumes this may reflect the actual use of private and public transportation modes in Khartoum State in the case of our study.

According to Kimpton (2017), there are at least three ways of conceptualising equity of distribution of UGSs: (1) accessibility, usually operationalised as distance to greenspace or proportion of the population within a given distance; (2) provision, usually operationalised as greenspace cover within a given area; and (3) population pressure, or the potential crowding of a greenspace. Accordingly, this paper addresses aspects of accessibility and provision of public green spaces in Khartoum State using the Service Area Analysis (SAA). According to ESRI ArcGIS Resources, a service area in the network analysis approach is a region that encompasses all accessible streets that are within a specific parameter (Marti, 2013). Boone et al. (2009) and Wendel et al. (2012) have argued that SAA can be considered as a base measure to indicate the green space available to residents within a specific distance.

In this regard, the two most common methods are usually used; circular buffer around location under investigation at a given radius and road network buffer to define areas within which an individual can walk as a base for measuring the built environment. There is a consent that network buffers are a new area of research and has the potential to more accurately define areas of the built environment that influence walking versus circular buffer analysis a circular buffer that does not correctly reflect the walking path and overestimates accessibility (Oliver et al., 2007; Mears & Brindley, 2019).

Another important aspect to consider is the users' opinion about green public spaces. Studies on environmental comfort and climate in public green spaces presume the acquisition of users' opinions on UGSs as important as other aspects such as environmental conditions and the physical characterization of spaces (Monteiro et al., 2018).

Climatic projections for 2060 anticipate that Khartoum State will face a significant increase in temperature between 1.1 and 3.1 °C and a decrease in precipitation (Higher Council for Environment and Natural Resources, 2003). Hence, the provision of public green spaces by city authorities can be a traditional measure to help increase the adaptive capacity of the dwellers to such climate extremities. Boultona et al. (2018) have argued that despite the rapid urbanisation process in South America and Africa there remains a significant research gap in UGSs studies in compact and densely populated cities in these two continents. Also, Boultona et al. (2018) found that almost two-thirds of the literature they systematically reviewed have identified and described issues related to accessibility or distance to greenspace areas and distributional equity. At Khartoum State level, green and open public spaces accessibility-related research is scarce. Only a few relevant studies were found while writing this paper, for instance, Awad (2018) has evaluated neighbourhoods developed open spaces in Khartoum State, the research promoted measurable development indicators such as availability, accessibility, safety, and management and then applied these indicators to analyze developed open spaces. Bashari et al. (2015) investigated the effects and interactions of different needs, perceptions, and experiences of gender-specific needs in planning public open spaces. This study falls into the category of applied research, as it is a practical application of scientific methods to investigate the issues of accessibility of green spaces and users' opinions about their quality as recreational destinations. As the authors of this study are attempting to provide a comprehensive picture of Khartoum's green spaces, this study addresses accessibility in conjunction with knowledge of the individual qualities of a place. Therefore, this study has tried to answer the question: How is Khartoum equipped with public green spaces and to what extent do the existing green spaces meet the expectations and needs of their users?

This study has two main objectives. First, it assesses the users' opinions and potential accessibility to 44 public parks, community gardens, and city forests in

Khartoum State based on a public survey and the integration of the GIS analysis. Second, it describes and explains the spatial disparities and potential accessibility of the available public parks between different residential areas and administrative units that compose Khartoum State.

2 Materials and Methods

2.1 Study Area

The study area for this research is defined by the territory of Khartoum State, the capital of Sudan (Fig. 1). It is situated at the confluence of the White and the Blue Niles, and it is located between latitudes 15°20' and 16°40' N and longitudes 31°40' and 34°20' E at an altitude of 405.6 m above sea level. Khartoum State has become increasingly urbanized during the past three decades, and its three main cities, Khartoum, Khartoum North and Omdurman, cover an area of 22,736 km^2. The land is generally flat with a gradual slope towards the north and the Niles (Agraa et al., 1985). Khartoum is typified by a desert and semi-desert climate. In the summer temperatures rise to over 40 °C while in winter they fall to 11 °C or less; winter and autumn sand-bearing winds expose the city to suspended sand for almost six months of the year (Higher Council for Environment, Urban and Rural Promotion, 2015). During the rainy season, from July to September, Khartoum receives between 110 and 200 mm of rainfall on average, while the remainder of the year is fair (Higher Council for Environment and Natural Resources, 2016). Sand storms occur regularly and river fluctuations threaten riverbank erosion and flooding. Rapid urban growth combined with rising temperatures, rainfall variability, and river fluctuations represent very serious problems for the urban sector of Sudan.

Sudan has a federal governance system that is a matter of continuous change. In 2011, following the separation of Southern Sudan, Sudan consisted of 15 federal states (*Wilayat*, singular *Wallayha*) (Fig. 1). In 2012 two new states were added—Central and East Darfur, and in 2013, West Kordufan State emerged. Khartoum State constitutes 1% of Sudan's area but hosts 21% of its population. According to the Central Bureau of Statistics (CBS, 2013), Khartoum State has 6,534,796 inhabitants with a general population density of 490 persons per km^2 at the outskirts and 1300 persons per km^2 in the inner city (Osman, 2009). This population is distributed throughout 7 localities/provinces and 36 administrative units (Districts) (Table 1). It has 864,400 households with an average household size of 6.3 persons (Higher Council for Strategic Planning, 2013). The age structure of the Khartoum State population is as follows: <15 years of age, 37%; 15–40 years of age, 45%, and >50 years, 18% (Central Bureau of Statistics, 2013). The conurbation area stretches over 5000 km^2 including rural areas. Urban plans follow the heritage of the colonial era according to which residential areas are divided into four classes defined by the level of income, plot sizes, services and construction standards: (1) The first class

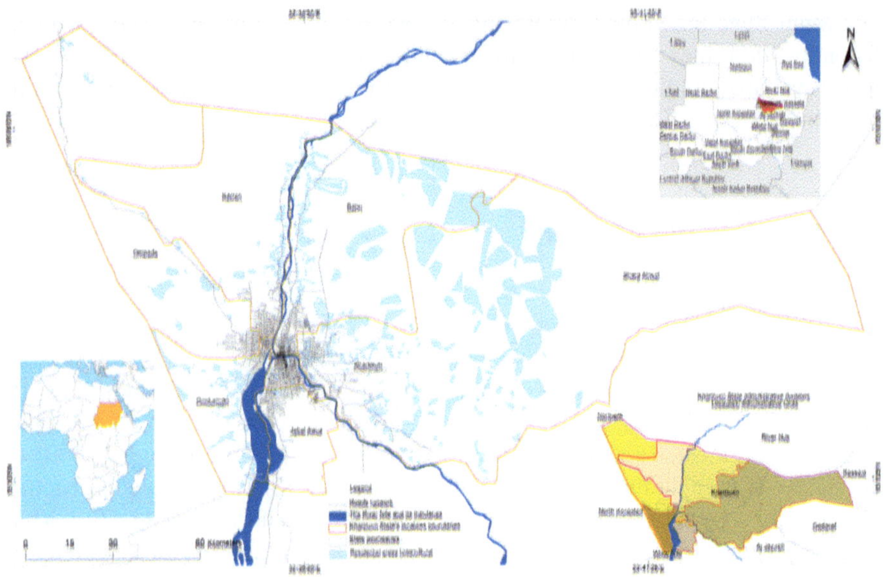

Fig. 1 Khartoum State (Wallayhat Alkhartoum). *Source* CBS (2008), OCHA (2020)

areas with land parcels ranging from 500 to 800 m^2 and up to 1,200 m^2; (2) The second class areas with plot areas ranging from 400 to 600 m^2; (3) The third class areas with plots ranging from 300 to 400 m^2 and (4) The fourth class areas with the plots of 200 m^2. Additionally, Babiker and Ibrahim (2016) listed five land use and land cover categories that shape Khartoum's surface. Water bodies represent 4.6%, the built-up area 59%, agricultural lands 15%, the rocky area 1.6% and bare land is 18.8%.

Table 1 Administrative units of the study area—Khartoum State

City	Locality/Province	No. of administrative units	No. of households
Omdurman	Karrari	3	115,907
	Ombada	4	182,586
	Omdurman	7	85,425
Khartoum North	Bahri	4	101,441
	Sharg Alneel	8	144,202
Khartoum	Khartoum	6	106,216
	Jabal Awlia	4	158,411

2.1.1 Road Network

The road network of Khartoum State stretches over 3684 km; it encompasses local roads, collector and distributor roads, motorways, and highways. Khartoum State speed limits are set at 40–60 km/h for local and collector roads, while for inter-city highways they are set at 80 km/h. The eight operating bridges that connect the three cities of Khartoum State form the actively interacting Greater Khartoum.

2.1.2 Green and Open Spaces in Khartoum State

In Khartoum State, green and non-green public spaces have been classified into four distinct classes, based on the numbers of served population and the radius of the service zone. The open public greened and non-green spaces of Khartoum State are as follows: (1) Neighbourhood courtyards serving 500–5000 people and covering a service zone of 300–500 m; (2) Residential gardens serving 10,000–20,000 people within 2–5 km; (3) Communal gardens serving 20,000–50,000 people with no defined service zone; (4) City Park serving 100,000–150,000 people with a service radius larger than 5 km (adapted from Nasifa Consultants, 2012).

Khartoum State possesses 300 km of waterfront, 44 residential and communal gardens, and 1000 non-greened open spaces (adapted from Nasifa Consultants, 2012). Riverine forests and private orchards on the banks of the River Nile and its tributaries are often used by the public to compensate for the lack of properly planned public green and recreational areas in Khartoum State. Formal statistics from the Forestry Administration show that Khartoum State has in total 224 forests, 190 of which are reserved forests. Two riverine city forests: Alsunut and Wad Alagaali forests which are popularly well-known as picnic attraction spots were included in the model of this study. Many governmental and private institutions and facilities, such as hospitals, sports facilities, schools, union clubs, etc., own green spaces varying in area and quality. Although they provide numerous ecosystem services, these green spaces are not accessible to the public for recreation, and therefore, they were not included in this study. Tree alleys and green corridors are also present in Khartoum State especially on the roads along the River Nile in the old parts of Khartoum and Omdurman cities. The most predominant tree species in Khartoum State streets are *Azadiracta indica*, *Khaya* and *Entandrophragma*, *Phoenix dactylifera* and most recently *Conocarpus laucifolius* in addition to different ornamental plants and bushes (Alhuseen, 2015).

2.2 Methods

Based on the definition established in the Introduction, a total number of 42 residential and communal gardens, parks, and two riverine city forests covering 859 hectares in total, were included in the GIS model of the study area. These modelled UGSs

Table 2 Base map-layers used in the GIS model and their sources

Shapefile	Type	Source of data	Date
Road network	Polyline	Ministry of infrastructure and transportation	2017
Residential areas	Polygon	Central bureau of statistics	2008
Residential classes	Polygon	Khartoum structural plan	2012
Parks and gardens	Point	Survey and google earth	2018
River Nile and its tributaries	Polygon	Khartoum structural plan	2012
Administrative boundaries	Polygon	Central bureau of statistics/ OCHA	2008/2020

elements are the well-known public UGSs accessible by the public for recreational purposes and for social and cultural gatherings and they are equipped with minimum facilities such as; benches, lights, green yards and trees, and access ways.

2.2.1 Data Collection and Software

Six GIS thematic vector shapefiles were obtained and used to construct the study area model. Table 2 illustrates the utilized shapefiles and their sources. In addition, secondary data relevant to the topic were reviewed and extracted from similar previous local and international studies, and from different governmental reports.

2.2.2 Data Processing

The six thematic map layers of the study area were stratified in the interface of ArcMap software using Universal Transverse Mercator (UTM) projection in zone 36. The main databases and analysis files were associated with the road network, public parks and gardens, residential areas, residential classes, and administrative boundaries. A new network dataset was created in ArcCatalog, and a topology procedure offered in the ArcMap was applied to the road network shapefile in order to rectify digitization and topology errors to guarantee an accurate network analysis. The network dataset consisted of junctions characterized by nodes, and streets and other routes represented by straight or curved polylines. Two columns were added to the database of the classified road network layer. The first column calculated the time needed to access the closest public green space based on the average human walking speed. This study considered 6.4 km h^{-1} as the average speed of a middle-aged person (Morris & Hardman, 1997). The second field computed the needed driving time between residential areas and green spaces. The speed of 60 km h^{-1} was assigned to local roads, collector and distributor roads, and 80 km h^{-1} was assigned to highways. The following formula was applied to both fields: road length/speed limit * 60. Time spent at traffic lights, the state of the road, and the elapsed time caused by frequent stops of public transportation or traffic jams were not included in this model. The 44 UGSs under study were represented with a centroid each.

To assess users' opinions on green spaces of Khartoum State, a simple random sample of households was selected using the following formula:

$$n = \frac{p(1-p)z^2}{e^2},$$

where p is the assumed prevalence of the event in the population under study. In this study, the value of 0.50 was used based on the results of the pilot survey that covered 20 households in Khartoum State to test the design of the questionnaire and the comprehension of the questions by the targeted respondents. Whereas, 10 households answered they visit UGSs frequently. z is the critical value obtained from a standard normal distribution. A confidence level of 95% and its corresponding z value 1.96 were used, and e is the maximum absolute error. The value for e was set to 0.50 as a maximum absolute error to give a larger sample size and then greater precision. The sample size given using the aforementioned formula was 385, taking into account non-response from some households, the sample size has been increased to 395.

The distributed questionnaires covered the seven localities and their subsequent thirty-six administrative units of Khartoum State. The selection of the simple random sample from the households in each administrative unit was done proportionally and then blocks and the first household in each block in all 36 administrative units were randomly selected. Keeping in consideration the absence of formal geographic names, geocodes, and house numbering information, the following neighbouring houses to the first household selected earlier were visited and interviewed. The questionnaire form consisted of questions which investigated the following: (1) demographic characteristics of the surveyed households (sex, age group, place of birth, education, and employment status), and (2) frequently- visited green spaces, reasons for choosing them, level of satisfaction and suggestions for improvement, and purpose of visiting UGSs. The questions aimed to investigate the usage of UGSs by the respondents were designed as open-ended questions to let respondents answer freely. The questionnaires were filled by a team of 10 graduate students from different academic backgrounds who were adequately trained for this purpose. From the 395 households questioned, only one household refused to respond.

2.2.3 Analysis

The buffer and network analysis embedded in GIS are two commonly used methods in green space accessibility studies. The network analysis approach is used in this study as residents are more likely to follow the actual road network to reach public parks or green spaces rather than use a straight-line buffer radius to the nearby public green space. Following the framework of Kimpton (2017), the analysis applied to this case study flips between area outcomes and population measures by applying the following procedures:

(1) Parameters to analyse service areas for both drivers and pedestrians' time were set using the three time intervals of 5, 10, and 15 min. The three interval values for real-time impedance were selected based on literature (Natural England, 2010; WHO, 2016; Le Texier et al., 2018).
(2) Facilities options—GIS offers three options to perform this step: overlapping, not overlapping, and merging by interval value. The merge by interval value option was selected in order to join the polygons of multiple facilities that have the same interval value. In addition, the ring option was ticked to exclude the area of smaller intervals.
(3) Three service area coverage polygons were generated based on the interval values determined earlier.

3 Results

The GIS model of the study area has generated a new layer representing the service zone of UGSs in Khartoum State. A geometry calculation revealed that the total service area of the public green space in Khartoum State accessible by drivers within 5–15 min corresponded to 616.70 km^2. This figure represents 12.3% of the conurbation area and 2.8% of the total area of Khartoum State. However, the UGSs service zone coverage accessible to pedestrians was only 115.5 km^2, i.e. 2.31% of the total conurbation area (Fig. 2a, b).

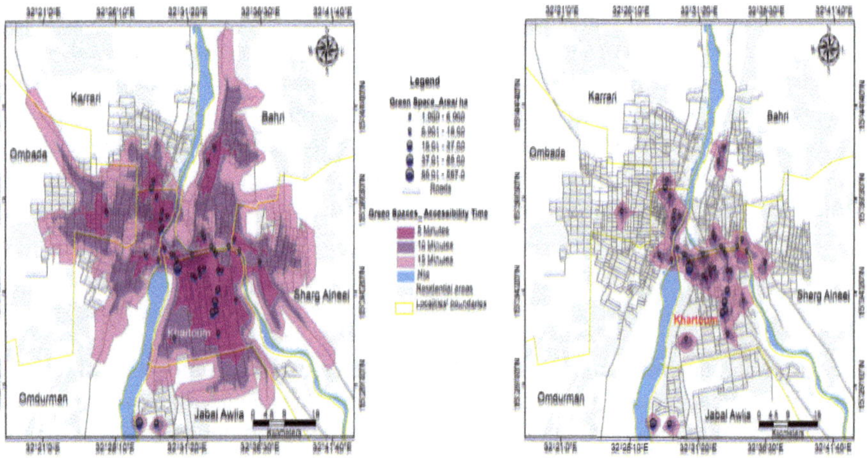

a UGSs accessible by driving

b UGSs accessible on foot

Fig. 2 **a** UGSs accessible by driving. **b** UGSs accessible on foot

Figure 3 shows the quota per each time break in (km^2) from the total service zone of public parks accessible by car/public transport and on foot. In addition, a clipping tool was applied on both stratified layers of our results and the layer of administrative boundaries to compare the allocation of accessible public green space across the different administrative units of Khartoum State.

Table 3 shows that four localities which host 96.5% of the households of Khartoum State have a service area coverage of UGSs of <1/km^2 each accessible within 5 min on foot. Table 4 reveals clearly the centralization of the urban parks within the Khartoum locality as the service area coverage of accessible UGSs decreases when the time break increases. Moreover, both tables show that the Khartoum locality, which contains many and the oldest first-class residential neighbourhoods, has the largest share of accessible green spaces on foot within 5–15 min intervals.

To estimate the population who have access to public green space in the study area, our result layer was clipped against the residential areas layer that carries population data. The total service zone of accessible UGSs within the sum of the three-time intervals covered 3,224,189 inhabitants, i.e., 40.3% of the total population of the

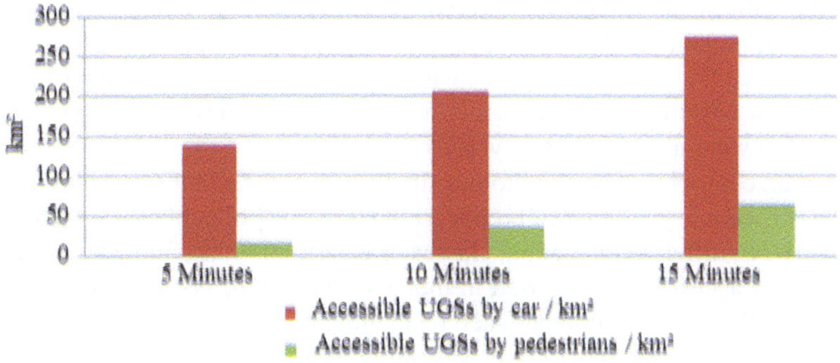

Fig. 3 UGSs service area (km^2) accessible by car and pedestrians in three-time intervals

Table 3 Allocation of accessible public green space to pedestrians per locality

Khartoum State		Locality area km^2	5 min km^2	% from total area	10 min km^2	% from total area	15 min km^2	% from total area
City	Locality							
Omdurman	Karrari	4780.45	0.27	0.006	1.018	0.021	2.057	0.043
	Ombada	2678.50	0.38	0.014	1.254	0.046	3.027	0.113
	Omdurman	891.55	2.98	0.334	6.394	0.717	9.673	1.085
Khartoum North	Bahri	4687.98	1.26	0.026	3.921	0.083	8.493	0.181
	Sharg Alneel	7997.04	0.84	0.010	2.988	0.037	6.850	0.085
Khartoum	Khartoum	186.06	8.62	4.637	18.321	9.846	31.229	0.087
	Jabal Awlia	612.46	0.67	0.109	2.027	0.331	3.232	0.527

Table 4 Allocation of accessible public green space by car per locality

Khartoum State		Locality area km^2	5 min km^2	% from total area	10 min km^2	% from Total area	15 min km^2	% from total area
City	Locality							
Omdurman	Karrari	4780.45	6.5	0.13	8.93	0.18	12.76	0.26
	Ombada	2678.50	10.1	0. 37	35.26	1.31	48.89	1.82
	Omdurman	891.55	18.8	2.1	32.68	3.66	38.57	4.32
Khartoum North	Bahri	4687.98	13.9	0.29	48.70	1.03	57.42	1.22
	Sharg Alneel	7997.04	11.7	0.14	29.87	0.37	60.51	0.75
Khartoum	Khartoum	186.06	72.3	38.8	33.25	17.87	16.33	8.78
	Jabal Awlia	612.46	4.1	0.66	16.50	2.69	39.01	6.37

study area. Nevertheless, only 8.7% of the total population fell within the 5-min service zone. On the other hand, the UGSs service zone that was accessible on foot within the sum of the three time intervals covered 303,989 inhabitants, i.e., 3.7% of the total population, while <1% of the total population could access on foot the existing public parks and green spaces within 5 min. Figures 4 and 5 show the details of population coverage of public green space service zones within the three-time intervals that were used in the analysis as per locality.

The intersect tool was applied to our final result layer and the residential classes layer, this procedure yielded results concerning the equity of public green space distribution between different residential classes of Khartoum State. Table 5 compares the two means of accessing UGSs used in the analysis and the corresponding potential accessible UGSs service zone per each residential class.

Another method used in this study was a public survey. The results of the questionnaires showed that among the 44 UGSs studied, respondents frequently mentioned 22 specific parks and public green spaces as the most appealing spots for recreation.

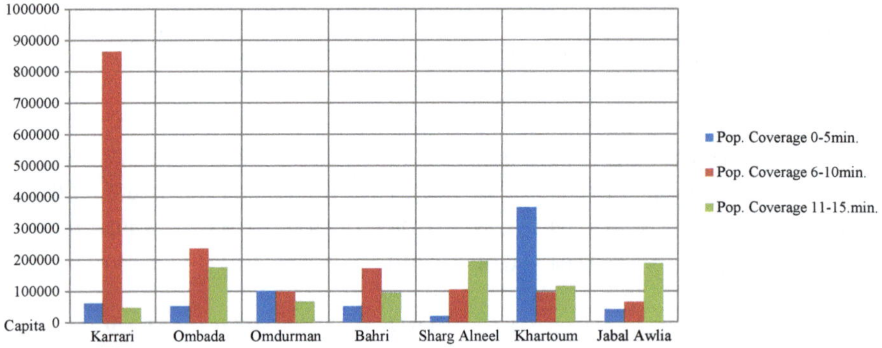

Fig. 4 UGSs population coverage within three-time intervals by car

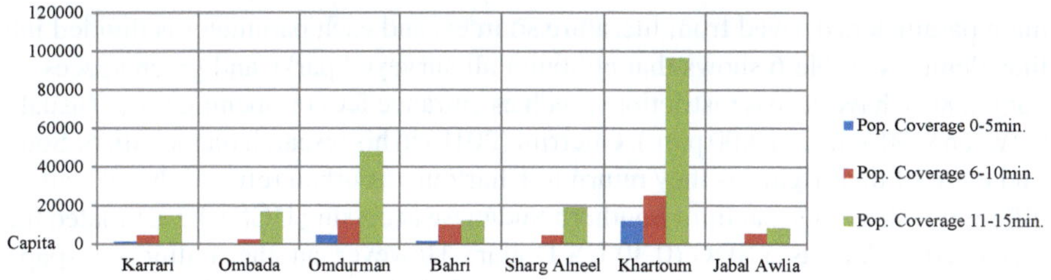

Fig. 5 UGSs population coverage within three time intervals by pedestrians

Table 5 Allocated UGSs service coverage per residential class

Pedestrian			Drive		
Minutes	Area/km^2	Class	Minutes	Area/km^2	Class
⟨0; 5⟩	7.04962	First	⟨0; 5⟩	49.198498	First
	5.78887	Second		49.567724	Second
	1.75038	Third		32.746469	Third
⟨5; 10⟩	16.6676	First	⟨5; 10⟩	43.339166	First
	10.2259	Second		55.936057	Second
	7.11965	Third		96.672152	Third
⟨10; 15⟩	27.9489	First	⟨10; 15⟩	33.114668	First
	15.8423	Second		55.055442	Second
	16.1889	Third		158.12278	Third

Figure 6 shows the features of the public green space that the respondents prefer. The graph clearly indicates that the most attractive features are well-equipped public gardens near the respondents' residences. Moreover, respondents show a remarkable willingness to bear extra travel costs and time in order to access UGSs well-equipped with recreational and educational facilities.

The users' opinion on UGSs assessment is based on the responses from the questionnaires and the GIS accessibility analysis. The assessment tool consists of six

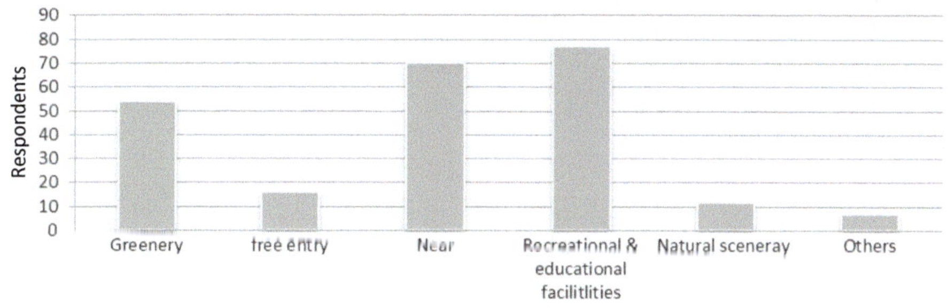

Fig. 6 Determinants of public green space selection among the surveyed sample

main parameters derived from literature sources, and each parameter is divided into fine elements. Table 6 shows that although all surveyed parks and green spaces are public, 80% have access restrictions, such as entrance fees or opening times (usually between 9:00 a.m. and 9:00 p.m.). Guerrini (2018) in his research on reclaim of South Sudanese of their rights as they return to Khartoum as urban refugees has discussed different restrictions that limit Southern Sudanese accessing UGSs, he estimated that a greened park costs 5 SDG (0.30 US Dollar). However, another online newspaper conducted an interview with UGSs visitors in 2013 has estimated the entrance costs of gated UGSs by 7 SDG (more than 1 US Dollar) (Mohamed, 2013). On the other hand, 95% of these parks are adjacent to and only accessible through collector roads and highways which limits their value as places of tranquillity. Although the safety parameter showed that 80% of the parks are safe according to respondents, in reality, public green space and parks in Khartoum lack basic safety precautions, aid facilities for the disabled, toilets, benches and playgrounds to suit all ages.

Table 6 Users' opinion on UGSs assessment results

Users' opinion parameter	Parameter's breakdown					
Maintenance of the parks	Cleanness	Seating	Walking tracks	Special designs for disability	Toilets	Children playgrounds
	30%	30%	2%	0%	8%	30%
Accessibility	Private	Public	Free	Restricted	Within 5 min. walking distance	Within 5 min car drive
	–	100%	20%	80%	<1%	2.5%
Proximity to heavy traffic road	Close to heavy traffic		Not close to heavy traffic			
	95%		5%			
Safety	Safe	Unsafe				
	80%	20%				
Natural vegetation	Natural	Not natural				
	1%	99%				
Peace & quiet	Yes	NO				
	6%	94%				

4 Discussion

The current study has examined the spatial disparities in the potential accessibility to and users' opinions on UGSs in Khartoum State. The results reveal two main findings: (1) an allocation deficit, and inequitable distribution of public green space in the study area that does not fulfil the quality of life standard for the majority of residents and users' expectations; (2) the current accessibility mode to green spaces and public parks in the study area favours residents from first-class areas at the expense of other residential classes.

The above-mentioned findings are outcomes of combined political, governance, urban planning, and socioeconomic factors. Unstable political and governance systems at both national and local levels in Sudan have led to the continuous changing and reshaping of management institutions and municipal jurisdictions. Such changes also affect the boundaries of these authorities. Consequently, the urban development plans and strategies (including the construction of new UGSs and maintenance of the existing ones) have been frequently interrupted. Further, meagre financial resources allocated to the state government and local councils can barely meet their main needs. In 2002, the state budget amounted to 36.7 billion Sudanese Pound (SDG) (US $143 million), and it was mostly spent on providing basic services, including the salaries of civil servants, e.g. teachers (Eltayeb, 2003). Presumably, the current situation has witnessed an increase in budgets allocated but no changes in priorities with regard to UGSs provision and maintenance.

Moreover, natural population growth, massive internal displacement, and asylum-seeking, due to internal and regional conflicts and imbalanced economic development in Sudan, have further burdened the urbanized area of Khartoum, which has had to expand horizontally in response to the influx of newcomers. According to Eltayeb (2003), Khartoum became 48 times horizontally larger during the period 1955–1998. Further, internally displaced people in Khartoum comprise between 18 and 23% of the population (Jacobsen, 2008). Weaver (1985) has estimated that in 1984 Khartoum hosted over 1 million Ethiopian refugee families. This situation has led to significant strains on the infrastructure and services of Khartoum State. Thus, the authorities have to deal with more urgent urban issues such as reducing poverty, planning slums, providing security, etc. Usually, in such a situation, providing and expanding green spaces could be considered a luxury among the other pressing municipal priorities. According to Fratini et al. (2009), in the period of economic recession municipalities tend to reduce maintenance expenditures and delay tasks concerning public park construction and maintenance (Tempesta, 2015).

The ill-conceived urban plans, which support a dividing conurbation area into different residential classes based on income, have created inequitable opportunities for residents to access the existing UGSs. Moreover, the current urban re-planning model of slums in Khartoum State was based on the assumption that the state government would lease land plots for long periods (from 50 to 90 years) so that inhabitants could build their houses and necessary services on their own. Later studies have proved the fallacy of such an assumption since less than 50% of land plots are occupied by families who can afford to build their houses and basic services (UN-Habitat, 2009). Formal statistics indicate that about 27% of Khartoum State inhabitants are living below the poverty line of US $2 per day (Urban and Water Unit, Sustainable Development Department, Africa-Region, 2011). Another prevailing physical planning concept is the service buffer zone as a basis for public services provision and distribution. The service buffer zone portrays accessibility to different services in straight lines without considering road blockages or the actual walking patterns and speed; therefore, this may have been another factor that deprived a vast majority of the inhabitants of having adequate and equitable access to UGSs in Khartoum State.

There are no universally accepted standards for providing public green space and each country sets its own guidelines concerning this matter, but there is a consensus that accessible public green space and urban parks that match the quality-of-life standard should be at the centre of the settlement area and not further than a five-minute walk (Etzioni, 1998). The results showed that only 15 km^2 out of 5,000 km^2 of the inhabited area of Khartoum, coincided with this standard. Moreover, 88% of modelled urban parks have an area range between 1 and 12 ha; thus, the per capita share of the existing urban parks is negligible. Nevertheless, Khartoum is not an exception in this regard. Rigolon et al. (2018) has found inequities in UGSs quantity and proximity mostly consistent across the cities of the Global South, i.e. across African, Asian, and Latin American cities.

Since inhabitants are the end-users of UGSs, they must be at the centre of concern. Study results showed that <1% of the population could potentially access nearby green public spaces on foot. Meanwhile, statistics show that 82% of Khartoum State population is aged between <15 and 40 years old (Central Bureau of Statistics, 2013). This young population needs to be provided with venues such as public green spaces for social interaction, sports, psychological relaxation, and recreation. A National Youth Strategy (2007) has been recently prepared and endorsed by the Youth and Sports Ministry, aiming to advocate and support involving youth in urban and rural affairs, however, most of its projects are unfinished or unrealized due to budgetary constraints (UN-Habitat, 2016).

Well-maintained UGSs have greater attraction factors and are expected to meet the users' needs and the purpose they were developed for. However, the maintenance of UGSs doesn't come at a cheap price. McDonough Maland (2012), has summarized the costs of urban parks as follows: (1) costs of acquiring the property; (2) costs associated with developing the property, including design and construction costs; (3) costs associated with operation and maintenance, including employee payroll and landscaping costs; (4) the opportunity costs associated with the loss of property income tax that communities would have received if the property had been developed for other purposes.

Referring to the results of the questionnaire, the majority of the respondents select green space based on the availability of recreational and educational facilities; this might be because most of the park visitors in the study area are parents accompanying their kids (UN-Habitat, 2016). The overall per capita share of public UGSs in Khartoum State is nominal (859 ha/7 million inhabitants). Very similar results are presented in the work of Rehačková and Pauditšová (2006). In Sudan, there is no reliable documentation of the expenditures on this category of municipal work; however, considering economic downturns, budget deficiencies, and very dry climate conditions, maintenance, operating costs and capital costs are assumed to be very high. Examples of the various maintenance costs needed for the upkeep of UGSs provide evidence on this matter. A study conducted in 40 municipalities in the Veneto Region highlighted that maintenance costs varied from 0.39 to 2.73 € per m^2 (constant price 2012) with an average of 1.10 € per m^2 (Tempesta, 2015). On average the cost per inhabitant was 10.08 € per year. In the 87 biggest USA cities the spending on parks and recreation per resident ranges from 10 to 287 € per capita per year (median = \$73) (TPL-CCPE, 2014). In 15 UK parks, the cost per inhabitant ranges from 10.61 to 44.12 € and the cost per m^2 from 0.28 to 1.34 € (constant price 2002) (Dunnett et al., 2002).

5 Limitations

This study suffers from number of limitations in different aspects that might affect the accuracy of the results. First of all, the method of network analysis approach used to estimate service coverage of public UGSs for both drivers and pedestrians didn't count for number of other barriers such as traffic jams, quality of the street and safety issues, treating different modes of transportation (private car and public transportation) equally without considering main differences like frequent stops of public transportation and waiting time. All these factors may affect the accuracy of the final output of the analysis. Another methodological limitation of this model is treating all UGSs as a center point instead of identifying access points like gates or even fence gaps that UGSs users may get into UGSs through it and may also affect the overall accuracy of the analysis outcome. Although Boone et al. (2009) have suggested to treat small parks (almost 80% of the parks modelled in this study have an area of 1–5 ha) as center points but still it is advised to model UGSs elements access points to obtain more accurate results. Moreover, there are many more methodological issues that we have not been addressed. Key amongst these is the explicit assumption that all greenspace goers are intending to visit the UGSs only for sitting and enjoying the scenery and atheistic value of the UGS (passive recreation) without considering other active usages of UGSs and parallelly assuming all UGSs elements are equally appealing and beneficial to visit. Matter of fact, greenspace characteristics including size, quality and amenities all contribute to usage preferences, this assumption is unlikely to be correct (Lee et al., 2015). There may also be bias introduced by assessing at household-level, rather than the level of the individual person, due to variation in average headcount per household across the city (Meers & Brindley, 2019). Besides the points mentioned above, the data such as population and street networks datasets related to different phases of time (ranging from 2008 to 2018) which may not represent precisely the real world now as urban centers are rapidly growing and changing. Future research on UGSs accessibility in Khartoum State should consider those gaps plus add more sophisticated analysis methods such as site suitability to locate green for more accurate results.

6 Conclusion

This study examines the spatial disparities in accessing UGSs in Khartoum State for both pedestrians and drivers. It has also evaluated UGSs functionality. The study has compared accessibility patterns among different administrative divisions of the study area and its population coverage. It explains the planning and socioeconomic factors responsible for the current situation. The study findings indicate that the existing pattern of accessing UGSs in Khartoum is far from fulfilling the minimum of internationally accepted standards of welfare and quality of life in the city. The prevailing urban planning methods and concepts have further marginalized the majority of the population in the study area and failed to provide them with adequate green spaces and equitable access to the existing ones. The economic situation also plays a major role in creating the current situation, where the provision of functional and fairly accessible green spaces has been offset by other more pressing municipal issues. In addition to all the above-mentioned factors, political instability and frequent changes in central and local governments have weakened local governance units, interrupted and delayed the execution of scheduled urban and strategic plans that might improve the current situation. This study suggests providing qualitatively, quantitatively and equitably accessed green spaces in the conurbation area of Khartoum State in order to improve its quality of life.

Acknowledgements This work was supported by the Ministry of Education, Youth and Sports of the Czech Republic, INTER-COST programme No LTC18072 and has received funding from the European Union's Horizon 2020 Research and Innovation Programme within Grant Agreement No 653522: RESIN—Climate Resilient Cities and Infrastructures.

References

Agraa, O. A., Elkheir, O., Haywood, I., & Ahmed. A. M. (1985). Human settlements in greater Khartoum (p. 251). Khartoum University Press.

Alhuseen, A. (2015). Sufficiency of open public and green spaces in Khartoum State-Sudan. *International Journal of Social Relevance & Concern (IJSRC), 3*(5), 15, 3–7.

Awad, Z. (2018). Evaluating neighborhoods developed open spaces in Khartoum-Sudan. *Civil Engineering and Architecture, 6*(6), 269–282, 2018 http://www.hrpub.org https://doi.org/10.13189/cea.2018.060601.

Babiker, M. E. A., & Ibrahim, A. M. (2016). Detecting land use and land cover. *Asian Journal of Geoinformatics, 2*(16), 9–17.

Balram, S., & Dragićević, S. (2005). Attitudes toward urban green spaces: Integrating questionnaire survey and collaborative GIS techniques to improve attitude measurements. *Landscape and Urban Planning, 71*(2–4), 147–162.

Bannaga, S. E. I. (2018). Revitalization of greater Khartoum urban transportation system. *Future Cities and Environment, 4*(1), 10, 1–15, https://doi.org/10.5334/fce.2.

Bashari, W., Merghani, A., Bahreldin, I. (2015). Spatial impact of gender variation on Khartoum city river side public open space, Conference: Sudanese Institute of Architects, 4th Scientific Conference, At Khartoum, Sudan.

Baycan-Levent, T., van Leeuwen, E., Rodenburg, C., & Nijkamp, P. (2002, September 21–26). Development and management of green spaces in European cities: A comparative analysis. In *38th International Planning Congress on "The Pulsar Effect" Planning with Peaks* (pp. 237–247). Glifada, Athens, Greece.

Bell, S., Montarzino, A., & Travlou, P. (2007). Mapping research priorities for green public urban space in UK. *Urban Forestry and Urban Greening, 6*(2), 103–115.

Boone, Ch. G., Buckley, G. L., Grove, M., Sister, Ch. (2009). Parks and people: An environmental justice inquiry in Baltimore, Maryland. *Annals of the Association of American Geographers, 99*(4), 767–787.

Boultona, C., Dedekorkut-Howesa, A., & Byrneb, J. (2018). Factors shaping urban greenspace provision: A systematic review of the literature. *Landscape and Urban Planning 178*, 82–101, 86–88.

Burgess, J., Limb, M., & Harrison, C. (1988). Exploring environmental values through the medium of small groups. Part 2. *Environment and Planning A, 20*, 457–476.

Central Bureau of Statistics (CBS), Republic of Sudan. (2013). Population projections. Annual statistical book. Central Bureau of Statistics, Republic of Sudan.

Cvejić, R., Eler, K., Pintar, M., Železnikar, Š. Haase, D., Kabisch, N., & Strohbach, M. (2015). A typology of urban green spaces, eco-system services provisioning services and demands, (ENV.2013.6.2-5-603567) GREEN SURGE project (2013–2017) (68 p.).

Dunnett, N., Swanwick C., & Woolley H. (2002). Improving urban parks, play areas and green spaces, February 2019. London (UK): Department for Transport, Local Government and the Regions. http://www.ocs.polito.it/biblioteca/verde/improving_full.pdf.

El Tayeb, S., Abdalla, S., Van den Bergh, G., & Heuch, I. (2015). Use of healthcare services by injured people in Khartoum State, Sudan. *International Health, 7*(3), 183–189. https://doi.org/10.1093/inthealth/ihu063.

Eltayeb, G. (2003). *Understanding slums: Case studies for the global report on human settlements.* Global report on human settlements. http://www.ucl.ac.uk/dpu-projects/Global_Report/pdfs/khartoum.pdf.

Etzioni, A. (1998). *The essential communitarian reader.* Rowman and Littlefield, Lanham.

Figueiredo, R., Gonçalves, A. B., & Ramos, I. L. (2016). Service areas of local urban green spaces: an explorative approach in arroyos, Lisbon. In *ISPRS Annals of the Photogrammetry, Remote Sensing and Spatial Information Sciences* (Vol. IV-4/W1).

Fratini, R., Marone, E., Riccioli, F., & Scozzafava, G. (2009). Green urban areas: Evaluation and analysis of public spending for management. *Geomatics and Environmental Engineering, 3*, 25–43.

Germann-Chiari, C., & Seeland, K. (2004). Are urban green spaces optimally distributed to act as places for social integration? Results of a geographical information system (GIS) approach for urban forestry research. *Forest Policy and Economics, 6*(1), 3–13.

Guerrini, D. (2018). *Return of South Sudanese to Khartoum: (Re) claiming their right to the city as urban refugees.* Utrecht University-6108210.

Higher Council for Environment and Natural Resources. (2003). *Sudan First National Communication Report to UNFCCC.* https://unfccc.int/resource/docs/natc/sudnc2.pdf.

Higher Council for Environment and Natural Resources. (2016). National adaption plan (105 p.). Republic of the Sudan, Ministry of Environment, Natural Resources & Physical Development Higher Council for Environment and Natural Resources, 2018. http://www4.unfccc.int/nap/Documents%20NAP.

Higher Council for Environment, Urban and Rural Promotion. (2015). Green Belt-Khartoum State. Republic of Sudan.

Higher Council for Strategic Planning. (2013). Annual strategic report, Khartoum State. Almazin Publisher Ltd. (In Arabic).

Jacobsen, K. (2008). Internal displacement to urban areas. The Tufts-IDMC profiling study, case 1 in Khartoum, Sudan, Feinstein International Centre, Tufts University, in collaboration with the Internal Displacement Monitoring Centre, Geneva.

Kimpton, A. (2017). A spatial analytic approach for classifying greenspace and comparing greenspace social equity. *Applied Geography., 82*, 129–142. https://doi.org/10.1016/j.apgeog.2017.03.016.

Le Texier M., Schiel K., & Caruso G. (2018). The provision of urban green space and its accessibility: Spatial data effects in Brussels. *PLoS ONE, 13*(10), e0204684.https://doi.org/10.1371/journal.pone.0204684.

Lee, A. C. K., Jordan, H. C., & Horsley, J. (2015). Value of urban green spaces in promoting healthy living and wellbeing: prospects for planning. Risk management and healthcare policy **8**, 131–137. Retrieved August 27, 2015, from https://doi.org/10.2147/RMHP.S61654. PMCID: PMC4556255.

Markevych, I., Schoierer, J., Hartig, T., Chudnovsky, A., Hystad, P., Dzhambov, A. M., De Vries, S., Triguero-Mas, M., Brauer, M., Nieuwenhuijsen, M. J., Lupp, G., Richardson, E. A., Astell-Burt, T., Dimitrova, D., Feng, X., Sadeh, M., Standl, M., Heinrich, J., & Fuertes, E. (2017). Exploring pathways linking greenspace to health: Theoretical and methodological guidance. *Environmental Research, 158*, 301–317.

Martí, R. M. (2013). Development of a pedestrian spatial network for the University Jaume I of Castellón (Spain). Master's Degree in Geospatial Technologies.

Mcdonough, M. M. (2012). The return on investment of parks and open space: How economics can inform design. Ph.D. dissertation, University of Georgia (USA).

Mears, M., & Brindley, P. (2019). Measuring urban greenspace distribution equity: The importance of appropriate methodological approaches. *ISPRS International Journal of Geo-Informatics, 8*, 286.

Mohamed, A. (2013). Khartoum's citizens flock to rare green spaces. https://www.theniles.org/en/articles/archive/2012/.

Monteiro, L. M., Okamoto Junior, J., Peixoto E.V., Assis J.S., Suzuki K.M., & Prata-Shimomura A. R. (2018). Assessment of outdoor comfort conditions based on the application of a participative model in open urban spaces. In W. Leal Filho, F. Frankenberger, P. Iglecias, R. Mülfarth (Eds.), *Towards Green Campus Operations*. World Sustainability Series. Springer.

Morris. J. N., & Hardman. A. E. (1997). Walking to health. *Sports Medicine, 23*(5), 306–332. https://doi.org/10.1007/BF03257359.

Nasifa Consultants. (2012). Khartoum State localities strategic plans, prepared for the benefit of Khartoum State government and the Chamber of Local Governance.

Natural England. (2010). *Nature nearby: Accessible natural greenspace guidance*. Sheffield: Natural England.

OCHA, United Nations Office for the Coordination of Humanitarian Affairs. (2020). *Sudan administrative level 0–2 boundaries*. https://data.humdata.org/dataset/sudan-administrative-boundaries-levels-0-2.

Oliver, L. N., Schuurman, N., & Hall, A. W. (2007). Comparing circular and network buffers to examine the influence of land use on walking for leisure and errands. *International Journal of Health Geographics, 6*, 41. https://doi.org/10.1186/1476-072X-6-41.

Osman, S. (2009). Making cities work—Khartoum state's new structural plan, the second lecture of the series at Khartoum International Community School Humanities Department Lecture Series 2009–10.

Reháčková, T., & Pauditšová, E. (2004). Evaluation of urban green, spaces in Bratislava. *Boreal Environment Research, 9*, 469–477.

Reháčková, T., & Pauditšová, E. (2006). *Vegetation in urban environment* (132 p.). Cicero Ltd. (in Slovak).

Rigolon, A., Browning, M., Lee, K., & Shin, S. (2018). Access to urban green space in cities of the global south: A systematic literature review. Licensee MDPI, Basel, Switzerland. This article is an open access article distributed under the terms and conditions of the Creative Commons Attribution (CC BY) license http://creativecommons.org/licenses/by/4.0/.

So, S. W. (2016). Urban green space accessibility and environmental justice: A GIS-based analysis in the city of Phoenix, Arizona. Master's thesis, University of Southern California. https://spatial.usc.edu/wp-content.

Swanwick, C., Dunnett, N., & Woolley, H. (2003). Nature, role and value of green space in towns and cities: An overview. *Built Environment, 29*(2), 94–106.

Tempesta, T. (2015, December). *Benefits and Costs of Urban Parks: A Review* (pp. 127–143). Firenze University Press. Aestimum 67. https://doi.org/10.13128/Aestimum-17943.

The International Organization for Standardization ISO/TS 37151:2015. Smart community infrastructures—Principles and requirements for performance metrics. https://www.iso.org/standard/61057.html.

TPL-CCPE, the Trust for Public Land Centre for City Park Excellence. (2014). 2014 city park facts, Washington, D.C. (USA). Retrieved March, 2018, from http://www.tpl.org/2014-city-park-facts.

UN-Habitat. (2009). Urban sector studies and capacity building for Khartoum State, United Nations Human Settlements Programme, International guidelines on decentralisation and access to basic services for all. https://www.academia.edu/5557474/Urban_sector_studies_and_capacity_building_for_Khartoum_States.

UN-Habitat. (2016). Sudan's report for United Nations' third conference on housing and sustainable urban development—Addressing urban youth needs. http://habitat3.org/wp-content/uploads/Sudan-National-Report.pdf.

Urban and Water Unit, Sustainable Development Department, Africa-Region. (2011). Report No. 63785-SD Sudan: Issues in urban development phase 1—Overview of the urban landscape.

Van Herzele, A., & Wiedemann, T. (2003). A monitoring tool for the provision of accessible and attractive urban green spaces. *Landscape and Urban Planning, 63*(2), 109–126.

Weaver, J. L. (1985). Sojourners along the Nile: Ethiopian Refugees in Khartoum. *Journal of Modern African Studies, 23*(1), 147–156.

Wendel, H., Zarger, R., & Mihelcic, J. (2012). Accessibility and usability: Green space preferences, perceptions, and barriers in a rapidly urbanizing city in Latin America. *Landscape and Urban Planning, 107*, 272–282.

WHO. (2016). Urban green spaces and health. Copenhagen: WHO Regional Office for Europe. http://www.euro.who.int/__data/assets/pdf_file/0005/321971/Urban-green-spaces-and-health-review-evidence.pdf?ua=1.

Yuan, L. H., & Xu, P. W. (2015). Accessibility of park green spaces in the central districts of Beijing. *Urban Environment & Urban Ecology, 28*(1), 22–25.

Policy Issues and Sustainable Development

Multifunctional Patterns and Governance Challenges of Blue–Green Infrastructure in Puebla, Mexico

María del Socorro Ramírez, Romeo A. Saldaña-Vázquez, and María E. Ibarrarán

Abstract Blue–Green infrastructure (BGI) consists of spaces with water bodies and vegetation inside the urban territory. Due to their characteristics, they have the potential to provide many functions and benefits to the inhabitants of the city, such as open space, places for social encounters and a healthy lifestyle, shady and humid environments for climate regulation and education, and economic well-being. The accumulation of these functions is denominated multifunctionality. The BGI multifunctionality depends on social and physical factors. In this chapter, two of these factors are reviewed: the size of the BGI and the urban legislation for the City of Puebla and its metropolitan area, in central Mexico. The measurement of multifunctionality is carried out with the satellite images of Google Maps Streetview and Google Earth Pro. Subsequently, a cluster analysis is done to classify those BGIs that were similar in their size and number of functions provided. Most of the BGI in Puebla are relatively small, on average 0.56 ha, and provide less than 3 functions. The larger ones (average 2.02 ha) can provide up to 5 functions, the largest number of functions observed. Additionally, there is a relationship between the size of the BGI and the number of functions provided, and the presence of water bodies. On the other hand, the instruments of urban legislation of different governmental scales are consulted, locating in their content the normativity related to BGI. According to the analysis of the urban legislation, the multifunctionality of the BGI is not addressed in a relevant way at the various levels of government, i.e., municipality, state, and federation. In conclusion, most of the Blue–Green Infrastructure (BGI) in the city of

M. S. Ramírez
M.A. in Habitat and Socio-Territorial Equity, Universidad Iberoamericana Puebla, Blvd. del Niño Poblano No. 2901, Col. Reserva Territorial Atlixcáyotl, San Andrés Cholula, 72820 Puebla, México

R. A. Saldaña-Vázquez · M. E. Ibarrarán (✉)
Instituto de Investigaciones en Medio Ambiente Xabier Gorostiaga, S.J. Universidad Iberoamericana Puebla, Blvd. del Niño Poblano No. 2901, Col. Reserva Territorial Atlixcáyotl, San Andrés Cholula, 72820 Puebla, México
e-mail: mariaeugenia.ibarraran@iberopuebla.mx

R. A. Saldaña-Vázquez
e-mail: romeoalberto.saldana@iberopuebla.mx

P. K. Joshi et al. (eds.), *Blue-Green Infrastructure for Sustainable Urban Settlements*,
https://doi.org/10.1007/978-3-031-62293-9_15

Puebla is relatively small and could enhance its functionality through redesign and modifications in urban legislation.

Keywords Atoyac basin · Built infrastructure · Care infrastructure · Ecosystem services · Heat waves · Nature-based solutions · Sponge city

1 Introduction

Climate extremes, particularly droughts, floods, and heatwaves, are growing challenges for anthropized and natural ecosystems (Aghakouchak et al., 2020; Hunt & Watkiss, 2011). Urban areas are experiencing increasing population density and expanding due to urban sprawl (Mahtta et al., 2022; Seto & Shepherd, 2009; Seto et al., 2011). To address immediate and medium-term needs, these areas rely on gray infrastructure. However, this approach leads to environmental impacts such as the heat island effect, soil waterproofing that reduces water collection and contributes to flooding, and the reduction of parks and green spaces to accommodate housing densification and transportation infrastructure (Deilami et al., 2018; Hibbs & Sharp, 2012; Tam et al., 2015).

The additional challenge posed by climate change is the increased frequency of extreme conditions like droughts and floods, which significantly impact living conditions and necessitate cities to adapt (Hunt & Watkiss, 2011). In this regard, Blue–Green Infrastructure (BGI) can serve as a nature-based solution (NbS) that relies on vegetation and water bodies to enhance resilience in cities amidst climate change (Siehr et al., 2022). This more diverse and comprehensive infrastructure has a significant impact on people's well-being, both in terms of physical and mental health, as it allows for greater interaction with green and blue spaces that protect urban ecosystems (Hansen & Pauleit, 2014) and allows for the enjoyment of nature. Benefits include improved social fabric, economic activity, and reduced costs associated with climate-related extremes (Weinbrenner et al., 2021), because BGI becomes part of the infrastructure necessary to mitigate such events, thereby freeing up resources for other development projects (Ghofrani et al., 2017).

The Puebla City Metropolitan Area (PCMA) is not an exception to environmental problems caused by urban expansion. Between 1980 and 2019, municipalities within this metropolitan area have witnessed a range of 100–500% increase in their urban surface (González & Aguilar, 2021; Sánchez Juárez, 2021). Consequently, natural vegetation and water bodies have been lost (González & Aguilar, 2021), and the per capita green area in the city of Puebla is merely 4.7 m^2 (Sánchez Juárez, 2021). Most urban water bodies have been covered, and the few remaining ones are heavily polluted (Pérez Castresana et al., 2018). PCMA faces challenges such as droughts, floods, water shortages, and the heat island effect. Therefore, there is an urgent need for BGI implementation to enhance resilience for both the population and ecosystems in the city.

This chapter is organized as follows: The first part introduces the concept of BGI and outlines its essential characteristics for contributing to population well-being and climate change adaptation within the framework of "resilience thinking." The second section focuses on the urban metropolitan area of Puebla, highlighting its existing green areas and their multifunctional attributes. The third part discusses the governance tools associated with BGI implementation. Finally, the fourth part concludes if the public policies within PCMA and their potential support to the establishment of a BGI system foster greater well-being for the PCMA and its inhabitants.

2 Definition and Characteristics of BGI

Nature serves as the foundation for of ecosystem services that sustain the well-being of people in both rural and urban ecosystems (Díaz et al., 2018). However, urban development repeatedly reduces green areas to promote higher population density, while also prioritizing the construction of durable and efficient road networks (Abass et al., 2018; Dobbs et al., 2017). When waterways obstruct the path of urban expansion, they are often covered or eliminated to facilitate the construction of streets and other urban infrastructure. As cities expand outward into the countryside, they encroach upon agricultural and forested lands, resulting in soil erosion and a decrease in water infiltration to aquifers (Sun & Caldwell, 2015; Venter et al., 2016). These processes contribute to the graying of cities, making them more susceptible to floods, droughts, and severe water shortages. Additionally, there is often a scarcity of clean and inviting open spaces for people to gather, leading them to seek refuge in shopping malls. Given these persistent conditions, there is an urgent need for a new approach to urban redevelopment.

Blue–Green Infrastructure (BGI) can be defined as open green spaces and water bodies, including ephemeral, intermittent, and perennial features, that serve multiple functions (Ghofrani et al., 2017; O'Donnell et al., 2021). BGI comprises an interconnected network of landscape components that effectively manage stormwater in urban areas, acting as a sponge-like system that enhances biodiversity, promotes food production, improves air quality, and facilitates climate adaptation and regulation (Siehr et al., 2022). Consequently, these types of implementations can also be classified as Nature-based Solutions (NbS). Furthermore, BGI provides urban populations with calm open spaces that foster social connections, offer areas for exercise and recreation, and contribute to improved mental and physical health.

BGI, however, faces certain obstacles, with cost being a significant factor. Nonetheless, as local environmental conditions change, BGI implementation can lead to reduced operating costs for cities. Studies conducted in the Netherlands and Norway have demonstrated cost savings resulting from decreased expenses in the use of chemical pesticides, water purification, and stormwater management (Ghofrani et al., 2017; Wilbers et al., 2022). Thus, BGI proves to be a more cost-effective alternative compared to expanding gray infrastructure, which, in addition, has a negative feedback effect by increasing water runoff and temperature. Furthermore, the benefits of BGI can also be assessed based on the damage avoided resulting from its implementation.

Cities embracing the BGI paradigm typically incorporate additional concepts within their framework, such as social inclusion, ensuring that this infrastructure is accessible to all rather than limited to those who can afford it through direct payment or taxes. As a result, BGI promises to promote social and environmental justice (Meenar et al., 2022).

In Mexico, both the country as a whole and its cities face significant challenges currently and in the future. Many cities experience periods of severe drought in Spring and extending into Summer, followed by flash floods in the Summer and Fall (Gleason & Flores, 2021). This leads to extreme conditions where water is either too scarce or abundant, often feeling uncontrollable (Soto-Montes-de-Oca & Alfie-Cohen, 2019). With climate change, these threats become insurmountable. Climate policy needs to expand beyond focusing solely on mitigation and address adaptation to effectively tackle the growing consequences of hydrometeorological events.

Although there is a Climate Change National Strategy (SEMARNAT, 2013) that can provide guidance, adaptation measures are mostly local, as one-size-fits-all solutions rarely work. Therefore, local adaptation processes in cities and peri-urban communities can leverage BGI to achieve significant local impacts, thereby harnessing nature as an ally in their development. There is a vast literature advocating for BGI, and numerous case studies have been conducted in countries such as the Netherlands, Belgium, the UK, the US, China, Japan, and India, among other countries (O'Donnell et al., 2021; Wilbers et al., 2022). The findings consistently reveal that there is potential for implementing BGI in all countries, with the main difference lying in the extent to which governments and communities engage in the process.

BGI operates within the framework of Resilience Thinking (Walker & Salt, 2012). This framework proposes that BGI interventions will not restore the environment to its original state but rather aim to create a trajectory of human and nature interaction, where key-system variables can return after abrupt or non-linear shifts, leading to multiple states of complex systems (Holling, 1973). There is no equilibrium, but a different state somewhere within a suitable trajectory. This is essential because, with BGI, the goal is not to restore but to foster new networks of ecosystems that can help cities cope with current and future conditions associated with climate variability and change. This implies that human actions play a role in resilience theory and influence resources through interventions, leading to unexpected and uncertain outcomes. Therefore, BGI should be designed to enhance the regions' capacity to resist stressors, both internal and external, related to climate change and other environmental challenges (Ghofrani et al., 2017).

Nature-based multifunctional infrastructure has already been studied in Mexico, albeit at a smaller scale within Puebla City Metropolitan Area (PCMA), particularly in San Pedro Cholula (Flores et al., 2019), where systemic and nature-based approaches have been incipiently implemented to make that location climate-proof.

The leapfrogging potential to accelerate a sustainable urban transition was evaluated. Findings indicate that water governance in the city is still weak and requires strengthening to enable leapfrogging.

To increase the knowledge about BGI in the PCMA, the objective of this chapter is to assess the extent of BGI´s multifunctionality in PCMA, as well as its institutional framework. Through the analysis, several recommendations will emerge to enhance PCMA's resilience to current conditions and climate change.

3 Puebla and Its Multifunctional Blue–Green Infrastructure

The ecological functionality and social benefits of BGI in cities have a wide range of environmental contributions due to the presence of vegetation, such as climate moderation, air filtration, groundwater recharge, noise reduction, soil stabilization, biodiversity conservation, and flood control, among others (Hansen & Pauleit, 2014; Lai et al., 2019; Manteghi et al., 2015; Sánchez Juárez, 2021). Regarding the social dimension, its functions include promoting both physical and mental health, providing suitable spaces for community interaction, recreation, and connecting people with nature, preserving, and fostering cultural, ethical, and spiritual values, as well as environmental education and their aesthetic value (Lee & Maheswaran, 2011; Riechers et al., 2018; Weinbrenner et al., 2021). On the economic front, benefits are the provision of agricultural and/or forestry products, reducing costs related to stormwater management, atmospheric pollution, and greenhouse gas emissions control. Additionally, by contributing to the health of the population, these services also benefit the healthcare sector by reducing its costs and enhancing the productivity of the workforce. It is essential to note that benefits and functions are often used interchangeably in the BGI literature. However, benefits encompass a range of goods and services that such an area can provide, whereas functions refer to the roles that BGIs play when recognized as elements of urban structure and are directly related to the benefits obtained from each space.

Cultural factors, urban form, aesthetic norms, and governance regimes can play a fundamental role in the distribution of BGI as well as the density of urban vegetation (Smart et al., 2020). For instance, the urban form determines the layout, size, and proportion of BGI. Aesthetic norms influence the choice of species based on their ornamental or functional characteristics, while governance regimes establish relevant regulations for their placement and features. The role of municipal policies, combined with resource allocation for maintenance practices, determines the overall condition of vegetation (Steinfort et al., 2020). This, in turn, affects the regulation of BGI care, as well as their design and vegetation, ideally ensuring diversity of species in open city spaces to shape BGI that provide socio-environmental benefits.

Although the need for BGI is emphasized due to the range of benefits they offer, which are crucial for urban living conditions, there are still many gaps regarding their suggested characteristics. While a recommended number of square meters of BGI has been mentioned to improve the quality of life, the appropriate dimensions, elements they should contain, their distribution, and biodiversity are not considered. This has resulted in a fragmented presence of BGI that provides few benefits and therefore fails to fulfill their potential functions. Spelling out the functions of BGI contributes to their recognition as an essential part of the urban environment. Just as roadways are indispensable because they serve a necessary function, that of communication, BGI should also be acknowledged. This can start with the identification of its characteristics, such as size, and its relationship to the functions they fulfill.

The objective of this chapter is to classify the Blue and Green Infrastructure of the Puebla City Metropolitan Area according to its level of multifunctionality. Additionally, it aims to evaluate if the multifunctionality is related to the size of such areas.

Publicly accessible BGIs were identified. However, most are green areas that can be addressed as Green Infrastructure (GI), given that the blue component is seldomly present. Thus, primarily parks and gardens were considered for this analysis, excluding private gardens. Similarly, other vegetated spaces such as median strips, planters, or small residual green areas are not included. The selected study area includes the conurbated area of the city of Puebla, which comprises 6 municipalities and 6 localities from nearby municipalities (Fig. 1). The topographic information for the conurbated area was obtained from the National Institute of Statistics and Geography of Mexico (INEGI, 2020). The codes of the maps used in digital formats are E14B43a, E14B43b, E14B43c, E14B43d, E14B43e, E14B43f, E14B53a, E14B53b, E14B53c, E14B53d at a scale of 1:20,000 and E14B43, E14B53 at a scale of 1:50,000.

To classify the blue–green infrastructure in the municipality of Puebla and its conurbation according to their level of functionality, a layer of geographic information in "shape" format was obtained from the National Institute of Statistics and Geography of Mexico (INEGI, 2020). This file includes 188 polygons corresponding to the perimeters of the green areas in Puebla, collected by the institute in 2018. Each polygon has a unique key that identifies it as an object, along with a name and type. The 188 objects that compose the layer were manually reviewed to ensure effective correspondence to each represented green space and that their perimeters matched appropriately. Those that did not correspond to a green area or had poorly drawn polygons have been either removed or updated. The removed objects were identified with the codes 10,446,870 and 202,506,615, while the updated ones were 10,446,872 and 10,446,863. To update the polygons, the QGIS program (ver. 3.16) was used to delete excess polygons and modify the size and shape by manipulating their vertices.

Once the layer of polygons was updated, the area of each polygon was calculated in square meters by accessing the attribute table and running the area function in the QGIS program (ver. 3.16). The information of the polygons was then exported to a Microsoft Excel database (ver. 16.48), which contained columns names of polygons such as "name," "objectid," "surface," "type," and an additional five fields to indicate the functionality of the BGI: 1: recreative, 2: environmental, 3: health, 4: social, and 5: culture, as well as a field to sum up these fields. To classify the BGI according to their social and environmental functions, the definitions proposed by Téllez Montes

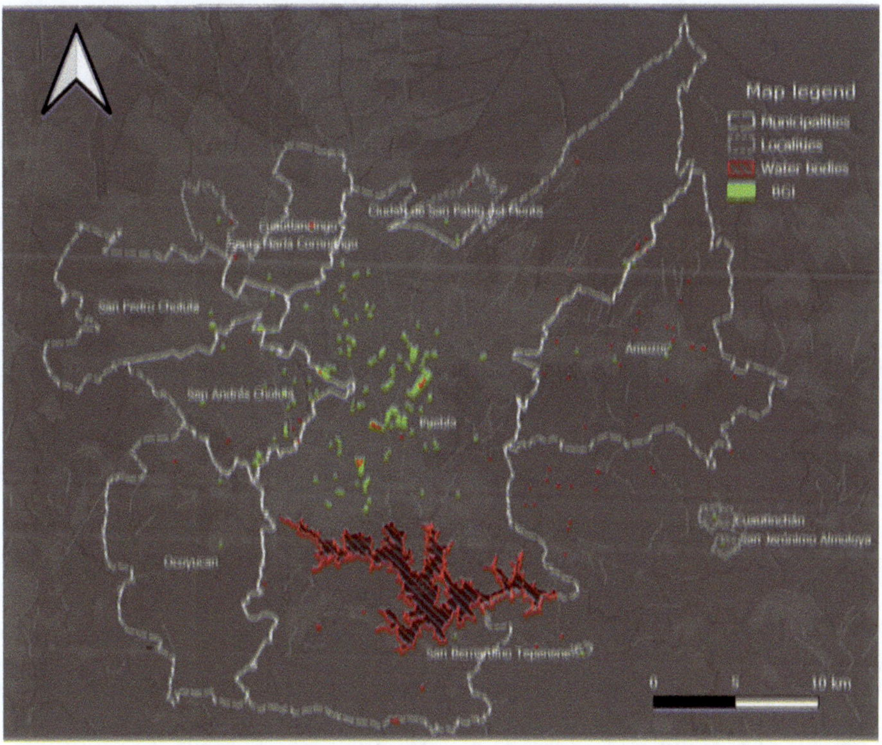

Fig. 1 Map showing the political delimitation of the municipalities and localities under study. BGI located within this zone are highlighted in red. Self-made, using data obtained from INEGI (2020)

Y de las and Castro Campos (2020) were followed. They define the functions of BGI as follows:

Recreational and leisure function: This function refers to activities related to leisure time, such as walking, pet walking, and individual or group recreation for children. Green areas with elements facilitating these activities, such as walking trails or playground equipment for children, were assigned a value of 1 in the associated column. In some cases, green areas without both elements still receive a value of 1 if they contain at least one of the mentioned elements. Parks lacking these elements received a value of 0.

Environmental function: This function considers parks as biotic areas that play a role in climate regulation, such as cooling microclimates in the urban environment. The green surfaces of parks act as filtering media for water recharge in aquifers. To fulfill this function, the presence of plant species in the parks was considered, with an emphasis on the number of species. Urban green areas were classified with an environmental function (value of 1) if the polygon predominantly consisted of areas with vegetation and was free of concrete.

Public health function: This function encompasses sports activities, both for individuals and groups, and the benefit of breathing higher-quality air in these spaces. Parks promote physical and mental activity, fostering relationships between people and tranquility. Like the first function, for the public health function, it was considered fulfilled if the park contained elements conducive to sports activities, such as sports fields, tracks, gym equipment, etc., hence assigned a value of 1.

Social cohesion function: Parks provide meeting points for residents, facilitating integration and social cohesion. This function was considered fulfilled when elements facilitating encounters and permanence of visitors, such as benches or picnic areas, statues, monuments, or other contemplative elements, were present.

Cultural and recreational function: This function aims to provide spaces for environmental education, appreciation, and care of nature by the visitors. To determine its application in the parks, activities or the presence of areas fostering this purpose were considered, along with the presence of animal species complementing the landscape or a more complex ecosystem. Generally, this function was related to larger parks or those with bodies of water.

Economic function: Several studies confirm that people prefer to buy homes in areas with nearby parks and are willing to pay more for the property as a result (Sandoval Chávez et al., 2021). However, this situation depends directly on the maintenance condition of the park. The analysis was done remotely and visually using the available data in the Google Earth Pro platform (ver. 7.3) and Google Maps. Therefore, the economic function could not be obtained with these tools and was not considered for the analysis.

Once this classification of park functionality was obtained, the matrix was imported into the R program, where descriptive analyses of the values in the "sum" column and a clustering analysis using the "k-means" method were performed to classify similar BGIs based on characteristics such as size and number of functions provided. All the analyses were conducted using the graphics, cluster, tibble, stats, factoextra, NbClust, and tydr packages of the R programming language (R Core Team, 2021).

Out of the 186 parks analyzed, 45 parks do not fulfill more than two functions. Less than 5 parks meet high functionality criteria, meaning they fulfill all five criteria. The most common scenario is parks that meet three functions. The parks vary in size from 1 to 500 km^2, and only a few show a surface area greater than 100 km^2 (Fig. 2).

A positive relationship was found between the cumulative number of functions and the area of the BGI (Pearson correlation, $r = 0.26$, $t = 3.78$, $d.f. = 184$, $P = 0.0002$; Fig. 3).

Finally, we found that the optimal number of BGI groups based on their functions and size was 5 (Fig. 4).

As a next step, the groups of BGIs were projected on a map according to their type of function in the conurbation area of the city of Puebla (Fig. 5). We can observe Group 1 (52 BGIs) where they fulfill two to three functions, highlighting recreational, environmental, and social functions. In Group 2 (49 BGIs), we find parks of different sizes, but they fulfill three or four functions, usually except for the cultural function. In Group 3 (51 BGIs), we find small-sized areas that fulfill zero to two functions;

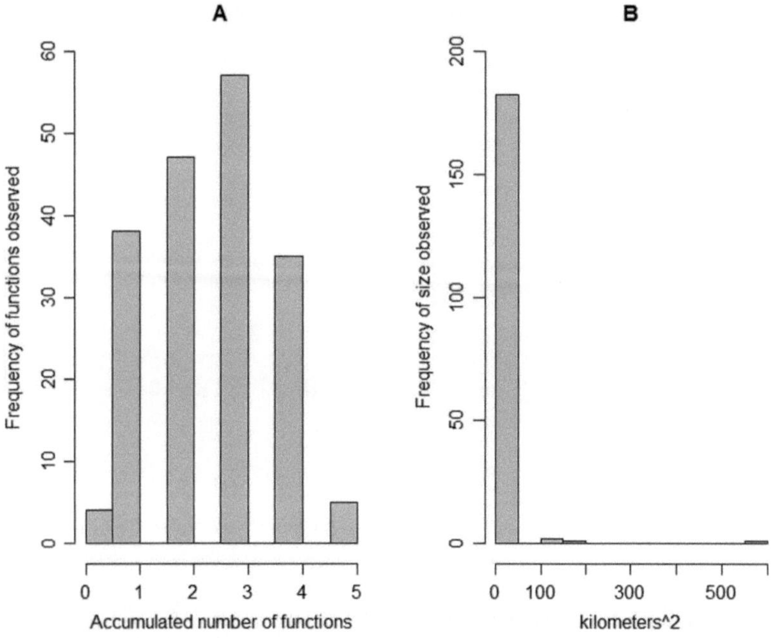

Fig. 2 Cumulative frequency functions and area of BGI in Puebla City and its conurbation. In graph A, we find the frequency of functions observed in each BGI (Y-axis) by cumulative number of functions (X-axis). In graph B, we find the frequency of size of BGI observed (Y-axis) according to their surface area in square kilometers (X-axis)

these functions can vary, but the environmental function generally predominates. Group 4 (30 BGIs) fulfills two to three functions, mainly recreational, health, and social functions. In Group 5 (4 BGIs), we find parks that fulfill all functions and have a large surface area.

Seventy-eight percent of urban green areas in Puebla fulfill less than 3 functions and have an average surface area of 4,391 m^2. The main function fulfilled is the environmental one (77.4%) due to the presence of vegetation, while the least fulfilled is the cultural function (2.6%). This demonstrates that in terms of the multifunctionality of BGIs in Puebla, there are opportunities for developing the potential of these existing spaces, which could improve the level of service and, consequently, the quality of life for users. Finally, we find that size is a factor that determines multifunctionality, as it provides additional space for the development of other functions.

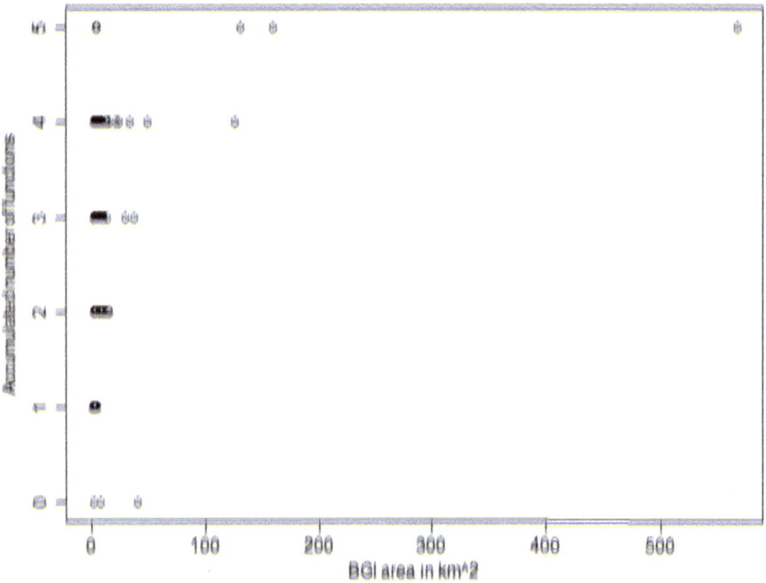

Fig. 3 Relationship between the surface area of BGI and the number of cumulative functions. On the X-axis, the size of BGI is observed in square kilometers, and on the Y-axis, the number of cumulative functions for each BGI

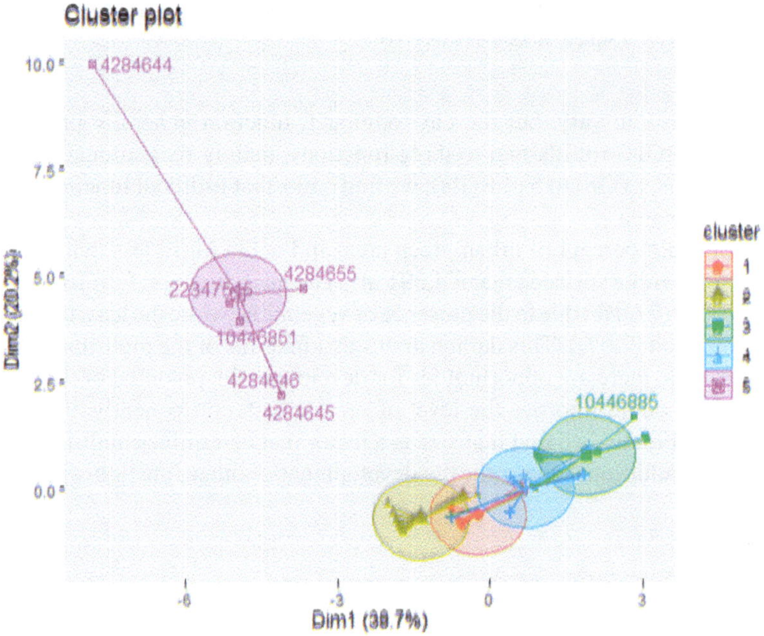

Fig. 4 Clustering plot of BGI in the city of Puebla and its conurbated area. Based on their similarity in functions and surface area, using the k-means method. Numbers near each group correspond to the ID of each BGI

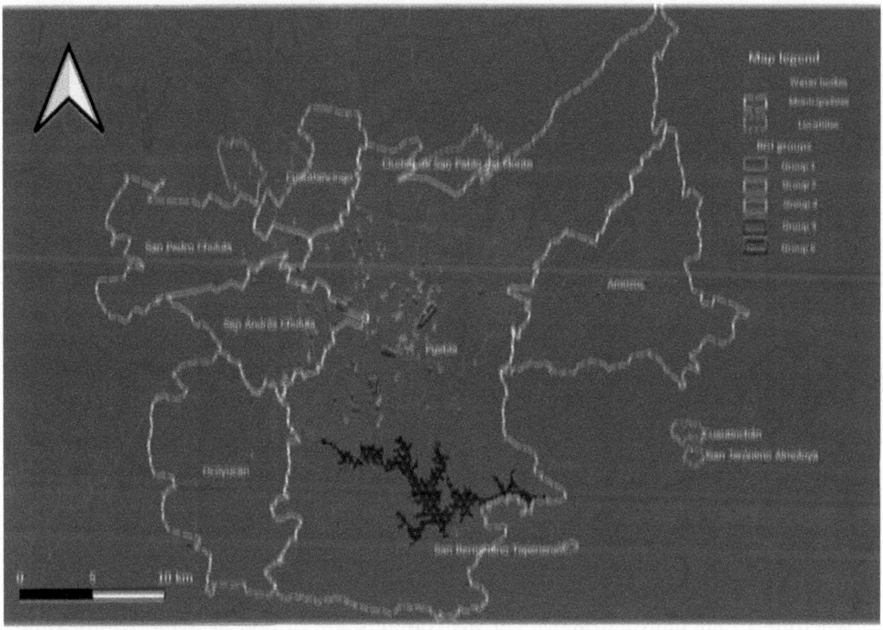

Fig. 5 Map of the BGI of Puebla City and their metropolitan area according to their functionality

4 Governance Instruments to Promote Multifunctionality of BGI in Puebla

The multifunctionality of blue and green infrastructure in developing countries' cities depends not only on their biophysical characteristics but also on the governance instruments used for their planning, implementation, and management (Reuvekamp, 2020). City or country regulations and laws can promote the multifunctionality of green and blue infrastructure if they are designed to encourage their use and protection. For example, land use regulations may require the inclusion of green areas or water bodies in new construction projects, or territorial planning policies may establish priority zones for biodiversity conservation and restoration. Additionally, governance instruments can facilitate citizen participation in decision-making and the management of blue and green infrastructure, enhancing its accessibility and use by the population. Therefore, it is essential to analyze the role of governance instruments in promoting the multifunctionality of blue and green infrastructure in developing countries' cities to identify opportunities for improvement and strengthen their impact on the population's quality of life.

In this section, we analyze the regulations related to BGI at different levels of government in Puebla, i.e., international, national, state, and municipal. The objective is to determine whether the regulations and public policy instruments for BGI management promote their multifunctionality.

Legislation can be defined as the set of current norms in a specific space and time, encompassing the rights and obligations of both governed and governing entities (Santos, 2008). In urban matters, we have urban law, which groups and systematizes all the legal norms that regulate the process of comprehensive administrative urban development, ranging from planning, organization, execution, control, and evaluation. This is to ensure that human settlements develop appropriately, with the greatest efficiency and effectiveness, and not simply grow in the least time and at the lowest cost (Ugalde, 2015).

The instruments of all governmental levels were consulted via Internet. Normative instruments tend to follow a hierarchy related to territorial scale, i.e., international, national, state, and municipal. Following this hierarchical order, the analysis began with international instruments such as the Sustainable Development Goals (SDGs) and the New Urban Agenda, which were consulted through their respective websites. To identify state-level instruments, the official website of the Government of Puebla was consulted under the section "Legislation of the State," where normative documents directly related to the topic of green areas were selected. For the selection of local regulations, the official website of the Government of Puebla was consulted under the section "Municipal Laws," where documents directly related to the topic of green areas were chosen.

We found that international-level instruments are aimed at establishing objectives and goals based on identified general problems. Therefore, they do not dictate specific paths to follow, and each country or region will be responsible for achieving them according to their conditions and resources. On the other hand, national-level instruments focus on organizing areas and on the organizational structure, granting powers and responsibilities to agencies, as well as general provisions to follow in the subject matter. State-level instruments dictate the responsibilities and attributions of agencies for the management of green areas. Finally, at the municipal level, we find laws and regulations that contain the responsibilities and attributions of administrative actors, along with specific processes to follow. Also, at this level, we find development plans and programs that set goals, objectives, actions, and strategies for their achievement.

4.1 Instruments at the International Level

Among the normative instruments at the international level, the Sustainable Development Goals (SDGs) and the New Urban Agenda stand out. The SDGs refer to access to BGI in a safe and inclusive manner, but they do not specifically mention their multifunctionality or characteristics. On the other hand, the New Urban Agenda does make a reference to multifunctionality and the design of these spaces to ensure human development. It even addresses their creation and maintenance for economic development and as an urban element for environmental resilience.

4.2 Instruments at the National Level

At this level, the most important instrument is the Constitution (Constitución Política de los Estados Unidos Mexicanos), in which the rights and individual guarantees are decreed. It specifically mentions the right to a healthy and quality environment. It also refers to the municipality's attributions and responsibilities, considering it as the basis of its territorial division and political and administrative organization. This means that the promotion, implementation, management, and maintenance of blue and green infrastructure falls under the responsibility of the municipality. Additionally, the municipal governments have the authority to approve regulations and administrative provisions that ensure citizen and neighborhood participation.

In the General Law of Ecological Balance and Environmental Protection (Ley General del Equilibrio Ecológico y Protección al Ambiente), the main emphasis is on the right and social responsibility to achieve a balanced environment. It highlights that the municipality is responsible for creating and managing green areas. It also emphasizes that the Federation, with the support of state and municipal governments, must promote the involvement and participation of different social groups in the creation and evaluation of environmental improvement programs.

The General Law of Human Settlements, Territorial Planning, and Urban Development (Ley General de Asentamientos Humanos, Ordenamiento Territorial y Desarrollo Urbano), refers to the right of all individuals to enjoy urban services and amenities, as well as the promotion by the State, through its governmental bodies, of a culture of civic and social co-responsibility. Therefore, one of its principles is to create habitable conditions for public spaces, which are fundamental elements for the right to a healthy life, social interaction, recreation, and citizen safety, considering the differentiated needs of individuals and groups. The law encourages the rescue, creation, and maintenance of public spaces, which may be expanded or improved but should never be destroyed or diminished. This law points out that the Ministry of Agrarian, Territorial, and Urban Development (Secretaría de Desarrollo Agrario, Territorial y Urbano) has the authority and responsibility to issue official norms for standardizing terminology for the prioritization of public spaces and facilities in urban development plans or programs. The law assigns to state legislation the provision of green areas, as well as the promotion and application of feasible and environmentally appropriate technologies for greater self-sufficiency, sustainability, and environmental protection. This includes the implementation of green roofs and vertical gardens. Similarly, it is the responsibility of municipalities to formulate, approve, and manage the zoning of population centers located within their territory. To do so, they issue applicable norms and technical provisions for the design or adaptation of green areas to ensure their consolidation, with sufficient and high-quality provisions. The law also stipulates that municipalities must define the best location and size of green areas according to the population served, maintain a balance between green areas and construction of infrastructure based on national norms regarding public spaces, and ensure their preservation. These spaces cannot be used for other purposes, and when developers and subdividers are obligated to cede plots to the municipality for green areas, they cannot be residual, located in flood-prone or risky zones, or have more complicated topographic conditions than the average subdivision or urban complex.

On the other hand, the objectives of the Official Mexican Standard for Public Spaces in Human Settlements (NOM-001-SEDATU-2021), are to standardize terminology, content, and methodologies regarding public spaces in urban development plans or programs. It aims to generate indicators and provide self-assessment tools. The standard presents a classification of public spaces, including urban green areas as public facilities, and specifies the terminology, dimensions, and some minimum design characteristics required for this classification. The standard indicates recommendations for proper provision, the importance of creating interconnections between public spaces and encouraging social participation through consultation, opinions, and deliberation to determine priorities and projects for public spaces. It also calls for monitoring the implementation of works, evaluating programs, and operating and managing these spaces, including all of it in urban development plans and programs.

4.3 Instruments at the State Level

At this level, one of the primary instruments related to urban green areas is the Organic Law of the Public Administration of the State of Puebla (Ley Orgánica de la Administración Pública del Estado de Puebla), that mentions the responsibilities and attributions of different instances of the state's public administration, such as the Secretariat of Environment, Sustainable Development, and Territorial Planning (Secretaría de Medio Ambiente, Desarrollo Sustentable y Ordenamiento Territorial), and the participation of Federal and Municipal Governments.

The Law of Subdivisions and Urban Actions (Ley de Fraccionamientos y Acciones Urbanísticas), distinguishes ecological areas, recognizing the former as recreation spaces, including parks, gardens, playgrounds, and sports fields. It also acknowledges "donation area," which is ceded to the municipality to be destined as ecological and equipment areas and should be equivalent to 20% of the total subdivided land, of which 50% must be allocated for urban facilities, recreation, and/or sports, and the rest for green areas. Once the donation is made, the promotion of projects for the construction of equipment, the management of spaces, as well as their maintenance and administration, falls under the municipality's guardianship.

The Law of Territorial Planning and Urban Development (Ley de Ordenamiento Territorial y Desarrollo Urbano), is aimed at determining the responsibilities and attributions of different levels of government in the field. Again, the creation, promotion, protection, and administration of urban green areas fall under the municipality's jurisdiction, as well as the development of norms, regulations, and other instruments for the application of the law. It contains all the regulations for this process, emphasizing green areas as part of public space. It stipulates that priority should be given to designing, adapting, maintaining, and protecting public spaces, always considering the city's evolution. Municipal plans or programs of urban development determine the provision of public space. Likewise, these instruments include the necessary aspects for their utilization and protection, defining the characteristics to ensure connectivity and adaptation to different densities over time. They define the best location and size in accordance with current norms and guidelines. These instruments contain the necessary processes and tools for the involvement of society in the entire planning process. The law includes principles of public policy, with a primary focus on the protection and progressivity of public space, recognizing the importance of these spaces for social interaction, recreation, and citizen safety, with the purpose of protecting, creating, and maintaining them.

One of the highlighted evaluation criteria is multifunctionality, indicating that projects related to space should be conceived for multiple uses in different times and areas. In general, for the processes and instruments applicable to territorial planning, this law establishes the provision of safe, sufficient, and high-quality green areas (public space), as well as their allocation in areas lacking such spaces to ensure universal access. It also contemplates the promotion and application of feasible and environmentally appropriate technologies, such as green roofs and vertical gardens, for greater self-sufficiency, sustainability, and environmental protection. The plans and programs determine the necessary measures to achieve the goals, as well as the applicable norms and technical provisions for the design or adaptation of green spaces.

Furthermore, this law enunciates basic norms for the organization of human settlements, highlighting, for example, that public spaces will be in central areas of each population center to correspond with the public character of such spaces, enhancing the surroundings' image. It also declares that reforestation of these spaces should consider native species to the area or others compatible with it to improve the environment and appearance.

On the other hand, within the diagnosis/forecast contained in the State Program of Sustainable Urban Development (Programa Estatal de Desarrollo Urbano Sustentable), green areas are classified as urban facilities, which are mentioned in the urban-social guidelines. The intention is to ensure their maintenance, improvement, expansion, and utilization, while regulating their distribution to guarantee a safe, productive, and healthy life for the inhabitants of population centers. Moreover, the allocation of facilities is recognized as a development catalyst. The territorial guidelines emphasize the need to adapt urban activities to the physiographic conditions by allocating the lands of the highest natural quality or vulnerability to parks, green areas, visual corridors, or for climate improvement. This entails the creation and reinforcement of instruments to control urban and environmental processes, as well as the expansion of regional facilities. Urban development strategies include strengthening facilities, particularly regulating the provision of regional facilities, for instance, the establishment of regional parks for recreation and sports.

In the realm of sustainable urban planning, it is recommended that urban facilities be preferably located along primary roads, close to urban development units, to facilitate access and increase options for their use, emphasizing the importance of creating or rehabilitating multipurpose green spaces throughout urban centers to meet recreation needs, thus contributing to biodiversity preservation, air pollution reduction, and improved microclimate. Regarding infrastructure, facilities, and services provision, updating the norms related to the coverage and characteristics of regional and urban facilities is stipulated. The promotion of social participation in urban matters is globally encouraged, although specific relations to urban green areas are not explicitly mentioned.

The Climate Change Law of the State of Puebla (Ley de Cambio Climático del Estado de Puebla), considers actions for climate change adaptation, such as increasing green areas, implementing measures to reduce the heat island effect, and protecting natural vegetation. The law generally expresses that state and municipal authorities regarding climate change must promote the society's joint and responsible participation in the planning, execution, and monitoring of the state's climate change policy.

The Law for the Protection of Natural Environment and Sustainable Development (Ley para la Protección del Ambiente Natural y el Desarrollo Sustentable) is to define principles of environmental public policy, asserting that every person has the right to enjoy a suitable natural environment for their development, health, and well-being. Thus, authorities will take the necessary measures to preserve this right, and consequently, they must assume responsibility for environmental protection. Another principle involves directing relations between society and nature through the concentration and collaboration of environmental actions between the state and people. Urban parks, public gardens, corridors, pathways, and other analogous areas planned by the jurisdiction are integrated into ecological preservation areas of population centers, which are designated for preserving essential natural elements for their well-being. Municipalities are responsible for establishing, regulating, and enforcing protection measures for municipal ecological preservation areas and urban parks in population centers.

Moreover, regulations for the Law for the Protection of Natural Environment and Sustainable Development, concerning Protected Natural Areas (Ley para la Protección del Ambiente Natural y el Desarrollo Sostenible), Prevention and Control of Air Pollution (Prevención y Control de la Contaminación Atmosférica), Environmental Impact Assessment and Risk Evaluation (Evaluación del Impacto y Riesgo Ambiental), and Prevention of Visual Pollution (Prevención de la Contaminación Visual) were reviewed. None of them contains specifications regarding the dimensions, design, or functions of BGI, nor the management or implementation processes, or social participation in these matters.

4.4 Instruments at the Municipal Level

The regulations of the municipalities that make up the Metropolitan Area of the City of Puebla were reviewed. The term "urban green area or blue–green infrastructure" and its classification are used differently in all the reviewed documents. Thus, there is no generalized concept of these spaces, nor a classification that tracks them at different scales within these instruments, leading to confusion and complications in their management. There is a concern about the issue at hand, as BGI is recognized to play an important role in the urban environment. Efforts are being made to incorporate it into urban legislation, especially at the international and national levels, but their inclusion is less pronounced at the state level and very scattered at the municipal level. While the international and national instruments recognize BGIs as part of the urban structure and city equipment, their multiple socio-environmental functions are not fully acknowledged, except for the Official Mexican Standard NOM-001-SEDATU-2021.

Overall, the regulations are quite lax, lacking specific guidelines for the regulation and management of these areas. While they mention what should be done, they do not provide detailed procedures for different specific cases, such as design requirements. The latter standard (NOM-001-SEDATU-2021) does specify some BGI implementation details and dimensions. In contrast, other municipal-level norms refer to SEDESOL's urban equipment guidelines. Some instruments do not clearly distinguish between the application of norms in natural or urban spaces, leading to generalized and sometimes ambiguous implementation. An exclusive instrument for improving the urban natural environment would be useful.

At the state level, secretariats and other government entities have the same denomination and functions, which differs at the municipal level, where departments, offices, etc., may have different names and functions. This lack of uniformity might not necessarily be a problem, but given the number of tasks assigned to municipalities, more order would be beneficial to ensure all functions are properly carried out. In this regard, it seems that these instances are insufficient. A positive aspect is the recognition of territorial challenges in terms of administration in the conurbated zones. Some instruments at this level are indicated, and the participation of municipal authorities that form part of these areas is specified for the entire planning process, from diagnosis to evaluation, as well as the formulation of necessary instruments for development and coordination in those areas.

Regarding social participation, most documents mention it, and some even propose instruments and actors for its development. However, except for the case of San Pedro Cholula, which proposes employing BGI users' labor for maintenance, none refers to how people can be involved in the management, creation, or maintenance processes. It is evident that the normative framework needs to be strengthened, especially at the municipal level, as there is repetitive information in its content. Each territory requires evaluation to formulate norms that respond to its unique context, as the problems are not common to all of them. Moreover, this legal framework must be comprehensive, including all the topics that affect the municipality, such as BGIs. The bottom line is therefore that even though some proxy for BGI is present in the national and state legislation, it is to be developed and operated by the municipal level. But as in many other cases, such responsibilities have not been translated into specific programs that can be implemented, and municipal governments are not granted the economic resources to truly instrument and operate BGI.

5 Conclusion and the Challenges Towards a BGI in Puebla

As shown, there is an incipient green infrastructure in Puebla with varying degrees of multifunctionality, but still limited in its coverage and functions. However, the rules and regulations for its governance are already being shaped. This is indisputably a first step.

The results of this research show that those urban facilities that, in addition to having green elements, also have bodies of water, have greater functionality. Although this has not yet been considered in depth for this chapter, from the reviewed literature, it is easy to notice the need to include blue components, that is, elements for water management, in view of the ever more frequent and intense rainfall events together with drier Summers.

Puebla and its metropolitan area, as well as most cities in Mexico, suffer all the difficulties mentioned above that are expected to become increasingly challenging to their operation and livability. At the same time, they are prone to heatwaves that affect its population. It is therefore required that green and blue infrastructure be integrated taking advantage of the existing green spaces with water bodies capable of addressing these extreme conditions, while at the same time allowing for adaptation to climate change.

This research identified green areas with low functionality. Specific actions may increase their level of service through community participation, as well as the modification of aspects of urban legislation, starting with the homologation and classification of the urban green area concept, the creation of instruments for their regulation, implementation, and use, in addition to improving the monitoring and control mechanisms for the application of these regulations. All these actions are municipal prerogatives, but municipalities have few resources. These mandates should come with sufficient resources so local authorities can develop BGI and creative ways to finance their operation. There is evidence for this in Mexico, where in some municipalities the operation of such infrastructure is funded through a fee charged by the local water operator. Other examples have developed in Ecuador and Argentina, receiving seed money from the Interamerican Development Bank.

In terms of physical infrastructure, the next step is scaling the green to blue infrastructure. This will require three conditions and collaborative planning. First, is finding and creating enough urban sponges for detaining, retaining, and delaying the flow of rainwater within the metropolitan area. Second, identify blue–green spots on which sponges can deposit excess water for discharge and storage. This will embed a close collaboration of state, municipal and regional water authorities for integrating and operating the grid. Third, pinpoint emergency valves for the blue–green network and overflow areas where water can temporarily go in extreme situations (Dolman, 2021).

Regarding institutional arrangements, another suggestion is to use the Government Assessment Tool (GAT) used by Flores et al. (2019), to assess the leapfrogging potential, based on contextual factors. As mentioned, this has been done for San Pedro Cholula and could be applied to consider expanding BGI.

Acknowledgements María del Socorro Ramírez Miranda expresses gratitude to the National Council of Humanities, Science and Technology (CONAHCyT) for the scholarship granted (CVU 710454) to carry out this research work within the framework of the Master's program in Habitat and Socio-Territorial Equity (MHEST) at the Universidad Iberoamericana Puebla. The authors would like to thank Jerónimo Chavarría Hernández for his support in searching for information on BGI polygons in INEGI and the guidance provided for their projection in GIS programs.

References

Abass, K., Afriyie, K., & Gyasi, R. M. (2018). From green to grey: The dynamics of land use/land cover change in urban Ghana. *Landscape Research 44*, 909–921. https://doi.org/10.1080/01426397.2018.1552251.

Aghakouchak, A., Chiang, F., Huning, L. S., Love, C. A., Mallakpour, I., Mazdiyasni, O., Moftakhari, H., Papalexiou, S. M., Ragno, E., & Sadegh, M. (2020). Climate extremes and compound hazards in a warming world. *The Annual Review of Earth and Planetary Sciences, 48*, 519–567. https://doi.org/10.1146/annurev-earth-071719.

Deilami, K., Kamruzzaman, M., & Liu, Y. (2018). Urban heat island effect: A systematic review of spatio-temporal factors, data, methods, and mitigation measures. *International Journal of Applied Earth Observation and Geoinformation, 67*, 30–42. https://doi.org/10.1016/J.JAG.2017.12.009.

Díaz, S., Pascual, U., Stenseke, M., Martín-López, B., Watson, R. T., Molnár, Z., Hill, R., Chan, K. M. A., Baste, I. A., & Brauman, K. A. (2018). KA (2018) Assessing nature's contributions to people. *Science, 359*, 270–272.

Dobbs, C., Nitschke, C., & Kendal, D. (2017). Assessing the drivers shaping global patterns of urban vegetation landscape structure. *Science of the Total Environment, 592*, 171–177. https://doi.org/10.1016/j.scitotenv.2017.03.058.

Dolman, N. (2021). Integration of water management and urban design for climate resilient cities. In: de Graaf-van Dinther, R (Ed.), *Climate resilient urban areas: Palgrave studies in climate resilient societies* (pp. 21–43). Palgrave Macmillan. https://doi.org/10.1007/978-3-030-57537-3_2

Flores, C. C., Crompvoets, J., Ibarraran Viniegra, M. E., & Farrelly, M. (2019). Governance assessment of the flood's infrastructure policy in San Pedro Cholula, Mexico: Potential for a leapfrog to water sensitive. *Sustainability, 11*, 7144. https://doi.org/10.3390/SU11247144.

Ghofrani, Z., Sposito, V., & Faggian, R. (2017). A comprehensive review of blue–green infrastructure concepts. *International Journal of Environment and Sustainability, 6*, 15–36.

Gleason, J. A., & Flores, C. C. (2021). Challenges of water sensitive cities in Mexico: The case of the metropolitan area of Guadalajara. *Water, 13*, 601. https://doi.org/10.3390/W13050601.

González, J. A. R., & Aguilar, A. G. (2021). Irregular urban expansion, change in land use and environmental deterioration in the northern periphery of the Puebla-Tlaxcala Metropolitan zone: The case of la malinche national park. *Cuadernos de Geografia: Revista Colombiana de Geografia, 30*, 441–458. https://doi.org/10.15446/rcdg.v30n2.89849.

Hansen, R., & Pauleit, S. (2014). From multifunctionality to multiple ecosystem services? A conceptual framework for multifunctionality in green infrastructure planning for Urban Areas. *Ambio, 43*, 516–529. https://doi.org/10.1007/S13280-014-0510-2/FIGURES/5.

Hibbs, B. J., & Sharp, J. M. (2012). Hydrogeological impacts of urbanization. *Environmental & Engineering Geoscience, 18*, 3–24. https://doi.org/10.2113/GSEEGEOSCI.18.1.3.

Holling, C. S. (1973). Resilience and stability of ecological systems. *Annual Review of Ecology and Systematics, 4*, 1–23. https://doi.org/10.1146/annurev.es.04.110173.000245.

Hunt, A., & Watkiss, P. (2011). Climate change impacts and adaptation in cities: A review of the literature. *Climate Change, 104*, 13–49. https://doi.org/10.1007/S10584-010-9975-6/METRICS.

INEGI. (2020). *Población rural y urbana*. Cuéntame de Mexico.

Lai, D., Liu, W., Gan, T., Liu, K., & Chen, Q. (2019). A review of mitigating strategies to improve the thermal environment and thermal comfort in urban outdoor spaces. *Science of the Total Environment, 661*, 337–353. https://doi.org/10.1016/J.SCITOTENV.2019.01.062.

Lee, A. C. K., & Maheswaran, R. (2011). The health benefits of urban green spaces: A review of the evidence. *J Public Health (Bangkok), 33*, 212–222. https://doi.org/10.1093/PUBMED/FDQ068.

Mahtta, R., Fragkias, M., Güneralp, B., Mahendra, A., Reba, M., Wentz, E. A., & Seto, K. C. (2022). Urban land expansion: the role of population and economic growth for 300+ cities. *NPJ Urban Sustainability, 2*, 1–11. https://doi.org/10.1038/s42949-022-00048-y.

Manteghi, G., bin Limit, H., & Remaz, D. (2015). Water bodies an urban microclimate: A review. *Mathematical Models and Methods in Applied Sciences, 9*, 1.

Meenar, M., Heckert, M., & Adlakha, D. (2022). "Green enough ain't good enough:" Public perceptions and emotions related to green infrastructure in environmental justice communities. *International Journal of Environmental Research and Public Health, 19*, 1448. https://doi.org/10.3390/IJERPH19031448.

O'Donnell, E. C., Netusil, N. R., Chan, F. K. S., Dolman, N. J., & Gosling, S. N. (2021). International perceptions of urban blue–green infrastructure: A comparison across four cities. *Water, 13*, 544. https://doi.org/10.3390/W13040544.

Pérez Castresana, G., Tamariz Flores, V., López Reyes, L., Hernández Aldana, F., Castelán Vega, R., Morán Perales, J., García Suastegui, W., Díaz Fonseca, A., & Handal Silva, A. (2018). Atoyac river pollution in the metropolitan area of Puebla, México. *Water (basel), 10*, 267. https://doi.org/10.3390/w10030267.

R Core Team. (2021). R: A language and environment for statistical computing.

Reuvekamp, T. (2020). *Effective policy instruments to successfully involve communities in the implementation of green-blue measures in the public space*. Master thesis, University of Twente.

Riechers, M., Barkmann, J., & Teja, T. (2018). Diverging perceptions by social groups on cultural ecosystem services provided by urban green. *Landscape and Urban Planning, 175*, 161–168.

Sánchez Juárez, C. J. (2021). *Áreas verdes en la ciudad de Puebla: Reflejo de la desigualdad del espacio público y componente del derecho a la ciudad*. Master thesis, BUAP.

Sandoval Chávez, D. A., Córdova y Vázquez, A., Cervantes Rendón, E., Cervera Gómez, L. E., Reyes Escalante, A. Y. (2021). Valoración económica de la multifuncionalidad de los parques urbanos. *Revista de Economía, Facultad de Economía, Universidad Autónoma de Yucatán, 38*, 93–123. https://doi.org/10.33937/reveco.2021.176.

Santos, H. (2008). La legislación y el derecho actualmente en México. In *Trabajo presentado en Congres virtual interinstitucional los grandes problemas nacionales*. Nuevo León.

SEMARNAT. (2013). Estrategia Nacional de Cambio Climático. México: Gobierno de la República.

Seto, K. C., Fragkias, M., Güneralp, B., & Reilly, M. K. (2011). A meta-analysis of global urban land expansion. *PLoS ONE, 6*, e23777. https://doi.org/10.1371/journal.pone.0023777.

Seto, K. C., & Shepherd, J. M. (2009). Global urban land-use trends and climate impacts. *Current Opinion in Environment Sustainability, 1*, 89–95. https://doi.org/10.1016/j.cosust.2009.07.012.

Siehr, S. A., Sun, M., & Aranda Nucamendi, J. L. (2022). Blue-green infrastructure for climate resilience and urban multifunctionality in Chinese cities. *Wiley Interdiscip Rev Energy Environ, 11*, e447. https://doi.org/10.1002/WENE.447.

Smart, N., Eisenman, T. S., & Karvonen, A. (2020). Street tree density and distribution: An international analysis of five capital cities. *Frontiers in Ecology and Evolution, 8*, 562646. https://doi.org/10.3389/FEVO.2020.562646/BIBTEX.

Soto-Montes-de-Oca, G., & Alfie-Cohen, M. (2019). Impact of climate change in Mexican peri-urban areas with risk of drought. *Journal of Arid Environments, 162*, 74–88. https://doi.org/10.1016/J.JARIDENV.2018.10.006.

Steinfort, U., Contreras, A., Albornoz, F., Reyes-Paecke, S., Guilleminot, P., Steinfort, U., Contreras, A., Albornoz, F., Reyes-Paecke, S., & Guilleminot, P. (2020). Vegetation survival and condition in public green spaces after their establishment: Evidence from a semi-arid metropolis. *International Journal of Agriculture and Natural Resources, 47*, 90–104. https://doi.org/10.7764/ijanr.V47I2.2045.

Sun, G., & Caldwell, P. (2015). Impacts of urbanization on stream water quantity and quality in the United States on JSTOR. *Water Resources IMPACT, 17*, 17–20. https://www.jstor.org/stable/wateresoimpa.17.1.0017.

Tam, B. Y., Gough, W. A., & Mohsin, T. (2015). The impact of urbanization and the urban heat island effect on day to day temperature variation. *Urban Clim, 12*, 1–10. https://doi.org/10.1016/J.UCLIM.2014.12.004.

Téllez Montes Y de las, N., & Castro Campos, J. (2020). Parques Urbanos de la Ciudad De México: ¿Son suficientes? In: S. D. L. V Estrada, R. E. R. Luter, G. D. C. H. Castillo (Eds.), *Factores críticos y estratégicos en la interacción territorial, desafíos actuales y escenarios futuros* (pp. 1125–1141). Universidad Nacional Autónoma de México y Asociación Mexicana de Ciencias para el Desarrollo Regional A.C, Ciudad de México.

Ugalde, V. (2015). Derecho a la ciudad, derechos en la ciudad. *Estud Demogr Urbanos Col Mex, 30*, 567–595.

Venter, O., Sanderson, E. W., Magrach, A., Allan, J. R., Beher, J., Jones, K. R., Possingham, H. P., Laurance, W. F., Wood, P., Fekete, B. M., Levy, M. A., & Watson, J. E. M. (2016). Sixteen years of change in the global terrestrial human footprint and implications for biodiversity conservation. *Nature Communications, 7*, 12558. https://doi.org/10.1038/ncomms12558.

Walker B, Salt D (2012) Resilience Thinking: Sustaining Ecosystems and People in a Changing World. Island Press.

Weinbrenner, H., Breithut, J., Hebermehl, W., Kaufmann, A., Klinger, T., Palm, T., & Wirth, K. (2021). "The forest has become our new living room"—The critical importance of urban forests during the COVID-19 pandemic. *Frontiers in Forests and Global Change, 4*, 68. https://doi.org/10.3389/ffgc.2021.672909.

Wilbers, G. J., de Bruin, K., Seifert-Dähnn, I., Lekkerkerk, W., Li, H., & Budding-Polo Ballinas, M. (2022). Investing in urban blue–green infrastructure—Assessing the costs and benefits of stormwater management in a Peri-urban catchment in Oslo, Norway. *Sustainability (switzerland), 14*, 1934. https://doi.org/10.3390/SU14031934/S1.

Blue–Green Infrastructure as a Stimulant of Growth: Case of Ahmedabad

Vidhu Bansal and Hareesh Haridasan

Abstract Blue Green Infrastructure (BGI), as a concept, utilises natural ecosystems and green cover in urban areas to create an environment which is sustainable. This includes the use of features such as water bodies, parks, urban forests, gardens, and wetlands, to name a few, to enhance the quality of life in cities. Green spaces in an urban environment create breathing space for citizens, thus creating a healthy environment. A healthy workforce with diversified economic contributors is essential for a fast-growing economy like India. This paper discusses the role of BGI as a stimulant in economic and social progress. The present study proposes to examine the BGI-induced growth in the city of Ahmedabad through examples of various waterfront/riverfront development projects. The study would incorporate the Sabarmati Riverfront Development Project (10.4 km stretch) in Ahmedabad as an example of implementing BGI, creating a vibrant public space for the city's residents. Similarly, the Kankaria and Vastrapur Lake Development Projects would also be investigated through the lens of BGI. The study will also incorporate another area of Ahmedabad city near the Isanpur Lake, which is currently dilapidated. In ancient times, it was a small village near Ahmedabad, now embedded into the main city. A few landmarks of historical importance, like Jethabhai Stepwell and an old mosque, make it an area which has the potential to become a vibrant hub. The inputs gathered from successful studies of Ahmedabad and example from around the world would be used to develop an academic understanding of how to include BGI strategies from the early design development stages to produce more environmentally conscious urban design in the Isanpur Lake.

Keywords Ahmedabad lakes · BGI · Urban design · Urban growth · Water sensitive urban design (WSUD)

V. Bansal (✉)
Anant National University, Ahmedabad 382115, India
e-mail: ar.vidhupandey@gmail.com

H. Haridasan
Vellore Institute of Technology, Vellore, India

P. K. Joshi et al. (eds.), *Blue-Green Infrastructure for Sustainable Urban Settlements*,
https://doi.org/10.1007/978-3-031-62293-9_16

1 Introduction

The environment that we live in has been altered to best suit the conditions of a particular time, ranging from ancient human beings to modern-day *Homo sapiens*. Throughout their evolution, humans have consistently modified the natural setting for their benefit. Agriculture is an initiative in this direction. Subsequently, the human society evolved with new technological advances. Human beings started transitioning from a herd-based community to a rationally thinking community. It is this rational thinking ability that has resulted in scientific prowess. On the one hand, gaining scientific prowess and causing destruction are sides of the same coin. In his book 'On the Future—Prospects for Humanity', Martin Rees comments that humanity has reached a critical moment and has a serious existential crisis. The future can be better defined based on how we ethically harness science and technology to address our challenges. The biggest challenges we face today are attributed predominantly to climate change (Bhadouria et al., 2022; Singh et al., 2022). There is a global rise in temperature and reduction of green cover, and riverine systems are tapped extensively for agriculture and as a source of drinking water for the urban zones, over-exploitation of non-renewable energy sources, extensive mining, etc. (Bhadouria et al., 2023).

Climate change, although a natural phenomenon, is accelerated by human activities (O'Brien et al., 2006). Floods in riverine delta and urban areas have been common in recent decades. The frequent occurrence of such disasters deems it essential to have a disaster management system. A new approach is needed to underpin the incorporation of risk management into work on climate change and the introduction of climate change into natural hazards and development planning (O'Brien et al., 2006). There exists a bidirectional relation between economic growth, energy consumption, population growth, and carbon emissions (Chen et al., 2022). Urban planning professionals are ideating on various means to mitigate issues about floods and reduce ecological footprints.

An integrative approach called Blue–Green Infrastructure development is a step in this direction. Infrastructure development is usually correlated with colours such as grey, brown, and green, to name a few (Dreiseitl & Wanschura, 2016). The term grey field refers to properties previously developed without serious environmental issues and are economically non-viable in their current state. The term brownfield refers to properties previously used for industrial purposes but contaminated with low concentrations of hazardous materials (EPA, 2023). Greenfield, in contrast, refers to an underdeveloped parcel of land in either an urban or rural setting with a negligible amount of buildings constructed.

On the other hand, the term green cities refers to the sustainable, efficient, adaptive, and resilient way to design whole cities (Brears, 2018). Green cities recognise connections between different sectors and development strategies that fulfill multiple functions and create multiple benefits for society and urban ecosystems (Brears, 2018). On the other hand, blue–green infrastructure demands holistic planning and management of water, wastewater, and stormwater across the city to ensure that populations are resilient to climate change and extreme weather events while ensuring the health of aquatic ecosystems (Brears, 2018). Blue–Green Infrastructure (BGI) initiatives use blue elements such as riverine systems, canals, lakes, and wetlands, to name a few, along with green elements such as trees, forests, farms, fields, and parks in the planning process (Dolman & O'Donnel, 2021). In comparison to the traditional approach of wastewater management in cities, especially in the Global South, the BGI uses a more globalised and integrated approach in urban ecology management. The traditional approach uses a more localised approach to address concerns of a neighbourhood or a city zone that faces water logging or flooding issues. In this approach, urban water managers also use various innovative fiscal and non-fiscal tools to encourage the implementation of BGI on public and private property to sustainably manage water resources and increase resilience to climate risks (Brears, 2018).

BGI plays a critical role in improving the economic and social conditions in urban areas. It provides a range of economic benefits, including increased property values, improved public health, and reduced infrastructure costs. In developing countries, BGI can help address social and environmental issues such as poverty, inequality, and climate change. In western India, particularly the states of Gujarat and Rajasthan, water forms a key element of social fabric and also holds historical value. The step-wells or '*vav*' today are admired for the invaluable lessons they offer on sustainable living. In Ahmedabad, water structures and the river Sabarmati have historical significance in urban development. The city has a rich history of using water structures such as Adalaj *Vav* and Kankaria Lake to cater to the demands of its populace during droughts and for irrigation. These infrastructures have played a significant role in the development of the city. The projects involved the renovation of the two lakes by creating new recreational spaces and promoting tourism in the city. The project created new job opportunities and boosted local businesses, contributing to the city's economic growth. The present chapter attempts to understand the role of BGI as a catalyst of growth and sustainable development in the cities around the world with a focus on an Indian city, i.e. Ahmedabad. The chapter illustrated component of BGI for more clarity before looking at various initiatives taken in India regarding it. After that brief cases studies are taken up on two cases in Germany, one case in Korea, and one case in USA to comprehend a global perspective. The chapter then focuses on the city of Ahmedabad and investigates few already completed project like Kankaria Lake and Sabarmati Riverfront. The chapter ends on the study, analysis and inferences for development on the lake of Isanpur which is also situated in Ahmedabad.

2 Components of Blue–Green Infrastructure (BGI)

The Blue–Green Infrastructure (BGI) has been in use in the recent decades. This use has become prominent with increased occurrences of floods in particularly urban areas. The origin of the term BGI is ambiguous. However, it found prominence in the later part of the 1990s. It is also the same phase where there was a paradigm shift in the discourse on sustainability and the environment. BGI includes the wider spectrum of natural, artificial, and semi-natural components of a multifunctional environmental system (Ghofrani et al., 2017).

The types of interventions in BGI are urban forests, rain gardens, bioswale, and constructed wetlands, to name a few. These interventions help in mitigating issues of flooding. Specific interventions, such as rain gardens, help in recharging the groundwater by percolation of water from the surface. Such interventions negate the effects of excessive nonporous hardscapes, a common scene in many urban landscapes. The BGI components can be classified based on function, position, and scale (Ghofrani et al., 2017). The classification based on function helps reduce stormwater runoffs, i.e., detention and retention. Detention systems can store water after heavy rainfall and slowly discharge it to the sewer system, while retention systems allow for slow systematic percolation (Ghofrani et al., 2017). The second categorisation is based on its position, i.e., on ground, above ground, or subsurface. These components could be green roofs, a meticulously planned rainwater drain system from buildings, or large storage tanks for rainwater harvesting. The last classification is about scale, which discusses the level at which the intervention happens. Interventions at an urban or regional scale, such as retention ponds, canals, and urban forests, fall under this category.

BGI is not limited to urban spaces, and its planning can be considered at multiple levels and in various planning contexts, such as urban, peri-urban, regional, and rural planning (Ghofrani et al., 2017). BGI can be integrated with the conventional 'grey' urban drainage systems, including conveyance (sewer) systems and storage and treatment facilities, to form blue–green-grey infrastructure (Almaaitah et al., 2021).

3 BGI Initiatives in India

In Southeast Asia, projections of rapid urban growth coupled with high water-related risks call for large investments in infrastructure, including in blue–green infrastructure (BGI) such as forests, parks, or vegetated engineered systems like rain gardens, bioretention units, blue–green roofs, etc. (Hammel & Tan, 2022). Post-economic liberalisation, India has been consistently growing both as an economy and in urban population density. Issues such as pollution, lack of safe public space, deforestation, depletion of green cover, etc., were perceived to be issues predominantly seen in developed countries. The accelerated developments in India and other countries in the Global South have manifested the issues mentioned above in the new developing block. Most of the development of BGI was initiated first in the Global North and then slowly moved to the Global South. Most information about the BGI has case studies from the global north. This limits the understanding of the geographies of the global south and its potential impact on BGI. The best practices from the global north have been adopted in the global south.

Flooding can have devastating effects on socio-economic as well as environmental systems (O'Donnell et al., 2017). India's rich biodiversity is at risk due to the constant upgradation of infrastructure. In such a scenario, initiatives such as BGI will help mitigate the effects of floods, better water management, and environmental sustainability, along with catalysing economic growth in the long run. Such initiatives maintain a balance between growth and sustainable living. India, in the past two decades, has increased its investment in infrastructure at an exponential rate. The environment ministry, since its inception in the 1980s, has been working persistently on issues related to controlling air and water pollution and preserving forests, mangroves and other natural resources (Udas-Mankikar & Driver, 2021). Indian cities like Delhi, Bhopal, Madurai and Bengaluru to name a few are including blue–green initiatives in their development plans with the aim of enhancing the existing natural features through a planned and phased manner (Udas-Mankikar & Driver, 2021). The above-mentioned cities have high-density development and predominantly used mixed-use type of land use. The BGI initiatives are either incorporated as part of the Smart City Mission Plan or through the new development plans. The goal for BGI in all these cities is in line with global goals such as mitigating pollution levels, adapting to climate change, improving health, adapting to urban flooding, reviving or accelerating economic growth, or moving towards a low-carbon economy. The BGI initiatives are perceived as a course correction in urban planning practices. The new plans or action plans are based on frameworks set up by the National Action Plan on Climate Change (NAPCC) in 2008.

Evaluating global and Indian practices points toward a multi-level and integrated approach for BGI. Local government bodies are being updated with knowledge on BGI, which is a paradigm shift in approach towards climate change mitigation. For executing any project of such importance, an institutional framework is equally essential to achieve the objectives. BGI projects are likely to succeed when they operate at a local level with the participation of authorities and communities (Udas-Mankikar & Driver, 2021). However, there are some key factors affecting citizens' willingness to implement green infrastructure are efficacy, aesthetics, and cost (Baptiste et al., 2015). The concept of BGI is in a relatively development stage wherein broader objectives are understood and accepted. A framework or an all-encompassing plan that the city and its sustainability have a symbiotic relationship with the environment must be managed. The existing blue–green resources can be better-managed assets for a sustainable future.

4 Case Studies Around the World

The studies taken here are from three continents: America, Europe, and Asia. The idea is to appraise how water-sensitive urban design is being used around the world to make more livable cities and habitats. Such ideas need to be explored to understand the application and replication in a variety of contexts. The case study of Philadelphia explores water-sensitive urban design (WSUD) to manage their stormwater runoff and its allied issues. The case of Germany explores the cases of urban spaces, stormwater, rainwater harvesting and green surfaces. They intend to use urban spaces as their point of application so that citizens become more aware of these solutions and get inspired to utilise them more. In South Korea, the case of Cheonggyecheon is studied, which takes up the renewal of an almost dried stream and restores the associated blue–green corridor. The cases are also taken from Ahmedabad, India and the development surrounding the various water bodies and lakes. The final case investigated in this paper is the lake of Isanpur in the eastern part of Ahmedabad city and how it can be benefitted if BGI features, which include WSUD, are applied.

4.1 Case of Greened Acre, Philadelphia

In many cities around the world, stormwater is not properly harvested, and hence, it goes to waste as surface runoff. It can also assimilate with the sewer water and then flow into the rivers, causing contamination of freshwater sources. Such is the case of Philadelphia in the USA. Sixteen billion gallons of untreated wastewater runoff was recorded annually. This affected stream banks, depleted fish habitats, and carried faecal coliform bacteria, petroleum products and other contaminants into the river. The condition used to get worse during the time of storms and hurricanes when water from wastewater treatment plants flooded the city and the river (Stutz, 2018).

The city administration decided there was a need to introduce a change in the infrastructure strategies and put more emphasis on strengthening the blue–green infrastructure. The idea of Green Acre evolved, which incorporated the various impervious surfaces all over the city to be converted into active blue–green infrastructure. These impervious surfaces are in the form of large parking lots, extensive rooftops, vacant plots, and others. In most of these green acres, a large part is converted to a green area by planting resilient native species of plants which can thrive with minimum care. The water from rain is collected using rain barrels, cisterns, or water channels. The collected water is then pumped to these green areas, where it gets cleaned by moving through the soil. The water also percolates to recharge the groundwater (Evergreen, 2020; Stutz, 2018).

4.2 Case of Zollhallen Plaza and Potsdamer Plaza, Germany

Most of the urban public places or plazas around the world are conventionally hardscapes. Generally, they are made of materials like concrete, brick, or stones, with raised planters and ineffectual seatings and have very little consideration for maintaining or restoring ecological balance. As a result, the delicate relationship between water and such urban places gets underplayed. Therefore, integrated solutions for sustainable use and efficient supply of water, improving quality of water, and flood risk management are required while designing such spaces. Such initiatives have already taken shape in Germany, for instance, the Zollhallen Plaza, Freiburg and the Potsdamer Plaza, Berlin (Grau & Dreiseitl, 2013).

The **Zollhallen Plaza** (Fig. 1) has been designed with a water-sensitive approach where disconnection from the sewer system of the city is achieved successfully. Much attention has been given to water infiltration such that all the water that falls on the surface infiltrates into the subsoil. This is accomplished by maintaining the slope so that water always reaches the designed permeable areas. Also, excess water during heavy rainfall is collected via a drainage gradient in the water storage boxes, from where it slowly infiltrates into the subsoil (Pearlmutter et al., 2020). Even during a "100-year flood" situation, the major portion of the plaza will be underwater, besides a safety zone that will stay dry and be useful during disaster management (Palazzo, 2019).

The **Postdamer Plaza** (Fig. 2) goes a step further by combining sustainable usage of water and recreational areas to make the issues related to water perceptible to the users. It is designed as an urban oasis with an interplay of nature and art. Around 60% roofs of the buildings on the plaza are green roofs, and they reduce the summer temperatures by 2 °C approximately (Nowobilska & Zaman, 2014). They also assist in rainwater harvesting, where harvested water is used for irrigation, toilets, and fire extinguishing systems. Excess rainwater is also stored in underground tanks and further discharged into narrow pools along the streets and walkways, where it is filtered by biotopes. The water features also moderate the microclimate as the moisture tempers the extreme temperature and also binds the dust particles in the air (Lehmann, 2012). The water bodies also act as pause points amidst the chaos of

Fig. 1 Zollhallen plaza (*Source* Google Earth)

Fig. 2 Postdamer plaza (*Source* Google Earth)

the commercial area. Both squares are examples of how ecological sensitivity can be attained by combining aesthetics and technology.

4.3 Case of Cheonggyecheon, South Korea

Cheonggyecheon in Seoul, South Korea, is a major urban renewal project that continues to be a massive success. Earlier, it was a 10-lane roadway and 4-lane elevated highway along the Cheonggyecheon stream and carried more than 170 thousand vehicles daily (Lee & Anderson, 2013). Besides the structural safety issues, the business district of Cheonggyecheon was on the downfall in almost all aspects (Fig. 3). In 2003, it was envisioned to restore the almost dry stream. The restoration led to a 10.9 km blue–green corridor for bicyclists, pedestrians, flora, and fauna. Nature-based landscape features like native willow swamps, marshes, shallows, fish spawning ground, terraced vertical walls, and natural stones to cross the stream make it not only a recreation spot for humans but also a habitat for amphibians, fish, birds, and insects (Jeon & Kang, 2019). With these interventions, the area has achieved flood protection for up to a 200-year event. While there has been an increase in the overall diversity by 639%, a significant reduction in the urban heat island effect, air pollution due to dust, and noise levels is also observed (Cho, 2010). The project not only revitalised the natural ecosystem of the area but also increased property values, improved air quality, and reduced traffic congestion. This, in turn, had an impact on social life.

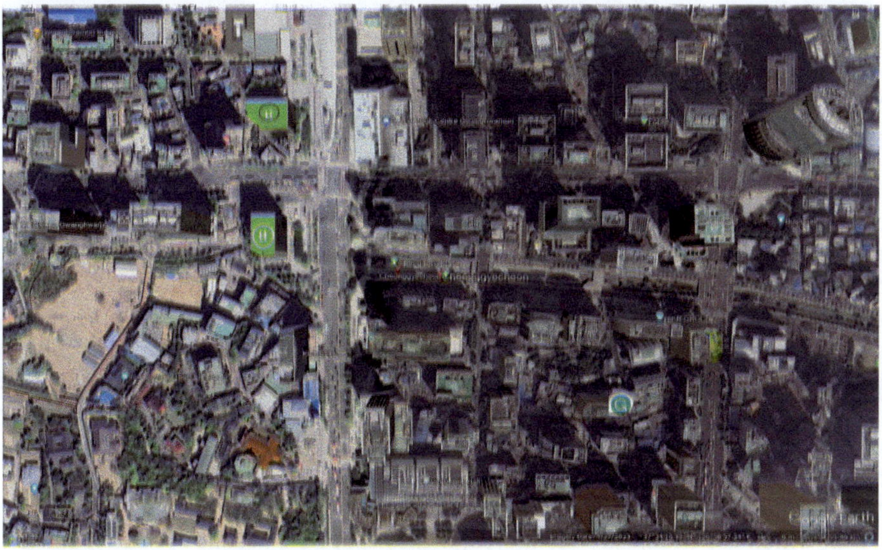

Fig. 3 Cheonggyecheon, South Korea (*Source* Google Earth)

5 Cases from Ahmedabad, India

The city of Ahmedabad, founded in the 15th century by Sultan Ahmad Shah on the banks of river Sabarmati, presents a rich urban heritage. For the cities of medieval eras, a river or a water source is an essential requirement since it is a basic necessity for life. Today, Ahmedabad is one of the fastest-growing urban centres. Since its founding, the city has had a close relationship with water sources. The river is one of them. The city is adorned with lakes and tanks. Kankaria Lake, Vastrapur Lake, Chandola Lake, Thaltej Lake, Chandlodia Lake and Isanpur Lake are some of the prominent ones within the urban limit.

Ahmedabad, with a population of over 5 million, is a commercial centre today. It is subject to a range of socio-economic and ecological pressures which influence the city's landscape (Mell, 2018). Ahmedabad's development plan outlines the importance of BGI or green infrastructure. Utilising ecological networks, the plan explores the value of integrating landscape projects to facilitate a spatially functional landscape resource base (Mell, 2018). Since the 2000s, local government agencies have been investing in mega projects such as the Kankaria Lake Redevelopment and the Sabarmati Riverfront Project. These projects leverage the existing ecological elements, namely lake and river, into an enhanced public space (Adhvaryu, 2011).

5.1 The Kankaria Lake Redevelopment Project

The historical records mention this site as *Hauz-e-Qutub*. The term *hauz* refers to a tank. It was built in 1451 during the reign of Sultan Qutb-ud-din Ahmad Shah. The lakefront was upgraded in 2006–08 at a cost of INR 36 crores (HUDCO, 2020). The project was undertaken by the Ahmedabad Municipal Corporation. The project created spaces for pedestrian infrastructure, such as seatings and pathways along the periphery of the lake. The existing traffic network was reworked along the periphery of the lake. Systematic entry points to the lakefront were defined for the users. The recreational potential of the site was fully explored by making provisions for toy trains, water rides, zoos, and eateries. The project also undertook the restoration of historic structures within the lake, such as Naginawadi, which is an island in the middle of the lake (Fig. 4). The lakefront development touched upon various facets pertaining to ecology as well as economy. The prime one is using the site for tourism. Since the tank also doubled up as space for collecting rainwater at an urban level, this redevelopment project helped improve the retention capacity. The quality of water, which is periodically monitored, makes space for aquatic species to make it its home. The recreational facilities along the lakefront contribute to the economic development.

Fig. 4 Satellite image of the Kankaria lake circa 2023 (*Source* Google Earth)

5.2 Sabarmati Riverfront Development

River Sabarmati is a non-perennial type. It depends on the monsoon for its supply of water. The river has been a witness to historical events in the city, right from its foundation to subsequent transitions. The city began to industrialise and grow during the 20th century and subsequently became a hub for the textile industry (Joshi, 2020). With this industrialisation and non-perennial nature, the riverbed was used for various activities such as laundry, dye textiles, and informal hutments that came along riverbanks, resulting in an unhygienic situation (Joshi, 2020). Industrial effluents and untreated domestic wastes were dumped into the river. To address the issue, efforts were first made in the 1960s in the form of feasibility studies and a proposal by French architect Bernard Kohn, along with architects Kamal Mangaldas, Anil Bakeri, and a few others. The proposal called for reclamation of about 30 hectares, of which a part was to be sold or leased for commercial development. A special-purpose vehicle called Sabarmati Riverfront Development Corporation (SRFDCL) was formed to execute this project. The Environmental Planning Collaborative (EPC) was tasked with the preparation of a comprehensive feasibility study, and the project effectively started its execution phase post-2004 (Joshi, 2020). The project had a multi-dimensional objective towards environmental improvement, social upliftment, and urban rejuvenation of Ahmedabad (Joshi, 2020) (Figs. 5, 6 and 7). Through this project, the risk of erosion and flooding in flood-prone neighbourhoods is controlled. It also provides public space with additional green space in the city. The additional parcels of land generated through the reclamation is used for commercial or cultural purpose. This project is seen as a landmark project which is known for its efficient execution.

Both the above-mentioned projects have an equal number of public criticisms. Both projects were criticised for their anti-poor narrative and displacements of communities. The criticism fails to discuss the legality of encroachment of public lands. In 2017, the lower promenade was submerged due to the release from upstream Dharoi Dam. The lower promenade getting submerged was seen as an anticipated scenario. Such flooding is linked to the level of rainfall, and it would soon breach the flooding level mark, which is set by the authorities. In such a scenario, the BGI initiates prove to be less flexible as well as accommodative to the changes.

5.3 Case of Isanpur Lake

There are around 142 lakes in the city of Ahmedabad that can be a potential sink for stormwater runoff. Each year, the city gets severely inundated during the rainy season. Around 33 such lakes are found in close to 42 waterlogging locations. More than five hectare-7 lakes, between 2 and 5 ha, 37 lakes, less than 2 hectares-98 lakes. Bureaucratic delays in recognising the water bodies as lakes with no record in revenue documents, which can legitimise their identity as lakes of the city, add to the

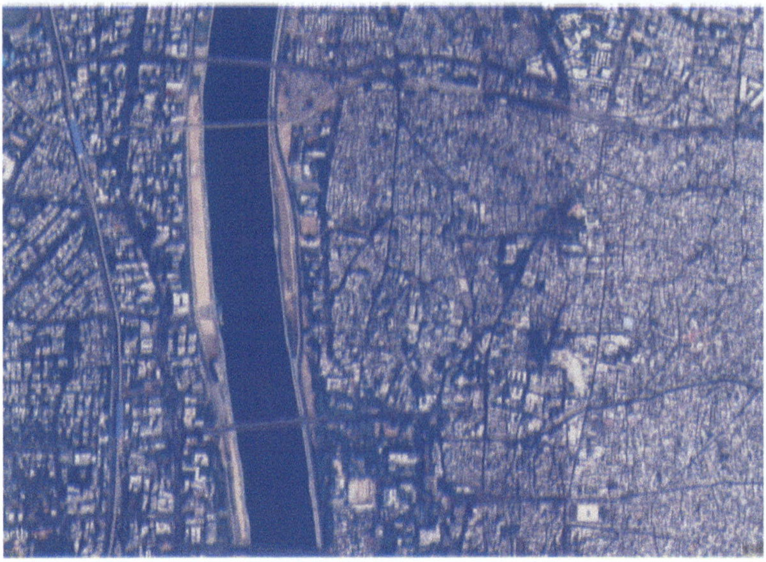

Fig. 5 Satellite image of the Riverfront circa 2023 (*Source* Google Earth)

Fig. 6 Satellite image of the Riverfront circa 2010 (*Source* Google Earth)

worries of the city. More than 360 acres of lakes (mostly unrecognised water bodies) have been reclaimed to build various amenities for government employees. The process disintegrated the natural hydrogeological lines, which help in the drainage

Fig. 7 The view of embankments of Sabarmati Riverfront Project (*Source* Author)

of stormwater during the rainy season. It is one of the major reasons why Ahmedabad becomes severely inundated during each rainy season.

One such example of a lake is the Isanpur lake in Ahmedabad (Figs. 8 and 9). The Isanpur area lies in the city of Ahmedabad in the eastern zone of the city. Isanpur was added to Ahmedabad in 2006, prior to which it was a village. The area has historical significance, with mentions in some of the historical records of the city. A historic stepwell called '*Jethbhai Ni Vav*' is located in the vicinity. Another important structure here is *Gumti Masjid,* also known as Hasan Shahid Masjid, which was built in the early 16th century and is in the vicinity. The architecture of this mosque is like that of Rani Sipri Mosque, located in the inner walled city of Ahmedabad.

The lake of Isanpur is one of the major features of the area. Lakes are generally located in the lowest areas of terrain and act as natural sinks for stormwater runoff. Isanpur Lake also fulfils a similar function. The lake is dotted by slum settlements at its edges (Fig. 9). The water is in a very degraded stage as the wastewater of nearby communities and the slum gets directly drained into the lake. Almost all the lanes leading towards the lake are filled with mud, debris, and human and animal waste. It impacts the livability condition of the communities dwelling on the edge.

The Isanpur lake is around 8–9 m deep at its lowest point (see Fig. 10). The area of the lake is 31000 sqm and has a perimeter of 770 m approximately. The dwellings situated on the edges get swamped with lake water during the rainy season as they are at the lowest side of the lake. The preliminary reconnaissance survey of the lake suggests strengthening of the water-land interface and repositioning of dwellings accordingly. It is important to understand the urban ecology of the area and

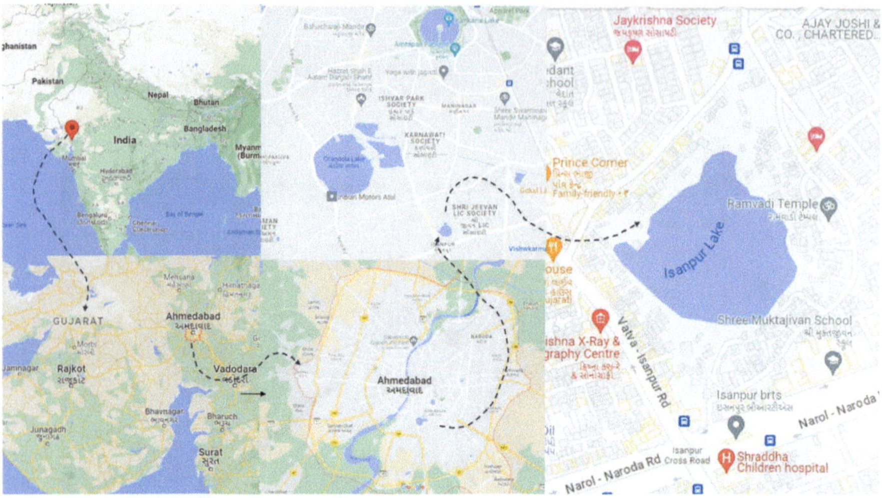

Fig. 8 Location of Isanpur Lake in Ahmedabad, Gujarat, India (*Source* Google Earth 2024)

Fig. 9 The Isanpur Lake satellite imagery of the vicinity (*Source* Google Earth 2023)

work on firming the blue–green infrastructure (BGI) of the lake. Here, BGI-based intervention, nature-based solutions, and environmentally conscious urban design or water-sensitive urban design (WSUD) can play a huge role.

The practice of water-sensitive urban design (WSUD) is a broad area where a wide range of objectives are taken into consideration, which include managing hydrology and water quality, improving urban amenities, and mitigating urban heat island impacts (Hoban, 2019). Nature-based solutions in WSUD usually provide

Fig. 10 A panorama of the Isanpur Lakeside (*Source* Author)

many benefits by augmenting recreational opportunities, holistic well-being, and biodiversity and improving aesthetics as well (Climate-ADAPT, 2023). The two key approaches which are followed in WSUD regarding stormwater runoff management are (Hoban, 2019):

1. Measures to reduce the generation of polluted stormwater
2. Measures to treat polluted stormwater.

6 Recommendations

In the case of Isanpur Lake, the areas on the edge of the lake are made up of a mix of permanent and temporary housings of different kinds of materials. All the stormwater runoff and the wastewater from the dwelling units is directly drained into the lake, making it unsuitable for human consumption. If WSUD is applied in the areas of Isanpur and especially around the edges of the lake, then the following measures are suggested:

6.1 Measures of Reduction

The measures of reduction of generation of polluted stormwater runoff can be done if the water itself can be slowed down and percolated into the subsoil. As the areas on the edge and around the lake do not have much green cover, much of the rainfall gets lost to the runoff. Therefore, the strategies can include green walls and roofs, permeable pavements, erosion and sediment control, rainwater tanks, and downpipe diverters (Hoban, 2019). Among all these strategies, green walls and roofs and permeable pavements can work best for the case of Isanpur.

6.2 Measures of Treatment

Most of the Isanpur lake is contaminated with wastewater. It is imperative to cleanse it such that a waterbody like Isanpur Lake can be restored for better livability of the area. The measures of treatment or mitigation can include gross pollutant traps, gully baskets, swales, bioretention, rain gardens, wetlands, floating wetlands, and street trees (Hoban, 2019). For the case of Isanpur, the WSUD strategies which can work well include swales, bioretention gardens, rain gardens, constructed wetlands, and street trees.

The swales can be designed on the existing hydrogeological lines of the site such that they provide better replenishment of the water and remove impurities during the passage of the water. Bioretention gardens and rain gardens can also be introduced at a variety of spots to act as sinks for rainwater as well as in times of excessive waterlogging. Constructed wetlands are typically shallow vegetated areas that retain water during excessive rain and then percolate it to the ground once the rains stop. They are useful for minimising water pollution and help in increasing the livability of the area.

A preliminary analysis of the Isanpur Lake site is shown in Figs. 11, 12, 13 and 14. Figure 11 shows one of the major hydrological lines where construction has happened, and the natural gradient has been disturbed. Such hydrological lines can be proposed as bioswales, which not only help in directing water and its percolation to sub-soil but also clean the water as it flows through it. Figure 12 shows the low-lying area around the lake, which gets filled up during the rainy season as the lake swells. These houses need to be relocated to the higher ground. Figure 13 suggests that the low-lying areas on the edge of the lake can be proposed as a continuous green belt. This green belt can be composed of constructed wetlands and other WSUD techniques, which can mitigate the flooding of the nearby areas, help maintain the water quality and help increase the biodiversity of the lake. Figure 14 suggests spots of green already existing in the nearby areas, which can be augmented by converting them into rain gardens and bioretention gardens on a case-to-case basis (Fig. 15).

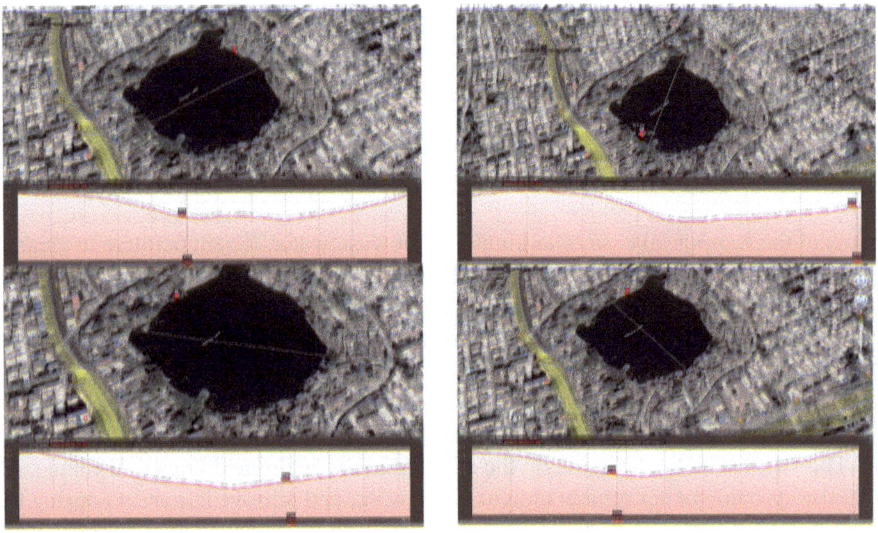

Fig. 11 Depth of Isanpur Lake at different locations (*Source* Google Earth)

Fig. 12 Isanpur Lake with one of the major hydrological lines, which can be reimagined like a bioswale (*Source* Author)

Fig. 13 Relocation of the slum areas from the low-lying area to a higher location (*Source* Author)

Fig. 14 Green belts based on WSUD along the edge of the Isanpur Lake (*Source* Author)

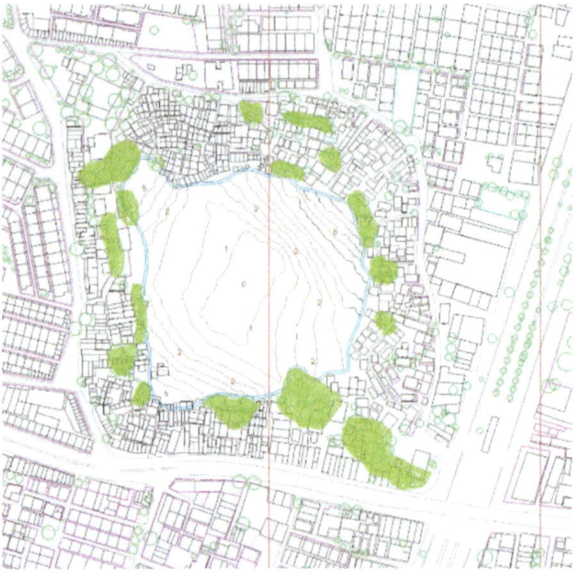

Fig. 15 Augmenting the green spots around the lake using bioretention, rain gardens, and swales (*Source* Author)

7 Conclusion

Water is an essential element for the survival of humankind. It dictates the way we live and govern the way we function. Water, in many forms, fulfils our day-to-day activities. Of these, one of the most important sources is the surface water. Rivers, streams (*nallah*), lakes, ponds, reservoirs, wetlands, etc., make up the surface water. With the rapid climate change affecting almost all the resources which are at the disposal of humankind, water has become one of the scarcest resources. The present study puts forward a similar case of surface water and how it gets exploited by the ignorance and sheer apathy of humankind. It is imperative to understand that water is a precious resource, and surface water needs to be conserved and preserved with utmost priority. These water bodies need to be managed more efficiently with potent systems of wastewater reduction and mitigation.

With the current state of urbanisation, people come to cities in anticipation of better opportunities. For those who cannot afford to get to some of the worst locations in the city, natural low-lying areas with risks of urban flooding are available, which become host to slums and unregulated development. The development is inevitable, yet it can be made better using WSUD-based design techniques. Such techniques can help these low-lying areas to develop more sensitively.

The lake of Isanpur discussed in the present study is one of the lakes of Ahmedabad, a city with its legacy of being the city of many lakes in the past. It is one of the low-lying areas of the city. The current state of Isanpur Lake can be understood as a degraded water body with wastewater being dumped from the nearby residential areas, including the slums on the edge of the lake. Understanding appropriate locations and current conditions is extremely important to do the required intervention. Here, techniques of water-sensitive urban design, if utilised appropriately, can help to change the state of the lake and nearby areas. WSUD makes up a huge part of BGI. Here, water is the essential component which has to be employed in conjunction with green surfaces to get a holistic result.

The chapter intends to make authorities, designers, developers, students, and citizens more aware of such issues and how a better study of the place and area-based solutions can be employed to make habitats more livable to a much larger section of society.

References

Adhvaryu, B. (2011). Analysing evolution of urban spatial structure: A case study of Ahmedabad, India. *Environment and Planning B: Planning and Design, 38*(5), 850–863.

Almaaitah, T., et al. (2021). The potential of Blue–Green infrastructure as a climate change adaptation strategy. *Blue Green Systems, 3*(1), 223–248.

Baptiste, A. K., Foley, C., & Smardon, R. (2015). Understanding urban neighborhood differences in willingness to implement green infrastructure measures: A case study of Syracuse, NY. *Landscape and Urban Planning, 136*, 1–12.

Bhadouria, R., Upadhyay, S., Tripathi, S., & Singh, P. (Eds.). (2022). *Urban Ecology and Global Climate Change*. Wiley.

Bhadouria, R., et al. (2023). Urban metabolism and global climate change: An overview. In *Urban Metabolism and Climate Change: Perspective for Sustainable Cities* (pp. 3–22). Springer.

Brears, R. (2018). *Blue and Green Cities*. Palgrave Macmillan.

Chen, H., et al. (2022). Does energy consumption, economic growth, urbanization, and population growth influence carbon emissions in the BRICS? Evidence from panel models robust to cross-sectional dependence and slope heterogeneity. *Environmental Science and Pollution Research, 29*, 19.

Cho, M.-R. (2010). The politics of urban nature restoration: The case of Cheonggyecheon restoration in Seoul. *Korea. International Development Planning Review, 32*(2), 145–165.

Climate-ADAPT. (2023). *Water sensitive urban and building design*. https://climate-adapt.eea.europa.eu/en/metadata/adaptation-options/water-sensitive-urban-and-building-design.

Dolman, N. & O'Donnel, E. (2021). *5 lessons learned from blue green infrastructure delivery*. https://www.ice.org.uk/news-and-insight/the-civil-engineer/july-2021/theory-and-practice-of-blue-green-infrastructure.

Dreiseitl, H., & Wanschura, B. (2016). *Strenghthening Blue–Green Infrastructure in our Cities*. Ramboll.

EPA. (2023). *Overview of EPA's brownfields program*. https://www.epa.gov/brownfields/overview-epas-brownfields-program.

Evergreen. (2020). *Five Blue–Green infrastructure projects making a splash*. https://www.evergreen.ca/blog/entry/five-blue-green-infrastructure-projects-making-a-splash/.

Ghofrani, Z., Sposito, V., & Faggian, R. (2017). A comprehensive review of blue–green infrastructure. *International Journal of Environment and Sustainability, 3*(3), 15–36.

Grau, D., & Dreiseitl, H. (2013). *Recent Waterscapes: Planning, Building and Designing with Water* (1st ed.). Birkhauser.

Hammel, P., & Tan, L. (2022). Blue-green infrastructure for flood and water quality management in Southeast Asia: Evidence and knowledge gaps. *Environmental Management, 69*, 699–718.

Hoban, A. (2019). Water sensitive urban design approaches and their description. In *Approaches to Water Sensitive Urban Design* (pp. 25–47). Elsevier.

HUDCO. (2020). *HUDCO best practice awards*. Retrieved August 19, 2023, from https://hudco.org.in/writereaddata/EB/HUDCO-BPA/HUDCO-BPA.html#p=34.

Jeon, C., & Kang, Y. (2019). Restoring and re-restoring the Cheonggyecheon: Nature, technology, and history in Seoul, South Korea. *Environmental History, 24*(4).

Joshi, A. (2020). *Reconnecting the city with its River—Sabarmati riverfront development project in Ahmedabad*. Retrieved August 19, 2023, from https://www.isocarp-institute.org/wp-content/uploads/2020/08/Review10_Reconnecting-the-City-with-its-River.pdf.

Lee, J. Y., & Anderson, C. D. (2013). The restored Cheonggyecheon and the quality of life in Seoul. *Journal of Urban Technology, 20*(4), 3–22.

Lehmann, S. (2012). Can rapid urbanisation ever lead to low carbon cities? The case of Shanghai in comparison to Potsdamer Platz Berlin. *Sustainable Cities and Society, 3*, 1–12.

Mell, I. C. (2018). Greening Ahmedabad—Creating a resilient Indian city using a green infrastructure approach to investment. *Landscape Research, 43*(3), 289–314.

Nowobilska, M., & Zaman, Q. M. (2014). *Potsdamer Platz: The Reshaping of Berlin* (1st ed.). Springer.

O'Brien, G., O'Keefe, P., Rose, J., & Wisner, B. (2006). Climate change and disaster management. *Disasters, 30*(1), 64–80.

O'Donnell, E., Lamond, J., & Thorne, C. (2017). Recognising barriers to implementation of blue green infrastructure: A newcastle case study. *Urban Water Journal, 14*(9), 964–971.

Palazzo, E. (2019). From water sensitive to floodable: Defining adaptive urban design for water resilient cities. *Journal of Urban Design, 24*(1), 137–157.

Pearlmutter, D., et al. (2020). Enhancing the circular economy with nature-based solutions in the built urban environment: Green building materials, systems and sites. *Blue–Green Systems, 2*(1), 46–72.

Singh, R., et al. (2022). Urban ecology and climate change: Challenges and mitigation strategies. In R. Bhadouria, S. Upadhyay, S. Tripathi, & P. Singh (Eds.), *Urban Ecology and Global Climate Change*. Wiley.

Stutz, B. (2018). *With a green makeover, Philadelphia is tackling its stormwater problem*. https://e360.yale.edu/features/with-a-green-makeover-philadelphia-tackles-its-stormwater-problem.

Udas-Mankikar, S., & Driver, B. (2021). *Blue–Green infrastructure: An opportunity for Indian cities—Occasional paper*. Retrieved August 19, 2023, from https://www.orfonline.org/research/blue-green-infrastructure-an-opportunity-for-indian-cities/.

Systemic Action Network for Improving Blue–Green Infrastructure Based on the Natural Capital Investigation: The Strategic Plan in Vlorë, Albania

Tao Dong, Angela Colucci, and Massimo Tadi

Abstract The natural capital (ecosystem services, good quality of ecological metabolic functionalities) is crucial in maintaining human well-being, as well as guaranteeing urban resilience for facing climate and Anthropocene challenges. The chapter reframes the Nature-based Solutions (NbS), assumed as an umbrella concept, contributing to activating the urban sustainable transformation processes integrating the NbS principles in the urban adaptive systems design approaches framework. This chapter takes the city Vlorë (Albania) as a representative study case of the Mediterranean context to explore the practicality of NbS by addressing the pivotal role of natural capital in long-term sustainable development through a holistic approach named Integrated Modification Methodology (IMM). By considering the city to be a complex adaptive system, the internal urban structure is pictured as a series of functioning patterns (permeability, effectiveness, proximity, and so on), among which urban diversity and continuity show great potential as the priority for driving Vlorë to transform towards a more resilient and sustainable status. The case study is built through a systematic network of interlinked actions for the improvement of blue–green infrastructure, synergizing the goals of UN SDG 11: Sustainable Cities and Communities with the regional guidelines of Vlorë, providing a comprehensive planning strategy. The project "Eco-breathing in Vlorë 2030" will be presented, highlighting the aspects of innovation of the urban model for the sustainable development of Vlorë for the next 10 years and beyond, where natural capital is assumed as a crucial Blue–Green infrastructure through NbS to address and overcome the challenges of the Anthropocene developing an attractive, sustainable, and resilient city,

T. Dong (✉) · M. Tadi
Department of Architecture, Built Environment and Construction Engineering, IMM Design Lab, Politecnico di Milano, Piazza Leonardo da Vinci 32, Milan, Italy
e-mail: tao.dong@polimi.it

M. Tadi
e-mail: massimo.tadi@polimi.it

A. Colucci
Department of Architecture, Built Environment and Construction Engineering, Politecnico di Milano, Piazza Leonardo da Vinci 32, Milan, Italy
e-mail: angela.colucci@polimi.it

P. K. Joshi et al. (eds.), *Blue-Green Infrastructure for Sustainable Urban Settlements*,
https://doi.org/10.1007/978-3-031-62293-9_17

meanwhile highlighting the potential advancement in connecting NbS and IMM to foster the innovation of transformative urban models toward more sustainable urban systems in developing countries. This Chapter explores the possibility of integrating Blue–Green Infrastructure as a promising solution when dealing with climate and urban challenges in the process of sustainable development and discusses the transferability of the IMM research model by the Vlorë case through a scientific process: Diagnosis-Synthesis-Strategies. The outcome demonstrates the significance of a systematic organization of design strategies and underscores the necessity for multidisciplinary interventions to propel sustainable urban transformation.

Keywords Climate change · Natural capital · SDG 11 · Systemic design · Urban regeneration

Abbreviations:

BGI	Blue–Green Infrastructure
BI	Blue Infrastructure
CAS	Complex Adaptive System
CC	Climate Change
DDR	Disaster Risk Reduction
DEM	Digital Elevation Model
DOP	Design Ordering Principle
EC	European Commission
GHG	Greenhouse Gas
GI	Green Infrastructure
GIS	Geographic Information System
KC	Key Categories
IMM	Integrated Modification Methodology
IUCN	International Union for Conservation of Nature
MEA	Millennium Ecosystem Assessment
NAPA	National Agency of Protected Areas
NbS	Nature-based Solutions
PEEN	Pan-European Ecological Network
SDG	Sustainable Development Goal
SOC	Soil Organic Carbon

1 Introduction

1.1 Anthropocene and Climate Change: Emerging Challenges for Urban Complex Systems

Urban areas are widely considered to be, contemporary: drivers significantly contributing to climate change and environmental pressures; places highly vulnerable and fragile to climatic-related events (Rosenzweig et al., 2010); and those are playing a pivotal role in proposing innovative strategies, and actions counterbalancing and facing the multiple phenomena and challenges of the Anthropocene.

Climate Change (CC) is now considered as one of the most challenging global issues mostly deriving from the past and ongoing urbanization processes (Kalnay & Cai, 2003) and urban lifestyles, with severe implications in environmental, social, economic, and organizational (institutional/political) components. In the last decade, CC has become a priority in the scientific debate and the global, national, and urban agendas. Cities increasingly demonstrate to be crucial actors in climate change counterbalancing policies implementation, assuming a pioneering role in developing strategies and initiatives aimed at reducing greenhouse gas (GHG) emissions and implementing actions for adaptation to the impacts of climate-related hazards (Dekker, 2019). The UN-Habitat Urban Agenda (United Nations, 2017) recognizes the crucial role of urban areas and urban policies in achieving global SDGs and, above all, facing environmental and climate challenges rooted in an integrated approach.

Alliances of networking and city collaboration, such as the World Mayors Council on Climate Change (WMCCC), the C40 Cities Climate Leadership Group, the Local Governments for Sustainability (ICLEI), and the Resilient Cities Network, attest to how cities have risen to the forefront of the climate change movement (Acuto & Leffel, 2021; Lee & Jung, 2018). Meanwhile, an extensive range of urban-centered visioning and strategies emerge from the institutional and scientific debate, like Smart City and Resilient City, providing a comprehensive and large frame for strategies and action definitions addressing mitigation and adaptation (Galderisi & Colucci, 2018, 2019).

In the climate approach the UK white paper "Our Energy Future: Creating a Low-Carbon Economy" introduced the concept of a "low-carbon economy" in 2003, leading to the widespread adoption of low-carbon practices and increased international cooperation (Tan et al., 2017). Cities contribute to over 70% of the global carbon dioxide (CO_2) emissions, predominantly originating from industrial activities and motorized transportation systems. These systems heavily rely on substantial quantities of fossil fuels and are associated with widespread infrastructure constructed using materials that yield a significant carbon footprint (Dasgupta et al., 2022). The term "Low-carbon City" derives from the growing need to address carbon reduction and climate change mitigation in urban areas. The transition towards the Low-carbon City calls for the adoption of a systemic approach that emphasizes the need to reconsider the traditional planning tools, too often ineffective in addressing the difficult relationship between urban development and environmental sustainability.

While recognizing the worldwide role of the cities, nevertheless, it is necessary to stress how cities in developing countries, which contribute significantly to global carbon dioxide (CO_2) emissions, are not acting fast enough toward carbon–neutral policies and in adaptation measures (Mukim & Roberts, 2023). Low or uncontrolled urbanization contributes to the general erosion of natural capital, ecological capacities, and functionalities (Mukim & Roberts, 2023).

Scientific evidence shows how urban areas in developing countries are the most impacted by the shock of climate change–related hazards suffering more severe negative impacts from extreme weather events. Considered as complex adaptive systems, these urban areas exhibit spatial and transformative potential to address the multifaceted challenges posed by climate change through optimized strategies leveraging natural resources.

1.2 Global–Local Sustainability and Integrated Approach

Supporting local actions taken by cities in developing countries to tackle climate change becomes urgently necessary. The UN-Habitat New Urban Agenda and the 2030 UN Agenda strongly invite actions recognizing the role of cities of developing countries for tangible changes toward sustainability scenarios (United Nations, 2017). The urban model shift must be rooted in an integrated approach facing multiple challenges jointly introducing environmental and climate, as well as social sustainability. The SDGs' achievement and the transition towards sustainable models must consider the universal values of justice and equity, integrating the just transition approach. In striving for a carbon emissions-free society, achieving a just transition becomes essential, marked by fairness and equity. Climate justice enables researchers to explicitly examine the distribution of risks and responsibilities (Barrett, 2013). Climate change implies inequalities in terms of the distribution of climate impacts and related risks and responsibilities.

Supporting the urban model shift towards more sustainable and Climate-neutral cities in developing countries, the Integrated Modification Methodology (IMM) research team developed several applied research in urban informal contexts and megacities of Africa and South America. The IMM (see Sect. 4) methodology, developed at the Department of Architecture, Built Environment and Construction Engineering (DABC) of Politecnico di Milano, is rooted in a holistic approach generating environmental performance diagnostic of the city and highlighting the actions able to generate synergies and multi-benefits impacts to the local urban system.

In the context of sustainability, developing countries encounter a multitude of intricate environmental and social issues when compared to developed countries. The uncontrolled and unregulated expansion of settlements has given rise to a range of physical issues, including inadequate drainage infrastructure and availability of clean water, restricted inclusion into the urban fabric, challenges with road systems and land topography, and the occupation of unsuitable areas with vulnerability to flooding or mudslides. On the social front, issues encompass the absence of proper occupancy registration, insufficient community cohesion, and the presence of crime and illicit activities, resulting in constrained social integration and the denial of residents' citizenship rights (Tadi et al. 2020).

The applied experiences developed in many urban contexts in developing countries, such as in the informal settlements of Rio de Janeiro (Masera & Tadi, 2020) or Maputo and Dakar, have confirmed, on the one hand, the urgency to act for a transformative shift in urban transformation approaches. On the other hand, the application of IMM has made it possible to understand integrated and synergic dynamics by proposing solutions and response scenarios to the challenges of the Anthropocene.

1.3 Nature-Human Synergies for Urban Resilience

International, European, and National policies and scientific reports assume how natural capital and natural habitats (ecological quality and functionality) support and sustain the long-term life and development of human (social and economic) components and urban areas. As assumed by the World Bank in several reports (World Bank Group, 2021), natural capital (ecosystem services, good quality of ecological metabolic functionalities) is crucial in maintaining human well-being, stabilizing and improving local economies as well as guaranteeing urban resilient for facing Climate and Anthropocene Challenges (Bhadouria et al., 2023).

It is essential to align and integrate the Disaster Risk Reduction (DDR) plans with international development and environmental objectives (such as the Sustainable Development Goals—SDGs) of the United Nations. The UN-SDG 11 on sustainable cities and UN-SDG 13 on climate policies provide actions for environmental deterioration counterbalance and ecosystem services enhancement for decrease of catastrophe risk. These measures also support SDGs 14 and 15 directly, which focus on terrestrial and marine life. Furthermore, Ecosystem-based Disaster Risk Reduction (Eco-DRR) provides measures directly addressing climate change adaptation goals and fulfilling commitments of the Convention on Biological Diversity, the Ramsar Convention on Wetlands, and the UN Convention to Combat Desertification (Browder et al., 2019).

The Nature-based Solution (NbS) emerges as a paramount model in Climate mitigation/adaptation and enhances the ecosystem capacities and functionalities, providing multiple benefits and assuring life on our planet. In particular, integrating NbS in the urban transformation processes is essential in developing countries' urban areas. NbS sustains the just transition and guarantees multiple benefits:

- provide tangible results in mitigating climate emissions (carbon emission) and in reducing the impacts (hazards and vulnerabilities) due to climate, improving the adaptation capacities;

- provide benefits health and well-being of citizens (e.g., reducing air pollution, water pollution, and related diseases);
- improve the quality of local communities' lives by improving the quality (quantity, liveability, accessibility/affordability) of public and private places;
- provide urban innovation opportunities, including social innovation and local circular economies promotion;
- support the transition to more sustainable (and economic) movement behavior (slow mobility).

In approaching the Vlorë strategic project, the relevance and the crucial role of natural capital are emphasized. The IMM methodology application focuses on maximizing the multi-benefit provided by ecological functions, implementing the urban complex adaptive system as a methodological advancement and addressing a strategic urban challenge.

2 Blue–Green Infrastructure as Core Object for Eco-Services Development

2.1 Umbrella Concepts: Nature-Based Solutions

Nature-based Solutions (NbS) play an essential component in international and European policies considering they are characterized by an integrated approach and give many advantages to complex adaptive systems that connect the ecological, social, and economic sectors (Eggermont et al., 2015; Roggema, 2020).

In the mid-2000s, the International Union for Conservation of Nature (IUCN) and the European Commission (EC) advocated NbS as a viable combo of measures to mitigate climate and catastrophe risks. Within the framework of sustainable development, NbS is an umbrella notion that encompasses a variety of ecosystem-based remedies to many societal concerns.

IUCN defines NbS as: "Actions to protect, sustainably manage and restore natural or modified ecosystems, that address societal challenges effectively and adaptively, simultaneously providing human wellbeing and biodiversity benefits" (IUCN, 2016). The EC defines NbS as: "Solutions that aim to help societies address a variety of environmental, social, and economic challenges in sustainable ways (Faivre et al., 2017)."

The EC considers NbS as nature-inspired and nature-supported solutions that provide environmental, social, and economic benefits while also promoting resilience capacities (Davies et al., 2021; European Commission. Directorate General for Research & Innovation, 2015), responding to various challenges, ranging from environmental emergencies to many features of complex adaptive systems. NbS has become a collective term for a diverse range of landscape design approaches and models, including green infrastructure, urban re-greening, and ecosystem services (Colucci, 2023; European Environment Agency., 2021).

Nature-based Solutions provide several benefits and action spheres by reducing the recurrence of hazards (e.g., Forests help to prevent landslides, which tend to be triggered by environmental degradation in combination with other reasons like high rainfall) and diminishing the severity of hazards' effects (e.g., sand dunes can act as a barrier for big waves). NbS aims to maintain ecosystem functions and biodiversity (Heymans et al., 2019).

2.2 *Nature-Based Solutions and Blue–Green Infrastructure: Approaches and Tools for Human-Nature Synergies*

The NbS concept is rooted and strongly linked to many existing green solutions, such as blue and green infrastructure (BGI). The EC defines the BGI planned network as composed of natural and semi-natural components providing ecological habitat diversity to guarantee the most comprehensive spectrum of ecosystem services provision.

Green Infrastructure (GI) originated in the 1990s as are planned solution that consciously and strategically preserves, restores, and enhances the natural/semi-natural components to improve higher-quality, more resilient, and lower-cost infrastructure generating a large range of related services. It is highly linked with the concept of Blue–Green Infrastructure (BGI). Being an advanced notion, BGI can exist on multiple scales, ranging from the region, urban, to water element and site, meanwhile across administrative boundaries (Ghofrani et al., 2017). BGI involves Blue Infrastructure (BI) to be a central object for sustainable development, aiming to draw attention to the water-based components, such as water courses, ponds, rivers, and lakes, supporting biological and hydrological processes such as evaporation, transpiration, drainage, infiltration, and short-term runoff and discharge storage.

BGI encompasses a comprehensive array of natural and semi-natural constituents, coalescing into a network of diverse landscape elements that operate across varying levels of spatial and organizational scales. BGI integrates local-scale elements, such as small local green spots (bushes, hedgerows, and ponds), and territorial/regional scale elements, such as urban parks, gardens, forests, green river/watercourse corridors, and wetlands habitats. Components of urban BGI could be realized as modular NbS interventions. However, ensuring the continuity of the ecological components of the network is crucial and essential by including the realization of semi-natural solutions in more dense settled areas, such as vegetated/tree lines, green roofs, or other re-green integrated into built urban components. In the case that the present BGI components are ineffective, they ought to be enlarged in size or upgraded with additional green and grey components, including semi-permeable surfaces, water collection tanks, infiltration planters, tree boxes, or hives (Somarakis et al., 2019).

A crucial role plays also in the restoration of wetlands by enhancing the ecological condition of coastal areas. Restoration of crucial ecological processes and services, such as sediment fixation, as well as a rise in biodiversity result in the stabilization of the whole system. Ecosystem functions that have been lost to people, such as defense against erosion and coastal risks, are restored through the ecological improvement of mangrove ecosystems. After the reforestation method has been successfully implemented, this occurs on its own accord without additional human interference. Another benefit is that NbS improves the conditions of living, human health, and well-being (Nguyen et al., 2022).

2.3 *Significance of Ecosystem Services in Sustainable Development*

The Ecosystem Services are "the direct and indirect benefits people derive from ecosystems". It offers a diverse array of services that are critically significant to the overall well-being, health, livelihoods, and survival of humans (Costanza & Limburg, 1997). According to the Convention on Biological Diversity (1992), an ecosystem is characterized as "a complex of living organisms and the abiotic environment with which they interact in a specified location." In other words, it is a small- and medium-network of beings that interact with one another and with the environment in which they reside. Humankind and ecosystems are intricately interconnected through a multitude of relationships.

There are several ways of categorizing ecosystem services. The Millennium Ecosystem Assessment (MEA) categorizes ecosystem services as A. Provisioning Services or the provision of food, fresh water, fuel, fiber, and other goods; B. Regulating Services, such as climate, water, and disease regulation as well as pollination; C. Supporting Services, which include soil formation, nutrient cycling, etc.; and D. Cultural Services, for example educational, aesthetic, and cultural heritage values as well as recreation and tourism (Millennium Ecosystem Assessment (Program) 2005).

Ecosystems furnish a diverse array of advantages to human populations, encompassing provisioning, regulating, supporting, and cultural services. Provisioning services encompass the tangible outputs derived from ecosystems, including sustenance, energy sources, textiles, freshwater, and genetic reservoirs. Regulating services involve the benefits accruing to people through the management of ecosystem processes, such as the maintenance of air quality, climate control, erosion mitigation, disease regulation, and water purification. Cultural services encompass intangible advantages arising from ecosystems, contributing to spiritual enrichment, cognitive advancement, contemplation, recreation, and aesthetic encounters. Supporting services constitute the foundational components essential for the generation of all other ecosystem services, encompassing primary production, oxygen generation, and soil formation (Alcamo and World Resources Institute 2003).

It is essential to ensure robust biodiversity and ecological principles to avoid unanticipated harmful repercussions by improving the status of ecosystem services through ecological engineering techniques that aim to achieve a very particular goal, such as protecting against floods by building retention zones or protecting against coastal erosion by constructing artificial dunes.

The challenge of the transition towards low-impact models requires, in fact, synthesizing interrelated mechanisms such as mobility, environmental protection, logistics, food security, energy, water and waste management, and in general, the transition to a society free of net carbon emissions by 2050.

Capital stock manifests in diverse identifiable forms, with physical manifestations, commonly referred to as "natural capital", comprising prominent examples (Costanza & Limburg, 1997). These physical forms encompass the planetary resources that sustain life and well-being (e.g., air, water, soils, minerals, plants, animals) and underpin water and energy security, shelter, clean air, medicine, and more.

3 Policies on Vlorë Natural Capital

3.1 Pan-European Ecological Network (from Natura2000 to Emerald Network)

The establishment of a Pan-European Ecological Network (PEEN) was supported by 54 European Ministers of the Environment on October 25, 1995, at the Environment for Europe Ministerial Conference in Sofia. To facilitate species dispersal and migration, they envisioned the creation of a physical network consisting of core areas and other relevant measures supported by buffer zones and connected by corridors. In proposing the establishment of PEEN, the Ministers responded to the reduction in biodiversity, the loss of precious ecosystems and landscapes, and the threat, that represented Europe's ability to maintain itself in the future. Biodiversity and ecosystems play a crucial role in fostering social and economic development and enhancing human well-being through the provision of vital goods and services. Moreover, the ecosystems' inherent ability for self-regulation significantly contributes to environmental stability at both global and regional scales. Europe exhibits notable richness in terms of species and natural and semi-natural habitats, boasting an estimated 2500 habitat types and 200,000 species of plants and animals, thereby representing a substantial repository of biological and landscape diversity. As such, Europe's diverse biological and landscape features are among the continent's most esteemed and distinctive heritage attributes.

3.2 The Emerald Network

Albania is engaged in the constitution of the Emerald Network, which is a contemporary network encompassing protected areas that hold significant importance in safeguarding natural habitats and biodiversity (Marushchak et al., 2019). These regions are designated as Areas of Special Conservation Interest (ASCI). The establishment of the EN is legally supported by the Berne Convention on the Conservation of European Wildlife and Natural Habitats (1979) in Europe and observer countries. Despite having been conceptualized as early as 1989, the Emerald Network took concrete form in 1996 as part of the Pan European Ecological Network and the Natura 2000 Network.

The Emerald network includes the Wetland protected under the Ramsar Convention (Ramsar, 2022), which seeks to save a particular kind of ecosystem that is biologically rich but faces grave threats on a global scale. Strives to preserve a particular type of environment that is biologically exceptionally rich yet is seriously threatened globally. These Ramsar sites hold significant importance in safeguarding the migration routes of water habitats and effectively managing the ecological processes and services offered by wetlands. Consequently, they contribute to establishing a cohesive spatial framework that plays a specific role in flood prevention and the mitigation of pollution effects, aligning with the sustainable development goals of the Pan-European Ecological Network.

3.3 Albania Ecological Network

Albania exhibits a notable abundance of biological diversity, evident in its diverse spectrum of ecosystems and habitats that support around 3,200 vascular plant species, 2,350 non-vascular plant species, and 15,600 animal species. A significant portion of these species encounter challenges at the global or European level. In acknowledgment of Albania's noteworthy contribution to European and global biodiversity, the European Union supports the implementation of the NaturAL project. Protected areas serve as pivotal elements in nature conservation and represent a reliable approach to preserving ecosystem biodiversity. Remarkably, Albania has made substantial progress in the enlargement of its protected area network, growing from 5.2% of the nation's land area in 2005 to 16% by 2014. However, notwithstanding this noteworthy advancement, investments in protected areas, which encompass infrastructure, equipment, and the technical capabilities of protected area personnel, continue to be insufficient. 799 protected areas make up around 16% (4,600 km2) of Albania's land surface. They are typically relatively tiny in size and the great majority of them have been classified as natural monuments (750). The National Agency of Protected Areas (NAPA) is responsible for directing Albania's national system of protected areas, while 12 Regional Administrations of Protected Areas (RAPAs), which are a component of NAPA, are in charge of administering protected areas on a daily basis.

4 Integrated Modification Methodology (IMM)

Today, it is well accepted that the sectoral approaches to sustainability are doomed to fail. The challenges that the world faces are highly interrelated, and the structure of the problem is too complex to attack to form a branched angle. Therefore, strategies for development, no matter local or global, should comply with frameworks that can address this integral structure. The city functions mostly by the rules of natural complexity rather than human logic, and just like nature, it is classified as a Complex Adaptive System (CAS). Therefore, to minimize the unwanted side effects, it is necessary to acknowledge the complexity practically and employ a systemic approach able to study the reactions as much as possible.

The Integrated Modification Methodology (IMM) stands out for its innovative approach, characterized by integration, multiscale analysis, data-driven methodologies, and a systems-oriented perspective in urban design and planning. It examines urban systems across various scales, employs principles of systems thinking, and relies on data-driven methodologies. By incorporating sustainability principles and fostering adaptability, IMM tackles the intricate challenges of urban environments in a transformative manner. It provides a forward-looking and evidence-based framework for the creation of sustainable and resilient cities. The Integrated Modification Methodology (IMM) has been developed by IMMdesignlab a research lab established at Politecnico di Milano's Department of Architecture, Built Environment and Construction Engineering (DABC) and relies on the use of advanced optimization techniques and data-driven analysis to identify optimal design solutions for complex systems. By integrating different domains of engineering, architecture and urban planning and using performance indicators to evaluate design alternatives, the methodology allows designers and planners to make informed decisions and achieve optimal design solutions that meet a wide range of performance criteria.

IMM serves as a theoretical framework for evaluating the behavior of existing cities and forecasting their potential performance under various intervention scenarios. It is an innovative design methodology that aligns with the UN SDG2030. Within this approach, cities are perceived as Complex Adaptive Systems, and the emphasis is placed on simulating their operational mechanisms, encompassing both internal subsystems and their interactions with external systems. The outcome of this methodology is a methodological interpretation of SDG Number 11, which proposes locally focused actions strongly connected to SDG indicators and evaluation methods.

System thinking is a fundamental principle embedded in IMM, guiding the comprehension, analysis, and design of urban systems. It enables a holistic understanding of the intricate interconnections, emergent properties, and complexities within urban environments, facilitating comprehensive and effective decision-making for sustainable and resilient urban development. This approach is instrumental in unraveling the intricate dynamics and interactions inherent in urban systems.

Cities exhibit life-like characteristics by acquiring energy, processing materials, and generating waste. They possess systems for digestion, respiration, and circulation. Instead of perceiving cities as disconnected components, we view them as unified systems. Similar to living organisms, cities have the capacity to adapt and transform, with the potential for positive growth through the collective efforts of citizens and leaders, enhancing their chances of long-term survival.

The IMM methodology is structured as a multi-stage process, consisting of integrated phases: Investigation, Formulation, Modification, Retrofitting, and Optimization. The built environment, being a complex system, is inherently dynamic, constantly undergoing changes and generating new transformation thresholds at various stages. In this context, the transformation of the built environment follows a non-linear pattern, resembling a spiral. Similarly, the proposed procedure is designed in a non-linear manner, enabling it to navigate different scales and maintain the system's capacity for self-learning. The arrangement of the methodology ensures both precision and flexibility in adapting to the evolving needs of the built environment.

4.1 Investigation—Vlorë's Natural Capital

The Vlorë Region encompasses diverse landscapes, including coastal areas, wetlands, mountains, canyons, and more, which represent a tremendous advantage for ecological services and a powerful tourist attractor. The natural capital is under stress as a result of unauthorized construction, unregulated littering, untreated wastewater, and other unmanaged human activities that have degraded certain places and are now endangering clean ones. The Albanian coastal ecosystems are under significant pressure, especially the risks related to the decline in biological diversity and natural habitats that play an important role in the health of humans, the food chain, and the availability of natural resources for economic development, as shown in Fig. 1a.

Two specific focuses were developed to understand the ecosystem services functionality level of the Vlorë area: the tree cover and its climate mitigation capacity, and Bio-diversity features and values.

In Fig. 1b, c, the Tree Cover Loss depicts the annual loss of tree cover, specifically referring to the replacement of vegetation taller than 5 m within the designated area. On the other hand, the Tree Cover Gain indicates the introduction of tree canopy at the Landsat pixel scale in a region previously devoid of tree cover. According to the information provided on the Global Forest Watch website (Global Forest watch, 2022), between 2001 and 2012, Vlorë witnessed the addition of 646 hectares of tree cover, accounting for 75% of the total tree cover gain across the region.

The Soil Carbon Density (Fig. 1d) integrates global soil organic carbon (SOC) data derived from natural factors (e.g., land cover, vegetation, topography, climate) and human factors (e.g., land use, management). Measuring global SOC is crucial for understanding carbon sequestration, agricultural productivity, and soil water levels in a changing climate (Poggio et al., 2021; Sanderman et al., 2018).

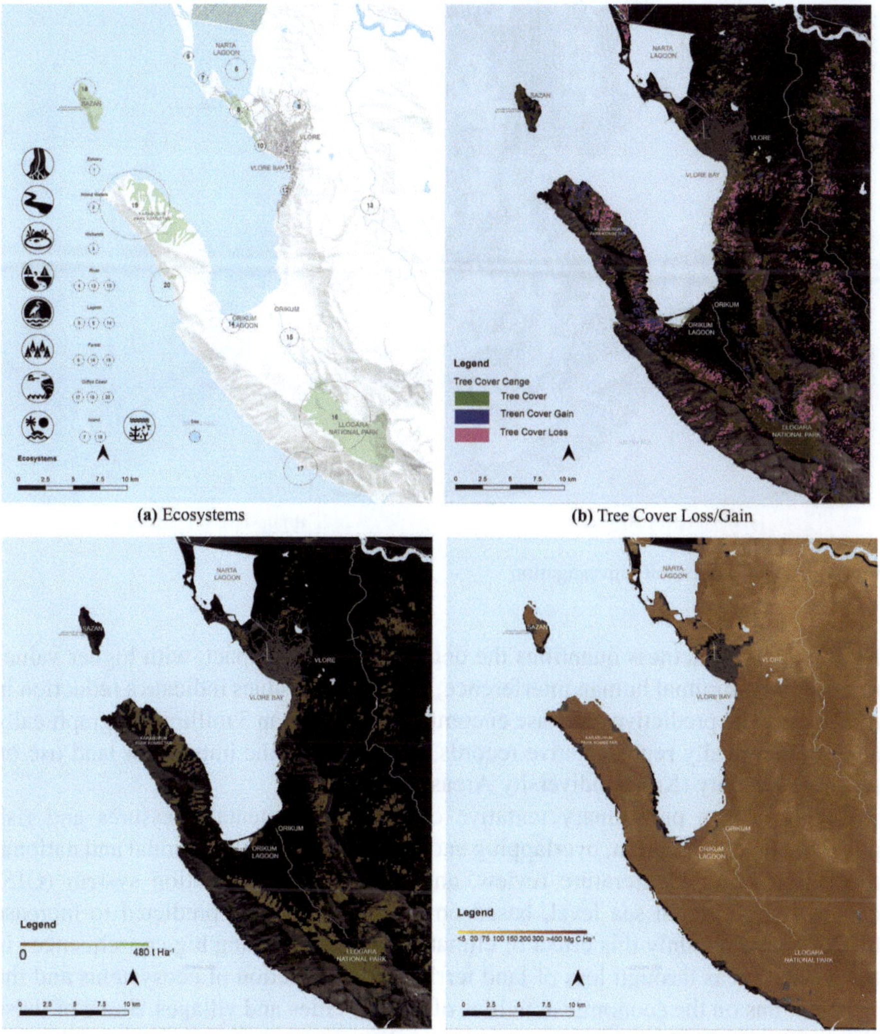

(a) Ecosystems

(b) Tree Cover Loss/Gain

(c) Tree Biomass Density

(d) Soil Carbon Density

Fig. 1 Vlorë's natural capital investigation

The Biodiversity Significance assesses the importance of each forest site in the context of biodiversity by evaluating the relative contribution of individual pixels to the worldwide distributions of forest-dependent mammals, birds, amphibians, and conifers (Fig. 2a). To assess it, forest-dependent species listed in the IUCN Red List are chosen, and their distribution maps are limited to their known altitudinal ranges (excluding amphibians, as their altitudinal range assessments are unavailable). This process involves utilizing a digital elevation model (DEM) dataset to clip the distribution maps and overlay them with the forest cover layer. In Fig. 2b

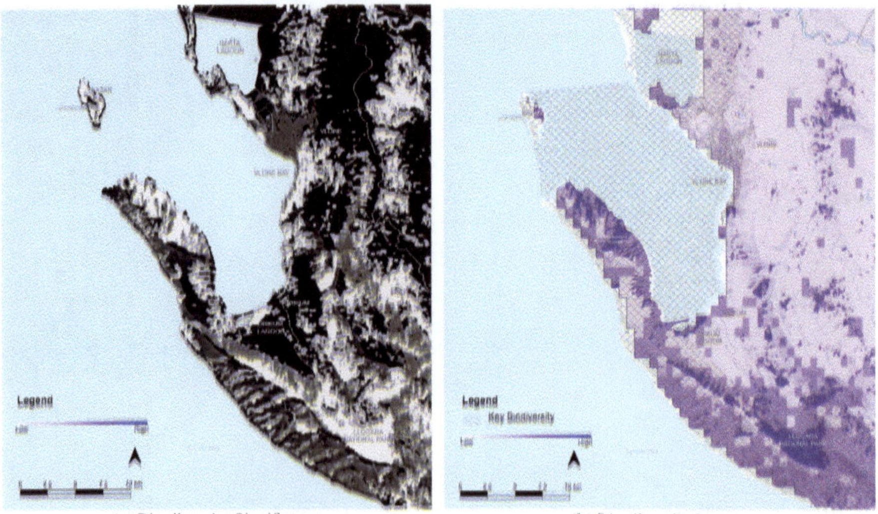

(a) Bio-diversity Significance

(b) Bio-diversity Intactness

Fig. 2 Vlorë's biodiversity investigation

Bio-diversity Intactness quantifies the degree of human impact, with higher values representing minimal human interference, while lower values indicate a reduction in intactness. The predictive database encompasses more than 3 million geographically and taxonomically representative records, documenting the impacts of land use on local biodiversity (Key Biodiversity Areas, 2022).

Figure 3 is a preliminary tentative of the environmental pressures and risk phenomena identification, overlapping and integrating the International and national reporting, scientific literature review, and geographic information system (GIS) database. The global sea level, based on the projections, is predicted to increase by 0.28–0.98 m. Only this effect of climatic change will bring big consequences in the coastal areas through loss of land territories, destruction of ecosystems and the impact it has on the economic activities of coastal cities and villages. Some of these changes are: an increase in air temperature and marine surface area; average changes and extreme rainfall; changes in the frequency and intensity of storms; an increase in sea level.

4.2 Diagnosis and Synthesis—Vlorë's Urban Systems

In the context of IMM, the urban diagnostic is a model-based process that involves systematically understanding, analyzing, and assessing the functioning mechanisms of urban systems (Tadi et al., 2020). It serves as the foundation for identifying key categories and indicators that capture the essential properties and behaviors of the urban environment. Through the use of quantitative data and metrics, the urban

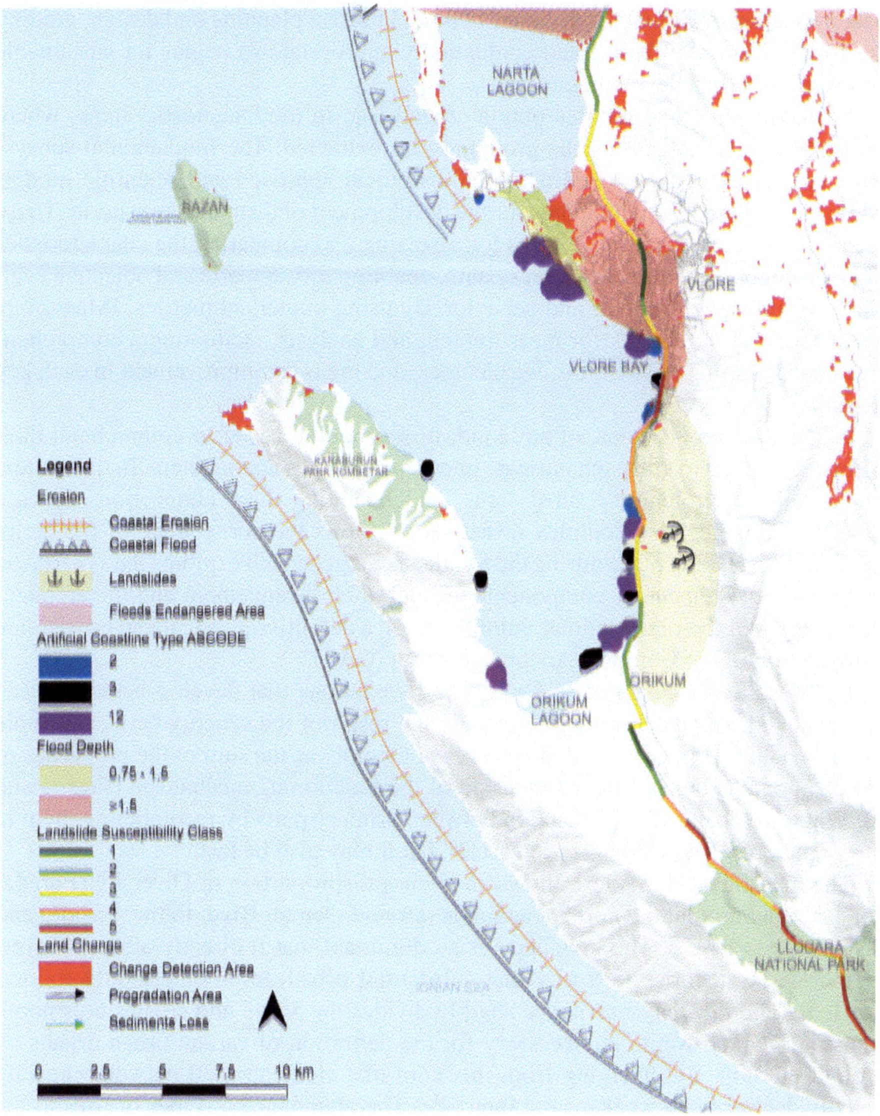

Fig. 3 Vlorë's Environmental Pressures and Risk Identification

diagnostic helps diagnose the strengths, weaknesses, and interconnections within the urban system.

The urban diagnostic in IMM employs a multi-scale and integrated approach, considering the complex interactions between subsystems and external systems. It helps uncover emergent properties and patterns, facilitating the identification of leverage points for transformative interventions. By combining scientific understanding, data analysis, and systems thinking principles, the urban diagnostic becomes a powerful tool for decision-making in urban planning and design, guiding the formulation of effective and contextually appropriate strategies for sustainable and resilient urban development.

In IMM, numerical metrics play a crucial role in the Diagnostic Stage, where measurable factors forming the groupings are extracted. The fundamental subsystems are separately examined using an analytical approach and scientific models while identifying key participants for the development of dismantling criteria. These metrics provide a quantitative basis for assessing and comparing the characteristics of key categories within the urban system, enabling a systematic and objective evaluation of their performance and behavior. By using numerical metrics, IMM offers a standardized framework for measurement and analysis, facilitating a comprehensive understanding of strengths, weaknesses, and areas for improvement in each key category.

When studying systems, of any kind, it is generally easier to comprehend their components than to grasp the intricate interdependencies among them. In the context of the built environment's subsystems, understanding these connections poses a greater challenge. This complex investigation forms the cornerstone of the entire methodology, taking place during the Synthesis Stage, which explores the synergistic integration of elementary components through their arrangement and the architectural aspects of their interactions, culminating in a definitive physical and provisional composition of the Complex Adaptive System (CAS).

In IMM, Key Categories (KC) are the interactions that develop between basic components to create a synergy. They are the result of the synergy between simple systems, a new organization that emerges not simply as the sum of the properties of the components but as a new dimension of the functioning mechanism (Figs. 4 and 5). When we change the connection between primary parts by recombining them in different ways, a new organization arises, but it may also be lost.

The investigation in Vlorë shows an intense disproportion of Diversity (Fig. 4a) concentrated around the north–south axis (around Qemali Blvd. to the seaside) and dispersed elsewhere. The imbalance is so dominant that it directly affects the rest of the Key Categories in a way that substantial urban development only follows this axis. The diversity imbalance sharply divides the Vlorë and creates no smooth transition, which would be necessary for the definition of variant urban organs all around the city. By diverging from this core-like element, road network capacity heavily declines and weakens the interface. The abundance of types of uses in the axis naturally leads to higher proximity indices while putting the rest of the area at risk of higher car dependency. The specified core easily attracts almost all capacities of the public transportation system and produces a significant imbalance of Accessibility (Fig. 4c) and Effectiveness of public transportation (Fig. 4d) in its favor.

In Fig. 4b, Green Continuity refers to the possibility of creating a connected and comprehensive green infrastructure network in urban areas. It highlights the potential for natural resources to interact in a harmonious and interconnected way. This connection is essential for the sustainability of our environment, as it allows resources to be used responsibly and thoughtfully. This can involve the careful management of resources to ensure they are not depleted or degraded, as well as finding ways

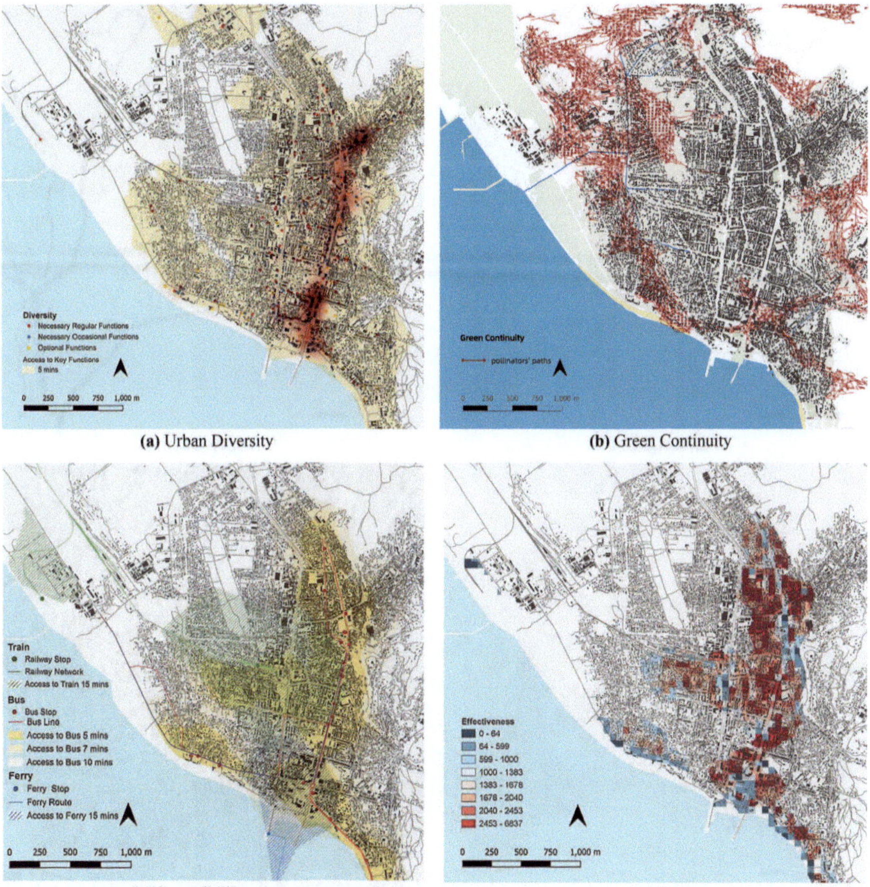

(a) Urban Diversity

(b) Green Continuity

(c) Accessibility

(d) Effectiveness

Fig. 4 Vlorë—urban diagnosis and synthesis of key urban properties

to ensure the health and diversity of natural ecosystems. This network consists of various green and blue spaces, such as street trees, wetlands, parks, and urban forests, but also green roofs integrated into the urban environment. The analysis simulates pollinators' paths to trace the covert connection between various natural elements.

The result shows that the suburb of Vlorë interacts actively with the surrounding farmlands, grasslands, and forests. However, in the city center, it's hard to recognize a connected system of large public open spaces. Despite the relevance of the surrounding natural system, made of natural reserves, forests, lagoons, and hills, the urban area lacks green spaces. Except for the east side, where there is a transition from the forest to the city center made of private and residual green pockets, the rest of the city mostly presents rows of trees on the main streets. This disruption of continuity implies a high potential for Nature-based Solutions intervention.

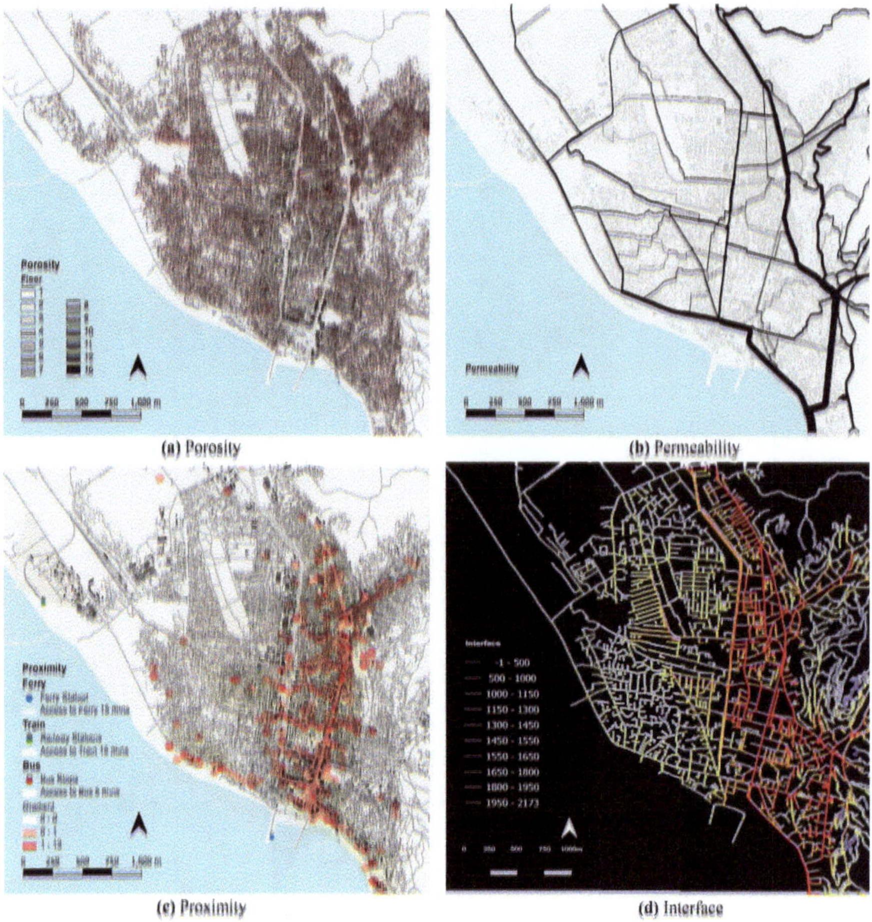

Fig. 5 Vlorë-urban diagnosis and synthesis of sub-level urban properties

Therefore, it is sensible to select Diversity and Green Continuity as the catalysts of transformation as the most preliminary systems to be modified. It is worth mentioning that Diversity and Green Continuity will only trigger the transformation and create a reaction change in other Key Categories to ensure an effective and balanced overall modification.

5 Eco-Breathing in Vlorë 2030, Strategic Plan by Systemic Action Network

5.1 *Design Ordering Principles and Systemic Actions Network*

Eco-breathing in Vlorë 2030 is an innovative urban project strictly based on the above diagnosis and synthesis, meanwhile rooted in the alliance between humans and nature. It is a vision for the sustainable development of the city for the next 10 years and beyond. Natural capital is assumed as a crucial infrastructure to address and overcome the challenges of the Anthropocene, developing an attractive, sustainable, and resilient city.

The Systemic Action Network is derived from The Design Ordering Principles (DOPs), which are universally accepted standards for modifying the built environment to perform sustainably. Rather than separate objectives, IMM defines them as an integrative and interactive system in which each DOP constantly affects the other ones. In this fashion, the non-linearity of the urban environment is addressed not only in investigation but also through modification.

As mentioned earlier, cities are not isolated entities but rather compounds composed of different systems. The interconnections between these systems are nonlinear. To enhance the overall performance of cities, it is essential to clarify the relationships between these systems and assess the impact of each design action on the systems. Similarly, the actions aimed at improving urban performance also play a role in different systems. A single design action not only achieves optimization in a specific Key Category and emphasizes a particular Design Ordering Principle but also, through the interlinkage of complex systems, simultaneously brings about improvements in other systems.

During the Diagnosis and Synthesis, Diversity and Green Continuity are selected as catalysts for triggering the modification. For this reason, the actions for Vlorë's retrofitting take DOP 3 (Balance the distribution of key types of uses) and DOP 4 (Make biodiversity an important part of urban life) as direction. To specify the task, the DOPs provide a comprehensive, integrated development framework that needs to be activated by specific Actions. In IMM, big efforts have been made to identify possible synergies, trade-offs, and interlinkages between DOPs and Actions and between the different Actions by themselves. Any specific DOP or any Action could generate impacts on other DOPs and on other Actions, with a large set of possible cascading effects. In particular, the interlinkage of specific Actions is a priority in IMM.

Categorized by DOPs, actions are divided into: A. Inner Actions; B. External Actions. The inner actions are actions fully included in a specific DOP, and they generally work aligned to achieve the aim of the DOP that includes them. They show a kind of priority for the activation of the transformation process concerning the DOP selected. Based on systemic thinking, the synergy of actions will bring a result more than the aggregation of single actions (Flood, 2010). This result not only achieves a singular DOP but also addresses other DOPs through actions that share similar objects and objectives. These actions are referred to as External Actions. The external actions could be seen as indirect actions that would be activated when the process starts due to the level of interlinkages, they have with any individual inner Action.

5.2 Vlorë 2030, Strategic Plan

For Diversity and Green Continuity defined as the catalysts of the transformation in the Vlorë project, the modification starts with DOP 3 and DOP 4, which are about balancing the typology of the types of uses and managing biodiversity resources, respectively. The Action Network offers clear guidelines for proceeding with the strategic plan by figuring out design options (Fig. 6). The inner actions in DOP 3 highlight the transformation of street facilities towards a pedestrian-friendly environment, by improving the green infrastructure like tree canopies and public parks. This is in coherence with the external actions like Action 2.6: implement eye-leveled greenery lines, which is under another DOP that emphasizes the permeability to facilitate the urban flow. Meanwhile, Action 3.5 calls for the vibrancy of public spaces, aligning with the objective of Action 5.5, which underscores the importance of creating interconnected open spaces. The inner actions of DOP 4 prioritize the significance of biodiversity. By activating vegetation and food production through Action 4.4, the cultivating fields are well preserved. Consequently, the local food production-supply-consumption chain will be shortened, and the internal regenerative capacity will be maximized according to external Action 1.6. Together with the maps of Diagnosis and Synthesis that pinpoint the problematic areas, the action network provides holistic and comprehensive instructions for specifying the transformation of Vlorë towards a more sustainable and resilient direction.

In the Strategic Plan (Fig. 7), Vlorë is enriched with new hybrid green infrastructures that offer multiple urban (eco)system services, innovating diversified urban functions and connecting new functional centralities and Eco-districts. Enhancing the existing natural capital, realizing the new green infrastructure network, and integrating Nature-based Solutions in the urban renewal Eco-breathing in Vlorë 2030 strategic plan maximizes the benefits of ecosystem services: green and blue infrastructure brings the richness of peri-urban and edge landscapes to the city of Vlorë by enabling effective water management and improving urban microclimates.

It is necessary to counterbalance the actual prevailing north–south urban axial development, centered on Rruga Sadik Zotaj and Rruga Gjergj Kastrioti, with new East–West connections in the form of a system of integrated green corridors. Hence, the plan introduces two main East–West Green hybrid infrastructures, two corridors with a strong green and nature-based component, able to catalyze new urban functions and to elect them as the foundation of a new integrated mobility system based on pedestrian and bicycle mobility fully integrated with public transportation.

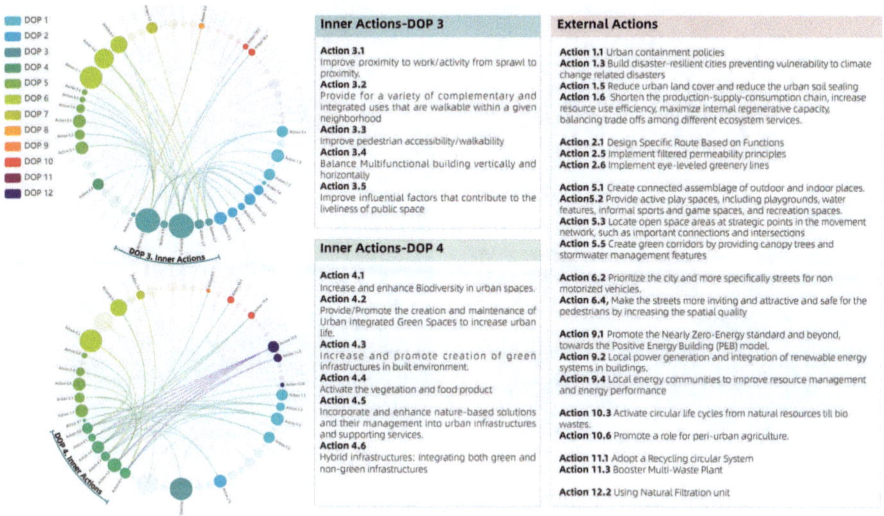

Fig. 6 Vlorë 2030: systemic action network (interlinkage of inner and external actions)

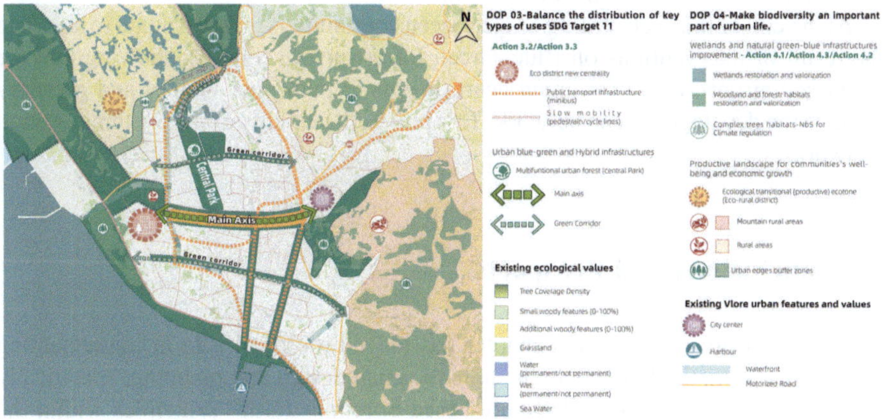

Fig. 7 Vlorë 2030: strategic plan

Urban productive landscapes are introduced to integrate with the hybrid green infrastructures to enhance and diversify tourist services. The new green network connects functionally, ecologically, and socio-economically with Soda Forest, Narta Lagoon, and agricultural areas. The new green heart in the Vlorë, on the dismissed airport site, becomes the green centrality providing ecosystem services for water and climate regulation and attractive places for Vlorë urban communities.

Along the new green boulevard are located opportunities for urban innovation and urban regeneration. The strategic plan makes the Eco-district composing the main green East–West corridors the most important area for the city development,

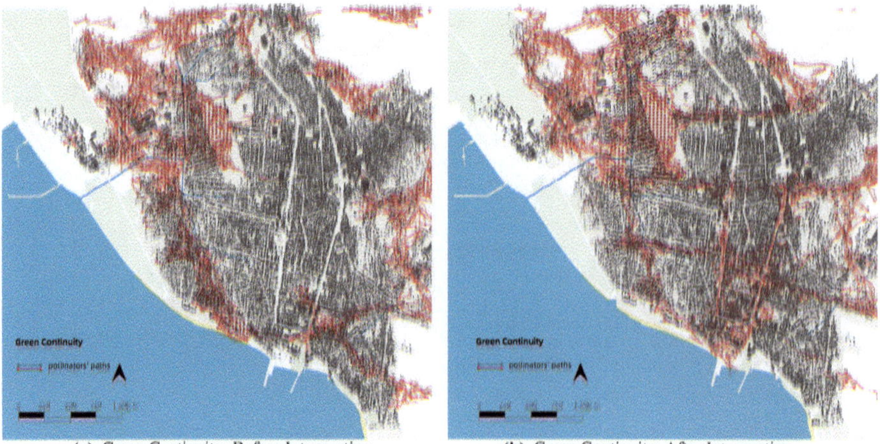

(a) Green Continuity: Before Intervention (b) Green Continuity: After Intervention

Fig. 8 Vlorë strategic plan: retrofitting

characterized by densification of the actual low-sprawled area, making it the blueprint of the new alliance between nature and the city.

At the end of this Retrofitting Stage, Green continuity is mapped again to simulate the result of interventions on Blue–Green Infrastructure (Fig. 8), showing that the dismissed airport, as a new green heart, begins to pump out vitality and greenery through the developed green network. This proves the validity of the Systemic Action Network.

6 Conclusions and Prospects

The research presented in this chapter has played a critical role in implementing new tools and methodologies in the field of urban design and planning, specifically focusing on the integration of Nature-based Solutions (NbS) principles within the context of the Integrated Modification Methodology (IMM). The study took the city of Vlorë in Albania as a representative case to explore the practicality of NbS and its role in achieving long-term sustainable development through a holistic approach.

The findings of the research emphasize the significance of natural capital, including ecosystem services and ecological functionalities, in maintaining human well-being and urban resilience in the face of climate and Anthropocene challenges. By viewing the city as a Complex Adaptive System (CAS), the study identified key functioning patterns, such as diversity and continuity, as essential drivers for transforming Vlorë into a more resilient and sustainable city.

The project "Eco-breathing in Vlorë 2030" was introduced as an innovative strategic plan, incorporating the principles of NbS in the frame of the IMM approach to foster sustainable development over the next decade and beyond. The plan aims to leverage natural capital as a crucial Blue–Green infrastructure to address the challenges of the Anthropocene and create an attractive, sustainable, and resilient city.

The strategic plan was developed based on the Design Ordering Principles (DOPs) and Systemic Action Network, which provide a comprehensive and integrated framework for urban transformation. By balancing the distribution of key types of uses and managing biodiversity resources, the plan seeks to create a more diverse and interconnected urban environment with improved green continuity.

Overall, this research demonstrates the potential of Nature-based Solutions and the Integrated Modification Methodology in driving sustainable urban transformation. The findings and strategic plan presented in this paper can serve as a valuable reference for other cities in the Mediterranean context and beyond, contributing to the advancement of transformative urban models towards more sustainable urban systems in developing countries. By prioritizing natural capital and leveraging the power of Nature-based Solutions, cities can work towards creating resilient and vibrant urban environments for the well-being of both humans and the environment.

Based on the achievements and findings presented in the chapter, several areas of further research and open questions will be implemented in the future to advance the integration of Nature-based Solutions (NbS) and IMM. One important aspect to explore is the Long-term Monitoring and Evaluation of NbS and the strategic plan. Establishing robust monitoring systems will allow us to assess the long-term effectiveness and impact of implemented Nature-based Solutions in urban areas. By conducting longitudinal studies, we can track changes in ecological, social, and economic indicators over time to understand the resilience and sustainability of urban transformations.

Another crucial area for further research is the transferability of Findings. It is essential to investigate how applicable the results and strategic plan developed for Vlorë are to other cities and regions with different contexts and challenges. Understanding the transferability of NbS and IMM principles will help identify best practices and adapt solutions to suit specific urban settings.

Community Engagement and Participation is another key aspect that requires exploration. Understanding how to effectively engage and involve local communities in the planning and implementation of Nature-based Solutions is crucial for successful and inclusive urban transformation. By considering community perceptions, needs, and preferences, cities can create more socially and environmentally sustainable solutions.

Climate change adaptation and mitigation are dual challenges that need to be addressed through NbS optimization. Investigating the synergies and trade-offs between NbS measures and their role in reducing greenhouse gas emissions while enhancing urban resilience is critical for climate-responsive urban planning. The Integration of NbS with Built Infrastructure is an area where further research can lead to more efficient and sustainable urban solutions. Understanding how to effectively integrate Nature-based Solutions with traditional built infrastructure to create hybrid green–blue systems can maximize urban benefits and improve urban living conditions.

Finally, Cross-Sectoral Collaboration is essential for successful NbS projects. Exploring multi-disciplinary and cross-sectoral partnerships can foster innovative solutions and maximize the positive impacts of Nature-based Solutions. By bringing together different sectors and stakeholders, cities can create more comprehensive and integrated urban development plans.

By addressing these open questions and conducting further research, cities can continuously improve their approach to sustainable urban development. The introduction of systemic thinking and systemic.

Leveraging the potential of Nature-based Solutions and integrated methodologies like IMM can lead to more resilient, vibrant, and ecologically responsible urban environments.

Acknowledgements and Attribution The authors are grateful to all the components of Politecnico di Milano's research team coordinated by Profs. Massimo Tadi and Gabriele Masera. Although the chapter is the result of a common reflection, paragraph 1 has been written by Angela Colucci, paragraph 6 by Massimo Tadi, paragraph 2 by Angela Colucci and Tao Dong, paragraph 4 by Massimo Tadi and Tao Dong, and paragraphs 3 and 5 by Tao Dong.

References

Acuto, M., & Leffel, B. (2021). Understanding the global ecosystem of city networks. *Urban Studies, 58*, 1758–1774. https://doi.org/10.1177/0042098020929261.

Alcamo, J., & Institute, W. R. (Eds.). (2003). *Ecosystems and Human Well-Being: A Framework for Assessment*. Island Press.

Barrett, S. (2013). Local level climate justice? Adaptation finance and vulnerability reduction. *Global Environmental Change, 23*, 1819–1829. https://doi.org/10.1016/j.gloenvcha.2013.07.015.

Bhadouria, R., Tripathi, S., Singh, P., Joshi, P. K., & Singh, R. (2023). Urban metabolism and global climate change: An overview. In R. Bhadouria, S. Tripathi, P. Singh, P. K. Joshi, & R. Singh (Eds.), *Urban Metabolism and Climate Change* (pp. 3–22). Springer International Publishing.

Browder, G., Ozment, S., Bescos, I. R., Gartner, T., & Lange, G.-M. (2019). Integrating green and gray.

Colucci, A. (2023). Resilience practices contribution enabling European landscape policy innovation and implementation. *Land, 12*, 637. https://doi.org/10.3390/land12030637.

Costanza, R., & Limburg, K. (1997). The value of the world's ecosystem.

Davies, C., Chen, W. Y., Sanesi, G., & Lafortezza, R. (2021). The European Union roadmap for implementing nature-based solutions: A review. *Environmental Science & Policy, 121*, 49–67. https://doi.org/10.1016/j.envsci.2021.03.018.

Dekker, S. (2019). *Cities Leading Climate Action: Urban Policy and Planning*. Routledge, Taylor & Francis Group.

Eggermont, H., Balian, E., Azevedo, J. M. N., Beumer, V., Brodin, T., Claudet, J., Fady, B., Grube, M., Keune, H., Lamarque, P., Reuter, K., Smith, M., Van Ham, C., Weisser, W. W., & Le Roux, X. (2015). Nature-based solutions: New influence for environmental management and research in Europe. In *GAIA—Ecological Perspectives for Science and Society* (Vol. 24, pp. 243–248). https://doi.org/10.14512/gaia.24.4.9.

European Commission. Directorate General for Research and Innovation. (2015). *Towards an EU research and innovation policy agenda for nature-based solutions & re-naturing cities: Final report of the Horizon 2020 expert group on "Nature based solutions and re naturing cities."* European Commission. Directorate General for Research and Innovation, LU.

European Environment Agency. (2021). *Nature-based solutions in Europe policy, knowledge and practice for climate change adaptation and disaster risk reduction*. Publications Office.

Faivre, N., Fritz, M., Freitas, T., De Boissezon, B., & Vandewoestijne, S. (2017). Nature-based solutions in the EU: Innovating with nature to address social, economic and environmental challenges. *Environmental Research, 159*, 509–518. https://doi.org/10.1016/j.envres.2017.08.032.

Flood, R. L. (2010). The relationship of 'systems thinking' to action research. *Syst Pract Action Res, 23*, 269–284. https://doi.org/10.1007/s11213-010-9169-1.

Galderisi, A., & Colucci, A. (2018). *Smart, Resilient and Transition Cities: Emerging Approaches and Tools for a Climate-Sensitive Urban Development*. Elsevier.

Galderisi, A., & Colucci, A. (2019). Cities addressing climate change: Hindering factors and seeds of innovation in current urban climate strategies. In *CRIOS* (pp. 85–94). https://doi.org/10.3280/CRIOS2019-017007.

Ghofrani, Z., Sposito, V., & Faggian, R. (2017). A comprehensive review of blue–green infrastructure concepts. *International Journal of Environment and Sustainability, 6*. https://doi.org/10.24102/ijes.v6i1.728.

Heymans, A., Breadsell, J., Morrison, G., Byrne, J., & Eon, C. (2019). Ecological urban planning and design: A systematic literature review. *Sustainability, 11*, 3723. https://doi.org/10.3390/su11133723.

Kalnay, E., & Cai, M. (2003). Impact of urbanization and land-use change on climate. *Nature, 423*, 528–531. https://doi.org/10.1038/nature01675.

Lee, T., & Jung, H. Y. (2018). Mapping city-to-city networks for climate change action: Geographic bases, link modalities, functions, and activity. *Journal of Cleaner Production, 182*, 96–104. https://doi.org/10.1016/j.jclepro.2018.02.034.

Marushchak, O., Nekrasova, O., Pupins, M., Tytar, V., & Ceirans, A. (2019). The role and importance of the protected areas' (emerald network) development for amphibians and reptiles on the example of Ukraine in the context of various factors' influence. In *ETR* (Vol. 1, p. 154). https://doi.org/10.17770/etr2019vol1.4100.

Masera, G., & Tadi, M. (Eds.). (2020). *Environmental Performance and Social Inclusion in Informal Settlements: A Favela Project Based on the IMM Integrated Modification Methodology*. Springer International Publishing.

Millennium Assessment, (Program) (Eds.). (2005). *Ecosystems and human well-being: Wetlands and water synthesis: A report of the Millennium Ecosystem Assessment*. World Resources Institute.

Mukim, M., & Roberts, M. (Ed.). (2023). *Thriving : Making Cities Green, Resilient, and Inclusive in a Changing Climate*. World Bank.

Nguyen, H., Chu, L., Harper, R. J., Dell, B., & Hoang, H. (2022). Mangrove-shrimp farming: A triple-win approach for communities in the Mekong River Delta. *Ocean & Coastal Management, 221*, 106082. https://doi.org/10.1016/j.ocecoaman.2022.106082.

Poggio, L., De Sousa, L. M., Batjes, N. H., Heuvelink, G. B. M., Kempen, B., Ribeiro, E., & Rossiter, D. (2021). SoilGrids 2.0: Producing soil information for the globe with quantified spatial uncertainty. *The Soil, 7*, 217–240. https://doi.org/10.5194/soil-7-217-2021.

Roggema, R. (Ed.). (2020). *Nature Driven Urbanism*. Springer.

Rosenzweig, C., Solecki, W., Hammer, S. A., & Mehrotra, S. (2010). Cities lead the way in climate–change action. *Nature, 467*, 909–911. https://doi.org/10.1038/467909a.

Sanderman, J., Hengl, T., Fiske, G., Solvik, K., Adame, M. F., Benson, L., Bukoski, J. J., Carnell, P., Cifuentes-Jara, M., Donato, D., Duncan, C., Eid, E. M., Ermgassen, P. Z., Lewis, C. J. E., Macreadie, P. I., Glass, L., Gress, S., Jardine, S. L., Jones, T. G., … Landis, E. (2018). A global map of mangrove forest soil carbon at 30 m spatial resolution. *Environmental Research Letters, 13*, 055002. https://doi.org/10.1088/1748-9326/aabe1c.

Somarakis, G., Stagakis, S., & Chrysoulakis, N. (2019). *ThinkNature/Nature-Based Solutions Handbook.* ThinkNature project funded by the EU Horizon 2020 research and innovation programme. https://doi.org/10.26225/JERV-W202.

Tadi, M., Zadeh, M. H., & Biraghi, C. A. (2020). The integrated modification methodology. In: G. Masera & M. Tadi (Eds.), *Environmental Performance and Social Inclusion in Informal Settlements. Research for Development*. Springer. https://doi.org/10.1007/978-3-030-44352-8_2.

Tan, S., Yang, J., Yan, J., Lee, C., Hashim, H., & Chen, B. (2017). A holistic low carbon city indicator framework for sustainable development. *Applied Energy, 185*, 1919–1930. https://doi.org/10.1016/j.apenergy.2016.03.041.

United Nations. (2017). New urban agenda (A/RES/71/256).

Websites

Dasgupta, S., Lall, S., & Wheeler, D. (2022). *Cutting global carbon emissions: Where do cities stand.* Retrieved July 05, 2022, from https://blogs.worldbank.org/sustainablecities/cutting-global-carbon-emissions-where-do-cities-stand.

Global Forest Watch. (2022). *Tree cover gain in Vlorës, Vlorë, Albania compared to other areas.* Retrieved February 05, 2022, from www.globalforestwatch.org.

IUCN. (2016). *WCC-2016-Res-069-EN. Defining nature-based solutions.* Retrieved May 07, 2023, from https://portals.iucn.org/library/node/46486.

Key Biodiversity Areas. (2022). *KBA data.* Retrieved February 01, 2022, from https://www.keybiodiversityareas.org/kba-data.

Ramsar. (2022). Global wetland outlook: Special edition 2021. In: *Ramsar convention on wetlands.* Retrieved February 01, 2022, from https://www.ramsar.org.

World Bank Group. (2021). *World Bank Group climate change action plan 2021–2025: Supporting green, resilient, and inclusive development.* Retrieved June 04, 2022, from http://hdl.handle.net/10986/35799.

Correction to: Role of Blue-Green Infrastructure in Achieving Sustainable Development Goals in Urban Centres of Developing countries

Kajal Patel, Meenakshi Chaurasia, Ranjana Singh, Rahul Bhadouria, and K. S. Rao

Correction to:
Chapter "Role of Blue-Green Infrastructure in Achieving Sustainable Development Goals in Urban Centres of Developing countries" in: P. K. Joshi et al. (eds.), *Blue-Green Infrastructure for Sustainable Urban Settlements*, https://doi.org/10.1007/978-3-031-62293-9_5

In the original version of chapter 5, the text "see for detail Well and Ludwig, 2020" has been added now at the end of some statements.

The updated version of this chapter can be found at
https://doi.org/10.1007/978-3-031-62293-9_5

P. K. Joshi et al. (eds.), *Blue-Green Infrastructure for Sustainable Urban Settlements*,
https://doi.org/10.1007/978-3-031-62293-9_18

MIX
Papier aus verantwortungsvollen Quellen
Paper from responsible sources
FSC® C105338

If you have any concerns about our products,
you can contact us on
ProductSafety@springernature.com

In case Publisher is established outside the EU,
the EU authorized representative is:
Springer Nature Customer Service Center GmbH
Europaplatz 3, 69115 Heidelberg, Germany

Printed by Libri Plureos GmbH
in Hamburg, Germany